FUNCTIONS MODELING CHANGE:
A Preparation for Calculus

FUNCTIONS MODELING CHANGE:
A Preparation for Calculus

Produced by the Consortium based at Harvard and funded by a National Science Foundation Grant.

Eric Connally
Wellesley College

Deborah Hughes-Hallett
University of Arizona

Andrew M. Gleason
Harvard University

Frank Avenoso
Nassau Community College

Philip Cheifetz
Nassau Community College

Andrew Pasquale
Chelmsford High School

Karen Rhea
University of Southern Mississippi

Pat Shure
University of Michigan

Carl Swenson
Seattle University

Katherine Yoshiwara
Los Angeles Pierce College

Ann Davidian
General Douglas MacArthur High School

Coordinated by
Srdjan Divac
S. Alex Mallozzi
Ann J. Ryu
Xianbao Xu

John Wiley & Sons, Inc.
New York Chichester Weinheim Brisbane Singapore Toronto

Dedicated to Maria, Ben, and Jonah

EXECUTIVE EDITOR	Ruth Baruth
ASSISTANT EDITOR	Mary Johenk
SENIOR MARKETING MANAGER	Peter Rytell
PRODUCTION EDITOR	Ken Santor
COVER DESIGNER	Madelyn Lesure
COVER PHOTO	Peter Mathis/Tony Stone Images, New York
OUTSIDE PRODUCTION SERVICES	Publication Services, Inc

This book was set in Times Roman by the Consortium based at Harvard using L&TEX, Mathematica, and the package AsTeX, which was written by Alex Kasman for this project. Special thanks to Srdjan Divac and Xianbao Xu for managing the process. It was printed and bound by Von Hoffman Press. The cover was printed by Lehigh Press.

This book is printed on acid-free paper.

The paper in this book was manufactured by a mill whose whose forest management programs include sustained yield harvesting of its timberlands. Sustained yield harvesting priciples ensure that the numbers of trees cut each year does not exceed the amount of new growth.

This material is based upon work supported by the National Science Foundation under Grant No. DUE-9352905. All royalties from the sale of this book will go toward the furtherance of the project. Opinions expressed are those of the authors and not necessarily those of the Foundation.

ISBN. 0-471 37918 2

Printed in the United States of America

10 9 8 7 6 5 4 3 2 1

PREFACE

Mathematics has the extraordinary power to reduce complicated problems to simple rules and procedures. Therein lies the danger in teaching mathematics: it is possible to teach the subject as nothing but the rules and procedures—thereby losing sight of both the mathematics and of its practical value. This new edition of *Functions Modeling Change: A Preparation for Calculus* continues our effort to refocus the teaching of mathematics on concepts as well as procedures.

A Focused Vision: Conceptual Understanding

These materials stress conceptual understanding and multiple ways of representing mathematical ideas. Our goal is to provide students with a clear understanding of the ideas of functions as a solid foundation for subsequent courses in mathematics and other disciplines. When we designed this curriculum under an NSF grant, we started with a clean slate. We focused on a small number of key concepts, emphasizing depth of understanding rather than breadth of coverage.

The Preliminary Edition: A New Curriculum

The preliminary edition of this book was the work of faculty at a consortium of ten institutions, generously supported by the National Science Foundation. It represents the first consensus between such a diverse group of faculty to have shaped a mainstream precalculus text.

The First Edition: A New Organization

The first edition has the same vision as the preliminary edition and provides instructors with a new organization. In particular:
- The first two chapters have been reorganized and divided into three chapters. Linear functions now come earlier in the text, since the new Chapter 1 is shorter than in the Preliminary Edition.
- An early and brief introduction to inverse functions has been added to the new Chapter 3.
- Chapter 4, Exponential and Logarithmic Functions, has been reorganized so that logarithmic functions are introduced as the inverse of exponential functions.
- There are now two chapters on trigonometry with additional material on identities and polar coordinates and more material on triangles.
- New material on conic sections and sequences and series has been added to Chapter 11.

Guiding Principles: Varied Problems and the Rule of Four

Since students usually learn most when they are active, we feel that the exercises in a text are of central importance. In addition, we have found that multiple representations encourage students to reflect on the meaning of the material. Consequently, we have been guided by the following principles.
- Our problems are varied and some are challenging. Most cannot be done by following a template in the text.
- The Rule of Four: Each function is represented symbolically, numerically, graphically, and verbally.
- Fewer topics are introduced than is customary, but each topic is treated in greater depth. The core syllabus of precalculus should include only those topics that are essential to the study of calculus.

- The components of a precalculus curriculum should be tied together by clearly defined themes. Algebra should be developed as needed, but should not serve as a central theme. Functions as models of change is our central theme.

- Problems involving real data are included to prepare students to use mathematics in other fields.

- To use mathematics effectively, students need skill in both symbolic manipulation and the use of technology. The exact proportions of each may vary widely, depending on the preparation of the student and the wishes of the instructor.

- Materials for precalculus should allow for a broad range of teaching styles. They should be flexible enough to use in large lecture halls, small classes, or in group or lab settings.

What Student Background is Expected?

Students using this book should have successfully completed a course in intermediate algebra or high school algebra II. The book is thought-provoking for well-prepared students while still accessible to students with weaker backgrounds. Providing numerical and graphical approaches as well as the algebraic gives students another way of mastering the material. This approach encourages students to persist, thereby lowering failure rates.

Content

The central theme of this course is functions as models of change. We emphasize that functions can be grouped into families and that functions can be used as models for real-world behavior. Because linear, exponential, power, and periodic functions are more frequently used to model physical phenomena, they are introduced before polynomial and rational functions. Once introduced, a family of functions is compared and contrasted with other families of functions.

A large number of the examples and problems that students see in this precalculus course are given in the context of real-world problems. Indeed, we hope that students will be able to create mathematical models that will help them understand the world in which they live. The inclusion of non-routine problems is intended to establish the idea that such problems are not only part of mathematics, but in some sense, are the point of mathematics.

The book does not require any specific software or technology. Test sites have used the material with graphing calculators and graphing software. Any technology with the ability to graph functions will suffice.

Chapter 1: Functions and Change

This chapter introduces the concept of a function and the possible representations. Proportionality, rates, and rates of change are covered to enable students to quantify change.

Chapter 2: Linear Functions

This chapter focuses exclusively on the development of linear functions. The chapter includes a section on fitting a linear function to data.

Chapter 3: Function Notation

Function notation is introduced in Chapter 1, and is studied in more detail in this chapter. Inside and outside changes, the concept of an inverse function and the domain and range are investigated. The chapter ends with a section on piecewise defined functions, including the absolute value function.

Chapter 4: Exponential and Logarithmic Functions

This chapter introduces the family of exponential functions and the number e. Logarithmic functions to base 10 and to the base e are introduced as inverses of exponential functions. Modeling with exponential functions, including compund interest, logarithmic scales, and a section on fitting exponential functions to data conclude the chapter.

Chapter 5: Transformations of Functions and Their Graphs

This chapter investigates transformations — shifting, reflecting, and stretching. These ideas are applied to the family of quadratic functions.

Chapter 6: Trigonometric Functions

This chapter, which focuses on modeling periodic phenomena, introduces the trigonometric functions: the sine, cosine, tangent, and briefly, the secant, cosecant, and cotangent. The inverse trigonometric functions are introduced.

Chapter 7: Trigonometry

This chapter focuses on triangles (both right and non-right triangles), identities, and polar coordinates. It includes a section on modeling using trigonometric functions.

Chapter 8: Compositions, Inverses, and Combinations of Functions

This chapter contains material on composition and combinations of functions. Inverse functions, which were introduced in Chapter 3, are studied in more detail in this chapter.

Chapter 9: Polynomial and Rational Functions

This chapter covers power functions, polynomials, and rational functions.

Chapter 10: Vectors

This chapter contains material on vectors and makes use of trigonometry.

Chapter 11: Other Ways of Defining Functions

This chapter contains material on arithmetic sequences and series, geometric series, parametric equations, implicit functions, complex numbers and hyperbolic functions.

Appendices

The appendices contain a review of algebra. Answers to all appendix problems are in the back of the book.

Our Experiences

The Preliminary Edition was used by over 200 schools around the country. In this diverse group of schools, the preliminary edition was class-tested with many different types of students in semester and quarter systems, in large lectures and small classes, as well as in full year courses in secondary schools. It has been used in computer labs, small groups, and traditional settings, and with a number of different technologies.

Changes from the Preliminary Edition

In preparing this edition, we solicited comments from a large number of mathematicians who had used the text. We continued to discuss with our colleagues in client disciplines the mathematical needs of their students. We were offered many valuable suggestions, which we have tried to incorporate, while maintaining our original commitment to a focused treatment of a limited number of topics.

General Changes: The problem sets in all chapters have been revised to include more easy conceptual problems and more drill problems. The discussions throughout the text have been rewritten for increased clarity

Chapter 1 Functions and Change and **Chapter 2 Linear Functions:** The first two chapters have been reorganized and broken into three chapters. The new Chapter 1 is easier than Chapter 1 in the Preliminary Edition, so it will ease students into the course. There is a new section (1.2) on proportionality, which is important in science and provides an introduction to rates. Linear functions now come earlier in the text, since the new Chapter 1 is shorter than in the Preliminary Edition.

Chapter 3 Function Notation: The remaining material from Chapter 1 in the Preliminary Edition is found in the new Chapter 3 in the First Edition. An early introduction to inverse functions has been added to this new Chapter 3. Inverse functions are discussed in detail in Chapter 8 (formerly Chapter 6), but are introduced briefly in Chapter 3 so that instructors who want to emphasize them when covering logarithmic and inverse trigonometric functions can do so.

Chapter 4 Exponential and Logarithmic Functions: (formerly Chapter 3) has been reorganized. Logarithms are now introduced as the inverse of the exponential function, rather than through log scales. The material on log scales is included, but is near the end of the chapter. The number e is now introduced earlier, thereby streamlining the presentation.

Chapter 5 Transformations of Functions and Their Graphs: Quadratic functions have been moved to the end of the chapter.

Chapter 6 Trigonometric Functions and **Chapter 7 Trigonometry:** There are now two chapters on trigonometry. There is additional material on identities and polar coordinates (from sections 8.5 and 9.5 in the Preliminary Edition), and more materials on triangles.

Chapter 8 Compositions, Inverses, and Combinations of Functions: Since inverse functions are now introduced in Chapter 3, this chapter (formerly Chapter 6) has been shortened to three sections.

Chapter 9 Polynomial and Rational Functions This was Chapter 7 in the Preliminary Edition.

Chapter 10 Vectors: This was Chapter 8 in the Preliminary Edition, without the section on polar coordinates.

Chapter 11 Other Ways of Defining Functions: The chapter starts with a new section on arithmetic sequences and series. New material on conic sections has been added to the sections on parametric functions and hyperbolic trigonometric functions.

Appendices Large numbers of additional straightforward problems have been added for students in need of a refresher.

Supplementary Materials

The following supplementary materials are available for the First Edition:

- **Instructor's Manual with Sample Exam Questions** containing teaching tips, calculator programs, some overhead transparency masters and test questions arranged according to section.

- **Instructor's Solution Manual** with complete solutions to all problems.

- **Student's Solution Manual** with complete solutions to half the odd-numbered problems.

- **Instructor's Resource CD-ROM** which contains the Instructor's Manual and Instructor's Solutions Manual as well as other valuable resources.

- **Student Study Guide** containing additional study aids for students. The topics are tied directly to the book.

- **Getting Started Graphing Calculator Manual** instructs students on how to utilize their TI-83/82 calculators with this book. The TI-86/85 and TI-89 are also discussed. Contains samples, tips, and trouble shooting sections to answer students' questions.

- **Wiley Web Tests for Precalculus and College Algebra** Customizable and expandable on-line testing system allows for administering and grading tests in a variety of modes. The software is platform independent and resides on the local server. Developed by John Orr and the University of Nebraska, Lincoln.

Acknowledgments

We would like to thank the many people who made this book possible. First, we would like to thank the National Science Foundation for their trust and their support; we are particularly grateful to Jim Lightbourne and Spud Bradley.

We are also grateful to our Advisory Board for their guidance: Benita Albert, Lida Barrett, Simon Bernau, Robert Davis, Lovenia Deconge-Watson, John Dossey, Ronald Douglas, Eli Fromm, Bill Haver, Don Lewis, Seymour Parter, John Prados, and Stephen Rodi. Working with Ruth Baruth, Lucille Buonocore, Ken Santor, Peter Ryttel, Pete Janzow, Mary Johenk, Maddy Lesure, and Jo Bakal at John Wiley is a pleasure. We appreciate their patience and imagination.

Many people have contributed significantly to this text. They include: Fahd Alshammari, David Arias, Tim Bean, Charlotte Bonner, Bill Bossert, Brian Bradie, Noah S. Brannen, Mike Brilleslyper, Donna Brouillette, Mauro Cassano, Ray E. Collings, Eva Demyan, Bob Dobrow, Ian Dowker, Carolyn Edmond, Mary Ehlers, Maryann Faller, Aidan Flanagan, Christie Gilliland, John Gerke, Seana Grey, Wynne Guy, Donnie Hallstone, Christine Healy, Brian Henderson, Larry Henly, Dean Hickerson, Bob Hoburg, Brian Hopkins, Phil Hotchkiss, Mike Huffman, Mac Hyman, Jerry Ianni, Rajini Jesudason, Loren Johnson, Bill Kiele, Mary Kilbride, Candace Kling, Rob LaQuaglia, Barbara Leasher, Andrew Lippai, Richard Little, Len Malinowski, Brad Mann, Nancy Marcus, Bob Megginson, Deborah Moore, Eric Motylinski, Bridget Neale Paris, Anne O'Sullivan, Ted Pyne, Mary Rack, Janet Ray, Jerry Resnick, Ken Richardson, Halip Saifi, Sharon Sanders, Ellen Schmierer, Mary Schumacher, John Screiber, Mike Seery, Mike Sherman, Donna Sherrill, Kanwal Singh, Fred Shure, Myra Snell, Natasha Speer, Sonya Stanley, Peggy Tibbs, Mike Totoro, Jerry Uhl, Pat Wagener, Maura Winkler, Dale Winter, Andre Yandl, and Victoria Yen.

Special thanks are owed to Kenny Ching, Dan Flath, Jo Ellen Hillyer, David Lovelock, Bill McCallum, Kate McGivney, Bill Mueller and Elias Toubassi for their work on the text, to "Suds" Sudholz for administering the project, to Alex Kasman for his software support, to Leonid Andreev and Bob Condon for their help with the computers.

Most of all, we owe our thanks and admiration to the fantastic team of people who did all the computer work: Ebo Bentil, Mike Esposito, Elliot Marks, and Xianbao Xu.

Eric Connally	Deborah Hughes-Hallett	Andrew M. Gleason	Frank Avenoso
Philip Cheifetz	Ann Davidian	Andrew Pasquale	Karen Rhea
Pat Shure	Carl Swenson	Katherine Yoshiwara	

To Students: How to Learn from this Book

- This book may be different from other math textbooks that you have used, so it may be helpful to know about some of the differences in advance. At every stage, this book emphasizes the *meaning* (in practical, graphical or numerical terms) of the symbols you are using. There is much less emphasis on "plug-and-chug" and using formulas, and much more emphasis on the interpretation of these formulas than you may expect. You will often be asked to explain your ideas in words or to explain an answer using graphs.

- The book contains the main ideas of precalculus in plain English. Success in using this book will depend on reading, questioning, and thinking hard about the ideas presented. It will be helpful to read the text in detail, not just the worked examples.

- There are few examples in the text that are exactly like the homework problems, so homework problems can't be done by searching for similar–looking "worked out" examples. Success with the homework will come by grappling with the ideas of precalculus.

- Many of the problems in the book are open-ended. This means that there is more than one correct approach and more than one correct solution. Sometimes, solving a problem relies on common sense ideas that are not stated in the problem explicitly but which you know from everyday life.

- This book assumes that you have access to a calculator or computer that can graph functions and find (approximate) roots of equations. There are many situations where you may not be able to find an exact solution to a problem, but can use a calculator or computer to get a reasonable approximation. An answer obtained this way can be as useful as an exact one. However, the problem does not always state that a calculator is required, so use your own judgement.

  If you mistrust technology, listen to this student, who started out the same way:

  > Using computers is strange, but surprisingly beneficial, and in my opinion is what leads to success in this class. I have difficulty visualizing graphs in my head, and this has always led to my downfall in calculus. With the assistance of the computers, that stress was no longer a factor, and I was able to concentrate on the concepts behind the shapes of the graphs, and since these became gradually more clear, I got increasingly better at picturing what the graphs should look like. It's the old story of not being able to get a job without previous experience, but not being able to get experience without a job. Relying on the computer to help me avoid graphing, I was tricked into focusing on what the graphs meant instead of how to make them look right, and what graphs symbolize is the fundamental basis of this class. By being able to see what I was trying to describe and learn from, I could understand a lot more about the concepts, because I could change the conditions and see the results. For the first time, I was able to see how everything works together . . . .

  That was a student at the University of Arizona who took calculus in Fall 1990. She was terrified of calculus, got a C on her first test, but finished with an A for the course.

- This book attempts to give equal weight to four methods for describing functions: graphical (a picture), numerical (a table of values), algebraic (a formula) and verbal (words). Sometimes it's easier to translate a problem given in one form into another. For example, you might replace the graph of a parabola with its equation, or plot a table of values to see its behavior. It is important to be flexible about your approach: if one way of looking at a problem doesn't work, try another.

- Students using this book have found discussing these problems in small groups helpful. There are a great many problems which are not cut-and-dried; it can help to attack them with the other perspectives your colleagues can provide. If group work is not feasible, see if your instructor can organize a discussion session in which additional problems can be worked on.

- You are probably wondering what you'll get from the book. The answer is, if you put in a solid effort, you will get a real understanding of functions as well as a real sense of how mathematics is used in the age of technology.

CONTENTS

Table of Contents

7 TRIGONOMETRY   287

8 COMPOSITIONS, INVERSES, AND COMBINATIONS OF FUNCTIONS 329

9 POLYNOMIAL AND RATIONAL FUNCTIONS   365

CHAPTER ONE

FUNCTIONS AND CHANGE

A function represents the relationship between two quantities and describes how the value of one quantity depends on the value of the other. A function can be represented in several ways: in words, by a graph, by a formula, or by a table of numbers. This chapter gives examples of all four representations and introduces the notation used to represent a function.

Proportional relationships are introduced. These represent an important type of function. The final section develops the idea of a rate of change which compares the changes in two quantities.

1.1 WHAT IS A FUNCTION?

In everyday language, the word *function* expresses the notion of dependence. For example, a person might say that election results are a function of the economy, meaning that the winner of an election is determined by how the economy is doing. Someone else might claim that car sales are a function of the weather, meaning that the number of cars sold on a given day is affected by the weather.

In mathematics, the meaning of the word *function* is more precise, but the basic idea is the same. A function is a relationship between two quantities. If the value of the first quantity determines exactly one value of the second quantity, we say the second quantity is a function of the first.

Example 1 In the early 1980s, the recording industry introduced the compact disc (CD). Table 1.1 gives the number of CDs (in millions) sold for the years 1982 through 1997.[1] The year determines exactly one sales figure—the number of CDs sold that year. Thus, we say the number of CDs sold is a function of the year.

TABLE 1.1 *Number of CDs sold (in millions), by year*

Year	1982	1983	1984	1985	1986	1987	1988	1989
Sales	0	0.8	5.8	23	53	102	150	207
Year	1990	1991	1992	1993	1994	1995	1996	1997
Sales	287	333	408	495	662	723	779	753

The quantities described by a function are called *variables*. In the previous example, the variables are the year and the sales. In the next example, the variables are named A and n.

Example 2 The number of gallons of paint needed to paint a house depends on the size of the house. A gallon of paint typically covers 250 square feet. Thus, the number of gallons of paint, n, is a function of the area to be painted, A. For example, if $A = 5000 \text{ ft}^2$, then $n = 5000/250 = 20$ gallons of paint. In general, n and A are related by the formula

$$n = \frac{A}{250}.$$

We make the following definition:

A **function** is a rule which takes certain values as inputs and assigns to each input value exactly one output value. The output is a function of the input.

In the CD example, the inputs are the years and the outputs are the sales given in Table 1.1. In the paint example, the input is area and the output is the number of gallons of paint, calculated by the formula $n = A/250$.

Representing Functions: Words, Tables, Graphs, and Formulas

A function can be described using words, data in a table, points on a graph, or a formula. In Example 1, the CD sales are represented by a table; in Example 2, the number of gallons of paint required is represented by a formula.

[1]Data from Recording Industry Association of America, Inc, 1998.

Example 3   It is a surprising biological fact that most crickets chirp at a rate that increases as the temperature increases. For the snowy tree cricket (*Oecanthus fultoni*), the relationship between temperature and chirp rate is so reliable that this type of cricket is called the thermometer cricket. We can estimate the temperature (in degrees Fahrenheit) by counting the number of times a snowy tree cricket chirps in 15 seconds and adding 40. For instance, if we count 20 chirps in 15 seconds, then a good estimate of the temperature is $20 + 40 = 60°$F.

The rule used to find the temperature T (in °F) from the chirp rate R (in chirps per minute) is an example of a function. Describe this function in four different ways: using words, a table, a graph, and a formula.

Solution   • **Words**: To estimate the temperature, we count the number of chirps in fifteen seconds and add forty. Alternatively, we can count R chirps per minute, divide R by four and add forty. This is because there are one-fourth as many chirps in fifteen seconds as there are in sixty seconds. For instance, 80 chirps per minute works out to $\frac{1}{4} \cdot 80 = 20$ chirps every 15 seconds, giving an estimated temperature of $20 + 40 = 60°$F.

• **Table**: Table 1.2 gives the estimated temperature, T, as a function of R, the number of chirps per minute. Notice the pattern in Table 1.2: each time the chirp rate, R, goes up by 20 chirps per minute, the temperature, T, goes up by 5°F.

• **Graph**: The data from Table 1.2 are plotted in Figure 1.1. For instance, the pair of values $R = 80, T = 60$ are plotted as the point P, which is 80 units along the horizontal axis and 60 units up the vertical axis. Data represented in this way are said to be plotted on the *Cartesian plane*. The precise position of P is shown by its coordinates, written $P = (80, 60)$.

TABLE 1.2   *Chirp rate and temperature*

R, chirp rate (chirps/minute)	T, predicted temperature (°F)
20	45
40	50
60	55
80	60
100	65
120	70
140	75
160	80

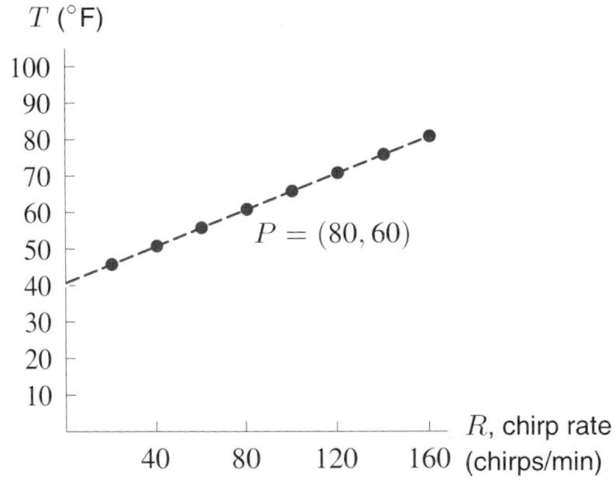

Figure 1.1: Chirp rate and temperature

• **Formula**: A formula is an equation giving T in terms of R. Dividing the chirp rate by four and adding forty gives the estimated temperature, so:

$$\underbrace{\text{Estimated temperature (in °F)}}_{T} = \frac{1}{4} \cdot \underbrace{\text{Chirp rate (in chirps/min)}}_{R} + 40.$$

Rewriting this using the variables T and R gives the formula:

$$T = \frac{1}{4}R + 40.$$

Let's check the formula. Substituting $R = 80$, we have

$$T = \frac{1}{4} \cdot 80 + 40 = 60$$

which agrees with point $P = (80, 60)$ in Figure 1.1.

The formula $T = \frac{1}{4}R + 40$ also tells us that if $R = 0$, then $T = 40$. Thus, the dashed line in Figure 1.1 crosses (or intersects) the T-axis at $T = 40$; we say the T-*intercept* is 40.

All of the descriptions given in Example 3 provide the same information, but each description has a different emphasis. A relationship between variables is often described in words, as at the beginning of Example 3. A table like Table 1.2 is useful because it shows the predicted temperature for various chirp rates. The graph in Figure 1.1 is more suggestive of a trend than the table, although it is harder to read exact values of the function. For example, you might have noticed that every point in Figure 1.1 falls on a straight line that slopes up from left to right. In general, a graph can reveal a pattern that might otherwise go unnoticed. Finally, the formula has the advantage of being both compact and precise. However, this compactness can also be a disadvantage since it is often harder to gain as much immediate insight from the brief notation of a formula as from a table or a graph.

When we use a function to describe an actual situation, the function is referred to as a **mathematical model**. The formula $T = \frac{1}{4}R + 40$ is a mathematical model of the relationship between the temperature and the cricket's chirp rate. Such models can be powerful tools for understanding phenomena and making predictions. For example, this model predicts that when the chirp rate is 80 chirps per minute, the temperature is 60°F. In addition, since $T = 40$ when $R = 0$, the model predicts that the chirp rate is 0 at 40°F. Whether the model's predictions are accurate for chirp rates down to 0 and temperatures as low as 40°F is a question that mathematics alone cannot answer; an understanding of the biology of crickets is needed. However, we can safely say that the model does not apply for temperatures below 40°F, because the chirp rate would then be negative. For the range of chirp rates and temperatures in Table 1.2, the model is remarkably accurate.

Function Notation

To indicate that a quantity Q is a function of a quantity t, we abbreviate

Q is a function of t

to

Q equals "f of t"

and, using function notation, to

$$Q = f(t).$$

Thus, applying the rule f to the input value, t, gives the output value, $f(t)$. In other words, $f(t)$ represents a value of Q. Here Q is called the *dependent variable* and t is called the *independent variable*. Symbolically,

$$\text{Output} = f(\text{Input})$$

or

$$\text{Dependent} = f(\text{Independent}).$$

We could have used any letter, not just f, to represent the rule.

Example 4   Example 1 describes the number of CDs sold in a year as a function of the year. Using C for the number of CDs and t for the year, we write $C = f(t)$. In this case, the function f is the rule which describes the number of CDs sold, C, during a particular year, t.

Example 5   Let $n = f(A)$ be the amount of paint (in gallons) needed to cover an area of A ft^2. Explain in words what the statement $f(10,000) = 40$ tells us about painting houses.

Solution   The input of the function $n = f(A)$ is an area and the output is an amount of paint. The statement $f(10,000) = 40$ tells us that an area of $A = 10,000$ ft^2 requires $n = 40$ gallons of paint.

The expressions "Q depends on t" or "Q is a function of t" do *not* imply a cause-and-effect relationship, as the snowy tree cricket example illustrates.

Example 6   Example 3 gives the following formula for estimating air temperature based on the chirp rate of the snowy tree cricket:

$$T = \frac{1}{4}R + 40.$$

In this formula, T depends on R. Writing $T = f(R)$ indicates that the relationship is a function. In everyday language, however, saying that T depends on R suggests that making the cricket chirp faster would somehow make the temperature go up. Clearly, the cricket's chirping doesn't cause the temperature to be what it is. In mathematics, saying that the temperature "depends" on the chirp rate means only that knowing the chirp rate is sufficient to tell us the temperature.

Functions Don't Have to Be Defined by Formulas

People sometimes think that functions are always represented by formulas. However, the next example shows a function which is not given by a formula.

Example 7   The average monthly rainfall, R, at Chicago's O'Hare airport is given in Table 1.3, where time, t, is in months and $t = 1$ is January, $t = 2$ is February, and so on. The rainfall is a function of the month, so we write $R = f(t)$. However there is no equation that gives R when t is known. Evaluate $f(1)$ and $f(11)$. Explain what your answers mean.

TABLE 1.3   *Average monthly rainfall at Chicago's O'Hare airport*

Month, t	1	2	3	4	5	6	7	8	9	10	11	12
Rainfall, R (inches)	1.8	1.8	2.7	3.1	3.5	3.7	3.5	3.4	3.2	2.5	2.4	2.1

Solution   The value of $f(1)$ is the average rainfall in inches at Chicago's O'Hare airport in a typical January. From the table, $f(1) = 1.8$. Similarly, $f(11) = 2.4$ means that in a typical November, there are 2.4 inches of rain at O'Hare.

When Is a Relationship Not a Function?

It is possible for two quantities to be related and yet for neither quantity to be a function of the other, because neither determines the other.

Example 8   A national park contains foxes that prey on rabbits. Table 1.4 gives the two populations, F and R, over a 12-month period, where $t = 0$ means January 1, $t = 1$ means February 1, and so on.

TABLE 1.4   *Number of foxes and rabbits in a national park, by month*

t, month	0	1	2	3	4	5	6	7	8	9	10	11
R, rabbits	1000	750	567	500	567	750	1000	1250	1433	1500	1433	1250
F, foxes	150	143	125	100	75	57	50	57	75	100	125	143

(a)   Is F a function of t? Is R a function of t?

(b)   Is F a function of R? Is R a function of F?

Solution   (a)   Both F and R are functions of t. For each value of t, there is exactly one value of F and exactly one value of R. For example, Table 1.4 shows that if $t = 5$, then $R = 750$ and $F = 57$. This means that on June 1 there are 750 rabbits and 57 foxes in the park. If we write $R = f(t)$ and $F = g(t)$, then $f(5) = 750$ and $g(5) = 57$.

(b)   No, F is not a function of R. For example, suppose $R = 750$, meaning there are 750 rabbits. This happens both at $t = 1$ (February 1) and at $t = 5$ (June 1). In the first instance, there are 143 foxes; in the second instance, there are 57 foxes. Since there are R-values which correspond to more than one F-value, F is not a function of R.

Similarly, R is not a function of F. At time $t = 5$, we have $R = 750$ when $F = 57$, while at time $t = 7$, we have $R = 1250$ when $F = 57$ again. Thus, the value of F does not uniquely determine the value of R.

How to Tell if a Graph Represents a Function: Vertical Line Test

For y to be a function of x, each value of x must be associated with exactly one value of y. What does this requirement mean graphically? In order for a graph to represent a function, each x-value must correspond to exactly one y-value. This means that the graph must intersect any vertical line at most once. If a vertical line cuts the graph twice, the graph would contain two points with different y-values but the same x-value. This violates the definition of a function, so we have the following:

> The **Vertical Line Test**: If there is a vertical line which intersects a graph in more than one point, then the graph does not represent a function.

Example 9   In which of the graphs in Figures 1.2 and 1.3 could y be a function of x?

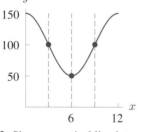

Figure 1.2: Since no vertical line intersects this curve at more than one point, the graph could represent y as a function of x

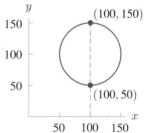

Figure 1.3: Since at least one vertical line intersects this curve at more than one point, the graph does not represent y as a function of x

Solution   The curve in Figure 1.2 could represent y as a function of x because no vertical line intersects this curve in more than one point. The curve in Figure 1.3 does not represent the graph of a function. The vertical line shown intersects the curve at two points.

A graph fails the vertical line test if at least one vertical line cuts the graph more than once, as in Figure 1.3. However, if a graph represents a function, then *every* vertical line must intersect the graph in at most one point.

Problems for Section 1.1

1.   Table 1.5 shows the daily low temperature for a one-week period in New York City during July.
    (a)   What was the low temperature on July 19?
    (b)   When was the low temperature 73°F?
    (c)   Is the daily low temperature a function of the date? Explain.
    (d)   Is the date a function of the daily low temperature? Explain.

TABLE 1.5

Date in July	17	18	19	20	21	22	23
Low temperature (°F)	73	77	69	73	75	75	70

2. Table 1.6 shows the number of calories used per minute as a function of body weight for three sports.[2]

    (a) Determine the number of calories that a 200-lb person uses in one half-hour of walking.
    (b) Who uses more calories, a 120-lb person swimming for one hour or a 220-lb person bicycling for a half-hour?
    (c) Does the number of calories used by a person walking increase or decrease as weight increases?

TABLE 1.6

Activity	100 lb	120 lb	150 lb	170 lb	200 lb	220 lb
Walking (3 mph)	2.7	3.2	4.0	4.6	5.4	5.9
Bicycling (10 mph)	5.4	6.5	8.1	9.2	10.8	11.9
Swimming (2 mph)	5.8	6.9	8.7	9.8	11.6	12.7

3. Use the data from Table 1.4 on page 5 to answer the following questions.

    (a) Plot R on the vertical axis and t on the horizontal axis. Use this graph to explain why you believe that R is a function of t.
    (b) Plot F on the vertical axis and t on the horizontal axis. Use this graph to explain why you believe that F is a function of t.
    (c) Plot F on the vertical axis and R on the horizontal axis. From this graph show that F is not a function of R.
    (d) Plot R on the vertical axis and F on the horizontal axis. From this graph show that R is not a function of F.

4. The following three tables represent the relationship between the button number, N, which you push, and the snack, S, delivered by three different vending machines. [3]

    (a) One of these vending machines is not a good one to use, because S is not a function of N. Which one? Explain why this makes it a bad machine to use.
    (b) For which vending machine(s) is S a function of N? Explain why this makes them user-friendly.
    (c) For which of the vending machines is N not a function of S? What does this mean to the user of the vending machine?

Vending Machine #1

N	S
1	m&ms
2	pretzels
3	dried fruit
4	Hersheys
5	fat free cookies
6	Snickers

Vending Machine #2

N	S
1	m&ms or dried fruit
2	pretzels or Hersheys
3	Snickers or fat free cookies

Vending Machine #3

N	S
1	m&ms
2	m&ms
3	pretzels
4	dried fruit
5	Hersheys
6	Hersheys
7	fat free cookies
8	Snickers
9	Snickers

5. Using Table 1.7, sketch a graph of $n = f(A)$, the number of gallons of paint needed to cover a house of area A. Identify the independent and dependent variables.

TABLE 1.7

A	0	250	500	750	1000	1250	1500
n	0	1	2	3	4	5	6

---

[2]From *1993 World Almanac.*

[3]For each N, vending machine #2 dispenses one or the other product at random.

6. Use Table 1.8 to fill in the missing values. (There may be more than one answer.)

    (a) $f(0) = ?$ (b) $f(?) = 0$ (c) $f(1) = ?$ (d) $f(?) = 1$

TABLE 1.8

x	0	1	2	3	4
$f(x)$	4	2	1	0	1

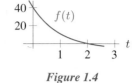

Figure 1.4

7. Use the graph in Figure 1.4 to fill in the missing values: (a) $f(0) = ?$ (b) $f(?) = 0$

8. (a) You are going to graph $p = f(w)$. Which variable goes on the horizontal axis?
   (b) If $10 = f(-4)$, give the coordinates of a point on the graph of f.
   (c) If 6 is a solution of the equation $f(w) = 1$, give a point on the graph of f.

9. (a) Which of the graphs in Figure 1.5 represent y as a function of x? (Note that an open circle indicates a point that is not included in the graph; a solid dot indicates a point that is included in the graph.)[4]

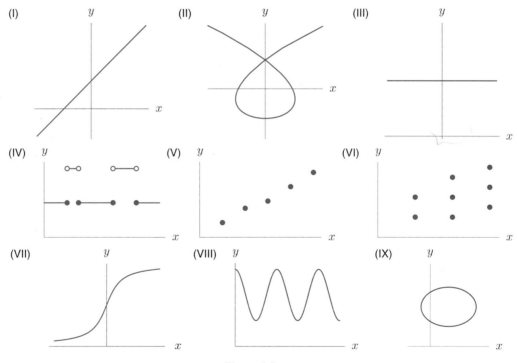

Figure 1.5

   (b) Which of the graphs in Figure 1.5 could represent the following situations? Give reasons.

      (i) SAT Math score versus SAT Verbal score for a small number of students.

      (ii) Total number of daylight hours as a function of the day of the year, shown over a period of several years.

   (c) Among graphs (I)–(IX) in Figure 1.5, find two which could give the cost of train fare as a function of the time of day. Explain the relationship between cost and time for both choices.

---

[4]See Appendix H, Example 11.

10. Consider the following stories about five different bike rides. Match each story to one of the graphs in Figure 1.6, where d represents distance from home and t is time in hours since the start of the ride. (A graph may be used more than once.)

    (a) Starts 5 miles from home and rides 5 miles per hour away from home.
    (b) Starts 5 miles from home and rides 10 miles per hour away from home.
    (c) Starts 10 miles from home and arrives home one hour later.
    (d) Starts 10 miles from home and is halfway home after one hour.
    (e) Starts 5 miles from home and is 10 miles from home after one hour.

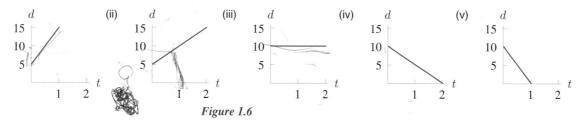

Figure 1.6

In Problems 11–14, label the axes for a sketch to illustrate the given statement.

11. "Over the past century we have seen changes in the population, P (in millions), of the city. . ."
12. "Sketch a graph of the cost of manufacturing q items. . ."
13. "Graph the pressure, p, of a gas as a function of its volume, v, where p is in pounds per square inch and v is in cubic inches."
14. "Graph D in terms of y. . ."
15. Suppose you are looking at the graph of y, a function of x.

    (a) What is the maximum number of times that the graph can intersect the y-axis? Explain.
    (b) Can the graph intersect the x-axis an infinite number of times? Explain.

16. A bug starts out ten feet from a light, flies closer to the light, then farther away, then closer than before, then farther away. Finally the bug hits the bulb and flies off. Sketch a possible graph of the distance of the bug from the light as a function of time.

17. A light is turned off for several hours. It is then turned on. After a few hours it is turned off again. Sketch a possible graph of the light bulb's temperature as a function of time.

18. The sales tax on an item is 6%. Express the total cost, C, in terms of the price of the item, P.

19. A cylindrical can is closed at both ends and its height is twice its radius. Express its surface area, S, as a function of its radius, r. [Hint: The surface of a can consists of a rectangle plus two circular disks.]

20. Suppose that $y = 3$ no matter what x is.

    (a) Is y a function of x? Explain.   (b) Is x a function of y? Explain.

21. Suppose that $x = 5$ no matter what y is.

    (a) Is y a function of x? Explain.   (b) Is x a function of y? Explain.

22. According to Charles Osgood, CBS news commentator, it takes about one minute to read 15 double-spaced typewritten lines on the air.[5]

    (a) Construct a table showing the time Charles Osgood is reading on the air in seconds as a function of the number of double-spaced lines read for $0, 1, 2, \ldots, 10$ lines. From your table, how long does it take Charles Osgood to read 9 lines?
    (b) Plot this data on a graph with the number of seconds on the vertical axis and the number of lines on the horizontal axis.
    (c) From your graph, estimate how long it takes Charles Osgood to read 9 lines. From your graph, estimate how many lines Charles Osgood can read in 30 seconds.
    (d) Construct a formula which relates the time T to n, the number of lines read.

---

[5]T. Parker, *Rules of Thumb*, (Boston: Houghton Mifflin, 1983).

23.   A chemical company spends $2 million to buy machinery before it starts producing chemicals. Then it spends $0.5 million on raw materials for each million liters of chemical produced.

   (a)   The number of liters produced ranges from 0 to 5 million. Make a table showing the relationship between the number of million liters produced, l, and the total cost, C, in millions of dollars, to produce that number of million liters.

   (b)   Find a formula that expresses C as a function of l.

24.   The distance between Cambridge and Wellesley is 10 miles. A person walks part of the way at 5 miles per hour, then jogs the rest of the way at 8 mph. Find a formula that expresses the total amount of time for the trip, $T(d)$, as a function of d, the distance walked.

25.   A person plans to leave home and walk due west for a time, and then to walk due north.

   (a)   Suppose the person will walk 10 miles in total. If w represents the (variable) distance west she walks, and D represents her (variable) distance from home at the end of her walk, is D a function of w? Why or why not?

   (b)   Suppose now that x is the distance that she walks in total. Is D a function of x? Why or why not?

1.2   PROPORTIONS AND RATES

Direct Proportionality

To tip a server in a restaurant, many people leave a gratuity that is *proportional to* the cost of the meal. The more expensive the meal, the larger the tip. Let's assume that a 15% tip is common, in which case

$$\text{Tip} = 15\% \times \text{Cost of meal}.$$

This formula gives the tip, T, as a function of the cost, c, of the meal, so we write $T = f(c)$. Table 1.9 reflects the fact that doubling the cost doubles the tip, tripling the cost triples the tip, and so on. The fact that the graph of T against c in Figure 1.7 is a straight line through the origin gives the same information.

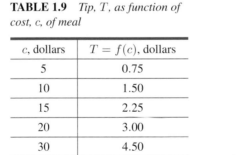

TABLE 1.9   *Tip, T, as function of cost, c, of meal*

c, dollars	$T = f(c)$, dollars
5	0.75
10	1.50
15	2.25
20	3.00
30	4.50

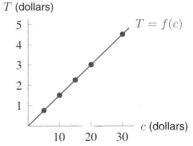

Figure 1.7: Tip, T, proportional to cost, c, of meal

In general, we say one quantity is *proportional*, or *directly proportional*, to another if

$$\text{Second quantity} = \text{Constant} \times \text{First quantity}.$$

The constant is known as the *constant of proportionality*. In symbols, the definition is as follows:

A quantity Q is **directly proportional**, or **proportional**, to a quantity t if

$$Q = k \cdot t,$$

where k is the constant of proportionality.

Example 1   (a)   Show that the circumference, C, of a circle is proportional to its diameter, d. What is the constant of proportionality?

(b)   Table 1.10 contains the diameter and circumference of the circular top of several types of tin cans. Graph the data and show that it satisfies the proportional relationship from part (a).

TABLE 1.10   *Diameter, d, and circumference, C, of tin cans*

Tin can	d (inches)	C (inches)
Tomato paste	2.1	6.6
Beef broth	2.6	8.2
Condensed milk	2.9	9.1
Chunk light tuna	3.4	10.7
Crushed tomatoes	4.0	12.6

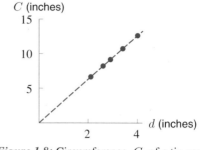

Figure 1.8: Circumference, C, of a tin can is proportional to its diameter, d

Solution   (a)   The formula giving the circumference of a circle in terms of its diameter is

$$C = \pi d,$$

where $\pi \approx 3.14$. Thus, C is proportional to d, with constant of proportionality π.

(b)   The data is graphed in Figure 1.8. Notice that the data points lie on a straight line through the origin. To check that the data satisfies the relationship $C = \pi d$, notice that Table 1.10 gives the diameter of a can of tomato paste as 2.1 inches and its circumference as 6.6 inches. After rounding,

$$6.6 = 3.14 \times 2.1.$$

Similarly, for the other cans, we see that, after rounding: $8.2 = 3.14 \times 2.6$, and $9.1 = 3.14 \times 2.9$, and $10.7 = 3.14 \times 3.4$, and $12.6 = 3.14 \times 4.0$. Thus,

$$C = 3.14d.$$

It is not always obvious that one quantity is proportional to another even when it is, as we see in the next example.

Example 2   For each of the following formulas, state whether y is directly proportional to x. If so, give the constant of proportionality, k.

(a) $y = 19x$ (b) $y = x/53$ (c) $y = x + 5.2\%(x)$
(d) $y = \sqrt{5}x$ (e) $y = x\pi^2$ (f) $y = 2 + 5x$

Solution   (a)   Yes, $y = kx$ with $k = 19$.
(b)   Yes, we have $y = (1/53)x$, so $k = 1/53$.
(c)   Yes. Since $5.2\% = 0.052$, we have $y = x + 0.052x = 1.052x$, so $k = 1.052$.
(d)   Yes, we have $k = \sqrt{5}$.
(e)   Yes, we have $k = \pi^2$.
(f)   No. It is not possible to find a value of k such that $y = kx$. The fact that we are adding 2 prevents this from being a direct proportion. However, $(y - 2)$ is directly proportional to x.

Proportionality to a Power of a Variable

The area, A, of a circle is a function of its radius, r. In fact, we know that $A = f(r) = \pi r^2$, so

$$\text{Area} = k \cdot (\text{Radius})^2.$$

TABLE 1.11   *Area, A, as a function of radius, r*

r	A
1	3.14
2	12.57
3	28.27
4	50.27
5	78.54
6	113.10

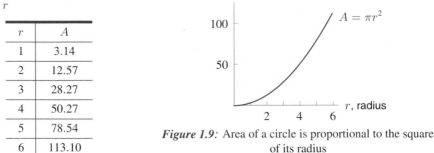

Figure 1.9: Area of a circle is proportional to the square of its radius

We say that A is proportional to the square of r. Values of f rounded to two decimal places are in Table 1.11; a graph f is in Figure 1.9.

Example 3   If we double the radius, r, of a circle, what happens to the corresponding value of the area, A?

Solution   Comparing the values of A for $r = 1$ and $r = 2$, for $r = 2$ and $r = 4$, or for $r = 3$ and $r = 6$ in Table 1.11 should convince you that the value of A does not double when the value of r is doubled. However, computing the ratios of these values does reveal a pattern:

$$\frac{f(2)}{f(1)} = \frac{12.57}{3.14} = 4.003, \qquad \frac{f(4)}{f(2)} = \frac{50.27}{12.57} = 3.999, \qquad \frac{f(6)}{f(3)} = \frac{113.10}{28.27} = 4.001.$$

All three ratios are close to 4. Taking into account the fact that the values in Table 1.11 are rounded to two decimal places suggests that doubling the value of r results in a four-fold increase in the value of A.

We can check this conjecture algebraically by comparing $f(r_2)$ with $f(r_1)$ for $r_2 = 2r_1$. First note that $f(r_1) = \pi(r_1)^2$. Then compute $f(r_2)$:

$$f(r_2) = \pi(r_2)^2 = \pi(2r_1)^2 = 4\pi(r_1)^2 = 4f(r_1).$$

Thus, the area of a circle of radius r_2 is four times the area of a circle of radius r_1.

The volume of a sphere of radius r is given by the formula $V = \frac{4}{3}\pi r^3$. This equation has the form

$$\text{Volume} = k \cdot (\text{Radius})^3.$$

We say that the volume of a sphere is proportional to the cube of its radius, and the constant of proportionality is $(4/3)\pi$. See Figure 1.10.

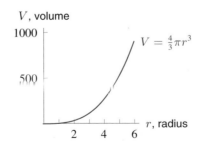

Figure 1.10: Volume of a sphere is proportional to the cube of its radius

The area of a circle and the volume of a sphere are examples of quantities that are directly proportional to a power of the variable r. The graphs of these functions in Figures 1.9 and 1.10 are both curves through the origin. Any function satisfying the following general definition with $n > 0$ has a graph through the origin.

A quantity y is **(directly) proportional to a power** of x if

$$y = kx^n, \qquad k \text{ and } n \text{ are constants.}$$

We usually take the power, n, to be positive. If n is negative, say $n = -m$, we often describe the relationship as *inverse proportionality* to x^m.

Inverse Proportionality

Provided n and k are positive, functions such as $y = kx^n$ that describe direct proportionality are examples of *increasing functions*. That is, as the input values increase, the output values increase also. The graph of an increasing function rises as we move from left to right. A *decreasing function* is one in which the output values decrease as the input values increase.

For example, suppose you take a part-time job to earn \$2400 for summer travel. The higher your rate of pay, the fewer hours you have to work. If you earn \$4 an hour, you have to work 600 hours to earn \$2400. On the other hand, if you find a job paying \$10 an hour, you only have to work 240 hours. If w is your hourly wage, the number of hours, h, you must work is given by

$$h = F(w) = \frac{2400}{w}.$$

Table 1.12 shows values of the function F; its graph is in Figure 1.11. Note that as the values of w increase, the values of $F(w)$ decrease.

TABLE 1.12 *Hours, h, as a function of wage, w*

w, dollars	h, hours
4	600
5	480
6	400
8	300
10	240

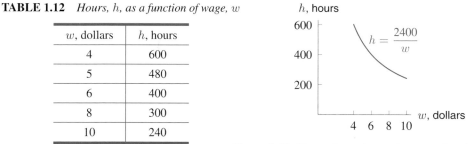

Figure 1.11: Hours, h, are inversely proportional to wages, w

The formula for the function F has the form

$$\text{Second quantity} = \frac{k}{\text{First quantity}}.$$

We say that the second quantity is *inversely proportional* to the first quantity.

It is also possible for one variable to be inversely proportional to a power of another variable. For example, if $y = k/x^2$, where k is a constant, then y is inversely proportional to x^2. In general, we make the following definition.

A quantity y is **inversely proportional** to x^n if

$$y = \frac{k}{x^n}, \qquad k \text{ and } n \text{ are constants, with } n > 0.$$

Example 4   If you drive 50 miles with your car set on cruise control, then the time, T, required to cover the distance is inversely proportional to your speed, R, because

$$T = \frac{50}{R}.$$

The graph of this function in Figure 1.12 shows that higher speeds lead to shorter travel times.

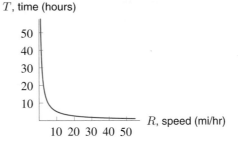

Figure 1.12: Time, T, is inversely proportional to speed, R

Example 5   An object's weight, w, is inversely proportional to the square of its distance, r, from the earth's center. This gives

$$w = \frac{k}{r^2}.$$

Here w is not inversely proportional to r, but to r^2.

It is important to realize that not every increasing function represents direct proportionality and not every decreasing function is an example of inverse proportionality.

Example 6   Which of the graphs in Figure 1.13-1.16 could describe direct proportionality to a power of x? Which could describe inverse proportionality to a power of x?

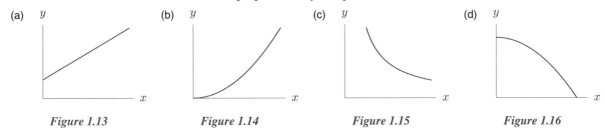

| Figure 1.13 | Figure 1.14 | Figure 1.15 | Figure 1.16 |

Solution   Graph (a) is increasing but cannot describe direct proportionality because it does not pass through the origin. Graph (b) could describe direct proportionality. Graph (c) could describe inverse proportionality. Graph (d) is decreasing but cannot describe inverse proportionality because it intersects the axes, which cannot happen with a function of the form $y = k/x^n$.

Proportionality and Rates

We can solve for the constant of proportionality, k, in the equation

$$\text{Second quantity} = k \times \text{First quantity}$$

to obtain

$$k = \frac{\text{Second quantity}}{\text{First quantity}}.$$

When k represents a commonly occurring ratio of two units, we often think of k as a rate.

Examples of rates include:

- Rate of pay ($7/hr)
- Rate of travel (60 miles/hour)
- Per capita rates ($300/person)
- Exchange rates ($1.60/£)

- Interest rates (6%/year)
- Rate of charge ($1.25/gallon)
- Rate of flow (15 gallons/min)
- Population density (200 people/square mile)

Example 7   At a constant speed, the distance a car travels is proportional to the time spent traveling:

$$\text{Distance traveled} = k \times \text{Time spent traveling.}$$

Solving for k gives

$$k = \frac{\text{Distance traveled}}{\text{Time spent traveling}}.$$

For instance, if a car goes 280 miles in 5 hours,

$$k = \frac{280 \text{ miles}}{5 \text{ hours}} = 56 \text{ mi/hr.}$$

In this example, k represents the car's rate of travel, or speed.

Example 8   The amount of interest a savings account pays in one year is proportional to the starting balance:

$$\text{Interest payment} = k \times \text{Starting balance.}$$

Solving for k gives

$$k = \frac{\text{Interest payment}}{\text{Starting balance}}.$$

For instance, if you deposit $1000 into an account that pays $55 interest after one year,

$$k = \frac{\$55/\text{year}}{\$1000} = 0.055 = 5.5\%/\text{year.}$$

In this example, the interest rate is 5.5% per year.

Example 9   For each situation, find the constant of proportionality and explain its meaning.

(a) The distance a car travels on the highway is proportional to the quantity of gas consumed. A car travels 225 miles on 5 gallons of gas.

(b) When you convert British pounds (£) into US dollars ($), the number of dollars you receive is proportional to the number of pounds you exchange. A traveler receives $400 in exchange for £250.

Solution   (a) Distance traveled is proportional to quantity of gas consumed, so

$$\text{Distance traveled} = k \times \text{Quantity of gas.}$$

Thus, the constant of proportionality is

$$k = \frac{225 \text{ miles}}{5 \text{ gallons}} = 45 \text{ miles/gallon.}$$

The constant k gives the rate of gas consumption in miles per gallon, or the fuel efficiency of the car.

(b) The number of dollars is proportional to the number of pounds, so

$$\text{Number of dollars} = k \times \text{Number of pounds.}$$

The constant of proportionality is

$$k = \frac{400 \text{ dollars}}{250 \text{ pounds}} = 1.60 \text{ dollars/pound.}$$

The constant k gives the conversion factor from pounds to dollars, or the rate of exchange in US dollars per British pound.

Example 10   In Example 9, the distance, d, in miles, traveled by a car is expressed as a function of the gas consumed, g, in gallons, so $d = f(g)$. The value in dollars, D, of currency is expressed as a function of its value in pounds, P, so $D = h(P)$. Write formulas for the functions f and h and graph them.

Solution   The car travels 45 miles/gallon, so

$$d = f(g) = 45g.$$

Each pound is worth $1.60, so

$$D = h(P) = 1.60P.$$

The graphs are in Figures 1.17 and 1.18. Note that both are straight lines through the origin.

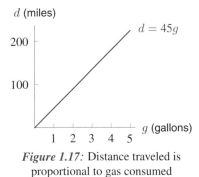

Figure 1.17: Distance traveled is proportional to gas consumed

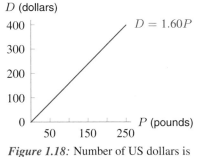

Figure 1.18: Number of US dollars is proportional to number of British pounds

Using Rates to Make Comparisons

We can use rates to make comparisons. For example, if a friend tells you that her new job has just paid her $200, you might be impressed. However, if she adds that the money was for 50 hours of work, you might be less impressed. Her

$$\text{Hourly rate of pay} = \frac{\text{Amount paid}}{\text{Hours worked}} = \frac{\$200}{50 \text{ hours}} = \$4/\text{hour},$$

less than minimum wage. Here, we are using a rate (your friend's rate of pay) to compare the amount of money she makes ($200) to the amount of time she works (50 hrs).

Rates make many types of comparisons easier. For instance, suppose a second friend tells you she has just been paid $100. This is less than $200, the amount your first friend made. However, the second friend explains that she earned the $100 in 10 hours, not 50. This means:

$$\text{Hourly rate of pay} = \frac{\text{Amount paid}}{\text{Hours worked}} = \frac{\$100}{10 \text{ hours}} = \$10/\text{hour}.$$

Even though she made less money than your first friend ($100 versus $200), the second friend's rate of pay is quite a bit higher ($10/hour versus $4/hour).

Crime Rates

The number of crimes in a city tends to be proportional to the size of its population. This leads to the notion of crime rate. Assuming that

$$\text{Number of crimes} = k \times \text{Number of people in city},$$

we solve for k by dividing:

$$k = \frac{\text{Number of crimes}}{\text{Number of people in city}} = \text{Crime rate}.$$

Example 11 In City A, 300 crimes are committed in one year. In City B, 400 crimes are committed in the same year. City A has 5000 people and City B has 8000 people. Which city is more dangerous?

Solution Since City B has a larger population than City A, we expect it to have more crimes than City A even if the two cities are equally dangerous. Calculating crime rates, we have

$$\text{City } A\text{'s crime rate} = \frac{\text{Number of crimes}}{\text{Number of people}} = \frac{300}{5000} = 0.06 = 6\%.$$

$$\text{City } B\text{'s crime rate} = \frac{\text{Number of crimes}}{\text{Number of people}} = \frac{400}{8000} = 0.05 = 5\%.$$

Thus, we conclude that City A is more dangerous, even though it has fewer crimes than City B.

Problems for Section 1.2

1. Suppose y is directly proportional to x. If $y = 6$ when $x = 4$, find the constant of proportionality and write the formula for y as a function of x. Use your formula to find x when $y = 8$.

2. Suppose y is inversely proportional to x. If $y = 6$ when $x = 4$, find the constant of proportionality and write the formula for y as a function of x. Use your formula to find x when $y = 8$.

3. Suppose c is directly proportional to the square of d. If $c = 45$ when $d = 3$, find the constant of proportionality and write the formula for c as a function of d. Use your formula to find c when $d = 5$.

4. Suppose c is inversely proportional to the square of d. If $c = 45$ when $d = 3$, find the constant of proportionality and write the formula for c as a function of d. Use your formula to find c when $d = 5$.

In Problems 5–12, state whether the function represents direct proportionality or inverse proportionality. Write each function in the form $y = kx^p$ and identify the values of k and p.

5. $y = -x$
6. $y = 2\pi x$
7. $y = x$
8. $y = mx$
9. $y = 0.34\left(\dfrac{x}{2}\right)$
10. $y = \dfrac{1}{x}$
11. $y = \dfrac{2}{3x}$
12. $y = \dfrac{-x}{2}$

13. The cost of denim fabric is directly proportional to the amount that you buy. Let $C(x)$ be the cost, in dollars, of x yards of denim fabric.

    (a) Write a formula for the cost, $C(x)$, in terms of x. Your answer will contain a constant, k.
    (b) A particular type of denim costs \$28.50 for 3 yards. Find the value of k and rewrite the formula for $C(x)$ using it.
    (c) Graph $C(x)$.
    (d) How much will it cost to buy 5.5 yards of denim?

14. A 30-second commercial during Super Bowl XXXIII in 1999 cost advertisers \$1.6 million. For the first Super Bowl in 1967, an advertiser could have purchased approximately 18.82 minutes of advertising time for the same amount of money.[6]

    (a) Assuming that cost is proportional to time, find the cost of advertising, in dollars/second, during the 1967 and 1999 Super Bowls.
    (b) How many times more expensive was Super Bowl advertising in 1999 than in 1967?

15. Three ounces of broiled ground beef contains 245 calories.[7] Are calories directly or inversely proportional to ounces? Explain your reasoning and write a formula for the proportion. How many calories are there in 4 ounces of broiled hamburger?

---

[6]From CNN.
[7]The World Almanac Book of Facts, 1999 p. 718

16.   A group of friends are planning to rent a house at the beach for spring break. If nine of them share the house, it costs $150 each. Is the cost to each person directly or inversely proportional to the number of people sharing the house? Explain your reasoning and write a formula for the proportion. How many people are needed to share the house if the students want to pay a maximum of $100 each?

17.   Driving at 55 mph, it takes approximately 3.5 hours to drive from Long Island to Albany, NY. Is the time the drive takes directly or inversely proportional to the speed? Explain your reasoning and write a formula for the proportion. To get to Albany in 3 hours, how fast would you have to drive?

18.   On a map, 1/2 inch represents 5 miles. Is the map distance between two locations directly or inversely proportional to the actual distance which separates the two locations? Explain your reasoning and write a formula for the proportion. How far apart are two towns if the distance between these two towns on the map is 3.25 inches?

19.   Throughout its long history, the annual number of crimes in a city has been directly proportional to the population. In 1900, the population was 23,000 and there were 415 crimes.

(a)   By 1925, the population had doubled. How many crimes were committed in the city that year?
(b)   In 1850, the population was only one-fourth as large as it was in 1900. How many crimes were committed in 1850?
(c)   By 1970, the population was about 160,000. How many crimes were there in 1970?
(d)   How large will the city be when there are 5000 crimes?

20.   On February 9, 1998, you could exchange $100 US for 143.27 Canadian dollars or for 847.46 Mexican pesos.

(a)   Fill in Table 1.13 with equivalent values in these three currencies.

TABLE 1.13

US $	200	50	10	1
Canadian dollars				
Mexican pesos				

(b)   What was the exchange rate from US currency to Canadian currency? (That is, how much was $1 US worth in Canadian dollars?)
(c)   What was the exchange rate from US dollars to Mexican pesos?
(d)   What was the exchange rate from Canadian dollars to US dollars? From Mexican pesos to US dollars?
(e)   What was the exchange rate from Canadian dollars to Mexican pesos?

In Problems 21–32, does the function represent proportionality to a power of the independent variable? That is, can the function be written in the form $y = kx^p$ for the variables given in the problem? If so, identify the constant, k, and the power, p.

21. $y = (2x)^5$

22. $y = \dfrac{3\sqrt{x}}{4}$

23. $y = \dfrac{3}{x^5}$

24. $y = \dfrac{3}{x^{2/3}}$

25. $y = \dfrac{-5}{t^{-4}}$

26. $s = \dfrac{x^{1/2}}{x^3}$

27. $y = \dfrac{\frac{1}{3}}{2x^7}$

28. $y = \dfrac{6}{-2/x^5}$

29. $z = 5(3)^4$

30. $C = 2q^3 \cdot 5$

31. $q = (5m + 1)^4$

32. $r = \dfrac{1}{2y}$

33.   The area of a circle is given by $A = \pi r^2$ and the volume of a sphere is given by $V = (4/3)\pi r^3$.

(a)   Find simplified formulas for A and V in terms of the circle's diameter, d, rather than its radius, r.
(b)   Is A proportional to a power of d? Is V proportional to a power of d?

34. The circulation time of a mammal—that is, the average time it takes for all the blood in the body to circulate once and return to the heart—is governed by the equation

$$t = 17.4m^{1/4},$$

where m is the body mass of the mammal in kilograms, and t is the circulation time in seconds.[8]

   (a) Complete Table 1.14 which shows typical body masses in kilograms for various mammals.[9]
   (b) If the circulation time of one mammal is twice that of another, what is the relationship between their body masses?

TABLE 1.14

Animal	Body mass (kg)	Circulation time (sec)
Blue whale	91000	
African elephant	5450	
White rhinoceros	3000	
Hippopotamus	2520	
Black rhinoceros	1170	
Horse	700	
Lion	180	
Human	70	

35. A volcano erupts in a powerful explosion. The sound from the explosion is heard in all directions for many hundreds of kilometers. The speed of sound is about 340 meters per second.

   (a) Fill in Table 1.15 showing the distance, d, in kilometers, that the sound of the explosion has traveled as a function of time. Write a formula for d as a function of time, t, in seconds.
   (b) How long after the explosion will a person living 200 km away hear the explosion?
   (c) Fill Table 1.15 showing the land area, A, in square kilometers, over which the explosion can be heard as a function of time. Write a formula for A as a function of t.
   (d) The average population density around the volcano is 31 people per square kilometer. Write a formula for P as function of t, where P is the number of people who have heard the explosion at time t.
   (e) Graph the function $P = f(t)$. How long will it take until 1 million people have heard the explosion?

TABLE 1.15

Time, t	5 sec	10 sec	1 min	5 min
Distance, d (km)				
Area, A (km^2)				

36. The thrust, T, in pounds delivered by a ship's propeller is proportional to the square of the propeller speed, R, in rotations per minute, times the fourth power of the propeller diameter, D, in feet.[10]

   (a) Write a formula for T in terms of R and D.
   (b) What happens to the thrust if the propeller speed is doubled?
   (c) What happens to the thrust if the propeller diameter is doubled?
   (d) If the propeller diameter is increased by 50%, by how much can the propeller speed be reduced to deliver the same thrust?

---

[8]K. Schmidt-Nielsen, *Scaling–Why is animal size so important?* (Cambridge, England: Cambridge University Press, 1984).

[9]R. McNeill Alexander, *Dynamics of Dinosaurs and Other Extinct Giants.* (New York: Columbia University Press, 1989).

[10]Gillner, Thomas C., *Modern Ship Design*, (US Naval Institute Press, 1972).

37.   Gold has the unusual property that it can be beaten into very thin sheets (called gold leaf), which can be used to ornament buildings, such as the dome of the Massachusetts State House. Table 1.16 shows the cost, C, of gold leaf needed to cover a hemispherical dome of various diameters.

(a)   Confirm that the cost appears to be proportional to the square of the diameter.

(b)   Assuming that cost is proportional to the square of diameter, find the constant of proportionality and write C as function of d.

(c)   Derive the formula you obtained in part (b) for C as a function of d. Use the fact that the surface area of a sphere of radius r is $S = 4\pi r^2$ and that one ounce of gold costs $400 and covers 17 square meters.

TABLE 1.16

Diameter, d (meters)	10	20	30	40
Cost, C (dollars)	3696	14784	33264	59136

38.   An average hailstone is a sphere of radius 0.3 centimeter. Severe thunderstorms can produce hailstones of radius 0.95 centimeter. The largest hailstone found in the US had radius 7.05 centimeters. Table 1.17 gives the masses of these hailstones in grams.[11]

(a)   Using the data given, check that mass, m, is proportional to the cube of the radius, r.

(b)   Find the constant of proportionality and write m as a function of r.

(c)   The largest recorded hailstone was found in India in 1939 and weighed 3.4 kilograms. What was its radius?

(d)   Calculate the density of ice in grams per cubic centimeter. [Hint: Density is mass per unit volume.]

TABLE 1.17

Radius, r (cm)	0.3	0.95	7.05
Mass, m (gm)	0.058	1.835	750

1.3   RATE OF CHANGE

In Section 1.2, we considered rates such as

$$\frac{\text{Number of dollars}}{\text{Number of hours}}, \quad \frac{\text{Number of miles}}{\text{Number of gallons}}, \quad \frac{\text{Number of crimes}}{\text{Number of people}}.$$

In this section, we consider a particular type of rate, called a *rate of change*. A rate of change is a rate of the form

$$\frac{\text{Change in one quantity}}{\text{Change in another quantity}}.$$

For example, in 1980 there were 63 million color TVs in use in the US, and in 1990 there were 90 million.[12] We say

$$\begin{aligned} \text{Rate of change in number of color TVs} &= \frac{\text{Change in number of color TVs}}{\text{Change in time}} \\ \text{between 1980 and 1990} \\ &= \frac{90 - 63}{10} \\ &= 2.7 \text{ million color TVs per year.} \end{aligned}$$

---

[11] Ahrens, C. Donald, *Essentials of Meteorology,* (Wadsworth: Belmont, CA, 1998).

[12] *Statistical Abstracts of the Unites States, 1998.* Table 915.

Let's compare the rate of change of color TVs before 1990 with a rate after 1990. In 1996 there were 95 million color TVs in use, so

$$\begin{aligned}\text{Rate of change in number of color TVs} \atop \text{between 1990 and 1996} &= \frac{\text{Change in number of color TVs}}{\text{Change in time}}\\ &= \frac{95 - 90}{6}\\ &= 0.83 \text{ million color TVs per year.}\end{aligned}$$

The smaller rate of change, 0.83 instead of 2.7 million color TVs per year, tells us that the rate of increase in the number of color TVs in use slowed in the early 1990s.

Average Speed

Two cyclists take a cross-country trip. Figure 1.19 is a graph of their distance traveled as a function of time, where $t = 0$ is the start of the trip. Amanda travels a constant speed throughout the trip, whereas Karim's speed varies.

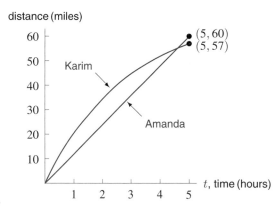

Figure 1.19: The distance traveled as a function of time

Over any time interval, we define:

$$\text{Average speed} = \frac{\text{Change in distance}}{\text{Change in time}} = \frac{\text{Distance traveled}}{\text{Time elapsed}}.$$

Thus, during the time interval $0 \le t \le 5$, Amanda went 60 miles, so

$$\text{Amanda's average speed} = \frac{\text{Distance traveled}}{\text{Time elapsed}} = \frac{60 \text{ miles}}{5 \text{ hours}} = 12 \text{ mph.}$$

Over the same interval $0 \le t \le 5$, Karim traveled 57 miles, so

$$\text{Karim's average speed} = \frac{\text{Distance traveled}}{\text{Time elapsed}} = \frac{57 \text{ miles}}{5 \text{ hours}} = 11.4 \text{ mph.}$$

Thus, for $0 \le t \le 5$, Amanda's average speed was greater than Karim's.

Figures 1.20 and 1.21 show that during the second hour of the trip, $1 \leq t \leq 2$,

$$\text{Karim's average speed} = \frac{15 \text{ miles}}{1 \text{ hour}} = 15 \text{ mph},$$

while

$$\text{Amanda's average speed} = \frac{12 \text{ miles}}{1 \text{ hour}} = 12 \text{ mph}.$$

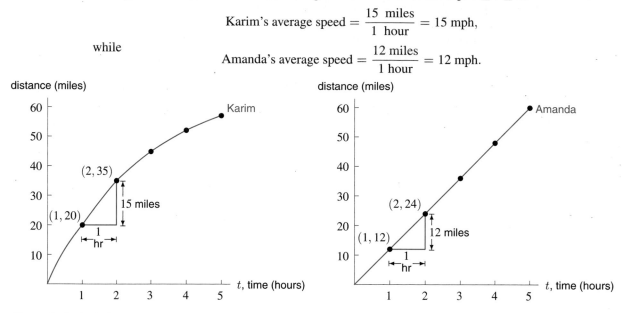

Figure 1.20: Karim's average speed is 15 mph for $1 \leq t \leq 2$ *Figure 1.21:* Amanda's average speed is 12 mph for $1 \leq t \leq 2$

Amanda's average speed remained a constant 12 mph over the entire trip. However, while Karim's speed over the entire 5-hour trip was 11.4 mph, during the second hour, his average speed was 15 mph. Average speed may be different on different time intervals.

Average Rate of Change of a Function

Average speed is a special case of the average rate of change of a function. In general, if $Q = f(t)$, we write ΔQ for a change in Q and Δt for a change in t. We define:[13]

> The **average rate of change**, or **rate of change**, of Q with respect to t over an interval is
>
> $$\frac{\text{Average rate of change}}{\text{over an interval}} = \frac{\text{Change in } Q}{\text{Change in } t} = \frac{\Delta Q}{\Delta t}.$$

The average rate of change of the function $Q = f(t)$ over an interval tells us how much Q changes, on average, for each unit change in t within that interval. On some parts of the interval, Q may be changing rapidly, while on other parts Q may be changing slowly. The average rate of change evens out these variations,

Example 1   Table 1.18 gives the annual sales (in millions) of compact discs (CDs) and vinyl long playing records (LPs) for some years between 1982 and 1994. What were the average rates of change of the annual sales of CDs and of LPs between 1982 and 1994?

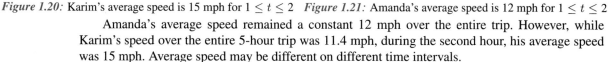

TABLE 1.18   *Annual sales of CDs and LPs (in millions) for selected years*

Year	1982	1984	1986	1988	1990	1992	1994
CD sales	0	5.8	53	150	287	408	662
LP sales	244	205	125	72	12	2.3	1.9

---

[13]The Greek letter Δ is often used in mathematics to represent change.

Solution   Let t be the number of years since 1982 and $s = C(t)$ be the sales (in millions) of CDs in year t. Then 1982 corresponds to $t = 0$ and 1994 corresponds to $t = 12$. Table 1.18 gives

$$\begin{array}{l} \text{Average rate of change of } s \\ \text{from 1982 to 1994} \end{array} = \frac{\text{Change in } s}{\text{Change in } t} = \frac{\Delta s}{\Delta t} = \frac{C(12) - C(0)}{12 - 0}$$

$$= \frac{662 - 0}{12}$$

$$\approx 55.2 \text{ million discs/year.}$$

Thus, CD sales increased on average by 55.2 million discs per year between 1982 and 1994.

Let $q = L(t)$ be the sales (in millions) of LPs during year t. Table 1.18 gives

$$\begin{array}{l} \text{Average rate of change of } q \\ \text{from 1982 to 1994} \end{array} = \frac{\text{Change in } q}{\text{Change in } t} = \frac{\Delta q}{\Delta t} = \frac{L(12) - L(0)}{12 - 0}$$

$$= \frac{1.9 - 244}{12}$$

$$\approx -20.2 \text{ million records/year.}$$

Thus, LP sales decreased on average by 20.2 million records per year between 1982 and 1994.

Increasing and Decreasing Functions

In the previous example, the average rate of change of CD sales is positive on the interval from 1982 to 1994 and we concluded that the sales of CDs increased over this interval. Similarly, the average rate of change of LP sales is negative on the same interval and we concluded that the sales of LPs decreased over this interval. The annual sales of CDs is an example of an *increasing function* and the annual sales of LPs is an example of a *decreasing function*. In general we say the following:

If $Q = f(t)$,
- f is an **increasing function** if the values of f increase as t increases.
- f is a **decreasing function** if the values of f decrease as t increases.

In the case of the CD sales, we observe that an increasing function has a positive rate of change. In the case of the LP sales, we see that a decreasing function has a negative rate of change. In general

---

If $Q = f(t)$,
- If f is an increasing function, then the average rate of change of f with respect to t is positive on every interval.
- If f is a decreasing function, then the average rate of change of f with respect to t is negative on every interval.

Example 2   Let $A = q(r)$ give the area, A, of a circle as a function of its radius, r, in centimeters. Sketch a graph of q. Explain how the fact that q is an increasing function and can be seen on a graph.

Solution   An increase in the radius results in an increase in the area, so $A = q(r)$ is an increasing function. Notice that the graph of $q(r)$ in Figure 1.22 climbs as we move from left to right and that the average rate of change, $\Delta A / \Delta r$ is positive on every interval.

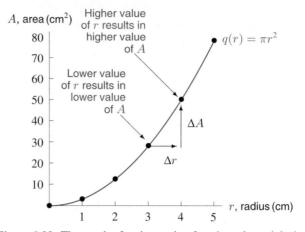

Figure 1.22: The graph of an increasing function: $A = q(r)$, the area of a circle of radius r, rises when read from left to right

Example 3   Carbon-14 is a radioactive element that exists naturally in the atmosphere and is absorbed by the tissues of living organisms. When an organism dies, carbon-14 is no longer absorbed and the amount present at death begins to decay (that is, to decrease).

Let $L = g(t)$ represent the quantity of carbon-14 (in micrograms, μg) in a tree t years after its death. A tree contains 200 μg (micrograms) of carbon-14 when it dies. Table 1.19 gives the amount of carbon-14 in the tree t years after its death. Explain why we expect g to be a decreasing function of t. How is this represented on a graph?

TABLE 1.19   *Quantity of carbon-14 as a function of time*

t, time (years)	0	1000	2000	3000	4000	5000
L, quantity of carbon-14 (μg)	200	177	157	139	123	109

Solution   Since the amount of carbon-14 is decaying over time, g is a decreasing function. A graph of $L = g(t)$ is shown in Figure 1.23. The graph falls as we move from left to right and the average rate of change in the level of carbon-14 with respect to time is negative on every interval.

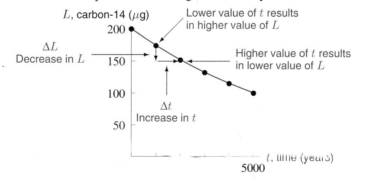

Figure 1.23: The graph of a deccreasing function: $L = g(t)$, the amount of carbon-14 remaining after t years, falls when read from left to right

In general, we can identify an increasing or decreasing function from its graph as follows:

- The graph of an increasing function rises when read from left to right.
- The graph of a decreasing function falls when read from left to right.

Many functions have some intervals on which they are increasing and other intervals on which they are decreasing. These intervals can be identified from the graph.

Example 4   On what intervals is the function graphed in Figure 1.24 increasing? Decreasing?

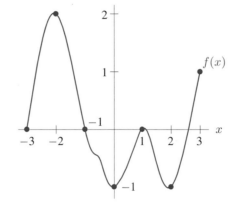

Figure 1.24: Graph of a function which is increasing on some intervals and decreasing on others

Solution   The function appears to be increasing for values of x between -3 and -2, for x between 0 and 1, and for x between 2 and 3. The function appears to be decreasing for x between -2 and 0 and for x between 1 and 2. Using inequalities, we say that f is increasing for $-3 < x < -2$, for $0 < x < 1$, and for $2 < x < 3$. Similarly, f is decreasing for $-2 < x < 0$ and $1 < x < 2$.

Function Notation for the Average Rate of Change

Suppose we want to find the average rate of change of a function $Q = f(t)$ over the interval $a \leq t \leq b$. On the interval from $t = a$ to $t = b$, the change in t is given by

$$\Delta t = b - a.$$

At $t = a$, the value of Q is $f(a)$, and at $t = b$, the value of Q is $f(b)$. Therefore, the change in Q is given by

$$\Delta Q = f(b) - f(a).$$

Using function notation, we express the average rate of change as follows:

$$\text{Average rate of change of } Q = f(t) \text{ over the interval } a \leq t \leq b = \frac{\text{Change in } Q}{\text{Change in } t} = \frac{\Delta Q}{\Delta t} = \frac{f(b) - f(a)}{b - a}.$$

In Figure 1.25, notice that the average rate of change is given by the ratio of the rise, $f(b) - f(a)$, to the run, $b - a$. This ratio is the *slope* of the dashed line segment.[14]

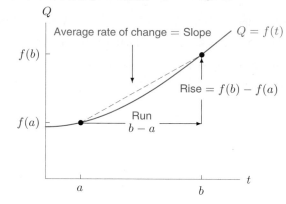

Figure 1.25: Visualizing the average rate of change of a function as a slope, or as the ratio Rise/Run

In previous examples we calculated the average rate of change from data. We now calculate average rates of change for functions given by formulas.

Example 5   Calculate the average rates of change of the function $f(x) = x^2$ between $x = 1$ and $x = 3$ and between $x = -2$ and $x = 1$. Show your results on a graph.

Solution   Between $x = 1$ and $x = 3$, we have

$$\begin{aligned}\text{Average rate of change of } f(x) \\ \text{over the interval } 1 \le x \le 3\end{aligned} = \frac{\text{Change in } f(x)}{\text{Change in } x} = \frac{f(3) - f(1)}{3 - 1}$$

$$= \frac{3^2 - 1^2}{3 - 1} = \frac{9 - 1}{2} = 4.$$

Between $x = -2$ and $x = 1$, we have

$$\begin{aligned}\text{Average rate of change of } f(x) \\ \text{over the interval } -2 \le x \le 1\end{aligned} = \frac{\text{Change in } f(x)}{\text{Change in } x} = \frac{f(1) - f(-2)}{1 - (-2)}$$

$$= \frac{1^2 - (-2)^2}{1 - (-2)} = \frac{1 - 4}{3} = -1.$$

The average rate of change between $x = 1$ and $x = 3$ is positive because $f(x)$ is increasing on this interval. See Figure 1.26. However, on the interval from $x = -2$ and $x = 1$, the function is partly decreasing and partly increasing. The average rate of change on this interval is negative because the decrease on the interval is larger than the increase.

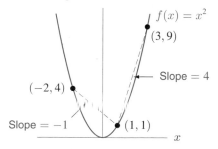

Figure 1.26: Average rate of change of $f(x)$ on an interval is slope of dashed line on that interval

---

[14]See Section 2.1 for further discussion of slope.

Problems for Section 1.3

1. Table 1.18 on page 22 gives the annual sales (in millions) of compact discs and vinyl long playing records. What was the average rate of change of annual sales of each of them between

   (a) 1982 and 1984?   (b) 1986 and 1988?

   (c) Interpret these results in terms of sales.

2. Table 1.18 on page 22 shows that CD sales are a function of LP sales. Is it an increasing or decreasing function?

3. Because scientists know how much carbon-14 a living organism should have in its tissues, they can measure the amount of carbon-14 present in the tissue of a fossil and then calculate how long it took for the original amount to decay to the current level, thus determining the time of the organism's death. A tree fossil is found to contain 130 μg of carbon-14, and scientists determine from the size of the tree that it would have contained 200 μg of carbon-14 at the time of its death. Using Table 1.19 on page 24, approximately how long ago did the tree die?

4. Figure 1.27 shows distance traveled as a function of time.

   (a) Find ΔD and Δt between:

   (i) $t = 2$ and $t = 5$ (ii) $t = 0.5$ and $t = 2.5$ (iii) $t = 1.5$ and $t = 3$

   (b) Compute the rate of change, $\Delta D / \Delta t$, and interpret its meaning.

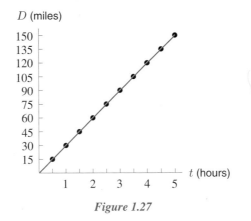

Figure 1.27

5. A classmate looks at the graph of the function $f(x) = x^2$ and says that f is an increasing function. Do you agree? Why or why not?

6. The most freakish change in temperature ever recorded was from $-4°$F to $45°$F between 7:30 am and 7:32 am on January 22, 1943 at Spearfish, South Dakota.[15] What was the average rate of change of the temperature for this time period?

7. Table 1.20 shows data for two populations for five different years. Find the average rate of change of each population over the following intervals.

   (a) 1980 to 1990   (b) 1985 to 1997   (c) 1980 to 1997

   TABLE 1.20

year	1980	1982	1985	1990	1997
P_1 (hundreds)	53	63	73	83	93
P_2 (hundreds)	85	80	75	70	65

---

[15] *The Guinness Book of Records.* 1995.

8. Table 1.21 gives the populations of two cities (in thousands) over a 17-year period.

    (a) Find the average rate of change of each population on the following intervals:

        (i)   1980 to 1990   (ii)   1980 to 1997   (iii)   1985 to 1997

    (b) What do you notice about the average rate of change of each population? Explain what the average rate of change tells you about each population.

    TABLE 1.21

Year	1980	1982	1985	1990	1997
P_1 (thousands)	42	46	52	62	76
P_2 (thousands)	82	80	77	72	65

9. (a) Let $f(x) = 16 - x^2$. Compute each of the following expressions, and interpret each as an average rate of change.

        (i) $\dfrac{f(2) - f(0)}{2 - 0}$ (ii) $\dfrac{f(4) - f(2)}{4 - 2}$ (iii) $\dfrac{f(4) - f(0)}{4 - 0}$

    (b) Sketch a graph of $f(x)$. Illustrate each ratio in part (a) by sketching the line segment with the given slope. Over which interval is the average rate of decrease the greatest?

10. Figure 1.28 shows the graph of the function $g(x)$.

    (a) Estimate $\dfrac{g(4) - g(0)}{4 - 0}$.

    (b) The ratio in part (a) is the slope of a line segment joining two points on the graph. Sketch this line segment on the graph.

    (c) Estimate $\dfrac{g(b) - g(a)}{b - a}$ for $a = -9$ and $b = -1$.

    (d) On the graph, sketch the line segment whose slope is given by the ratio in part (c).

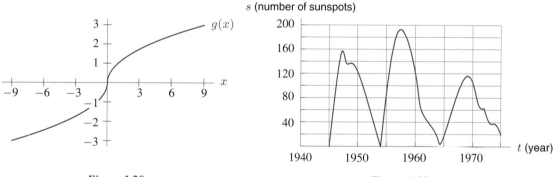

s (number of sunspots)

Figure 1.28		*Figure 1.29*

11. The surface of the sun has dark areas known as sunspots. Sunspots are cooler than the rest of the sun's surface, although they are still quite hot. The number of visible sunspots fluctuates with time, as shown in Figure 1.29. Is the number of sunspots, s, observed during year t a function of t? If so, is s an increasing or a decreasing function of t?

12. Using Figure 1.29, approximate the time intervals on which s is an increasing function of t.

13. Table 1.22 is a data set corresponding to the times achieved every 10 meters by Carl Lewis in the 100 meter final of the World Championship in Rome in 1987.[16]

    (a) For each successive time interval, calculate the average rate of change of distance. What is a common name for the average rate of change of distance?

---

[16]W. G. Pritchard, "Mathematical Models of Running", *SIAM Review.* 35, 1993, p.359 - 379.

(b)   Where did Carl Lewis attain his maximum speed during this race? Some runners are running their fastest as they cross the finish line. Does that seem to be true in this case?

TABLE 1.22

Time (sec)	0.00	1.94	2.96	3.91	4.78	5.64	6.50	7.36	8.22	9.07	9.93
Distance (meters)	0	10	20	30	40	50	60	70	80	90	100

14.   Table 1.23 gives the amount of garbage produced in the US per year as reported by the EPA.[17]

(a)   What is the value of Δt for consecutive entries in this table?
(b)   Calculate the value of ΔG for each pair of consecutive entries in this table.
(c)   Are all the values of ΔG you found in part (b) the same? What does this tell you?

TABLE 1.23

t (year)	1960	1965	1970	1975	1980	1985	1990
G (millions of tons of garbage per year)	90	105	120	130	150	165	180

15.   Match each story with the table and graph which best represent it.

(a)   When you study a foreign language, the number of new verbs you learn increases rapidly at first, but slows almost to a halt as you approach your saturation level.
(b)   You board an airplane in Philadelphia heading west. Your distance from the Atlantic Ocean, in kilometers per minute, increases at a constant rate.
(c)   The interest on your savings plan is compounded annually. At first your balance grows slowly, but its rate of growth continues to increase.

(E)

x	0	5	10	15	20	25
y	20	275	360	390	395	399

(F)

x	0	5	10	15	20	25
y	20	36	66	120	220	400

(G)

x	0	5	10	15	20	25
y	20	95	170	245	320	395

(I)   (II)   (III)

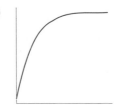

---

[17] Adapted from *Characterization of Solid Waste in the United States.* 1992, EPA.

REVIEW PROBLEMS FOR CHAPTER ONE

1. The nautilus, a multi-chambered mollusk, the canal of the inner ear, and petals of certain flowers have the spiral shape shown in Figure 1.30.

   (a) Is y a function of x? (b) Is x a function of y?

   (c) Is there any interval on the x-axis for which y is a function of x?

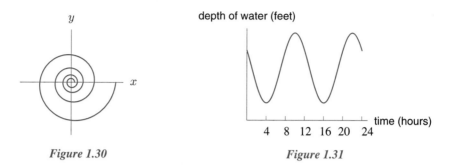

   Figure 1.30   **Figure 1.31**

2. Figure 1.31 gives the depth of the water at Montauk Point, New York, for a day in November.

   (a) How many high tides took place on this day?
   (b) How many low tides took place on this day?
   (c) How much time elapsed in between high tides?

3. According to Jill Phillips, a home economist, the ideal size refrigerator for a family is 8 cubic feet for two people, plus 1 cubic foot for each additional member of the family for families with 3 through 10 members.[18]

   (a) Construct a table showing the size of the ideal refrigerator as a function of the number of family members. From your table, what size refrigerator should a family of 5 own?
   (b) Plot a graph with refrigerator size on the vertical axis and the number of family members on the horizontal axis.
   (c) From your graph, estimate the ideal sized refrigerator for a family of 6.
   (d) Construct a formula which relates the size S to n, the number of family members.

4. (a) Is the area, A, of a square a function of the length of one of its sides, s?
   (b) Is the area, A, of a rectangle a function of the length of one of its sides, s?

5. A person's blood sugar level at a particular time of the day is partially determined by the time of the most recent meal. After a meal, blood sugar level increases rapidly, then slowly comes back down to a normal level. Sketch a graph showing a person's blood sugar level as a function of time over the course of a day. Label the axes to indicate normal blood sugar level and the time of each meal.

6. Many people think that hair growth is stimulated by haircuts. In fact, there is no difference in the rate hair grows after a haircut, but there *is* a difference in the rate at which hair's ends break off. A haircut eliminates dead and split ends, thereby slowing the rate at which hair breaks. However, even with regular haircuts, hair will not grow to an indefinite length. The average life cycle of human scalp hair is 3-5 years, after which the hair is shed.[19]

   Suppose Judy trims her hair once a year, when its growth is slowed by split ends. She cuts off just enough to eliminate dead and split ends, and then lets it grow another year. After 5 years, she realizes her hair won't grow any longer. Give a possible graph of the length of her hair as a function of time. Indicate when she receives her haircuts.

---

[18]T. Parker, *Rules of Thumb.* (Boston: Houghton Mifflin, 1983).
[19]*Britannica Micropedia* vol. 5. (Chicago: Encyclopaedia Britannica, Inc., 1989).

7. At the end of semester, students' math grades are listed in a table which gives each student's ID number in the left column and the student's grade in the right column. Let N represent the ID number and the G represent the grade. Which quantity, N or G, must necessarily be a function of the other?

8. Match each of the following descriptions with an appropriate graph and table of values.

    (a) The weight of your jumbo box of Fruity Flakes decreases by an equal amount every week.
    (b) The machinery depreciated rapidly at first, but its value declined more slowly as time went on.
    (c) In free fall, your distance from the ground decreases at an increasing rate.
    (d) For a while it looked like the decline in profits was slowing down, but then they began declining ever more rapidly.

(E)

x	0	1	2	3	4	5
y	400	384	336	256	144	0

(F)

x	0	1	2	3	4	5
y	400	320	240	160	80	0

(G)

x	0	1	2	3	4	5
y	400	184	98	63	49	43

(H)

x	0	1	2	3	4	5
y	412	265	226	224	185	38

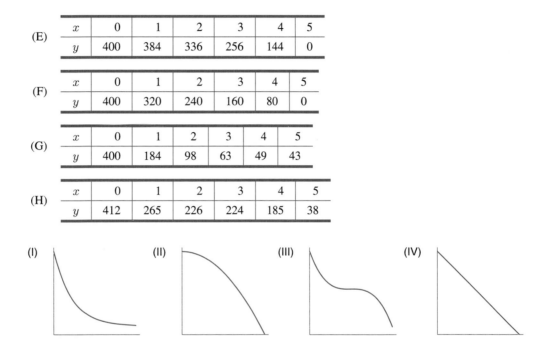

(I)   (II)   (III)   (IV)

9. A price increases 5% due to inflation and is then reduced 10% for a sale. Express the final price as a function of the original price, P.

10. An 8-foot tall cylindrical water tank has a base of diameter 6 feet.

    (a) How much water can the tank hold?
    (b) How much water is in the tank if the water is 5 feet deep?
    (c) Write a formula for the volume of water as a function of its depth in the tank.

For each of the formulas in Problems 11–19, state whether y is directly proportional to x and, if so, give the constant of proportionality.

11. $y = 5x$ 12. $y = x \cdot 7$ 13. $y = x \cdot x$ 14. $y = \sqrt{5} \cdot x$

15. $y = x/9$ 16. $y = 9/x$ 17. $y = x + 2$ 18. $y = 3(x + 2)$

19. $y = 6z$ where $z = 7x$

Identify each of the relationships in Problems 20–21 as a direct or inverse proportion. Explain your reasoning. Find the missing quantity.

20. If two typists can type a manuscript in 8 days, how many typists (working at the same rate) would be needed to type the manuscript in 5 days?

21. A cookie recipe which calls for one and a half cups of sugar makes about 2 dozen cookies. Is the number of cookies directly or inversely proportional to the number of cups of sugar? How many cookies can be made with two cups of sugar?

22. A meal in a restaurant costs M. The tax on the meal is 5%. You decide to tip the server 15%.

    (a)   Write an algebraic expression in terms of M for the amount you pay the server. Include the cost of the meal, tax, and tip and assume that you tip the waiter 15% of the cost of the meal:

        (i)   Not including the tax.   (ii)   Including the tax.

        (Opinions differ as to which method is appropriate.)

    (b)   In each case in part (a), is the total amount paid proportional to the cost of the meal?

23. The number of crimes, N, in a city is proportional to its population, P. Year after year, the constant of proportionality has remained fixed even though the population has grown.

    (a)   Find a formula for N in terms of P. Your answer will involve a constant, k.
    (b)   If the population P doubles, what happens to N?
    (c)   If the population P goes up by 1,000, what happens to N?

24. Carbon-14 is a radioactive isotope of carbon that decays into other elements over time. It is often used to determine the age of fossils. The amount of carbon-14 that decays in 1,000 years is proportional to the starting amount and is given by the formula

$$\text{Amount that decays in 1,000 years} = k \times \text{Starting amount}$$

where $k = 0.883$. A bone contains 200 mg of carbon-14. How much carbon-14 will it contain after:
    (a) 1,000 years?   (b) 2,000 years?   (c) 5,000 years?

25. In Example 4 on page 14 we saw that the time, T, needed to travel a fixed distance, D, is inversely proportional to the (constant) rate, R.

    (a)   Find D if it takes 20 hours to travel this distance at a constant rate of 60 miles per hour.
    (b)   Graph T as a function of R for $0 < R < 600$.
    (c)   If $T = f(R)$, calculate $f(300)$ and label the corresponding point on your graph. What does $f(300)$ mean in the context of the problem?

26. An astronaut's weight, w, is inversely proportional to the square of his distance, r, from the earth's center. Suppose that he weighs 180 pounds at the earth's surface and that the radius of the earth is approximately 3960 miles.

    (a)   Find the constant of proportionality, k.
    (b)   Graph w as a function of r for $3960 < r < 6960$.
    (c)   If $w = f(r)$, find $f(5000)$ and label the corresponding point on the graph. What does $f(5000)$ mean in the context of the problem?

27. The radius, r, of a sphere is proportional to the cube root of its volume, V.

    (a)   A spherical tank has radius 10 centimeters and volume 4188.79 cubic centimeters. Find the constant of proportionality and write r as a function of V, that is, $r = f(V)$.
    (b)   Graph $r = f(V)$ for $0 < V < 10,000$.
    (c)   The volume of the sphere in part (a) is doubled. What is the new radius? Write your answer using function notation and label the corresponding point on the graph.

28.   The energy, E, in foot-pounds delivered by an ocean wave is proportional[20] to the length, L, of the wave times the square of its height, h.

   (a)   Write a formula for E in terms of L and h.
   (b)   A 30-foot high wave of length 600 feet delivers 4 million foot-pounds of energy. Find the constant of proportionality.
   (c)   Sketch a graph of E as a function of L for 10-foot high waves.
   (d)   Sketch a graph of E as a function of h for 100-foot long waves.

29.   For n a positive integer, the product $1 \times 2 \times 3 \times \cdots \times (n-1) \times n$ is called n *factorial*, and is denoted by $n!$.

   (a)   Let $p(n) = n!$ for n a positive integer. Evaluate $p(n)$ for $n = 1, 2, 3, \ldots, 10$. Compile your results in a table.
   (b)   Most calculators can evaluate $n!$. What is the largest value of n for which your calculator will compute $n!$?

30.   The period of a pendulum is the amount of time required for the pendulum to make one full swing. The period, p, of a pendulum is proportional to the square root of its length, l.

   (a)   The pendulum in a grandfather clock is 3 feet long and has a period of 1.924 seconds. Find the constant of proportionality, and write p as a function of l, that is, $p = f(l)$.
   (b)   Graph $p = f(l)$ for $0 < l < 200$.
   (c)   The period of Foucault's pendulum, built in 1851 in the Pantheon in Paris, was 15.59 seconds. Find the length of the pendulum, and write your answer using function notation. Locate the corresponding point on your graph.

31.   Figure 1.32 shows the average monthly temperature in Albany, New York, over a twelve-month period. (January is month 1.)

   (a)   Make a table showing average temperature as a function of the month of the year.
   (b)   What is the warmest month in Albany?
   (c)   Over what interval of months is the temperature increasing? Decreasing?

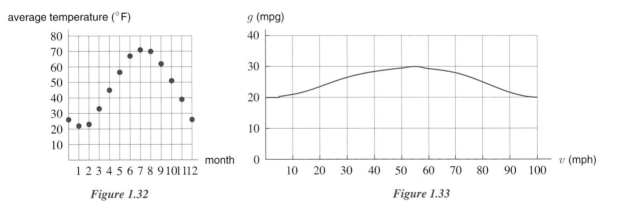

Figure 1.32   Figure 1.33

32.   Figure 1.33 shows the fuel consumption (in miles per gallon, mpg) of a car traveling at various speeds.

   (a)   How much gas is used on a 300 mile trip at 40 mph?
   (b)   How much gas is saved by traveling 60 mph instead of 70 mph on a 200 mile trip?
   (c)   According to this graph, what is the most fuel-efficient speed to travel? Explain.

[20] Cillner, Thomas C., *Modern Ship Design*, (US Naval Institute Press, 1972).

33. Academics have suggested that loss of worker productivity can result from sleep deprivation. An article in the Sunday, September 26, 1993, *New York Times* quotes David Poltrack, the senior vice president for planning and research at CBS, as saying that seven million Americans are staying up an hour later than usual to watch talk show host David Letterman. The article goes on to quote Timothy Monk, a professor at the University of Pittsburgh School of Medicine, as saying "... my hunch is that the effect [on productivity due to sleep deprivation among this group] would be in the area of a 10 percent decrement." The article next quotes Robert Solow, a Nobel prize-winning professor of economics at MIT, who suggests the following procedure to estimate the impact that this loss in productivity will have on the US economy – an impact he dubbed "the Letterman loss." First, Solow says, we find the percentage of the work force who watch the program. Next, we determine this group's contribution to the gross domestic product (GDP). Then we reduce the group's contribution by 10% to account for the loss in productivity due to sleep deprivation. The amount of this reduction is "the Letterman loss."

(a) The article estimated that the GDP is $6.325 trillion, and that 7 million Americans watch the show. Assume that the nation's work force is 118 million people and that 75% of David Letterman's audience belongs to this group. What percentage of the work force is in Dave's audience?

(b) What percent of the GDP would be expected to come from David Letterman's audience? How much money would they have contributed if they hadn't watched the show?

(c) How big is "the Letterman Loss"?

34. While white-water rafting, the guide in a kayak accompanies the raft. At points where the river is narrow, the kayak moves ahead of the raft, then waits for the raft to catch up. Where the river is wide, the kayak stays alongside the raft. The raft moves faster in regions where the river is narrow than in regions where the river is wide. Suppose that a kayak and a raft travel down the Reventazón River in Costa Rica. The river is narrow for the first half-mile, then wide for the next three quarters of a mile, then narrow for another two miles, and lastly wide for a mile.

(a) Sketch the graph of the raft's distance from its starting point as a function of time. Label the narrow and wide regions of the river on your graph.

(b) On the same graph as in part (a), sketch the position of the kayak as a function of time.

35. An incumbent politician running for reelection declared that the number of violent crimes is no longer rising and is presently under control. Does the graph shown in Figure 1.34 support this claim? Why or why not?

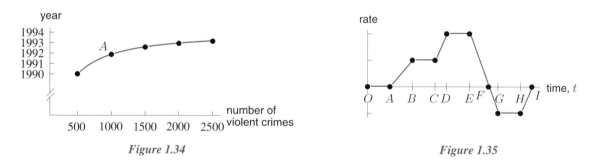

Figure 1.34   *Figure 1.35*

36. The rate at which water is entering a reservoir is given for time $t > 0$ by the graph in Figure 1.35. A negative rate means that water is leaving the reservoir. For each of the following statements, give the largest interval on which:

(a) The volume of water is increasing.

(b) The volume of water is constant.

(c) The volume of water is increasing fastest.

(d) The volume of water is decreasing.

CHAPTER TWO

LINEAR FUNCTIONS

Many functions can be grouped into families. Functions belonging to the same family have formulas that are alike; their graphs share a characteristic shape and their tables display similar patterns. In this chapter, we study linear functions — functions whose average rates of change are the same over every interval and whose graphs are straight lines.

2.1 WHAT MAKES A FUNCTION LINEAR?

Constant Rate of Change

In Chapter 1 we introduced the average rate of change of a function on an interval. For many functions, the average rate of change is different on different intervals. In this chapter, we consider functions which have the same average rate of change on every interval. Such a function has a graph which is a line and is called *linear*.

Population Growth

Mathematical models of population growth are used by city planners to project the growth of towns and by federal policy makers to project the growth of states. Biologists and ecologists model the growth of animal populations and physicians model the spread of an infection in the bloodstream. One model, a linear growth model, assumes that the population changes at the same average rate on every interval.

Example 1   A town of 30,000 people grows by 2000 people every year. If P gives the town's population, then P is a linear function of the time, t, measured in years, because the population is growing at the constant rate of 2000 people per year.

  (a)   What is the average rate of change of P over every time interval?
  (b)   Make a table that gives the town's population every five years over a 20-year period. Sketch a graph of the population over time.
  (c)   Find a formula for P in terms of t.

Solution   (a)   The average rate of change of population with respect to time is 2000 people per year.
  (b)   The initial population in year $t = 0$ is $P = 30,000$ people. Since the town grows by 2000 people every year, after five years it has grown by

$$\frac{2000 \text{ people}}{\text{year}} \times 5 \text{ years} = 10,000 \text{ people}.$$

Thus, in year $t = 5$ the population is given by

$$P = \text{Initial population} + \text{New population} = 30,000 + 10,000 = 40,000.$$

Similarly, in year $t = 10$ the population is given by

$$P = 30,000 + \underbrace{2000 \text{ people/year} \times 10 \text{ years}}_{20,000 \text{ new people}} = 50,000.$$

In year $t = 15$ the population is given by

$$P = 30,000 + \underbrace{2000 \text{ people/year} \times 15 \text{ years}}_{30,000 \text{ new people}} = 60,000,$$

and in year $t = 20$,

$$P = 30,000 + \underbrace{2000 \text{ people/year} \times 20 \text{ years}}_{40,000 \text{ new people}} = 70,000.$$

The results are in Table 2.1 and Figure 2.1; the dashed line shows the trend in the data.

TABLE 2.1   *Population over a 20-year period*

t, years	P, population
0	30,000
5	40,000
10	50,000
15	60,000
20	70,000

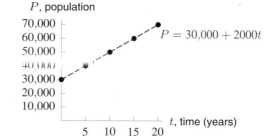

Figure 2.1: The town's population over 20 years

(c)   From part (b), we see that the size of the population is given by

$$P = \text{Initial population} + \text{Number of new people}$$
$$= 30{,}000 + 2000 \text{ people/year} \times \text{Number of years,}$$

so a formula for P in terms of t is

$$P = 30{,}000 + 2000t.$$

The graph of the population data in Figure 2.1 is a straight line. The average rate of change of the population over every interval is the same, namely 2000 people per year. Any linear function has the same average rate of change over every interval. Thus, we talk about *the* rate of change of a linear function. We say that a linear function changes at a constant rate. In contrast, a nonlinear function has different rates of change over different intervals. In general:

- A **linear function** has a constant rate of change.
- The graph of any linear function is a straight line.

Financial Models

Economists and accountants use linear functions for *straight-line depreciation*. For tax purposes, the value of certain equipment is considered to decrease, or depreciate, over time. For example, computer equipment may be state-of-the-art today, but after several years it is outdated. Straight-line depreciation assumes that the rate of change of value with respect to time is constant.

Example 2   A small business spends $20,000 on new computer equipment and, for tax purposes, chooses to depreciate it to $0 at a constant rate over a five-year period. Make a table and a graph showing the value of the equipment over the five-year period.

Solution   After five years the equipment is valued (for tax purposes) at $0. If V is the value and t the number of years, we see that

$$\begin{array}{c}\text{Average rate of change of value} \\ \text{from } t = 0 \text{ to } t = 5\end{array} = \frac{\text{Change in value}}{\text{Change in time}} = \frac{\Delta V}{\Delta t} = \frac{-\$20{,}000}{5 \text{ years}} = -\$4000 \text{ per year.}$$

Thus, the value drops at the constant rate of $4000 per year. (Notice that ΔV is negative because the value of the equipment decreases.) Table 2.2 gives the value of the equipment each year. The data are plotted in Figure 2.2. Since V changes at a constant rate, $V = f(t)$ is a linear function and its graph is a straight line. The rate of change, $-\$4000$ per year, is negative because the function is decreasing and the graph slopes down.

TABLE 2.2   *Value of equipment depreciated over a 5-year period*

t, year	V, value ($)
0	20,000
1	16,000
2	12,000
3	8,000
4	4,000
5	0

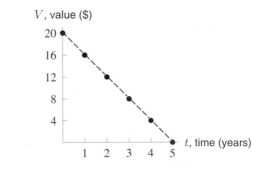

Figure 2.2: Value of computer equipment depreciated over a 5-year period

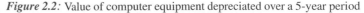

The total cost of production is another application of linear functions in economics.

Example 3   A woodworker goes into business selling rocking horses. His start-up costs, including tools, plans, and advertising, total $5000. Labor and materials for each horse cost $350.

(a)   Calculate the woodworker's total cost, C, to make 1, 2, 5, 10, and 20 rocking horses. Sketch a graph of C versus n, the number of rocking horses that he carves.
(b)   Find a formula for C in terms of n.
(c)   What is the rate of change of the function C? What does the rate of change tell us about the woodworker's expenses?

Solution   (a)   One horse costs the woodworker $5000 in start-up costs plus $350 for labor and materials, a total of $5350. Thus, if $n = 1$, we have

$$C = \underbrace{5000}_{\text{Start-up costs}} + \underbrace{350}_{\text{Extra cost for 1 horse}} = 5350.$$

Similarly, for 2 horses

$$C = \underbrace{5000}_{\text{Start-up costs}} + \underbrace{350 \cdot 2}_{\text{Extra cost for 2 horses}} = 5700,$$

and for 5 horses

$$C = \underbrace{5000}_{\text{Start-up costs}} + \underbrace{350 \cdot 5}_{\text{Extra cost for 5 horses}} = 6750.$$

Similar calculations show that for 10 horses, $C = 5000 + 350 \cdot 10 = 8500$, and for 20 horses, $C = 12{,}000$. See Table 2.3 and Figure 2.3.

TABLE 2.3   *Total cost of carving n horses*

n, number of horses	C, total cost ($)
0	5000
1	5350
2	5700
5	6750
10	8500
20	12,000

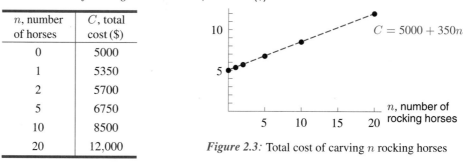

Figure 2.3: Total cost of carving n rocking horses

Notice that it costs the woodworker $5000 to carve 0 horses since he buys the tools, plans, and advertising even if he never carves a single horse.

(b)   From part (a), a formula for C, as a function of n, is

$$C = \underbrace{5000}_{\text{Start-up costs}} + \underbrace{350 \cdot n}_{\text{Extra cost for } n \text{ horses}} = 5000 + 350n.$$

(c)   The average rate of change of this function is

$$\text{Rate of change} = \frac{\Delta C}{\Delta n} = \frac{\text{Change in cost}}{\text{Change in number of horses carved}}.$$

Each additional horse costs an extra $350, so

$$\Delta C = 350 \qquad \text{if} \qquad \Delta n = 1.$$

Thus, the rate of change is given by

$$\frac{\Delta C}{\Delta n} = \frac{\$350}{1 \text{ horse}} = \$350 \text{ per horse.}$$

The rate of change of C tells us the additional cost to carve one additional horse. Since the total cost increases at a constant rate (\$350 per horse), the graph of C against n is a straight line sloping upward.

A General Formula for the Family of Linear Functions

Example 1 involved a town whose population is growing at a constant rate. A formula for this function is

$$\underset{\text{population}}{\text{Current}} = \underset{\underset{30,000 \text{ people}}{\underbrace{}}}{\underset{\text{population}}{\text{Initial}}} + \underset{\underset{2000 \text{ people per year}}{\underbrace{}}}{\underset{\text{rate}}{\text{Growth}}} \times \underset{\underset{t}{\underbrace{}}}{\underset{\text{years}}{\text{Number of}}}$$

so

$$P = 30{,}000 + 2000t.$$

In Example 3, the woodworker's total cost, C, as a function of n is given by

$$\underset{\text{cost}}{\text{Total}} = \underset{\underset{\$5000}{\underbrace{}}}{\underset{\text{cost}}{\text{Start-up}}} + \underset{\underset{\$350 \text{ per horse}}{\underbrace{}}}{\underset{\text{horse}}{\text{Cost per}}} \times \underset{\underset{n}{\underbrace{}}}{\underset{\text{horses}}{\text{Number of}}}$$

so

$$C = 5000 + 350n.$$

The formulas for both of these linear functions follow the same pattern:

$$\underset{\underset{y}{\underbrace{}}}{\text{Output}} = \underset{\underset{b}{\underbrace{}}}{\text{Initial value}} + \underset{\underset{m}{\underbrace{}}}{\text{Rate of change}} \times \underset{\underset{x}{\underbrace{}}}{\text{Input}}.$$

We rewrite this equation using the symbols x, y, b, m to get the following result:

If $y = f(x)$ is a linear function, then for some constants b and m:

$$y = b + mx.$$

- m is called the **slope**, and gives the rate of change of y with respect to x. Thus,

$$m = \frac{\Delta y}{\Delta x}.$$

If (x_0, y_0) and (x_1, y_1) are any two distinct points on the graph of f, then

$$m = \frac{\Delta y}{\Delta x} = \frac{y_1 - y_0}{x_1 - x_0}.$$

- b is called the **vertical intercept**, or **y-intercept**, and gives the value of y for $x = 0$. In mathematical models, b typically represents an initial, or starting, value of the output.

It turns out that every linear function has a formula of the form $y = b + mx$. Different linear functions have different values for m and b. These constants are known as *parameters*.

Example 4   In Example 1, the population is given by $P = 30,000 + 2000t$. The slope, $m = 2000$, tells us that the town grows by 2000 people per year. The vertical intercept, $b = 30,000$, tells us that when $t = 0$ the town had 30,000 people. In Example 3, the rocking-horse function is $C = 5000 + 350n$. The slope, $m = 350$, indicates that the total cost of production increases at a rate of \$350 per horse. The value $b = 5000$ indicates that the carpenter's start-up costs are \$5000.

Example 5   The value of new computer equipment is \$20,000 and the value drops at a constant rate so that it is worth \$0 after five years. If V is the value of computer equipment t years after the equipment is purchased, find a formula for V in terms of t.

Solution   Since the rate of change of V is constant, $V = f(t)$ is a linear function. Therefore $V = b + mt$. Since b is the initial value, $b = \$20,000$. Also,

$$\text{Rate of change} = m = \frac{\Delta V}{\Delta t} = \frac{-\$20,000}{5 \text{ years}} = -4000 \text{ dollars per year.}$$

Thus,

$$V = 20,000 - 4000t.$$

Tables for Linear Functions

A table of values could represent a linear function if the rate of change is constant, that is, if the

$$\frac{\text{Change in output}}{\text{Change in input}} = \text{Constant.}$$

This means that if the value of x goes up by equal steps in a table for a linear function, then the value of y goes up (or down) by equal steps as well. We say that changes in the value of y are *proportional* to changes in the value of x.

Example 6   Table 2.4 gives values of two functions, p and q. Could either of these functions be linear?

TABLE 2.4   *Values of two functions p and q*

x	50	55	60	65	70
$p(x)$	0.10	0.11	0.12	0.13	0.14
$q(x)$	0.01	0.03	0.06	0.14	0.15

Solution   The value of x goes up by equal steps of $\Delta x = 5$. The value of $p(x)$ also goes up by equal steps $\Delta p = 0.01$—from 0.10 to 0.11 to 0.12 and so on. See Table 2.5. Thus, p could be a linear function.

TABLE 2.5   *Values of $\Delta p / \Delta x$*

x	$p(x)$	$\Delta p / \Delta x$
50	0.10	
		0.002
55	0.11	
		0.002
60	0.12	
		0.002
65	0.13	
		0.002
70	0.14	

TABLE 2.6   *Values of $\Delta q / \Delta x$*

x	$q(x)$	$\Delta q / \Delta x$
50	0.01	
		0.004
55	0.03	
		0.006
60	0.06	
		0.016
65	0.14	
		0.002
70	0.15	

In contrast, the value of $q(x)$ does not go up by equal steps. The value climbs from 0.01 to 0.03 to 0.06, and so on. See Table 2.6. This means that Δq is not constant, even though Δx is. Thus, q could not be a linear function.

It is possible to have data from a linear function in which neither the x-values nor the y-values go up by equal steps. However the rate of change ratio must be constant, as in the following example.

Example 7   The former Republic of Yugoslavia began exporting cars called Yugos in 1985. Table 2.7 gives the quantity of Yugos sold, Q, and the price, p, for each year from 1985 to 1988.

TABLE 2.7   *Price and sales of Yugos*

Year	Price in $, p	Number sold, Q
1985	3990	49,000
1986	4110	43,000
1987	4200	38,500
1988	4330	32,000

(a)   Based on the data in Table 2.7, show that Q could be a linear function of p.

(b)   What does the rate of change of this function tell you about Yugos?

Solution   (a)   We are interested in Q as a function of p, so we plot a graph with Q on the vertical axis and p on the horizontal axis. The data points in Figure 2.4 appear to fall along a straight line, suggesting a linear function.

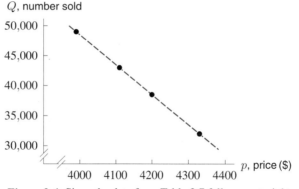

Figure 2.4: Since the data from Table 2.7 falls on a straight line, the table could represent a linear function

To provide further evidence that Q is a linear function, we check that the rate of change of Q with respect to p is constant for the data points given. When the price of a Yugo rose from $3990 to $4110, the number sold fell from 49,000 to 43,000. Thus,

$$\Delta p = 4110 - 3990 = 120,$$

and

$$\Delta Q = 43,000 - 49,000 = -6000.$$

Notice that ΔQ is negative because the number of Yugos sold decreased. Thus, as the price increased from $3990 to $4110,

$$\text{Rate of change of quantity as price increases} = \frac{\Delta Q}{\Delta p} = \frac{-6000}{120} = -50 \text{ cars per dollar.}$$

Next, we calculate the rate of change as the price increased from \$4110 to \$4200 to see if the rate remains constant:

$$\text{Rate of change} = \frac{\Delta Q}{\Delta p} = \frac{38{,}500 - 43{,}000}{4200 - 4110} = \frac{-4500}{90} = -50 \text{ cars per dollar},$$

and as the price increased from \$4200 to \$4330

$$\text{Rate of change} = \frac{\Delta Q}{\Delta p} = \frac{32{,}000 - 38{,}500}{4330 - 4200} = \frac{-6500}{130} = -50 \text{ cars per dollar}.$$

In each case, the rate of change is -50. Since the rate of change is a constant, Q could be a linear function of p. (It is still possible that, given additional data, $\Delta Q/\Delta p$ may not remain constant. However, based on the data in the table, it appears that the function is linear.)

(b) ΔQ is the change in the number of cars sold while Δp is the change in price. The rate of change is -50 cars per dollar, meaning that the number of Yugos sold decreased by 50 each time the price increased by \$1.

Warning: Not All Graphs That Look Like Lines Represent Linear Functions

The graph of any linear function is a line. However, a function's graph can look like a line without actually being one. Consider the following example.

Example 8   The function $P = 67.38(1.026)^t$ models the population of Mexico in the early 1980s. Here P is the population (in millions) and t is the number of years since 1980. Table 2.8 and Figure 2.5 show values of P over a six-year period. Is P a linear function of t?

TABLE 2.8 *Population of Mexico t years after 1980*

t (years)	P (millions)
0	67.38
1	69.13
2	70.93
3	72.77
4	74.67
5	76.61
6	78.60

Figure 2.5: Graph of $P = 67.38(1.026)^t$ over 6-year period: Looks linear (but isn't)

Solution   The formula $P = 67.38(1.026)^t$ is not of the form $P = b + mt$, so P is not a linear function of t. However, the graph of P in Figure 2.5 appears to be a straight line. We check P's rate of change in Table 2.8. When $t = 0$, $P = 67.38$ and when $t = 1$, $P = 69.13$. Thus, between 1980 and 1981,

$$\text{Rate of change of population} = \frac{\Delta P}{\Delta t} = \frac{69.13 - 67.38}{1 - 0} = 1.75.$$

For the interval from 1981 to 1982, we have

$$\text{Rate of change} = \frac{\Delta P}{\Delta t} = \frac{70.93 - 69.13}{2 - 1} = 1.80.$$

and for the interval from 1985 to 1986, we have

$$\text{Rate of change} = \frac{\Delta P}{\Delta t} = \frac{78.60 - 76.61}{6 - 5} = 1.99.$$

Thus, P's rate of change is not constant. In fact, P appears to be increasing at a faster and faster rate. Table 2.9 and Figure 2.6 show values of P over a longer (60-year) period. As you can see on this larger scale, these points do not fall on a straight line. However, the graph of P curves upward so gradually at first that over the short interval shown in Figure 2.5, it barely curves at all. The graphs of many non-linear functions, when viewed on a small scale, appear to be linear.

TABLE 2.9   *Population of Mexico over 60 years*

t (years since 1980)	P (millions)
0	67.38
10	87.10
20	112.58
30	145.53
40	188.12
50	243.16
60	314.32

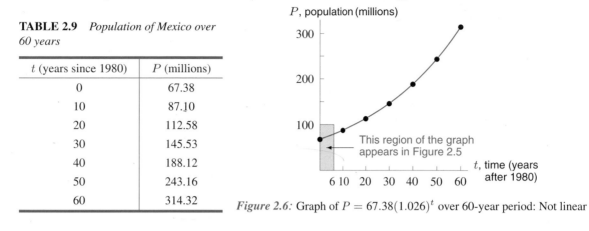

Figure 2.6: Graph of $P = 67.38(1.026)^t$ over 60-year period: Not linear

Problems for Section 2.1

1. Which of the following tables could represent a linear function? Check by determining if equal increments in x correspond to equal increments in the function values.

x	$f(x)$
0	10
5	20
10	30
15	40

x	$g(x)$
0	50
100	100
300	150
600	200

x	$h(x)$
0	20
10	40
20	50
30	55

x	$j(x)$
-3	5
-1	1
0	-1
3	-7

2. Table 2.10 shows the cost of selling various amounts of coffee per day from a cart.

    (a) Using the table, show that the relationship appears to be linear.
    (b) Plot the data in the table.
    (c) Find the slope of the line. Explain what this means in the context of the given situation.
    (d) Why should it cost $50 to serve zero cups of coffee?

    TABLE 2.10

x (cups per day)	0	5	10	50	100	200
C (dollars)	50.00	51.25	52.50	62.50	75.00	100.00

3. The population, $P(t)$, in millions, of a country in year t, is given by the formula $P(t) = 22 + 0.3t$.

    (a) Construct a table of values for $t = 0, 10, 20, \ldots, 50$.
    (b) Plot the points you found in part (a).
    (c) What is the country's initial population?
    (d) What is the average rate of change of the population, in millions of people/year?

4. In each case, graph a linear function with the given rate of change. Label and put scales on the axes.

    (a) Increasing at 2.1 inches/day   (b) Decreasing at 1.3 gallons/mile

5. Table 2.11 gives the area and perimeter of a square as a function of the length of its side.

   (a) From the table, decide if either area or perimeter is a linear function of side length.
   (b) From the data make two graphs, one showing area as a function of side length, the other showing perimeter as a function of side length. Connect the points.
   (c) If you find a linear relationship, give its corresponding rate of change and interpret its significance.

   TABLE 2.11

Length of side	0	1	2	3	4	5	6
Area of square	0	1	4	9	16	25	36
Perimeter of square	0	4	8	12	16	20	24

6. Make two tables, one comparing the radius of a circle to its area, the other comparing the radius of a circle to its circumference. Repeat parts (a), (b), and (c) from Problem 5, this time comparing radius with circumference, and radius with area.

7. Table 2.12 gives the fine $r = f(v)$ imposed on a motorist for speeding, where v is the motorist's speed and 55 mph is the speed limit.

   (a) Decide whether f is linear.
   (b) What does the rate of change represent in practical terms for the motorist?
   (c) Plot the data points.

   TABLE 2.12

v (mph)	60	65	70	75	80	85
r (dollars)	75	100	125	150	175	200

8. A new Toyota RAV4 costs $21,000. The car's value depreciates linearly to $10,500 in three years time. Write a formula which expresses its value, V, in terms of its age, t, in years.

9. In 1999, the population of a town was 18,310 and growing by 58 people per year. Find a formula for P, the town's population, in terms of t, the number of years since 1999.

10. Sri Lanka is a tropical island in the Indian Ocean which experienced approximately linear population growth from 1970 to 1990. On the other hand, Afghanistan, a country in Asia, was torn by warfare in the 1980s and did not experience linear nor near-linear growth.

    (a) In Table 2.13 which of these two countries is Country A and which is Country B? Explain.
    (b) What is the approximate rate of change of the linear function? What does the rate of change represent in practical terms?
    (c) Estimate the population of Sri Lanka in 1988.

    TABLE 2.13

Year	1970	1975	1980	1985	1990
Population of Country A (millions)	13.5	15.5	16	14	16
Population of Country B (millions)	12.2	13.5	14.7	15.8	17

11. Tuition cost T (in dollars) for part-time students at Stonewall College is given by $T = 300 + 200C$, where C represents the number of credits taken.

    (a) Find the tuition cost for eight credits.
    (b) How many credits were taken if the tuition was $1700?
    (c) Make a table showing costs for taking from one to twelve credits. For each value of C, give both the tuition cost, T, and the cost per credit, T/C. Round to the nearest dollar.
    (d) Which of these values of C has the smallest cost per credit?
    (e) What does the 300 represent in the formula for T?
    (f) What does the 200 represent in the formula for T?

12. Outside the US, temperature readings are usually given in degrees Celsius; inside the US, they are often given in degrees Fahrenheit. The exact conversion from Celsius, C, to Fahrenheit, F, uses the formula

$$F = \frac{9}{5}C + 32.$$

An approximate conversion is obtained by doubling the temperature in Celsius and adding $30°$ to get the equivalent Fahrenheit temperature.

(a) Write a formula using C and F to express the approximate conversion.
(b) How far off is the approximation if the Celsius temperature is $-5°, 0°, 15°, 30°$?
(c) For what temperature (in Celsius) does the approximation agree with the actual formula?

13. A company has found that there is a linear relationship between the amount of money that it spends on advertising and the number of units it sells. If it spends no money on advertising, it sells 300 units. For each additional $5000 spent on advertising, an additional 20 units are sold.

(a) If x is the amount of money that the company spends on advertising, find a formula for y, the number of units sold as a function of x.
(b) How many units does the firm sell if it spends $25,000 on advertising? $50,000?
(c) How much advertising money must be spent to sell 700 units?
(d) What is the slope of the line you found in part (a)? Give an interpretation of the slope that relates units sold and advertising costs.

14. Figure 2.7 shows the graph of $y = x^2/100 + 5$ in the window $-10 \le x \le 10, -10 \le y \le 10$. Discuss whether this is a linear function.

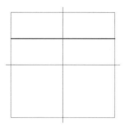

Figure 2.7

15. Graph the following function in the window $-10 \le x \le 10, -10 \le y \le 10$. Is this graph a line? Explain.

$$y = -x\left(\frac{x - 1000}{900}\right)$$

16. Graph $y = 2x + 400$ using the window $-10 \le x \le 10, -10 \le y \le 10$. Describe what happens, and how you can fix it by using a better window.

17. Graph $y = 200x + 4$ using the window $-10 \le x \le 10, -10 \le y \le 10$. Describe what happens and how you can fix it by using a better window.

18. The cost of a cab ride is given by the function $C = 1.50 + 2d$, where d is the number of miles traveled and C is in dollars. Choose an appropriate window and graph the cost of a ride for a cab that travels no farther than a 10 mile radius from the center of the city.

19. Let $f(x) = 0.003 - (1.246x + 0.37)$.

(a) Calculate the following average rates of change:

(i) $\dfrac{f(2) - f(1)}{2 - 1}$ (ii) $\dfrac{f(1) - f(2)}{1 - 2}$ (iii) $\dfrac{f(3) - f(4)}{3 - 4}$

(b) Rewrite $f(x)$ in the form $f(x) = b + mx$.

20. The graph of a linear function $y = f(x)$ passes through the two points $(a, f(a))$ and $(b, f(b))$, where $a < b$ and $f(a) < f(b)$.

(a) Sketch a possible graph of the function labeling the two points.
(b) Find the slope of the line in terms of f, a, and b.

2.2 FORMULAS FOR LINEAR FUNCTIONS

To find a formula for a linear function we find values for the slope, m, and the vertical intercept, b.

Finding a Formula for a Linear Function from a Table of Data

If a table of data represents a linear function, we use the data to calculate m. Having found m, we can determine the value of b.

Example 1   A grapefruit is thrown into the air. Its velocity, v, is a linear function of t, the time since it was thrown. A positive velocity indicates the grapefruit is rising and a negative velocity indicates it is falling. Check that the data in Table 2.14 corresponds to a linear function. Find a formula for v in terms of t.

TABLE 2.14   *Velocity of a grapefruit t seconds after being thrown into the air*

t, time (sec)	1	2	3	4
v, velocity (ft/sec)	48	16	-16	-48

Solution   Figure 2.8 shows the data in Table 2.14. The points appear to fall on a line. To check that the velocity function could be linear, calculate the rates of change of v and see that they are constant. From time $t = 1$ to $t = 2$, we have

$$\text{Rate of change of velocity with time} = \frac{\Delta v}{\Delta t} = \frac{16 - 48}{2 - 1} = -32.$$

For the next second, from $t = 2$ to $t = 3$, we have

$$\text{Rate of change} = \frac{\Delta v}{\Delta t} = \frac{-16 - 16}{3 - 2} = -32.$$

You can check that the rate of change from $t = 3$ to $t = 4$ is also -32.

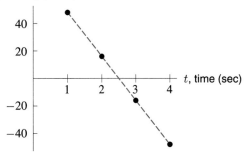

Figure 2.8: Velocity of a grapefruit is a linear function of time

A formula for v is of the form $v = b + mt$. Since m is the rate of change, we have $m = -32$ so $v = b - 32t$. The initial velocity (at $t = 0$) is represented by b. We are not given the value of v when $t = 0$, but we can use any data point to calculate b. For example, $v = 48$ when $t = 1$, so

$$48 = b - 32 \cdot 1,$$

which gives

$$b = 80.$$

Thus, a formula for the velocity is $v = 80 - 32t$.

What does the rate of change in Example 1 tell us about the grapefruit? Think about the units of m, the slope. We have

$$m = \frac{\Delta v}{\Delta t} = \frac{\text{Change in velocity}}{\text{Change in time}} = \frac{-32 \text{ ft/sec}}{1 \text{ sec}} = -32 \text{ ft/sec per second.}$$

Thus, the value of m is -32 ft/sec per second. This tells us that the grapefruit's velocity is decreasing by 32 ft/sec for every second that goes by. We say the grapefruit is accelerating at -32 ft/sec per second. (Negative acceleration is also called deceleration. The units ft/sec per second are often written ft/sec^2.)[1]

Finding a Formula for a Linear Function from a Graph

We can calculate the slope, m, of a linear function by using the coordinates of two points on its graph. Having found m we can use either of the points to calculate b, the vertical intercept.

Example 2   Figure 2.9 shows graphs of the oxygen consumption as a function of heart rate of two people.
(a)   Assuming linearity, find formulas for these two functions.
(b)   Interpret the slope of each graph in terms of oxygen consumption.

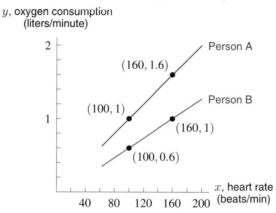

Figure 2.9: Oxygen consumption of two people running on treadmills

Solution   (a)   Let x be the heart rate and y the corresponding oxygen consumption. Since we are assuming linearity, we have $y = b + mx$. Two points on person A's line are given: $(100, 1)$ and $(160, 1.6)$. The first point indicates that when person A's heart rate is 100 beats/minute, person A consumes oxygen at a rate of 1 liter/minute. For person A,

$$m = \frac{\Delta y}{\Delta x} = \frac{1.6 - 1}{160 - 100} = 0.01.$$

Thus $y = b + 0.01x$. To determine b, use the fact that $y = 1$ when $x = 100$:

$$1 = b + 0.01(100)$$
$$1 = b + 1$$
$$b = 0.$$

Alternatively, b can be found using the fact that $x = 160$ if $y = 1.6$. Either way leads to the formula $y = 0.01x$.

For person B, we again begin with the formula $y = b + mx$. In Figure 2.9, we see that

---

[1]The notation ft/sec^2 is shorthand for ft/sec per second; it does not mean a "square second" in the same way that areas are measured square feet or square meters.

two points on the line are $(100, 0.6)$ and $(160, 1)$, so

$$m = \frac{\Delta y}{\Delta x} = \frac{1 - 0.6}{160 - 100} = \frac{0.4}{60} \approx 0.0067.$$

To determine b, use the fact that $y = 1$ when $x = 160$:

$$1 = b + 0.0067 \cdot 160$$
$$1 = b + 1.067$$
$$b = -0.067.$$

Thus, for person B, we have $y = -0.067 + 0.0067x$.

(b)   The slope for person A is $m = 0.01$, or

$$m = \frac{\text{Change in oxygen consumption}}{\text{Change in heart rate}} = \frac{\text{Change in liters/min}}{\text{Change in beats/min}} = 0.01\frac{\text{liters}}{\text{heart beat}}.$$

Every additional heart beat (per minute) for person A translates to an additional 0.01 liters (per minute) of oxygen consumed.

The slope for person B is $m = 0.0067$. Thus, for every additional beat per minute, person B consumes an additional 0.0067 liter of oxygen per minute. Since the slope for person B is smaller than for person A, person B consumes less additional oxygen than person A for the same increase in pulse.

What do the y-intercepts of the functions in Example 2 say about oxygen consumption? Often the y-intercept of a function is a starting value. In this case, the y-intercept would be the oxygen consumption of a person whose pulse is zero (i.e. $x = 0$). Since a person running on a treadmill must have a pulse, it makes no sense to interpret the y-intercept this way. The formula for oxygen consumption is useful only for realistic values of the pulse.

Finding a Formula for a Linear Function from a Verbal Description

Sometimes the verbal description of a linear function is less straightforward than those we saw in Section 2.1. Consider the following example.

Example 3   We have $24 to spend on soda and chips for a party. A six-pack of soda costs $3 and a bag of chips costs $2. The number of six-packs we can afford, y, is a function of the number of bags of chips we decide to buy, x.

(a)   Find an equation relating x and y.

(b)   Graph the equation. Interpret the intercepts and the slope in the context of the party.

Solution   (a)   If we spend all $24 on soda and chips, then we have the following equation:

$$\text{Amount spent on chips} + \text{Amount spent on soda} = \$24.$$

If we buy x bags of chips at $2 per bag, then the amount spent on chips is $2x$. Similarly, if we buy y six-packs of soda at $3 per six-pack, then the amount spent on soda is $3y$. Thus,

$$2x + 3y = 24.$$

We can solve for y, giving

$$3y = 24 - 2x$$
$$y = 8 - \frac{2}{3}x.$$

This is a linear function with slope $m = -2/3$ and y-intercept $b = 8$.

(b)   The graph of this function is a discrete set of points, since the number of bags of chips and the number of six-packs of soda must be (non-negative) integers.

To find the y-intercept, we set $x = 0$, giving

$$2 \cdot 0 + 3y = 24.$$

So $3y = 24$, giving $y = 8$.

Substituting $y = 0$ gives the x-intercept,

$$2x + (3 \cdot 0) = 24.$$

So $2x = 24$, giving $x = 12$. Thus the points $(0, 8)$ and $(12, 0)$ are on the graph.

The point $(0, 8)$ indicates that we can buy 8 six-packs of soda if we buy no chips. The point $(12, 0)$ indicates that we can buy 12 bags of chips if we buy no soda. The other points on the line describe affordable options between these two extremes. For example, the point $(6, 4)$ is on the line, because

$$2 \cdot 6 + 3 \cdot 4 = 24.$$

This means that if we buy 6 bags of chips, we can afford 4 six-packs of soda.

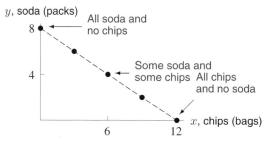

Figure 2.10: Relation between the number of six-packs, y, and the number of bags of chips, x

The graph is shown in Figure 2.10. The points marked represent affordable options. All affordable options lie on or below the line $2x + 3y = 24$. Not all points on the line are affordable options. For example, suppose we purchase one six-pack of soda for \$3.00. That leaves \$21.00 to spend on chips, meaning we would have to buy 10.5 bags of chips, which is not possible. Therefore, the point $(10.5, 1)$ is not an option, although it is a point on the line $2x + 3y = 24$.

To interpret the slope, notice that

$$m = \frac{\Delta y}{\Delta x} = \frac{\text{Change in number of six-packs}}{\text{Change in number of bags of chips}},$$

which means that the units of m are six-packs per bags of chips. The fact that $m = -2/3$ means that for each additional 3 bags of chips purchased, we can purchase 2 fewer six-packs of soda. This occurs because 2 six-packs cost \$6, the same as 3 bags of chips. Thus, $m = -2/3$ is the rate at which the amount of soda we can buy decreases as we buy more chips.

Alternative Forms for the Equation of a Line

In Example 3, the equation $2x + 3y = 24$ represents a linear relationship between x and y even though the equation is not in the form $y = b + mx$. The following equations represent lines.

- The *slope-intercept form* is
  $$y = b + mx \qquad \text{where } m \text{ is the slope and } b \text{ is the } y\text{-intercept.}$$
- The *point-slope form* is
  $$y - y_0 = m(x - x_0) \qquad \text{where } m \text{ is the slope and } (x_0, y_0) \text{ is a point on the line.}$$
- The *standard form* is
  $$Ax + By + C = 0 \qquad \text{where } A, B, \text{ and } C \text{ are constants.}$$

If we know the slope of a line and the coordinates of a point on the line, it is convenient to use the point-slope form of the equation.

Example 4   Use the point-slope form to find the equation of the line for the oxygen consumption of Person A in Example 2.

Solution   Since we know the slope of the line, $m = 0.01$, and the point $(100, 1)$, we use the point-slope form of the equation, which gives

$$y - 1 = 0.01(x - 100).$$

To check that this is just a different form of the equation we got in Example 2, we multiply out and simplify:

$$y - 1 = 0.01x - 1$$
$$y = 0.01x.$$

Alternatively, we could have used the point $(160, 1.6)$ instead of $(100, 1)$, giving

$$y - 1.6 = 0.01(x - 160).$$

Multiplying out again gives $y = 0.01x$.

Problems for Section 2.2

Graphing a linear equation is often easier if the equation is in slope-intercept form, $y = b + mx$. If possible, rewrite the equations in Problems 1–9 in this form.

1. $3x + 5y = 20$
2. $0.1y + x = 18$
3. $y - 0.7 = 5(x - 0.2)$
4. $5(x + y) = 4$
5. $5x - 3y + 2 = 0$
6. $\dfrac{x + y}{7} = 3$
7. $3x + 2y + 40 = x - y$
8. $x = 4$
9. $y = 5$

Find formulas for the linear functions with the properties in Problems 10–15.

10. Slope 3 and y-intercept 8
11. Passes through the points $(-1, 5)$ and $(2, -1)$
12. Slope -4 and x-intercept 7
13. Has x-intercept 3 and y-intercept -5
14. Slope $2/3$ and passes through the point $(5, 7)$
15. Slope 0.1, passes through $(-0.1, 0.02)$

16. If $y = f(x)$ is a linear function and if $f(-2) = 7$ and $f(3) = -3$, find a formula for f.

17. Describe a linear (or nearly linear) relationship that you have encountered outside the classroom. Determine the rate of change and interpret it in practical terms.

Problems 18–19 give data from a linear function. Find a formula for the function.

18.

t	1.2	1.3	1.4	1.5
$f(t)$	0.736	0.614	0.492	0.37

19.

x	200	230	300	320	400
$g(x)$	70	68.5	65	64	60

20. John wants to buy a dozen rolls. The local bakery sells sesame and poppy seed rolls for the same price.

   (a) Make a table of all the possible combinations of rolls if he buys a dozen, where s is the number of sesame seed rolls and p is the number of poppy seed rolls.
   (b) Find a formula for p as a function of s.
   (c) Graph this function.

21. In a college meal plan you pay a membership fee; then all your meals are at a fixed price per meal.

    (a)   If 30 meals cost $152.50 and 60 meals cost $250, find the membership fee and the price per meal.
    (b)   Write a formula for the cost of a meal plan, C, in terms of the number of meals, n.
    (c)   Find the cost for 50 meals.
    (d)   Find n in terms of C.
    (e)   Use part (d) to determine the maximum number of meals you can buy on a budget of $300.

22. A theater manager graphed weekly profits as a function of the number of patrons and found that the relationship was linear. One week the profit was $11,328 when 1324 patrons attended. Another week 1529 patrons produced a profit of $13,275.50.

    (a)   Find a formula for weekly profit, y, as a function of the number of patrons, x.
    (b)   Interpret the slope and the y-intercept.
    (c)   What is the break-even point (the number of patrons for which there is zero profit)?
    (d)   Find a formula for the number of patrons as a function of profit.
    (e)   If the weekly profit was $17,759.50, how many patrons attended the theater?

23. An empty champagne bottle is tossed from a hot-air balloon. Its upward velocity is measured every second and recorded in Table 2.15.

    (a)   Describe the motion of the bottle in words. What do negative values of v represent?
    (b)   Find a formula for v in terms of t.
    (c)   Explain the physical significance of the slope of your formula.
    (d)   Explain the physical significance of the t-axis and v-axis intercepts.

TABLE 2.15

t (sec)	0	1	2	3	4	5
v (ft/sec)	40	8	−24	−56	−88	−120

24. A bullet is shot straight up into the air from ground level. After t seconds, the velocity of the bullet, in meters per second, is approximated by the formula

$$v = f(t) = 1000 - 9.8t.$$

    (a)   Evaluate the following: $f(0)$, $f(1)$, $f(2)$, $f(3)$, $f(4)$. Compile your results in a table.
    (b)   Describe in words what is happening to the speed of the bullet. Discuss why you think this is happening.
    (c)   Evaluate and interpret the slope and both intercepts of $f(t)$.
    (d)   The gravitational field near the surface of Jupiter is stronger than that near the surface of the earth, which, in turn, is stronger than the field near the surface of the moon. How is the formula for $f(t)$ different for a bullet shot from Jupiter's surface? From the moon?

25. Find the equation of the line l, shown in Figure 2.11, if its slope is $m = 4$.

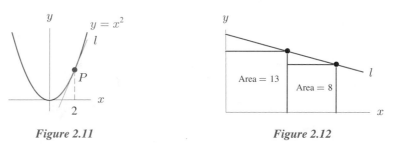

        Figure 2.11   *Figure 2.12*

26. Find the equation of the line l in Figure 2.12. The shapes under the line are squares.

27.   Find the equation of line l in Figure 2.13.

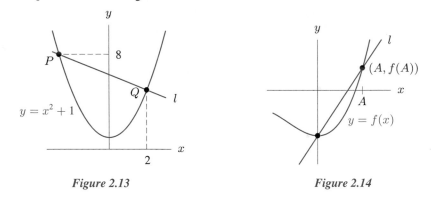

<table>
<tr><td>Figure 2.13</td><td>Figure 2.14</td></tr>
</table>

28.   Find an equation for the line l in Figure 2.14 in terms of the constant A and values of the function f.

29.   According to one economic model, the demand for gasoline is a linear function of price. If the price of gasoline is $p = \$1.10$ per gallon, the quantity demanded in a fixed period is $q = 65$ gallons. If the price rises to $\$1.50$ per gallon, the quantity demanded falls to 45 gallons in that period.

    (a)   Find a formula for q in terms of p.

    (b)   Explain the economic significance of the slope of your formula.

    (c)   Explain the economic significance of the q-axis and p-axis intercepts.

30.   A business consultant works 10 hours a day, 6 days a week. She divides her time between meetings with clients and meetings with co-workers. A client meeting requires 3 hours while a co-worker meeting requires only 2 hours. Let x be the number of co-worker meetings the consultant holds during a given week. If y is the number of client meetings for which she has time remaining, then y is a function of x. Assume this relationship is linear and that meetings can be split up and continued on different days.

    (a)   Sketch a graph that represents the relationship between y and x. [Hint: Consider the maximum number of client and co-worker meetings that can be held.]

    (b)   Find a formula for y as a function of x.

    (c)   Explain what the slope and the x- and y-intercepts represent in the context of the consultant's meeting schedule.

    (d)   Suppose that, due to a procedural change, co-worker meetings now take 90 minutes instead of two hours. Sketch the graph for this situation. Describe those features of this graph that have changed from the one sketched in part (a) and those that have remained the same.

2.3 GEOMETRIC PROPERTIES OF LINEAR FUNCTIONS

Interpreting the Parameters of a Linear Function

The slope-intercept form for a linear function is $y = b + mx$, where b is the y-intercept and m is the slope. The parameters b and m can be used to compare linear functions.

Example 1   The populations of four different towns, P_A, P_B, P_C and P_D, are given by the following formulas, where t is the year:

$$P_A = 20{,}000 + 1600t, \quad P_B = 50{,}000 - 300t, \quad P_C = 650t + 45{,}000, \quad P_D = 15{,}000(1.07)^t$$

    (a)   Which populations are represented by linear functions?

    (b)   Describe in words what each linear model tells you about that town's population. Which town starts out with the most people? Which town is growing fastest?

Solution (a) The populations of towns A, B, and C are represented by linear functions. Each of these formulas can be written in the form $y = b + mt$. For example, the equation for town A is $P_A = 20{,}000 + 1600t$. This is a linear equation with $b = 20{,}000$ and $m = 1600$. Town B also has linear population growth with $b = 50{,}000$ and $m = -300$, since we have

$$P_B = \underbrace{50{,}000}_{b} + \underbrace{(-300)}_{m} \cdot t.$$

Town C's growth is linear with $b = 45{,}000$ and $m = 650$, since the formula for P_C can be written as

$$P_C = \underbrace{45{,}000}_{b} + \underbrace{650}_{m} \cdot t.$$

Town D's population does not grow linearly since its formula, $P_D = 15{,}000(1.07)^t$, cannot be expressed in the form $P_D = b + mt$.

(b) For town A, we have $b = 20{,}000$ and $m = 1600$. This means that in year $t = 0$, town A has 20,000 people. It grows by 1600 people per year.

For town B, we have $b = 50{,}000$ and $m = -300$. This means that town B starts with 50,000 people. The negative slope indicates that the population is decreasing at the rate of 300 people per year.

For town C, we have $b = 45{,}000$ and $m = 650$. This means that town C begins with 45,000 people and grows by 650 people per year.

Town B starts out with the most people, (50,000), but town A, with a rate of change of 1600 people per year, grows the fastest of the three towns that grow linearly.

The Effect of the Parameters on the Graph of a Linear Function

The graph of a linear function is a line. Varying the values of b and m gives different members of the family of linear functions. In summary:

> Let $y = b + mx$. Then the graph of y against x is a line.
> - The y-intercept, b, tells us where the line crosses the y-axis.
> - If the slope, m, is positive, the line climbs from left to right. If the slope, m, is negative, the line falls from left to right.
> - The slope, m, tells us how fast the line is climbing or falling.
> - The larger the value of m (either positive or negative), the steeper the graph of f.

Example 2 (a) Graph the three linear functions P_A, P_B, P_C from Example 1 and show how to identify the values of b and m from the graph.

(b) Graph P_D from Example 1 and explain how the graph shows P_D is not a linear function.

Solution (a) Figure 2.15 gives graphs of the three functions:

$$P_A = 20{,}000 + 1600t, \qquad P_B = 50{,}000 - 300t, \quad \text{and} \quad P_C = 45{,}000 + 650t.$$

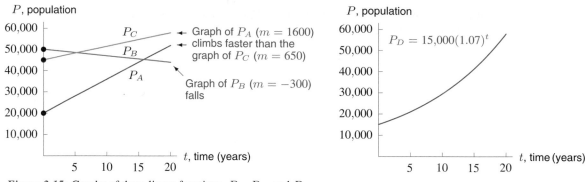

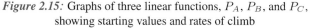

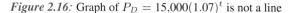

Figure 2.15: Graphs of three linear functions, P_A, P_B, and P_C, showing starting values and rates of climb

Figure 2.16: Graph of $P_D = 15{,}000(1.07)^t$ is not a line

The values of b identified in Example 1 tell us the vertical intercepts. Figure 2.15 shows that the graph of P_A crosses the P-axis at $P = 20{,}000$, the graph of P_B crosses at $P = 50{,}000$, and the graph of P_C crosses at $P = 45{,}000$.

Notice that the graphs of P_A and P_C are both climbing and that P_A climbs faster than P_C. This corresponds to the fact that the slopes of these two functions are positive ($m = 1600$ for P_A and $m = 650$ for P_C) and the slope of P_A is larger than the slope of P_C.

The graph of P_B falls when read from left to right, indicating that population decreases over time. This corresponds to the fact that the slope of P_C is negative ($m = -300$).

(b) Figure 2.16 gives a graph of P_D. Since it is not a line, P_D is not a linear function.

Equations of Horizontal and Vertical Lines

An increasing linear function has positive slope and a decreasing linear function has negative slope. What about a line with slope $m = 0$? If the rate of change of a quantity is zero, then the quantity does not change. Thus, if the slope of a line is zero, the value of y must be constant. Such a line is horizontal.

Example 3 Explain why the equation $y = 4$ represents a horizontal line and the equation $x = 4$ represents a vertical line.

Solution The equation $y = 4$ represents a linear function with slope $m = 0$. To see this, notice that this equation can be rewritten as $y = 4 + 0 \cdot x$. Thus, the value of y is 4 no matter what the value of x is. See Figure 2.17. Similarly, the equation $x = 4$ means that x is 4 no matter what the value of y is. Every point on the line in Figure 2.18 has x equal to 4, so this line is the graph of $x = 4$.

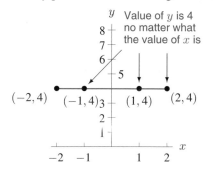

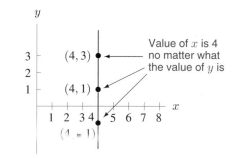

Figure 2.17: The horizontal line $y = 4$ has slope 0

Figure 2.18: The vertical line $x = 4$ has an undefined slope

What is the slope of a vertical line? Figure 2.18 shows three points, $(4, -1)$, $(4, 1)$, and $(4, 3)$ on a vertical line. Calculating the slope, gives

$$m = \frac{\Delta y}{\Delta x} = \frac{3 - 1}{4 - 4} = \frac{2}{0}.$$

The slope is undefined because the denominator, Δx, is 0. The slope of every vertical line is undefined for the same reason. All the x-values on such a line are equal, so Δx is 0, and the denominator of the expression for the slope is 0. A vertical line is not the graph of a function, since it fails the vertical line test. It does not have an equation of the form $y = b + mx$.

In summary,

For any constant k:
- The graph of the equation $y = k$ is a horizontal line and its slope is zero.
- The graph of the equation $x = k$ is a vertical line and its slope is undefined.

Slopes of Parallel and Perpendicular Lines

Figure 2.19 shows two parallel lines. These lines are parallel because they have equal slopes.

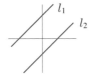

Figure 2.19: Parallel lines: l_1 and l_2 have equal slopes

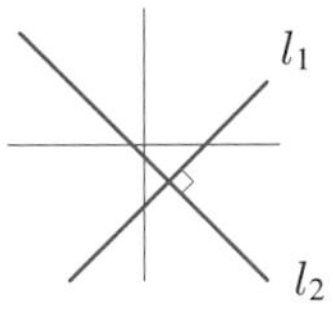

Figure 2.20: Perpendicular lines: l_1 has a positive slope and l_2 has a negative slope

What about perpendicular lines? Two perpendicular lines are graphed in Figure 2.20. We can see that if one line has a positive slope, then any perpendicular line must have a negative slope. Perpendicular lines have slopes with opposite signs.

It can be shown[2] that if l_1 and l_2 are two perpendicular lines with slopes, m_1 and m_2, then m_1 is the negative reciprocal of m_2. If m_1 and m_2 are not zero, we have the following result:

Let l_1 and l_2 be two lines having slopes m_1 and m_2, respectively. Then:
- These lines are parallel if and only if $m_1 = m_2$.
- These lines are perpendicular if and only if $m_1 = -\dfrac{1}{m_2}$.

In addition, any two horizontal lines are parallel and $m_1 = m_2 = 0$. Any two vertical lines are parallel and m_1 and m_2 are undefined. A horizontal line is perpendicular to a vertical line. See Figures 2.21–2.23.

Figure 2.21: Any two horizontal lines are parallel

Figure 2.22: Any two vertical lines are parallel

Figure 2.23: A horizontal line and a vertical line are perpendicular

---

[2]For a justification of the relationship between slopes of perpendicular lines, see page 58.

Intersection of Two Lines

To find the point at which two lines intersect, notice that the (x, y)-coordinates of such a point must satisfy the equations for both lines. Thus, in order to find the point of intersection algebraically, solve the equations simultaneously.[3]

Example 4   Find the point of intersection of the lines $y = 3 - \frac{2}{3}x$ and $y = -4 + \frac{3}{2}x$.

Solution   Since the y-values of the two lines are equal at the point of intersection, we have

$$-4 + \frac{3}{2}x = 3 - \frac{2}{3}x.$$

Notice that we have converted a pair of equations into a single equation by eliminating one of the two variables. This equation can be simplified by multiplying both sides by 6:

$$6\left(-4 + \frac{3}{2}x\right) = 6\left(3 - \frac{2}{3}x\right)$$
$$-24 + 9x = 18 - 4x$$
$$13x = 42$$
$$x = \frac{42}{13}.$$

We can evaluate either of the original equations at $x = \frac{42}{13}$ to find y. For example, $y = -4 + \frac{3}{2}x$ gives

$$y = -4 + \frac{3}{2}\left(\frac{42}{13}\right) = \frac{11}{13}.$$

Therefore, the point of intersection is $\left(\frac{42}{13}, \frac{11}{13}\right)$. You can check that this point also satisfies the other equation. The lines and their point of intersection are shown in Figure 2.24.

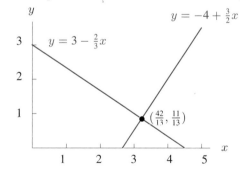

Figure 2.24: Intersection of lines is solution to simultaneous equations

---

If linear functions are modeling real quantities, their points of intersection often have practical significance. Consider the next example.

Example 5   The cost in dollars of renting a car for a day from three different rental agencies and driving it d miles is given by the following functions:

$$C_1 = 50 + 0.10d, \qquad C_2 = 30 + 0.20d, \qquad C_3 = 0.50d.$$

---

[3]If you have questions about the algebra in this section, please see Appendix G.

(a) Describe in words the daily rental arrangements made by each of these three agencies.

(b) Which agency is cheapest?

Solution (a) Agency 1 charges $50 plus $0.10 per mile driven. Agency 2 charges $30 plus $0.20 per mile. Agency 3 charges $0.50 per mile driven.

(b) The answer depends on how far we want to drive. If we aren't driving far, agency 3 may be cheapest because it only charges for miles driven and has no other fees. If we want to drive a long way, agency 1 may be cheapest (even though it charges $50 up front) because it has the lowest per-mile rate.

The three functions are graphed in Figure 2.25. The graph shows that for d up to 100 miles, the value of C_3 is less than C_1 and C_2 because its graph is below the other two. For d between 100 and 200 miles, the value of C_2 is less than C_1 and C_3. For d more than 200 miles, the value of C_1 is less than C_2 and C_3.

By graphing these three functions on a calculator, we can estimate the coordinates of the points of intersection by tracing. To find the exact coordinates, we solve simultaneous equations. Starting with the intersection of lines C_1 and C_2, we set the costs equal, $C_1 = C_2$, and solve for d:

$$50 + 0.10d = 30 + 0.20d$$
$$20 = 0.10d$$
$$d = 200.$$

Thus, the cost for driving 200 miles is the same for agencies 1 and 2. Solving $C_2 = C_3$ gives

$$30 + 0.20d = 0.50d$$
$$0.30d = 30$$
$$d = 100,$$

which means the cost of driving 100 miles is the same for agencies 2 and 3.

Thus, agency 3 is cheapest up to 100 miles. Agency 1 is cheapest for more than 200 miles. Agency 2 is cheapest between 100 and 200 miles. See Figure 2.25. Notice that the point of intersection of C_1 and C_3, $(125, 62.5)$, does not influence our decision as to which agency is the cheapest.

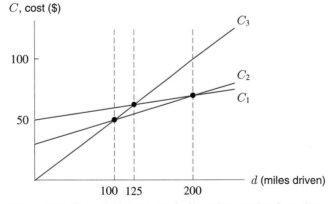

Figure 2.25: Cost of driving a car d miles when renting from three different agencies. Cheapest agency corresponds to the lowest graph for a given d value

Justification of Formula for Slopes of Perpendicular Lines

Figure 2.26 shows l_1 and l_2, two perpendicular lines with slope m_1 and m_2. Neither line is horizontal or vertical, so m_1 and m_2 are both defined and nonzero. We will show that

$$m_1 = -\frac{1}{m_2},$$

We use the two triangles, ΔPQR and ΔSPR. We show that ΔPQR and ΔSPR are similar by showing that corresponding angles have equal measure. The line PR is horizontal, so $\angle QRP = \angle SRP$ since both are right angles. Since ΔQPS is a right triangle, $\angle S$ is complementary to $\angle Q$ (that is, $\angle S$ and $\angle Q$ add to $90°$). Since ΔQRP is a right triangle, $\angle QPR$ is complementary to $\angle Q$. Therefore $\angle S = \angle QPR$. Since two pairs of angles in ΔPQR and ΔSPR have equal measure, the third must be equal also; the triangles are similar.

Corresponding sides of similar triangles are proportional. (See Figure 2.27.) Therefore,

$$\frac{\|RS\|}{\|RP\|} = \frac{\|RP\|}{\|RQ\|},$$

where $\|RS\|$ means the length of side RS.

Next, we calculate m_1 using points S and P, and we calculate m_2 using points Q and P. In Figure 2.26, we see that

$$\Delta x = \|RP\|, \quad \Delta y_1 = \|RS\|, \quad \text{and} \quad \Delta y_2 = -\|RQ\|,$$

where Δy_2 is negative because y-values of points on l_2 decrease as x increases. Thus,

$$m_1 = \frac{\Delta y_1}{\Delta x} = \frac{\|RS\|}{\|RP\|} \quad \text{and} \quad m_2 = \frac{\Delta y_2}{\Delta x} = -\frac{\|RQ\|}{\|RP\|}.$$

Therefore, using the result obtained from the similar triangles, we have

$$m_1 = \frac{\|RS\|}{\|RP\|} = \frac{\|RP\|}{\|RQ\|} = -\frac{1}{m_2}.$$

Thus, $m_1 = -1/m_2$.

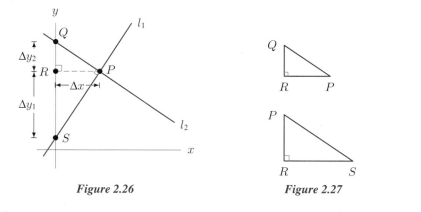

Figure 2.26   *Figure 2.27*

Problems for Section 2.3

1. Solve the system of equations given by $3x - y = 17$ and $-2x - 3y = -4$ algebraically.

2. Using the window $-10 \le x \le 10$, $-10 \le y \le 10$, graph $y = x$, $y = 10x$, $y = 100x$, and $y = 1000x$.

   (a) Explain what happens to the graphs of the lines as the slopes become large.

   (b) Write an equation of a line that passes through the origin and is horizontal.

3. Figure 2.28 gives five different lines, A, B, C, D, and E. Without using a calculator, match each line to one of the following functions f, g, h, u and v:

$$f(x) = 20 + 2x$$
$$g(x) = 20 + 4x$$
$$h(x) = 2x - 30$$
$$u(x) = 60 - x$$
$$v(x) = 60 - 2x$$

Figure 2.28

4. Graph $y = x + 1$, $y = x + 10$, and $y = x + 100$ in the window $-10 \leq x \leq 10, -10 \leq y \leq 10$.

    (a) Explain what happens to the graph of a line, $y = b + mx$, as b becomes large.
    (b) Write a linear equation whose graph cannot be seen in the window $-10 \leq x \leq 10, -10 \leq y \leq 10$ because all its y-values are less than the y-values shown.

5. The graphical interpretation of the slope is that it shows steepness. Using a calculator or a computer, graph the function $y = 2x - 3$ in the following windows:

    (a) $-10 \leq x \leq 10$ by $-10 \leq y \leq 10$
    (b) $-10 \leq x \leq 10$ by $-100 \leq y \leq 100$
    (c) $-10 \leq x \leq 10$ by $-1000 \leq y \leq 1000$
    (d) Write a sentence about how steepness is related to the window being used.

6. Without using a calculator, match the functions (a) – (c) to the graphs (i) – (iii).

    (a) $f(x) = 3x + 1$ (b) $g(x) = -2x + 1$ (c) $h(x) = 1$

7. Without using a calculator, match the following functions to the lines in Figure 2.29:

$$f(x) = 5 + 2x$$
$$g(x) = -5 + 2x$$
$$h(x) = 5 + 3x$$
$$j(x) = 5 - 2x$$
$$k(x) = 5 - 3x$$

Figure 2.29

8. Without using a calculator, match the equations (a) – (g) to the graphs (I) – (VII).

    (a) $y = x - 5$ (b) $-3x + 4 = y$ (c) $5 = y$ (d) $y = -4x - 5$
    (e) $y = x + 6$ (f) $y = x/2$ (g) $5 = x$

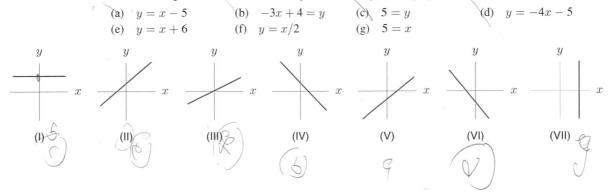

9. The total costs, in dollars, of producing four goods follow. The number of goods produced is represented by n. For each good, state the fixed and unit costs.

   (a) $C_1 = 8000 + 200n$ (b) $C_2 = 5000 + 200n$

   (c) $C_3 = 10{,}000 + 100n$ (d) $C_4 = 50n$

10. Graph the functions defined in Problem 9 on a single coordinate system.

11. Fill in the missing coordinates for the points in the following figures.

    (a) The triangle in Figure 2.30.   (b) The parallelogram in Figure 2.31.

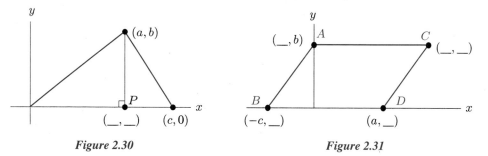

Figure 2.30   Figure 2.31

12. (a) Sketch, by hand, the graphs of $y = 3$ and $x = 3$.

    (b) Can the equations in part (a) be written in slope-intercept form?

13. Line l is given by $y = 3 - \frac{2}{3}x$ and point P has coordinates $(6, 5)$.

    (a) Find the equation of the line containing P and parallel to l.

    (b) Find the equation of the line containing P and perpendicular to l.

    (c) Graph the equations in parts (a) and (b).

14. Find the equation of the line l_2 in Figure 2.32.

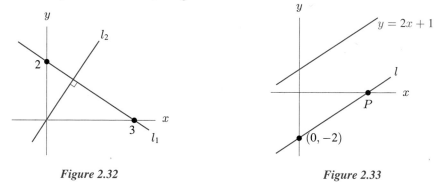

Figure 2.32   Figure 2.33

15. Line l in Figure 2.33 is parallel to the line $y = 2x + 1$. Find the coordinates of the point P.

16. Estimate the slope of the line in Figure 2.34 and find an approximate equation for the line.

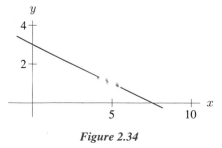

Figure 2.34

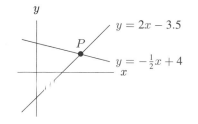

Figure 2.35: Axes not necessarily to scale

17. Find the coordinates of point P in Figure 2.35.

18. The cost of a Frigbox refrigerator is $950, and it depreciates $50 each year. The cost of an Arctic Air refrigerator is $1200, and it depreciates $100 per year.

(a) If a Frigbox and an Arctic Air are bought at the same time, when do the two refrigerators have equal value?

(b) If the depreciation continues at the same rate, what happens to the values of the refrigerators in 20 years time? What does this mean?

19. You need to rent a car and compare the charges of three different companies. Company A charges 20 cents per mile plus $20 per day. Company B charges 10 cents per mile plus $35 per day. Company C charges $70 per day with no mileage charge.

(a) Find formulas for the cost of driving cars rented from companies A, B, and C, in terms of x, the distance driven in miles in one day.

(b) Graph the costs for each company for $0 \leq x \leq 500$. Put all three graphs on the same set of axes.

(c) What do the slope and the vertical intercept tell you in this situation?

(d) Use the graph in part (b) to find under what circumstances would company A be the cheapest? Company B? Company C? Explain why your results make sense.

20. You want to choose one long-distance telephone company from the following options.

- Company A charges $0.37 per minute.
- Company B charges $13.95 per month plus $0.22 per minute.
- Company C charges a fixed rate of $50 per month.

Let Y_A, Y_B, Y_C represent the monthly charges using Company A, B, and C, respectively. Let x denote the number of minutes spent on long distance calling during a month.

(a) Find formulas for Y_A, Y_B, Y_C as functions of x.

(b) Figure 2.36 gives the graphs of the functions in part (a). Which function corresponds to which graph?

(c) Find the values of x for which Company B is the cheapest.

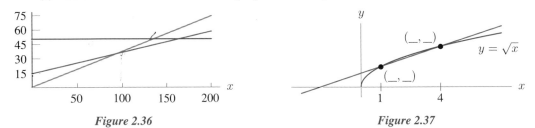

Figure 2.36   Figure 2.37

21. Fill in the missing coordinates in Figure 2.37. Write an equation for the line connecting the two points. Check your work by solving the system of two equations.

22. The solid waste generated each year in the cities of the US is increasing.[4] The solid waste generated, in millions of tons, was 82.3 in 1960 and 139.1 in 1980. The trend appears linear during this time.

(a) Construct a mathematical model of the amount of municipal solid waste generated in the US by finding the equation of the line through these two points.

(b) Use this model to predict the amount of municipal solid waste generated in the US, in millions of tons, in the year 2020.

23. Two lines are given by $y = b_1 + m_1 x$ and $y = b_2 + m_2 x$, where b_1, b_2, m_1, and m_2 are constants.

(a) What conditions are imposed on b_1, b_2, m_1, and m_2 if the two lines have no points in common?

(b) What conditions are imposed on b_1, b_2, m_1, and m_2 if the two lines have all points in common?

(c) What conditions are imposed on b_1, b_2, m_1, and m_2 if the two lines have exactly one point in common?

(d) What conditions are imposed on b_1, b_2, m_1, and m_2 if the two lines have exactly two points in common?

---

[4]*Statistical Abstracts of the US*, 1988, p. 193, Table 333.

2.4 FITTING LINEAR FUNCTIONS TO DATA

When real data are collected in the laboratory or the field, they are often subject to experimental error. Even if there is an underlying linear relationship between two quantities, real data may not fit this relationship perfectly. However, even if a data set does not perfectly conform to a linear function, we may still be able to use a linear function to help us analyze the data.

Laboratory Data: The Viscosity of Motor Oil

The viscosity of a liquid, or its resistance to flow, depends on the liquid's temperature. Pancake syrup is a familiar example: straight from the refrigerator, it pours very slowly. When warmed on the stove, its viscosity decreases and it becomes quite runny.

The viscosity of motor oil is a measure of its effectiveness as a lubricant in the engine of a car. Thus, the effect of engine temperature is an important determinant of motor-oil performance. Table 2.16 gives the viscosity, v, of motor oil as measured in the lab at different temperatures, T.

TABLE 2.16 *The measured viscosity, v, of motor oil as a function of the temperature, T*

T, temperature (°F)	v, viscosity (lbs·sec/in^2)
160	28
170	26
180	24
190	21
200	16
210	13
220	11
230	9

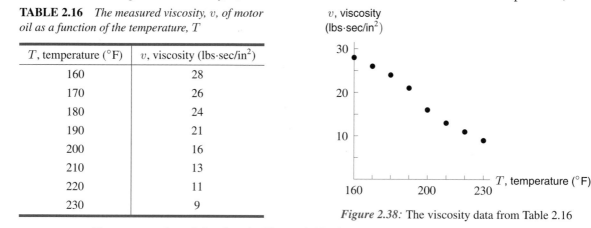

Figure 2.38: The viscosity data from Table 2.16

The *scatter plot* of the data in Figure 2.38 shows that the viscosity of motor oil decreases, approximately linearly, as its temperature rises. To find a formula relating viscosity and temperature, we fit a line to these data points.

Fitting the best line to a set of data is called *linear regression*. One way to fit a line is to draw a line "by eye." Alternatively, many computer programs and calculators compute regression lines. Figure 2.39 shows the data from Table 2.16 together with the computed regression line,

$$v = 75.6 - 0.293T.$$

Notice that none of the data points lie exactly on the regression line, although it fits the data well.

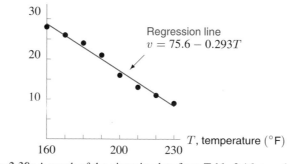

Figure 2.39: A graph of the viscosity data from Table 2.16, together with a regression line (provided by a calculator)

The Assumptions Involved In Finding a Regression Line

When we find a regression line for the data in Table 2.16, we are assuming that the value of v is related to the value of T. However, there may be experimental errors in our measurements. For example, if we measure viscosity twice at the same temperature, we may get two slightly different values. Alternatively, something besides engine temperature could be affecting the oil's viscosity (the oil pressure, for example). Thus, even if we assume that the temperature readings are exact, the viscosity readings include some degree of uncertainty.

Interpolation and Extrapolation

The formula for viscosity can be used to make predictions. Suppose we want to know the viscosity of motor oil at $T = 196°$F. The formula gives

$$v = 75.6 - 0.293(196) \approx 18.2 \text{ lb} \cdot \text{sec/in}^2.$$

To see that this is a reasonable estimate, compare it to the entries in Table 2.16. At $190°$F, the measured viscosity was 21, and at $200°$F, it was 16; the predicted viscosity of 18.2 is between 16 and 21. See Figure 2.40. Of course, if we measured the viscosity at $T = 196°$F in the lab, we might not get exactly 18.2.

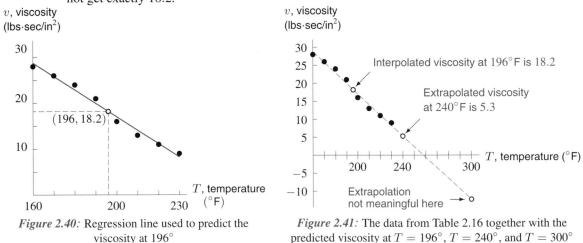

Figure 2.40: Regression line used to predict the viscosity at $196°$

Figure 2.41: The data from Table 2.16 together with the predicted viscosity at $T = 196°$, $T = 240°$, and $T = 300°$

Since the temperature $T = 196°$ is between two temperatures for which v is known ($190°$ and $200°$), the estimate of 18.2 is said to be an *interpolation*. If instead we estimate the value of v at a temperature outside the values for T in Table 2.16, our estimate is called an *extrapolation*.

Example 1   Predict the viscosity of motor oil at $240°$F and at $300°$F.

Solution   At $T = 240°$F, the formula for the regression line predicts that the viscosity of motor oil is

$$v = 75.6 - 0.293(240) = 5.3 \text{ lb} \cdot \text{sec/in}^2.$$

This is reasonable. Figure 2.41 shows that the predicted point—represented by an open circle on the graph—is consistent with the trend in the data points from Table 2.16.

On the other hand, at $T = 300°$F the regression-line formula gives

$$v = 75.6 - 0.293(300) = -12.3 \text{ lb} \cdot \text{sec/in}^2.$$

This is unreasonable because viscosity cannot be negative. To understand what went wrong, notice that in Figure 2.41, the open circle representing the point $(300, -12.3)$ is far from the plotted data points. By making a prediction at $300°$F, we have assumed—incorrectly—that the trend observed in laboratory data extended as far as $300°$F.

In general, interpolation tends to be more reliable than extrapolation because we are making a prediction on an interval we already know something about instead of making a prediction beyond the limits of our knowledge.

How Regression Works

How does a calculator or computer decide which line fits the data best? We assume that the value of y is related to the value of x, although other factors could influence y as well. Thus, we assume that we can pick the value of x exactly but that the value of y may be only partially determined by this x-value.

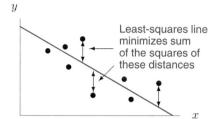

Figure 2.42: A given set of data and the corresponding least-squares regression line

One way to fit a line to the data is shown in Figure 2.42. The line shown was chosen to minimize the sum of the squares of the vertical distances between the data points and the line. Such a line is called a *least-squares line*. There are formulas which a calculator or computer uses to calculate the slope, m, and the y-intercept, b, of the least-squares line.

Correlation

When a computer or calculator calculates a regression line, it also gives a *correlation coefficient*, r. This number lies between -1 and $+1$ and measures how well a particular regression line fits the data. If $r = 1$, the data lie exactly on a line of positive slope. If $r = -1$, the data lie exactly on a line of negative slope. If r is close to 0, the data may be completely scattered, or there may be a non-linear relationship between the variables. (See Figure 2.43.)

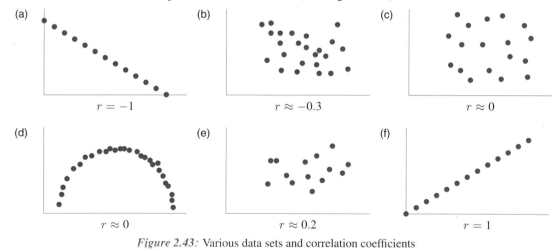

Figure 2.43: Various data sets and correlation coefficients

Example 2 The correlation coefficient for the viscosity data in Table 2.16 on page 62 is $r \approx -0.99$. The fact that r is negative tells us that the regression line has negative slope. The fact that r is close to -1 tells us that the regression line fits the data well.

The Difference between Relation, Correlation, and Causation

It is important to understand that a high correlation (either positive or negative) between two quantities does *not* imply causation. For example, there is a high correlation between children's reading level and shoe size.[5] However, large feet do not cause a child to read better (or vice versa). Larger feet and improved reading ability are both a consequence of growing older.

Notice also that a correlation of 0 does not imply that there is no relationship between x and y. For example, in Figure 2.43(d) there is a relationship between x and y-values, while Figure 2.43(c) exhibits no apparent relationship. Both data sets have a correlation coefficient of $r \approx 0$. Thus a correlation of $r = 0$ usually implies there is no linear relationship between x and y, but this does not mean there is no relationship at all.

Problems for Section 2.4

1. Match the r values with scatter plots in Figure 2.44.
$$r = -0.98, \quad r = -0.5, \quad r = -0.25, \quad r = 0, \quad r = 0.7, \quad r = 1$$

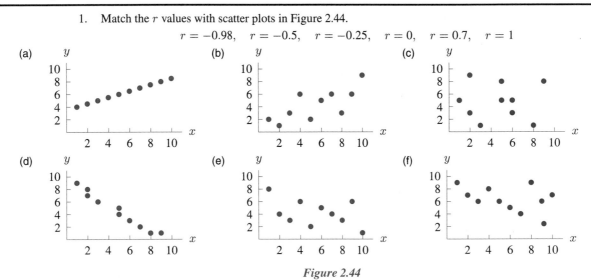

Figure 2.44

2. Table 2.17 shows the number of calories burned per minute by a person walking at 3 mph.
   (a) Make a scatter plot of this data.
   (b) Draw a regression line by eye.
   (c) Estimate the correlation coefficient by eye.

   TABLE 2.17

Body weight (lb)	100	120	150	170	200	220
Calories	2.7	3.2	4.0	4.6	5.4	5.9

3. The rate of oxygen consumption for Colorado beetles increases with temperature. See Table 2.18.
   (a) Make a scatter plot of this data.
   (b) Draw an estimated regression line by eye.
   (c) Use a calculator or computer to find the equation of the regression line. (Alternatively, find the equation of your line in part (b).) Round constants in the equation to the nearest integer.
   (d) Interpret the slope and each intercept of the regression equation.
   (e) Is there a correlation between temperature and oxygen rate?

   TABLE 2.18

°C	10	15	20	25	30
Oxygen consumption rate	90	125	200	300	375

[5]From *Statistics*, 2ed, by David Freedman. Robert Pisani, Roger Purves, Ani Adhikari, p. 142 (New York: W.W.Norton, 1991).

4. An ecologist tracked 145 deer that were born in 1992. The number of deer, d, living each subsequent year is recorded in Table 2.19.

   (a) Make a scatter plot of this data. Let $t = 0$ represent 1992.
   (b) Draw by eye a good fitting line and estimate its equation. (Round the coefficients to integers.)
   (c) Use a calculator or computer to find the equation of the least square line. (Round the coefficients to integers.)
   (d) Interpret the slope and each intercept of the line.
   (e) Is there a correlation between the year and the number of deer born in 1992 that are still alive?

TABLE 2.19

Year	1992	1993	1994	1995	1996	1997	1998	1999	2000
Deer	145	144	134	103	70	45	32	22	4

5. Table 2.20 shows the IQ of ten students and the number of hours of TV each watches per week.

   (a) Make a scatter plot of the data.
   (b) By eye, make a rough estimate of the correlation coefficient.
   (c) Use a calculator or computer to find the least squares regression line and the correlation coefficient. Your values should be correct to four decimal places.

TABLE 2.20

IQ	110	105	120	140	100	125	130	105	115	110
Hours of TV	10	12	8	2	12	10	5	6	13	3

6. The data on hand strength in Table 2.21 was collected from college freshman using a grip meter.

   (a) Make a scatter plot of these data treating the strength of the preferred hand as the independent quantity and the strength of the nonpreferred hand as the dependent quantity.
   (b) Draw a line on your scatter plot that is a good fit for these data and use it to find an approximate regression line equation.
   (c) Using a graphing calculator or computer find the least squares line equation.
   (d) What would the predicted grip strength in the nonpreferred hand be for a student with a preferred hand strength of 37?
   (e) Discuss interpolation and extrapolation using specific examples in relation to this regression line.
   (f) Discuss why r, the correlation coefficient, is both positive and close to 1.
   (g) Why do the points tend to "cluster" into two groups on your scatter plot?

TABLE 2.21 *Hand strength for 20 students*

Preferred hand (kg)	28	27	45	20	40	47	28	54	52	21
Nonpreferred hand (kg)	24	26	43	22	40	45	26	46	46	22
Preferred hand (kg)	53	52	49	45	39	26	25	32	30	32
Nonpreferred hand (kg)	47	47	41	44	33	20	27	30	29	29

7. In baseball, Henry Aaron holds the record for the greatest number of home-runs hit in the major leagues. Table 2.22 shows his cumulative yearly record [6] from the start of his career, 1954, until 1973.

   (a) Plot Aaron's cumulative number of home runs H on the vertical axis, and the time t in years along the horizontal axis, where $t = 1$ corresponds to 1954.
   (b) By eye draw a straight line that fits these data well and find its equation.
   (c) Use a calculator or computer to find the equation of the regression line for these data. What is the correlation coefficient, r, to 4 decimal places? To 3 decimal places? What does this tell you?

---

[6] Adapted from "Graphing Henry Aaron's home-run output" by H. Ringel, The Physics Teacher, January 1974, page 43.

(d)   What does the slope of the regression line mean in terms of Henry Aaron's home-run record?

(e)   From your answer to part (d), how many home-runs do you estimate Henry Aaron hit in each of the years 1974, 1975, 1976, and 1977? If you were told that Henry Aaron retired at the end of the 1976 season, would this affect your answers?

TABLE 2.22   *Henry Aaron's cumulative home-run record from 1954 to 1973*

Year	1	2	3	4	5	6	7	8	9	10	11	12	13	14	15	16	17	18	19	20
Home-runs	13	40	66	110	140	179	219	253	298	342	366	398	442	481	510	554	592	639	673	713

8.   Table 2.23 shows men's and women's world records for swimming distances from 50 meters to 1500 meters.[7]

(a)   What values would you add to Table 2.23 to represent the time taken by both men and women to swim 0 meters?

(b)   Plot the men's time against distance, with time t in seconds on the vertical axis and distance d in meters on the horizontal axis. It is claimed that a straight line models this behavior well. What is the equation for that line? What does its slope represent? On the same graph, plot the women's time against distance and find the equation of the straight line that models this behavior well. Is this line steeper or flatter than the men's line? What does that mean in terms of swimming? What are the values of the vertical intercepts? Do these values have a practical interpretation?

(c)   On another graph plot the women's times against the men's times, with women's times, w, on the vertical axis and men's times, m, on the horizontal axis. It should look linear. How could you have predicted this linearity from the equations you found in part (b)? What is the slope of this line and how can it be interpreted? A newspaper reporter claims that the women's records are about 8% slower than the men's. Do the facts support this statement? What is the value of the vertical intercept? Does this value have a practical interpretation?

TABLE 2.23   *Men's and women's world records for freestyle swimming*

Distance (meters)	50	100	200	400	800	1500
Men (seconds)	21.81	48.21	106.69	223.80	466.00	881.66
Women (seconds)	24.51	54.01	116.78	243.85	496.22	952.10

REVIEW PROBLEMS FOR CHAPTER TWO

1.   Find the equation of the line parallel to $3x + 5y = 6$ and passing through the point $(0, 6)$.

2.   Find the equation of the line passing through the point $(2, 1)$ and perpendicular to the line $y = 5x - 3$.

3.   Find the equations of the lines parallel to and perpendicular to the line $y + 4x = 7$, and through the point $(1, 5)$.

4.   A linear function, $f(t)$, generated the data in the following table. Find a formula for $f(t)$.

t	1.2	1.3	1.4	1.5
$f(t)$	0.736	0.614	0.492	0.37

[7]Accurate as of August 1995. Data from "The World Almanac and Book of Facts: 1996" edited by R. Famighetti, Funk and Wagnalls, New Jersey, 1995.

5.   Without using a calculator, match the equations (a) – (f) to the graphs (I) – (VI).

    (a) $y = -2.72x$ (c) $y = 27.9 - 0.1x$ (e) $y = -5.7 - 200x$

    (b) $y = 0.01 + 0.001x$ (d) $y = 0.1x - 27.9$ (f) $y = x/3.14$

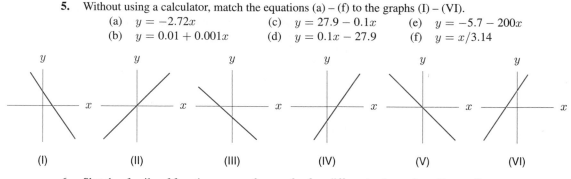

    (I)   (II)   (III)   (IV)   (V)   (VI)

6.   Sketch a family of functions $y = -2 - ax$ for five different values of x with $a < 0$.

7.   Assume A, B, C are constants with $A \neq 0$, $B \neq 0$. Consider the equation

$$Ax + By = C.$$

    (a)   Show that $y = f(x)$ is linear. State the slope and the x- and y-intercepts of $f(x)$.

    (b)   Graph $y = f(x)$, labeling the x- and y-intercepts in terms of A, B, and C, assuming

        (i) $A > 0, B > 0, C > 0$ (ii) $A > 0, B > 0, C < 0$ (iii) $A > 0, B < 0, C > 0$

8.   There are x male job-applicants at a certain company and y female applicants. Suppose that 15% of the men are accepted and 18% of the women are accepted. Write an expression in terms of x and y representing each of the following quantities:

    (a)   The total number of applicants to the company.

    (b)   The total number of applicants accepted.

    (c)   The percentage of all applicants accepted.

9.   You start 60 miles east of Pittsburgh and drive east at a constant speed of 50 miles per hour. (Assume that the road is straight and permits you to do this.) Find a formula for d, your distance from Pittsburgh as a function of t, the number of hours of travel.

10.   A small café sells coffee for $0.95 per cup. On average, it costs the café $0.25 to make a cup of coffee (for grounds, hot water, filters). The café also has a fixed daily cost of $200 (for rent, wages, utilities).

    (a)   Let R, C, and P be the café's daily revenue, costs, and profit, respectively, for selling x cups of coffee in a day. Find formulas for R, C, and P as a function of x. [Hint: The revenue, R, is the total amount of money that the café brings in. The cost, C, includes the fixed daily cost as well as the cost for all x cups of coffee sold. P is the café's profit after costs have been accounted for.]

    (b)   Plot P against x. For what x-values is the graph of P below the x-axis? Above the x-axis? Interpret your results.

    (c)   Interpret the slope and both intercepts of your graph in practical terms.

11.   Owners of an inactive quarry in Australia have decided to resume production. They estimate that it will cost them $1000 per month to maintain and insure their equipment and that monthly salaries will be $3000. It costs $80 to mine a ton of rocks. Write a formula that expresses the total cost each month, c, as a function of r, the number of tons of rock mined per month.

12.   You can type four pages in 50 minutes and nine pages in an hour and forty minutes.

    (a)   Find a linear function for the number of pages typed, p, as a function of time, t. If time is measured in minutes, what values of t make sense in this example?

    (b)   How many pages can be typed in two hours?

    (c)   Interpret the slope of the function in practical terms.

    (d)   Use the result in part(a) to solve for time as a function of the number of pages typed.

    (e)   How long does it take to type a 15 page paper?

    (f)   Write a short paragraph explaining why it is useful to know both of the formulas obtained in part(a) and part(d).

13. Hooke's Law states that the force in pounds, $F(x)$, necessary to keep a spring stretched x units beyond its natural length is directly proportional to x, so $F(x) = kx$. The positive constant k is called the spring constant. Stiffer springs have larger values of k.

(a) Do you expect $F(x)$ to be an increasing function or a decreasing function of x? Explain.

(b) A force of 2.36 pounds is required to hold a certain spring stretched 1.9 inches beyond its natural length. Find the value of k for this spring and rewrite the formula for $F(x)$ using this value of k.

(c) How much force is needed to stretch this spring 3 inches beyond its natural length?

Table 2.24 gives the cost, $C(n)$, of producing a certain good as a linear function of n, the number of units produced. Use the table to answer Problems 14–16.

TABLE 2.24

n (units)	100	125	150	175
$C(n)$ (dollars)	11000	11125	11250	11375

14. Evaluate the following expressions. Give economic interpretations for each.

(a) $C(175)$ (b) $C(175) - C(150)$ (c) $\dfrac{C(175) - C(150)}{175 - 150}$

15. Estimate $C(0)$. What is the economic significance of this value?

16. The *fixed cost* of production is the cost incurred before any goods are produced. The *unit cost* is the cost of producing an additional unit. Find a formula for $C(n)$ in terms of n, given that

$$\text{Total cost} = \text{Fixed cost} + \text{Unit cost} \cdot \text{Number of units}$$

17. Wire is sold by gauge size, where the diameter of the wire is a decreasing linear function of gauge. Gauge 2 wire has a diameter of 0.2656 inches and gauge 8 wire has a diameter of 0.1719 inches. Find the diameter for wires of gauge 12.5 and gauge 0. What values of the gauge do not make sense in this model?

18. A rock is thrown into the air. The rock's velocity in feet per second after t seconds is given by $v = 80 - 32t$

(a) Construct a table of values of v for $t = 0, 0.5, 1, 1.5, 2, 2.5, 3, 3.5, 4$.

(b) Describe the motion of the rock. How can you interpret negative values of v?

(c) At what time t is the rock highest above the ground?

(d) Interpret the slope and both intercepts of the graph of v against t.

(e) How would the slope of the graph of v be different on the moon, whose gravitational pull is less than the earth's? How would the slope be different on Jupiter, which has a greater gravitational pull than the earth?

19. One of the problems that challenged early Greek mathematicians was whether it was possible to construct a square whose area was equal to that of a given circle.

(a) If the radius of a given circle is r, give an expression, in terms of r, for the side, s, of the square with area equal to that of the circle.

(b) The side, s, of such a square is a function of the radius of the circle. What kind of function is it? How do you know?

(c) For what values of r is the side, s, equal to zero? Explain why your answer makes sense.

20. A dose-response function can be used to describe the increase in risk associated with the increase in exposure to various hazards. For example, the risk of contracting lung cancer depends, among other things, on the number of cigarettes a person smokes per day. This risk can be described by a linear dose-response function. For example, it is known that smoking 10 cigarettes per day increases a person's probability of contracting lung cancer by a factor of 25, while smoking 20 cigarettes a day increases the probability by a factor of 50.

(a) Find a formula for $i(x)$, the increase in the probability of contracting lung cancer for a person who smokes x cigarettes per day as compared to a non-smoker.

(b) Evaluate $i(0)$.

(c) Interpret the slope of the function i.

21. In economics, the *demand* for a product is the amount of that product that consumers are willing to buy at a given price. The quantity demanded of a product usually decreases if the price of that product increases. Suppose that a company believes there is a linear relationship between the demand for its product and its price. The company knows that when the price of its product was $3 per unit, the quantity demanded weekly was 500 units, and that when the unit price was raised to $4, the quantity demanded weekly dropped to 300 units. Let D represent the quantity demanded weekly at a unit price of p dollars.

    (a) Calculate D when $p = 5$. Interpret your result.
    (b) Find a formula for D in terms of p.
    (c) Suppose that the company raises the price of the good and that the new quantity demanded weekly is only 50 units. What is the new price?
    (d) Give an economic interpretation of the slope of the function you found in part (b).
    (e) Find D when $p = 0$. Find p when $D = 0$. Give economic interpretations of both these results.

22. In economics, the *supply* of a product is the quantity of that product suppliers are willing to provide at a given price. In theory, the quantity supplied of a product increases if the price of that product increases. Suppose that there is a linear relationship between the quantity supplied, S, of the product described in Problem 21 and its price, p. The quantity supplied weekly is 100 when the price is $2 and the quantity supplied rises by 50 units when the price rises by $0.50.

    (a) Find a formula for S in terms of p.
    (b) Interpret the slope of your formula in economic terms.
    (c) Is there a price below which suppliers will not provide this product?
    (d) The *market clearing price* is the price at which supply equals demand. According to theory, the free-market price of a product is its market clearing price. Using the demand function from Problem 21, find the market clearing price for this product.

23. When economists graph demand or supply equations, they place quantity on the horizontal axis and price on the vertical axis.

    (a) On the same set of axes, graph the demand and supply equations you found in Problems 21 and 22, with price on the vertical axis.
    (b) Indicate how you could estimate the market clearing price from your graph.

24. An initial investment of $12,467.00 grows according to Table 2.25.

    (a) Let $M(t)$ be the value of the investment after t months. Make a similar table showing t and $M(t)$.
    (b) Plot the data points you got from part (a).
    (c) Find an equation for a regression line for the data points.
    (d) Use the equation of the line to predict $M(9)$ and $M(100)$.
    (e) Are these reasonable extrapolations?

TABLE 2.25

Month	1	2	3	4	5	6	7	8
Interest	$95.48	$96.21	$96.95	$97.70	$98.47	$99.23	$99.99	$100.74

25. A student exercised for 30 minutes and then measured her ten-second pulse count at one minute intervals as she rested. The data are shown in Table 2.26, where time is in minutes after exercise.

    (a) Make a scatter plot of this data.
    (b) Because the pulse values are the same after 4 and 5 minutes, these data are clearly not linear. Are there values of t for which a regression line is a good model for these data? If so, what values?
    (c) Discuss the correlation between minutes after exercising and pulse rate.

TABLE 2.26

Time (min)	0	1	2	3	4	5
Pulse	22	18	15	12	10	10

26. Repeat Problem 25 after collecting your own data in a table like Table 2.26.

27. Record the height and shoe size of at least five females or five males in your class and make a data table.

   (a) Make a scatter plot of these data, with height on the y-axis and shoe size on the x-axis.
   (b) By eye, draw a line that fits this data and find its equation.
   (c) Use a graphing calculator or computer to find the equation of the least-squares line.
   (d) Discuss interpolation and extrapolation using specific examples in relation to this regression line. Check the interpolation with a student in the class whose shoe size is close to the one you have chosen.
   (e) Find and interpret r, the correlation coefficient.

28. You spend c dollars on x apples and y bananas. In Figure 2.45, line l gives y as a function of x.

   (a) If apples cost p dollars each and bananas cost q each, label the x- and y-intercepts of l. [Note: Your labels will involve the constants p, q or c.]
   (b) What is the slope of l?

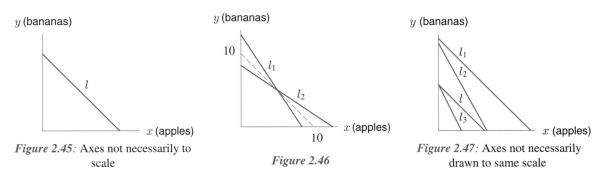

Figure 2.45: Axes not necessarily to scale

Figure 2.46

Figure 2.47: Axes not necessarily drawn to same scale

29. The apples in Problem 28 cost more than bananas, so $p > q$. Which of the two lines, l_1 or l_2, in Figure 2.46 could represent $y = f(x)$?

30. Figure 2.47 shows line l from Problem 28 and three other lines, l_1, l_2, l_3. If possible, match the following stories to lines l_1, l_2, l_3.

   (a) Your total budget, c, has increased, but the prices of apples and bananas have not changed.
   (b) Your total budget and the cost of bananas do not change, but the cost of apples increases.
   (c) Your budget increases, as does the price of apples, but the price of bananas stays fixed.
   (d) Your budget increases, but the price of apples goes down, and the price of bananas stays fixed.

CHAPTER THREE

FUNCTION NOTATION

In this chapter, we investigate properties and notation common to all functions. We begin with a closer look at function notation, including an introduction to inverse functions. The ideas of domain, range, and piecewise defined functions are addressed.

3.1 INPUT AND OUTPUT

Finding Output Values: Evaluating a Function

Evaluating a function means figuring out the value of a function's output from a particular value of the input.

In the housepainting example on page 2, the notation $n = f(A)$ indicates that n is a function of A. The expression $f(A)$ represents the output of the function—specifically, the amount of paint required to cover an area of A ft^2. For example $f(20{,}000)$ represents the number of gallons of paint required to cover a house of 20,000 ft^2.

Example 1 Using the fact that 1 gallon of paint will cover 250 ft^2, evaluate the expression $f(20{,}000)$.

Solution To evaluate $f(20{,}000)$, calculate the number of gallons required to cover 20,000 ft^2:

$$f(20{,}000) = \frac{20{,}000 \text{ ft}^2}{250 \text{ ft}^2/\text{gallon}} = 80 \text{ gallons of paint.}$$

Evaluating a Function Using a Formula

If we know the radius of a circle, we can find its area. We can write area as a function of radius:

$$A = q(r) = \pi r^2.$$

Example 2 Use the formula $q(r) = \pi r^2$, with r in cm, to evaluate $q(10)$ and $q(20)$. What do your results tell you about circles?

Solution In the expression $q(10)$, the value of r is 10, so

$$q(10) = \pi \cdot (10)^2 = 100\pi \approx 314.$$

Similarly, substituting $r = 20$, we have

$$q(20) = \pi \cdot (20)^2 = 400\pi \approx 1257.$$

The statements $q(10) \approx 314$ and $q(20) \approx 1257$ tell us that a circle of radius 10 cm has an area of approximately 314 cm^2 and a circle of radius 20 cm has an area of approximately 1257 cm^2.

Evaluating functions given by a formula can involve algebraic simplification, as the following example shows.

Example 3 Let $h(x) = x^2 + bx + c$, where b and c are constants. Evaluate and simplify the following expressions.

(a) $h(2)$ (b) $h(b)$ (c) $h(2b)$ (d) $h(2x)$

Solution Notice that x is the input and $h(x)$ is the output. It is helpful to rewrite the formula as

$$\text{Output} = h(\text{Input}) = (\text{Input})^2 + b \cdot (\text{Input}) + c.$$

(a) For $h(2)$, we have Input $= 2$, so
$$h(2) = (2)^2 + b \cdot (2) + c,$$
which simplifies to $h(2) = 2b + c + 4$.

(b) In this case, Input $= b$. Thus,

$$h(b) = (b)^2 + b \cdot (b) + c.$$

Simplifying gives $h(b) = 2b^2 + c$.

(c)   For the input $2b$, we have

$$h(2b) = (2b)^2 + b \cdot (2b) + c$$
$$= 4b^2 + 2b^2 + c$$
$$= 6b^2 + c.$$

(d)   To find $h(2x)$, we have

$$h(2x) = (2x)^2 + b \cdot (2x) + c$$
$$= 4x^2 + 2bx + c.$$

Example 4   Let $g(x) = \dfrac{x^2 + 1}{5 + x}$. Evaluate the following expressions.

(a) $g(3)$ (b) $g(-1)$ (c) $g(a)$ (d) $g(a - 2)$ (e) $g(a) - 2$ (f) $g(a) - g(2)$

Solution   (a)   To evaluate $g(3)$, replace every x in the formula with 3:

$$g(3) = \frac{(3)^2 + 1}{5 + (3)} = \frac{10}{8} = 1.25.$$

(b)   To evaluate $g(-1)$, replace every x in the formula with (-1):

$$g(-1) = \frac{(-1)^2 + 1}{5 + (-1)} = \frac{2}{4} = 0.5.$$

(c)   To evaluate $g(a)$, replace every x in the formula with a:

$$g(a) = \frac{a^2 + 1}{5 + a}.$$

(d)   To evaluate $g(a - 2)$, replace every x in the formula with $(a - 2)$:

$$g(a - 2) = \frac{(a - 2)^2 + 1}{5 + (a - 2)}$$
$$= \frac{a^2 - 4a + 4 + 1}{5 + a - 2}$$
$$= \frac{a^2 - 4a + 5}{3 + a}.$$

(e)   To evaluate $g(a) - 2$, first evaluate $g(a)$ (as we did in part (c)), and then subtract 2:

$$g(a) - 2 = \frac{a^2 + 1}{5 + a} - 2$$
$$= \frac{a^2 + 1}{5 + a} - \frac{2}{1} \cdot \frac{5 + a}{5 + a} \qquad \text{Creating common denominator}$$
$$= \frac{a^2 + 1}{5 + a} - \frac{10 + 2a}{5 + a}$$
$$= \frac{a^2 + 1 - 10 - 2a}{5 + a}$$
$$= \frac{a^2 - 2a - 9}{5 + a}.$$

(f)   To evaluate $g(a) - g(2)$, first find $g(2)$:

$$g(2) = \frac{2^2 + 1}{5 + 2} = \frac{5}{7}.$$

From part (c), we have $g(a) = \dfrac{a^2 + 1}{5 + a}$. Subtracting $g(2)$ from $g(a)$ gives:

$$g(a) - g(2) = \frac{a^2 + 1}{5 + a} - \frac{5}{7}$$
$$= \frac{7(a^2 + 1) - 5(5 + a)}{7(5 + a)}$$
$$= \frac{7a^2 - 5a - 18}{7(5 + a)}.$$

Finding Input Values: Solving Equations

Given an input, we evaluate the function to find the output. Often the situation is reversed; we know the output value and we want to find the corresponding input value(s). If the function is given by a formula, the input values are solutions to an equation.

Example 5   Use the cricket function $T = \frac{1}{4}R + 40$, introduced on page 3, to find how fast the snowy tree cricket chirps when the temperature is 76°F.

Solution   We want to find R when $T = 76$. Substitute $T = 76$ into the formula and solve the equation

$$76 = \frac{1}{4}R + 40$$

$$36 = \frac{1}{4}R \qquad \text{subtract 40 from both sides}$$

$$144 = R. \qquad \text{multiply both sides by 4}$$

The cricket chirps at a rate of 144 chirps per minute when the temperature is 76°F.

Example 6   Suppose $y = \dfrac{1}{\sqrt{x - 4}}$.

(a)   Find an x-value that results in $y = 2$.
(b)   Is there an x-value that results in $y = -2$?

Solution   (a)   To find an x-value that results in $y = 2$, solve the equation

$$2 = \frac{1}{\sqrt{x - 4}}.$$

Square both sides

$$4 = \frac{1}{x - 4}.$$

Now multiply by $(x - 4)$

$$4(x - 4) = 1$$
$$4x - 16 = 1$$
$$x = \frac{17}{4} = 4.25.$$

The x-value is 4.25. (Note that the simplification $(x - 4)/(x - 4) = 1$ in the second step was valid because $x - 4 \neq 0$.)

(b)   Since $\sqrt{x - 4}$ is nonnegative if it is defined, its reciprocal, $y = \dfrac{1}{\sqrt{x - 4}}$ is also nonnegative if it is defined. Thus, y is not negative for any x input, so there is no x-value that results in $y = -2$.

If we solve an equation involving a function which is being used to model a physical quantity, we must choose the solutions that make sense in the context of the model. Consider the next example.

Example 7   Let $A = q(r)$ be the area of a circle of radius r, where r is in cm. What is the radius of a circle whose area is 100 cm²?

Solution   The output $q(r)$ is an area. Solving the equation $q(r) = 100$ for r gives the radius of a circle whose area is 100 cm². Since the formula for the area of a circle is $q(r) = \pi r^2$, we solve

$$q(r) = \pi r^2 = 100$$
$$r^2 = \frac{100}{\pi}$$
$$r = \pm\sqrt{\frac{100}{\pi}} \approx \pm 5.64.$$

We found two solutions for r, one positive and one negative. Since a circle cannot have a negative radius, we take $r \approx 5.64$ cm. Thus, a circle of area 100 cm² has a radius of approximately 5.64 cm.

Finding Output and Input Values From Tables and Graphs

The following two examples use function notation with a table and a graph respectively.

Example 8   Table 3.1 shows the revenue, $R = f(t)$, received, and to be received, by the National Football League,[1] NFL, from network TV as a function of the year, t, since 1975.

(a) Evaluate and interpret $f(25)$.   (b) Solve and interpret $f(t) = 1159$.

TABLE 3.1

Year, t (since 1975)	0	5	10	15	20	25	30
Revenue, R (million $)	201	364	651	1075	1159	2200	2200

Solution   (a) The table shows $f(25) = 2200$. Since $t = 25$ in the year 2000, we know that NFL's revenue from TV is $2200 million in the year 2000.

(b) Solving $f(t) = 1159$ means finding the year in which TV revenues were $1159 million; it is $t = 20$. In 1995, NFL's TV revenues were $1159 million.

Example 9   A man drives from his home to a store and and back. The entire trip takes 30 minutes. Figure 3.1 gives his velocity $v(t)$ (in mph) as a function of the time t (in minutes) since he left home. A negative velocity indicates that he is traveling away from the store back to his home.

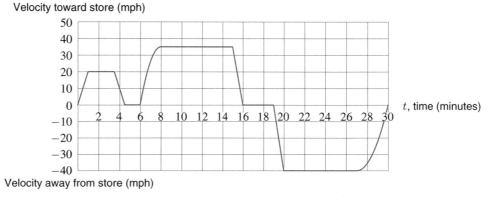

Figure 3.1: Velocity of a man on a trip to the store and back

[1]*Newsweek*, January 26, 1998.

Evaluate and interpret:

(a) $v(5)$ (b) $v(24)$ (c) $v(8) - v(6)$ (d) $v(-3)$

Solve for t and interpret:

(e) $v(t) = 15$ (f) $v(t) = -20$ (g) $v(t) = v(7)$

Solution   (a)   To evaluate $v(5)$, look on the graph where $t = 5$ minutes. Five minutes after he left home, his velocity is 0 mph. Thus, $v(5) = 0$. Perhaps he had to stop at a light.

(b)   The graph shows that $v(24) = -40$ mph. After 24 minutes, he is traveling at 40 mph away from the store.

(c)   From the graph, $v(8) = 35$ mph and $v(6) = 0$ mph. Thus, $v(8) - v(6) = 35 - 0 = 35$. This shows that the man's speed increased by 35 mph in the interval between $t = 6$ minutes and $t = 8$ minutes.

(d)   The quantity $v(-3)$ is not defined since the graph only gives velocities for nonnegative times.

(e)   To solve for t when $v(t) = 15$, look on the graph where the velocity is 15 mph. This occurs at $t \approx 0.75$ minute, 3.75 minutes, 6.5 minutes, and 15.5 minutes. At each of these four times the man's velocity was 15 mph.

(f)   To solve $v(t) = -20$ for t, we see that the velocity is -20 mph at $t \approx 19.5$ and $t \approx 29$ minutes.

(g)   First we evaluate $v(7) \approx 27$. To solve $v(t) = 27$, we look for the values of t making the velocity 27 mph. One such t is of course $t = 7$; the other t is $t \approx 15$ minutes.

Problems for Section 3.1

1.   Let $F = g(t)$ be the number of foxes in a park as a function of t, the number of months since January 1. Evaluate $g(9)$ using Table 1.4 on page 5. What does this tell us about the fox population?

2.   Let $F = g(t)$ be the number of foxes in month t in the national park described in Example 8 on page 5. Solve the equation $g(t) = 75$. What does your solution tell you about the fox population?

3.   Chicago's average monthly rainfall, $R = f(t)$ inches, is given as a function of month, t, in Table 3.2. (January is $t = 1$.) Solve and interpret: (a) $f(t) = 3.7$ (b) $f(t) = f(2)$

TABLE 3.2

Month, t	1	2	3	4	5	6	7	8	9	10	11	12
Rainfall, R (inches)	1.8	1.8	2.7	3.1	3.5	3.7	3.5	3.4	3.2	2.5	2.4	2.1

4.   If $f(x) = 2x + 1$, (a) Find $f(0)$ (b) Solve $f(x) = 0$.

5.   If $f(t) = t^2 - 4$, (a) Find $f(0)$ (b) Solve $f(t) = 0$.

6.   If $g(x) = x^2 - 5x + 6$, (a) Find $g(0)$ (b) Solve $g(x) = 0$.

7.   If $g(t) = \dfrac{1}{t+2} - 1$, (a) Find $g(0)$ (b) Solve $g(t) = 0$.

8.   Use the graph of $f(x)$ in Figure 3.2 to estimate:

(a) $f(0)$ (b) $f(1)$ (c) $f(b)$ (d) $f(c)$ (e) $f(d)$

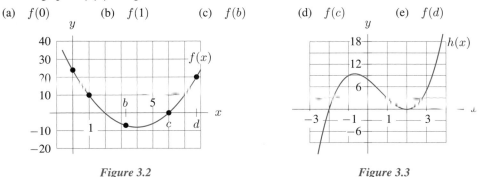

Figure 3.2   *Figure 3.3*

9. (a) Using the graph of $h(x)$ in Figure 3.3, fill in Table 3.3:

TABLE 3.3

x	-2	-1	0	1	2	3
$h(x)$						

   (b) Evaluate $h(3) - h(1)$ (c) Evaluate $h(2) - h(0)$
   (d) Evaluate $2h(0)$ (e) Evaluate $h(1) + 3$

10. Let $f(x) = \sqrt{x^2 + 16} - 5$.

   (a) Find $f(0)$.   (b) For what values of x is $f(x)$ zero?
   (c) Find $f(3)$.   (d) What is the vertical intercept of the graph of $f(x)$?
   (e) Where does the graph cross the x-axis?

11. Use the letters a, b, c, d, e, h in Figure 3.4 to answer the following questions.

   (a) What are the coordinates of the points P and Q?
   (b) Evaluate $f(b)$.
   (c) Solve $f(x) = e$ for x.
   (d) Suppose $c = f(z)$ and $z = f(x)$. What is x?
   (e) Suppose $f(b) = -f(d)$. What additional information does this give you?

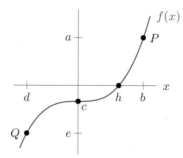

Figure 3.4

12. (a) Complete Table 3.4 using

$$f(x) = 2x(x - 3) - x(x - 5) \quad \text{and} \quad g(x) = x^2 - x.$$

   What do you notice? Graph these two functions. Are the two functions the same? Explain.
   (b) Complete Table 3.5 using

$$h(x) = x^5 - 5x^3 + 6x + 1 \quad \text{and} \quad j(x) = 2x + 1.$$

   What do you notice? Graph these two functions. Are the two functions the same? Explain.

TABLE 3.4

x	-2	-1	0	1	2
$f(x)$					
$g(x)$					

TABLE 3.5

x	-2	-1	0	1	2
$h(x)$					
$j(x)$					

13. If a ball is thrown up from the ground with an initial velocity of 64 ft/sec, its height as a function of time, t, is given by

$$h(t) = -16t^2 + 64t.$$

   (a) Evaluate $h(1)$ and $h(3)$. What does this tell us about the height of the ball?
   (b) Sketch this function. Using a graph, determine when the ball hits the ground and the maximum height of the ball.

14. Suppose $v(t) = t^2 - 2t$ gives the velocity, in ft/sec, of an object at time t, in seconds.
    (a) What is the initial velocity, $v(0)$?
    (b) When does the object have a velocity of zero?
    (c) What is the meaning of the quantity $v(3)$? What are its units?

15. Let $s(t) = 11t^2 + t + 100$ be the position, in miles, of a car driving on a straight road at time t, in hours. The car's velocity at any time t is given by $v(t) = 22t + 1$.
    (a) Use function notation to express the car's position after 2 hours. Where is the car then?
    (b) Use function notation to express the question, "When is the car going 65 mph?"
    (c) Where is the car when it is going 67 mph?

16. New York state income tax is based on what is called taxable income. A person's taxable income is part of his/her total income; the tax owed to the state is calculated using the taxable income (not total income). For a person with a taxable income between $65,000 and $100,000, the tax owed is $4635 plus 7.875% of the taxable income over $65,000.
    (a) Compute the tax owed by a lawyer whose taxable income is $68,000.
    (b) Consider a lawyer whose taxable income is 80% of her total income, x, where x is between $85,000 and $120,000. Write a formula for $T(x)$, the taxable income.
    (c) Write a formula for $L(x)$, the amount of tax owed by the lawyer in part (b).
    (d) Use $L(x)$ to evaluate the tax liability for $x = 85,000$ and compare your results to part (a).

17. In bowling, ten pins are arranged in a triangular fashion as shown in Figure 17. If a fifth row were added, the total number of pins would be fifteen. Let $s(n)$ be the sum of the pins in rows 1 to n inclusive. For example, $s(3) = 1 + 2 + 3 = 6$.

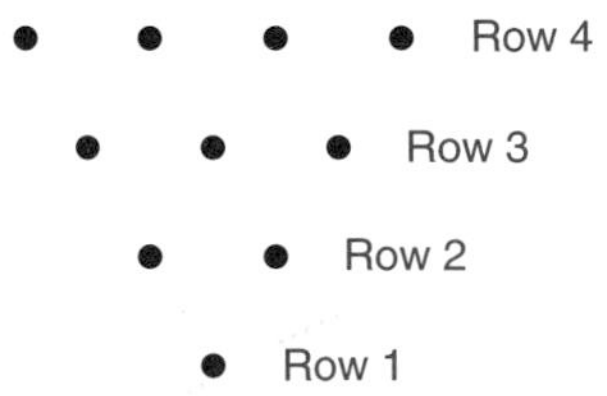

Figure 3.5

TABLE 3.6

n	1	2	3	4	5
$s(n)$					

    (a) Complete Table 3.6.
    (b) Using Table 3.6, check that the following formula holds for $1 \leq n \leq 5$:
$$s(n) = \frac{n(n+1)}{2}.$$
    (c) Assuming the formula for $s(n)$ holds for all n, calculate how many pins would be used if there were 100 rows.

18. The Fibonacci sequence is a sequence of numbers that begins $1, 1, 2, 3, 5 \ldots$.
    (a) Notice that each term in the sequence is the sum of the two preceding terms. For example,
$$2 = 1 + 1, \quad 3 = 2 + 1, \quad 5 = 2 + 3, \ldots.$$
    Based on this observation, complete the following table of values for $f(n)$, the nth term in the Fibonacci sequence.

n	1	2	3	4	5	6	7	8	9	10	11	12
$f(n)$	1	1	2	3	5							

    (b) The table of values in part (a) can be completed even though we don't have a formula for $f(n)$. Does the fact that we don't have a formula mean that $f(n)$ is not a function?
    (c) Are you able to evaluate the following expressions in a way that is consistent with the observations from parts (a) and (b)? If so, do so; if not, explain why not.
$$f(0), \quad f(-1), \quad f(-2), \quad f(0.5).$$

3.2 CHANGES IN INPUT AND OUTPUT

Inside and Outside Changes

In this section, we consider changes to the input and output of a function. For the function

$$Q = f(t),$$

a change inside the function's parentheses can be called an "inside change" and a change outside the function's parentheses can be called an "outside change."

Example 1   If $n = f(A)$ gives the number of gallons of paint needed to cover a house of area A ft^2, explain the meaning of the expressions $f(A + 10)$ and $f(A) + 10$ in the context of painting.

Solution   These two expressions are similar in that they both involve adding 10. However, for $f(A + 10)$, the 10 is added on the inside, so 10 is added to the area, A. Thus,

$$n = f(\underbrace{A + 10}_{\text{Area}}) = \begin{array}{c}\text{Amount of paint needed} \\ \text{to cover an area of } (A+10) \text{ ft}^2\end{array} = \begin{array}{c}\text{Amount of paint needed to cover} \\ \text{an area 10 ft}^2 \text{ larger than } A.\end{array}$$

The expression $f(A) + 10$ represents an outside change. We are adding 10 to $f(A)$, which represents an amount of paint, not an area. We have

$$n = \underbrace{f(A)}_{\substack{\text{Amount} \\ \text{of paint}}} + 10 = \begin{array}{c}\text{Amount of paint needed} \\ \text{to cover a region of area } A\end{array} + 10 \text{ gals} = \begin{array}{c}\text{10 gallons more paint} \\ \text{than the amount needed} \\ \text{to cover an area } A.\end{array}$$

In $f(A + 10)$, we added 10 square feet on the inside of the function, which means that the area to be painted is now 10 ft^2 larger. In $f(A) + 10$, we added 10 gallons to the outside, which means that we have 10 more gallons of paint than we need.

Example 2   Let $s(t)$ be the average weight (in pounds) of a baby at age t months. The weight, V, of a particular baby named Jonah is related to the average weight function $s(t)$ by the equation

$$V = s(t) + 2.$$

Find Jonah's weight at ages $t = 3$ and $t = 6$ months. What can you say about Jonah's weight in general?

Solution   At $t = 3$ months, Jonah's weight is

$$V = s(3) + 2.$$

Since $s(3)$ is the average weight of a 3-month old boy, we see that at 3 months, Jonah weighs 2 pounds more than average. Similarly, at $t = 6$ months we have

$$V = s(6) + 2,$$

which means that, at 6 months, Jonah weighs 2 pounds more than average. In general, Jonah weighs 2 pounds more than average for babies of his age.

Example 3   The weight, W, of another baby named Ben is related to $s(t)$ by the equation

$$W = s(t + 4).$$

What can you say about Ben's weight at age $t = 3$ months? At $t = 6$ months? Assuming that babies increase in weight over the first year of life, decide if Ben is of average weight for his age, above average, or below average.

Solution   Since $W = s(t + 4)$, at age $t = 3$ months Ben's weight is given by

$$W = s(3 + 4) = s(7).$$

We defined $s(7)$ to be the average weight of a 7-month old baby. At age 3 months, Ben's weight is the same as the average weight of 7-month old babies. Since, on average, a baby's weight increases as the baby grows, this means that Ben is heavier than the average for a 3-month old. Similarly, at age $t = 6$, Ben's weight is given by

$$W = s(6 + 4) = s(10).$$

Thus, at 6 months, Ben's weight is the same as the average weight of 10-month old babies. In both cases, we see that Ben is above average in weight.

Notice that in Example 3, the equation

$$W = s(t + 4)$$

involves an inside change, or a change in months. This equation tells us that Ben weighs as much as babies who are 4 months older than he is. However in Example 2, the equation

$$V = s(t) + 2$$

involves an outside change, or a change in weight. This equation tells us that Jonah is 2 pounds heavier than the average weight of babies his age. Although both equations tell us that the babies are heavier than average for their age, they vary from the average in different ways.

Interchanging Input and Output: Inverse Functions

The roles of input and output are not necessarily fixed. For example, in the house painting example, we write $n = f(A)$, with input A and output n, because knowing the area, A, determines the amount of paint, n, needed. However, knowing the amount of paint enables us to calculate the area which can be painted. Thus we can define a new function, $A = g(n)$, which tells us the value of A given the value of n instead of the other way around. For this function, n is the input and A is the output. The functions f and g are called *inverses* of each other.

Example 4   The function f has formula $n = f(A) = A/250$. Find a formula for the inverse function $A = g(n)$.

Solution   Since

$$n = \frac{A}{250},$$

solving for A gives

$$A = 250n.$$

Thus, $A = g(n) = 250n$.

The fact that f and g are inverse functions means that they go in "opposite directions." The function f takes A as input and returns n, while g takes n as input and returns A.

Inverse Function Notation

In the preceding discussion, there was nothing about the names of the two functions that stressed their special relationship. If we want to emphasize that a function is the inverse of f we call it f^{-1} (read "f-inverse").

For example, to express the fact that the number of gallons of paint, n, is a function of the area, A, we write

$$n = f(A).$$

To express the additional fact that the area A is also determined by n, so that A is a function of n, we write

$$A = f^{-1}(n).$$

The symbol f^{-1} is used to represent the function that gives the output A for a given input n.

Warning: The -1 which appears in inverse function notation is not an exponent.

Unfortunately, the notation $f^{-1}(x)$ might lead us to interpret it as $\frac{1}{f(x)}$. The two expressions are not the same in general: $f^{-1}(x)$ is the output when x is fed into the inverse of f, while $\frac{1}{f(x)}$ is the reciprocal of the number we get when x is fed into f.

Example 5 Interpret and evaluate $f(100)$ and $f^{-1}(100)$ for the house-painting function, $n = f(A) = A/250$, in Example 1 on page 81.

Solution Since $n = f(A)$, in $f(100)$ we have $A = 100$ ft². Evaluating $f(100)$ tells us how much paint is needed for 100 ft². Since

$$n = f(100) = \frac{100}{250} = 0.4,$$

we know that it takes 0.4 gallon of paint to cover 100 ft².

In $f^{-1}(100)$, the 100 is the number of gallons, so $f^{-1}(100)$ represents the area which can be painted by 100 gallons:

$$A = f^{-1}(100) = 250 \cdot (100) = 25{,}000 \text{ ft}^2.$$

Notice that $f(100) = 0.4$ and $f^{-1}(100) = 25{,}000$, so $(f(100))^{-1} = \frac{1}{f(100)} = \frac{1}{25{,}000} = 0.00004$. Thus, in the previous example, the reciprocal of the function is not the same as the inverse function:

$$f^{-1}(100) \neq (f(100))^{-1}.$$

Example 6 The population of birds on an island is given, in thousands, by $P(t)$, where t is the number of years since 1999. Assume that population is a function of time and vice versa.

(a) What does $P(4)$ represent?
(b) What does $P^{-1}(4)$ represent?

Solution (a) $P(4)$ is the bird population (in thousands) in the year 2003.

(b) Since P^{-1} is the inverse function, P^{-1} is a function which takes population as input and returns time as output. Therefore, $P^{-1}(4)$ is the year (after 1999) in which there were 4,000 birds on the island.

In the next example, we find the formula for an inverse function.

Example 7 The cricket function, which gives temperature, T, in terms of chirp rate, R, is

$$T = f(R) = \frac{1}{4} \cdot R + 40.$$

Find a formula for the inverse function, $R = f^{-1}(T)$.

Solution The inverse function gives the chirp rate in terms of the temperature, so we solve the following equation for R:

$$T = \frac{1}{4} \cdot R + 40,$$

giving

$$T - 40 = \frac{1}{4} \cdot R$$
$$R = 4(T - 40).$$

Thus, $R = f^{-1}(T) = 4(T - 40)$.

Problems for Section 3.2

1. The following table shows values for functions $f(x)$ and $g(x)$:

x	0	1	2	3	4	5	6	7	8	9
$f(x)$	-10	-7	4	29	74	145	248	389	574	809
$g(x)$		-6	-7	6	33	74	129	198	281	378	489

(a) Evaluate
    (i) $f(x)$ for $x = 6$.   (ii) $f(5) - 3$.   (iii) $f(5 - 3)$.
    (iv) $g(x) + 6$ for $x = 2$.   (v) $g(x + 6)$ for $x = 2$.   (vi) $3g(x)$ for $x = 0$.
    (vii) $f(3x)$ for $x = 2$.   (viii) $f(x) - f(2)$ for $x = 8$. (ix) $g(x+1) - g(x)$ for $x = 1$.

(b) Solve
    (i) $g(x) = 6$.   (ii) $f(x) = 574$.   (iii) $g(x) = 281$.

(c) The values in the table were obtained using the formulas $f(x) = x^3 + x^2 + x - 10$ and $g(x) = 7x^2 - 8x - 6$. Use the table to find two solutions to the equation $x^3 + x^2 + x - 10 = 7x^2 - 8x - 6$.

2. Using the graph of $v(t)$ in Figure 3.1 on page 77, solve the equation $v(t + 2) = -10$ and interpret your result.

3. Let $f(x) = \left(\dfrac{x}{2}\right)^3 + 2$. The graph of $y = f(x)$ is in Figure 3.6.

    (a) Calculate $f(-6)$.
    (b) Solve $f(x) = -6$.
    (c) Find points that correspond to parts (a) and (b) on the graph of the function.
    (d) Calculate $f(4) - f(2)$. Draw a vertical line segment on the y-axis that illustrates this calculation.
    (e) If $a = -2$, compute $f(a + 4)$ and $f(a) + 4$.
    (f) In part (e) above, what x-value corresponds to $f(a + 4)$? To $f(a) + 4$?

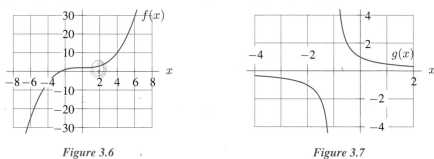

Figure 3.6   Figure 3.7

4. Let $g(x) = \dfrac{1}{x + 1}$. The graph of $y = g(x)$ is in Figure 3.7.

    (a) Calculate $g(-2)$.
    (b) Solve $g(x) = -2$.
    (c) Find points that correspond to parts (a) and (b) on the graph of the function.
    (d) Calculate $g(0) - g(2)$. Draw a vertical line segment on the y-axis that illustrates this calculation.
    (e) If $a = -3$, compute $g(a - 1)$ and $g(a) - 1$.
    (f) In part (e) above, what x-value corresponds to $g(a - 1)$? To $g(a) - 1$?

5. Let $h(x) = \sqrt{x + 4}$.

    (a) Find a point on the graph of $h(x)$ whose x-coordinate is 5.
    (b) Find a point on the graph whose y-coordinate is 5.
    (c) Sketch the graph and locate the points in parts (a) and (b).
    (d) Let $p = 2$. Calculate $h(p + 1) - h(p)$.

6. Let $f(x) = 1 - x$. Evaluate the following and simplify your answers.

    (a) $f(2x)$ (b) $f(x+1)$ (c) $f(1-x)$ (d) $f(x^2)$ (e) $f(1/x)$ (f) $f(\sqrt{r})$

7. Let $g(x) = x^2 + x$. Find formulas for the following functions. Simplify your answers.

    (a) $g(-3x)$ (b) $g(1-x)$ (c) $g(x+\pi)$
    (d) $g(\sqrt{x})$ (e) $g(1/(x+1))$ (f) $g(x^2)$

8. Let $f(x) = \dfrac{x}{x-1}$.

    (a) Find and simplify

        (i) $f\left(\dfrac{1}{t}\right)$ (ii) $f\left(\dfrac{1}{t+1}\right)$

    (b) Solve $f(x) = 3$.

9. The number of gallons of paint, n, needed to cover a house is a function of the surface area, measured in ft^2, of the house. That is, $n = f(A)$. Match each story below to one expression.

    (a) I figured out how many gallons I needed and then bought two extra gallons just in case.
    (b) I bought enough paint to cover my house twice.
    (c) I bought enough paint to cover my house and my welcome sign, which measures 2 square feet.

    (i) $2f(A)$ (ii) $f(A+2)$ (iii) $f(A)+2$

10. Suppose every day I take the same taxi over the same route from home to the train station. The trip is x miles, so the cost for the trip is $C = f(x)$. Match each story in (a)–(d) to a function in (i)–(iv) which represents the amount paid to the taxi driver.

    (a) I received a raise yesterday, so today I gave my driver a five dollar tip.
    (b) I had a new driver today and he got lost. He drove five extra miles and charged me for it.
    (c) I haven't paid my driver all week. Today is Friday and I'll pay what I owe for the week.
    (d) The meter in the taxi went crazy and showed five times the number of miles I actually traveled.

    (i) $C = 5f(x)$ (ii) $C = f(x)+5$ (iii) $C = f(5x)$ (iv) $C = f(x+5)$

11. Let $R = P(t)$ be the number of rabbits living in the national park in month t. (See Example 8 on page 5.)

    (a) What does the expression $P(t+1)$ represent?   (b) What does the expression $2P(t)$ represent?

12. The perimeter of a square is the distance around its sides. If s is the length of one side of a square, then $P = 4s$. Let $P = h(s)$.

    (a) Evaluate $h(3)$.   (b) Evaluate $h(s+1)$. What does this expression represent?

13. Suppose $P(t)$ is the US population in millions today and t is in years. Match each statement (I) – (IV) with one of the formulas (a) – (h).

    I. The population 10 years before today.

    II. Today's population plus 10 million immigrants.

    III. Ten percent of the population we have today.

    IV. The population after $100,000$ people have emigrated.

    (a) $P(t)-10$ (b) $P(t-10)$ (c) $0.1P(t)$ (d) $P(t)+10$
    (e) $P(t+10)$ (f) $P(t)/0.1$ (g) $P(t)+0.1$ (h) $P(t)-0.1$

14. Let $A = f(r)$ be the area of a circle as a function of radius.

    (a) Write a formula for $f(r)$.
    (b) Which expression represents the area of a circle whose radius is increased by 10%? Explain.
        (i) $0.10f(r)$ (ii) $f(r + 0.10)$ (iii) $f(0.10r)$ (iv) $f(1.1r)$ (v) $f(r) + 0.10$
    (c) By what percent does the area increase if the radius is increased by 10%?

15. Use the graph in Figure 3.8 to fill in the missing values:

    (a) $f(0) = ?$ (b) $f(?) = 0$ (c) $f^{-1}(0) = ?$ (d) $f^{-1}(?) = 0$

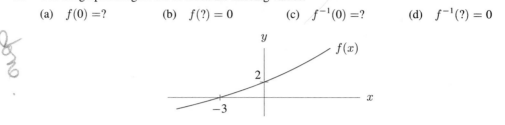

Figure 3.8

16. Values of an invertible function, f, are given in Table 3.7.

    (a) Using the table, fill in the missing values:
        (i) $f(0) = ?$ (ii) $f(?) = 0$ (iii) $f^{-1}(0) = ?$ (iv) $f^{-1}(?) = 0$
    (b) How do the answers to (i)–(iv) in part (a) relate to one another? In particular, how could you have obtained the answers to (iii) and (iv) from the answers to (i) and (ii)?

TABLE 3.7

x	-2	-1	0	1	2
$f(x)$	5	4	2	0	-3

TABLE 3.8

m	0	1	2	3	4	5
$C(m)$	0	2.50	4.00	5.50	7.00	8.50

17. Table 3.8 shows the cost, $C(m)$, of a taxi ride as a function of the number of miles, m, traveled.

    (a) What does $C(3.5)$ mean in practical terms? Estimate $C(3.5)$.
    (b) Assume C is invertible. What does $C^{-1}(3.5)$ mean in practical terms? Estimate $C^{-1}(3.5)$.

18. The area, in square centimeters, of a circle whose radius is r cm is given by $A = \pi r^2$.

    (a) Write this formula using function notation.
    (b) Evaluate $f(0)$.
    (c) Evaluate and interpret $f(r + 1)$.
    (d) Evaluate and interpret $f(r) + 1$.
    (e) What are the units of $f^{-1}(4)$?

19. The perimeter, in meters, of a square whose side is s meters is given by $P = 4s$.

    (a) Write this formula using function notation, where f is the name of the function.
    (b) Evaluate $f(s + 4)$ and interpret its meaning.
    (c) Evaluate $f(s) + 4$ and interpret its meaning.
    (d) What are the units of $f^{-1}(6)$?

20. Let $D(p)$ represent the number of iced cappuccinos sold each week by a coffeehouse when the price is set at p cents each.

    (a) What does the expression $D(225)$ represent?
    (b) Do you think that $D(p)$ is an increasing function or a decreasing function? Why?
    (c) What does the following equation tell you about p? $D(p) = 180$
    (d) The coffeehouse sells n iced cappuccinos when they charge the average price in their area, t cents. Thus, $D(t) = n$. What is the meaning of the following expressions: $D(1.5t)$, $1.5D(t)$, $D(t+50)$, $D(t) + 50$?

3.3 DOMAIN AND RANGE

In Example 7 on page 5, we defined R to be the average monthly rainfall at Chicago's O'Hare airport in month t. Although R is a function of t, the value of R is not defined for every possible value of t. For instance, it makes no sense to consider the value of R for $t = -3$, or $t = 8.21$, or $t = 13$ (since a year has 12 months). Thus, although R is a function of t, this function is defined only for certain values of t. Notice also that R, the output value of this function, takes only the values $\{1.8, 2.1, 2.4, 2.5, 2.7, 3.1, 3.2, 3.4, 3.5, 3.7\}$.

A function is often defined only for certain values of the independent variable. Also, the dependent variable often takes on only certain values. This leads to the following definitions:

> If $Q = f(t)$, then
> - the **domain** of f is the set of input values, t, which yield an output value.
> - the **range** of f is the corresponding set of output values, Q.

Thus, the domain of a function is the set of input values, and the range is the set of output values.

If the domain of a function is not specified, we usually assume that it is as large as possible—that is, all numbers that make sense as inputs for the function. For example, if there are no restrictions, the domain of the function $f(x) = x^2$ is the set of all real numbers, because we can substitute any real number into the formula $f(x) = x^2$. Sometimes, however, we may restrict the domain to suit a particular application. If the function $f(x) = x^2$ is used to represent the area of a square of side x, we restrict the domain to positive numbers.

If a function is being used to model a real-world situation, the domain and range of the function are often determined by the constraints of the situation being modeled, as in the next example.

Example 1   The house painting function $n = f(A)$ in Example 2 on page 2 has domain $A > 0$ because all houses have some positive area. There is a practical upper limit to A because houses cannot be infinitely large, but in principle, A can be as large or as small as we like, as long as it is positive. Therefore we take the domain of f to be $A > 0$.

The range of this function is $n \geq 0$, because we cannot use a negative amount of paint.

Choosing Realistic Domains and Ranges

When a function is used to model a real situation, it may be necessary to modify the domain and range.

Example 2   Algebraically speaking, the formula
$$T = \frac{1}{4}R + 40$$
can be used for all values of R. If we know nothing more about this function than its formula, its domain is all real numbers. The formula for $T = \frac{1}{4}R + 40$ returns any value of T we like when we choose an appropriate R-value (See Figure 3.9.) Thus, the range of the function is also the set of all real numbers.

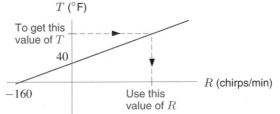

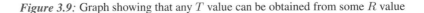

Figure 3.9: Graph showing that any T value can be obtained from some R value

However, if we use this formula to represent the temperature, T, as a function of a cricket's chirp rate, R, as we did in Example 3 on page 3, some values of R cannot be used. For example, it doesn't make sense to talk about a negative chirp rate. Also, there is some maximum chirp rate $R_{\max}$ that no cricket can physically exceed. Thus, to use this formula to express T as a function of R, we must restrict R to the interval $0 \leq R \leq R_{\max}$ shown in Figure 3.10.

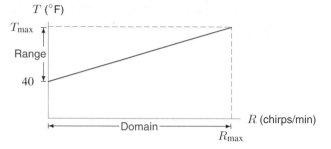

Figure 3.10: Graph showing that if $0 \leq R \leq R_{\max}$, then $40 \leq T \leq T_{\max}$

The range of the cricket function is also restricted. Since the chirp rate is nonnegative, the smallest value of T occurs when $R = 0$. This happens at $T = 40$. On the other hand, if the temperature gets too hot, the cricket won't be able to keep chirping faster. If the temperature $T_{\max}$ corresponds to the chirp rate $R_{\max}$, then the values of T are restricted to the interval $40 \leq T \leq T_{\max}$.

Using a Graph to Find the Domain and Range of a Function

A good way to estimate the domain and range of a function is to examine its graph. The domain is the set of input values on the horizontal axis which give rise to a point on the graph; the range is the corresponding set of output values on the vertical axis.

Example 3   A sunflower plant is measured every day t, for $t \geq 0$. The height, $h(t)$ centimeters, of the plant[2] can be modeled by using the *logistic function*

$$h(t) = \frac{260}{1 + 24(0.9)^t}.$$

  (a) Using a graphing calculator or computer, sketch a graph of the height over 80 days.

  (b) What is the domain of this function? What is the range? What does this tell you about the height of the sunflower?

Solution   (a)   The logistic function is graphed in Figure 3.11.

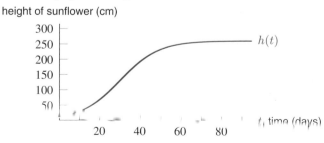

Figure 3.11: Height of sunflower as a function of time

---

[2] Adapted from H.S. Reed and R.H. Holland, "Growth of Sunflower Seeds" *Proc. Nat. Acad. Sci.*, 5, 1919.

(b) The domain of this function is $t \geq 0$. If we consider the fact that the sunflower dies at some point, then there is an upper bound on the domain, $0 \leq t \leq T$, where T is the day on which the sunflower dies.

To find the range, notice that the smallest value of h occurs at $t = 0$. Evaluating gives $h(0) = 10.4$ cm. This means that the plant was 10.4 cm high when it was first measured on day $t = 0$. Tracing along the graph, $h(t)$ increases. As t-values get large, $h(t)$-values approach, but never reach, 260. This suggests that the range is $10.4 \leq h(t) < 260$. This information tells us that sunflowers typically grow to a height of about 260 cm.

Using a Formula to Find the Domain and Range of a Function

When a function is defined by a formula, its domain and range can often be determined by examining the formula algebraically.

Example 4   State the domain and range of g, where

$$g(x) = \frac{1}{x}.$$

Solution   The domain is all real numbers except those which do not yield an output value. The expression $1/x$ is defined for any real number x except 0 (division by 0 is undefined). Therefore,

Domain: all real x, $x \neq 0$.

The range is all real numbers that the formula can return as output values. It is not possible for $g(x)$ to equal zero, since 1 divided by a real number is never zero. All real numbers except 0 are possible output values, since all nonzero real numbers have reciprocals. Thus

Range: all real values, $g(x) \neq 0$.

The graph in Figure 3.12 indicates agreement with these values for the domain and range.

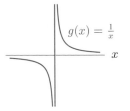

Figure 3.12: Domain and range of $g(x) = 1/x$

Example 5   Find the domain of the function $y = \dfrac{1}{\sqrt{x-4}}$ by examining its formula.

Solution   The domain is all real numbers except those for which the function is undefined. The square root of a negative number is undefined (if we restrict ourselves to real numbers), and so is division by zero. Therefore we need

$$x - 4 > 0.$$

Thus, the domain is all real numbers greater than 4.

Domain: $x > 4$.

In Example 6 on page 76, we saw that for $y = 1/\sqrt{x-4}$, the output, y, cannot be negative. Note that y cannot be zero either. (Why?) The range of $y = 1/\sqrt{x-4}$ is $y > 0$. See Problem 22.

Example 6   Find the range of the function $y = 2 + \dfrac{1}{x}$ algebraically.

Solution   The range is all real numbers which the function can return as output values.

We determine which output values are possible by solving for the input. To see why this works, suppose we want to check whether $y = -\frac{1}{2}$ is a possible y-value. This means we want to know whether there is an x-value corresponding to $y = -\frac{1}{2}$. To decide, we substitute $y = -\frac{1}{2}$ and solve for x:

$$-\frac{1}{2} = 2 + \frac{1}{x},$$

$$\frac{-5}{2} = \frac{1}{x},$$

or

$$x = \frac{-2}{5}.$$

This means that when we substitute $x = -\frac{2}{5}$, the result is $y = -\frac{1}{2}$. Thus, $y = -\frac{1}{2}$ is in the range. Now we apply the same reasoning to a general y-value. We start with $y = 2 + 1/x$ and solve for x:

$$y = 2 + \frac{1}{x}.$$

Subtracting 2 from both sides gives

$$y - 2 = \frac{1}{x}$$

and taking the reciprocal of both sides gives

$$x = \frac{1}{y - 2}.$$

From this formula we see that for every y-value, except $y = 2$, there is a corresponding x-value. Therefore the range of this function is all real numbers except 2.

$$\text{Range: all } y \text{ values,} \quad y \neq 2.$$

The graph in Figure 3.13 agrees with the result we obtained algebraically.

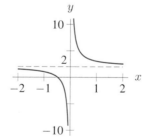

Figure 3.13: Range of $y = 2 + 1/x$ is all y-values except $y = 2$

Problems for Section 3.3

1. What is the domain of the function f giving average monthly rainfall at Chicago ????? ??????? (See Table 1.3 on page 5)

2. A movie theater seats 200 people. For any particular show, the amount of money the theater makes is a function of the number of people, n, in attendance. If a ticket costs $4.00, find the domain and range of this function. Sketch its graph.

3. A car gets the best mileage at intermediate speeds. Sketch a possible graph of the gas mileage as a function of speed. Determine a reasonable domain and range for the function and justify your reasoning.

4. Let f be defined by
$$f(x) = \frac{1}{x^2 - 5x + 6}.$$
   (a) Use algebra to find the domain of f.
   (b) Estimate the range of f using a graphing calculator or computer.

5. (a) What is the domain of the function $f(x) = \sqrt{4 - x^2}$?
   (b) Use your answer to part (a) to help you sketch a graph of $f(x)$.
   (c) What is the range of $f(x)$?

Sketch a graph of each of the functions in Problems 6–17. Then state the domain and range of the function.

6. $y = \sqrt{x}$ 7. $y = \sqrt{x - 3}$ 8. $y = \sqrt{8 - x}$ 9. $y = \dfrac{1}{x^2}$

10. $y = \dfrac{1}{(x - 2)^2}$ 11. $y = \dfrac{-1}{(x + 1)^2}$ 12. $y = x^2 + 1$ 13. $y = x^2 - 4$

14. $y = 9 - x^2$ 15. $y = x^3$ 16. $y = x^3 + 2$ 17. $y = (x - 4)^3$

In Problems 18–21, use a graph to help you find the range of each function on the given domain.

18. $y = \dfrac{1}{x^2}, \quad -1 \le x \le 1$ 19. $y = \dfrac{1}{x}, \quad -2 \le x \le 2$

20. $y = x^2 - 4, \quad -2 \le x \le 3$ 21. $y = \sqrt{9 - x^2}, \quad -3 \le x \le 1$

22. Find the range of $y = \dfrac{1}{\sqrt{x - 4}}$ algebraically.

23. Find the domain and range for each of the following functions algebraically:
   (a) $m(x) = 9 - x$ (b) $n(x) = 9 - x^4$ (c) $q(x) = \sqrt{x^2 - 9}$

24. (a) Use Table 3.9 to determine the number of calories that a person weighing 200 lb uses in a half–hour of walking.[3]
   (b) Table 3.9 illustrates a relationship between the number of calories used per minute walking and a person's weight in pounds. Describe in words what is true about this relationship. Identify the dependent and independent variables. Specify whether it is an increasing or decreasing function.
   (c) (i) Graph the linear function for walking, as described in part (b), and estimate its equation.
       (ii) Interpret the meaning of the vertical intercept of the graph of the function.
       (iii) Specify a meaningful domain and range for your function.
       (iv) Use your function to determine how many calories per minute a person who weighs 135 lb uses per minute of walking.

TABLE 3.9   *Calories used per minute as a function of weight*

Activity	100 lb	120 lb	150 lb	170 lb	200 lb	220 lb
Walking(3 mph)	2.7	3.2	4.0	4.6	5.4	5.9
Bicycling(10 mph)	5.4	6.5	8.1	9.2	10.8	11.9
Swimming(2 mph)	5.8	6.9	8.7	9.8	11.6	12.7

---

[3] Source: 1993 World Almanac

25. The last digit, d, of a phone number is a function of n, its position in the phone book. Table 3.10 gives d for the first 10 listings in the 1998 Boston telephone directory. The table shows that the last digit of the first listing is 3, the last digit of the second listing is 8, and so on. In principle we could use a phone book to figure out other values of d. For instance, if $n = 300$, we could count down to the 300th listing in order to determine d. So we write $d = f(n)$.

 (a) What is the value of $f(6)$?
 (b) Explain how you could use the phone book to find the domain of f.
 (c) What is the range of f?

TABLE 3.10

n	1	2	3	4	5	6	7	8	9	10
d	3	8	4	0	1	8	0	4	3	5

26. In month $t = 0$, a small group of rabbits escapes from a ship onto an island where there are no rabbits. The island rabbit population, $p(t)$, in month t is given by

$$p(t) = \frac{1000}{1 + 19(0.9)^t}, \quad t \geq 0.$$

 (a) Evaluate $p(0)$, $p(10)$, $p(50)$, and explain their meaning in terms of rabbits.
 (b) Sketch a graph of $p(t)$, $0 \leq t \leq 100$, and describe the graph in words. Does it suggest the growth in population you would expect among rabbits on an island?
 (c) Estimate the range of $p(t)$. What does this tell you about the rabbit population?
 (d) Explain how you can find the range of $p(t)$ from its formula.

27. Bronze is an alloy or mixture of the metals copper and tin. The properties of bronze depend on the percentage of copper in the mix. A chemist decides to study the properties of a given alloy of bronze as the proportion of copper is varied. She starts with 9 kg of bronze that contain 3 kg of copper and 6 kg of tin and either adds or removes copper. Let $f(x)$ be the percentage of copper in the mix if x kg of copper are added ($x > 0$) or removed ($x < 0$).

 (a) State the domain and range of f. What does your answer mean in the context of bronze?
 (b) Find a formula in terms of x for $f(x)$.
 (c) If the formula you found in part (b) was not intended to represent the percentage of copper in an alloy of bronze, but instead simply defined an abstract mathematical function, what would be the domain and range of this function?

3.4 PIECEWISE DEFINED FUNCTIONS

The Absolute Value Function

The absolute value of a number x, written $|x|$, is its distance from 0, measured along the number line. For example, $|4| = 4$, because the number 4 is 4 units away from 0 on the number line. Similarly, $|-3| = 3$, because the number -3 is 3 units away from 0 along the number line. (See Figure 3.14.)

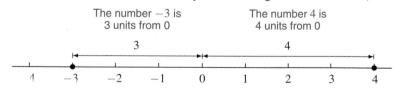

Figure 3.14: Diagram showing $|4| = 4$ and $|-3| = 3$.

Notice that for $x \geq 0$, we have $|x| = x$. For $x < 0$, we have $|x| = -x$. (Remember that $-x$ is a positive number if x is a negative number.) For example, if $x = -3$, then

$$|-3| = -(-3) = 3.$$

So the function which gives the absolute value of a function uses two different formulas depending on the value of x:

$$y = x \text{ whenever } x \geq 0, \text{ and } y = -x \text{ whenever } x < 0.$$

This leads to the following two-part definition:

The **Absolute Value Function** is defined by

$$f(x) = |x| = \begin{cases} x & \text{for} \quad x \geq 0 \\ -x & \text{for} \quad x < 0 \end{cases}.$$

Table 3.11 gives values of $f(x) = |x|$ and Figure 3.15 shows a graph of $f(x)$.

TABLE 3.11   *Absolute value function*

x	$|x|$
-3	3
-2	2
-1	1
0	0
1	1
2	2
3	3

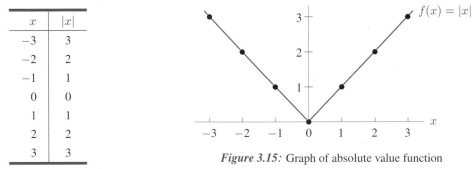

Figure 3.15: Graph of absolute value function

Piecewise Defined Functions

Because different formulas are used on different parts of the domain of f, the absolute value function is said to be *piecewise defined*. The next example is another such function.

Example 1   Graph the function $y = g(x)$ given by two different formulas depending on the value of x:

$$g(x) = x + 1 \text{ for } x \leq 2 \qquad \text{and} \qquad g(x) = 1 \text{ for } x > 2.$$

Using bracket notation, this function is written:

$$g(x) = \begin{cases} x + 1 & \text{for } x \leq 2 \\ 1 & \text{for } x > 2 \end{cases}$$

Solution   For $x \leq 2$, graph the line $y = x + 1$. The solid dot at the point $(2, 3)$ shows that it is included in the graph. For $x > 2$, graph the horizontal line $y = 1$. See Figure 3.16. The open circle at the point $(2, 1)$ shows that it is not included in the graph. (Note that $g(2) = 3$, and $g(2)$ cannot have more than one value.)

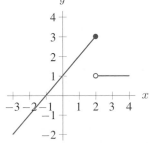

Figure 3.16: Graph of the piecewise defined function g

Example 2   The Ironman Triathlon is a race that consists of three parts: a 2.4 mile swim followed by a 112 mile bike race and then a 26.2 mile marathon. A participant swims steadily at 2 mph, cycles steadily at 20 mph, and then runs steadily at 9 mph.[4] Assuming that no time is lost during the transition from one stage to the next, find a formula for the distance d, covered in miles, as a function of the elapsed time t in hours, from the beginning of the race. Graph the function.

Solution   For each leg of the race, we use the formula Distance = Rate · Time. First, we calculate how long it took for the participant to cover each of the three parts of the race. The first leg took $2.4/2 = 1.2$ hours, the second leg took $112/20 = 5.6$ hours, and the final leg took $26.2/9 \approx 2.91$ hours. Thus, the participant finished the race in $1.2 + 5.6 + 2.91 = 9.71$ hours.

During the first leg, $t \leq 1.2$ and the speed is 2 mph, so

$$d = 2t \quad \text{for} \quad 0 \leq t \leq 1.2.$$

During the second leg, $1.2 < t \leq 1.2 + 5.6 = 6.8$ and the speed is 20 mph. The length of time spent in the second leg is $(t - 1.2)$ hours. Thus, by time t,

$$\text{Distance covered in the second leg} = 20(t - 1.2) \quad \text{for } 1.2 < t \leq 6.8.$$

When the participant is in the second leg, the total distance covered is the sum of the distance covered in the first leg (2.4 miles) plus the part of the second leg that has been covered by time t.

$$d = 2.4 + 20(t - 1.2)$$
$$= 20t - 21.6 \quad \text{for } 1.2 < t \leq 6.8.$$

In the third leg, $6.8 < t \leq 9.71$ and the speed is 9 mph. Since 6.8 hours were spent on the first two parts of the race, the length of time spent on the third leg is $(t - 6.8)$ hours. Thus, by time t,

$$\text{Distance covered in the third leg} = 9(t - 6.8) \quad \text{for } 6.8 < t \leq 9.1.$$

When the participant is in the third leg, the total distance covered is the sum of the distances covered in the first leg (2.4 miles) and the second leg (112 miles), plus the part of the third leg that has been covered by time t:

$$d = 2.4 + 112 + 9(t - 6.8)$$
$$= 9t + 53.2 \quad \text{for } 6.8 < t \leq 9.71.$$

The formula for d is different on different intervals of t:

$$d = \begin{cases} 2t & \text{for} \quad 0 \leq t \leq 1.2 \\ 20t - 21.6 & \text{for} \quad 1.2 < t \leq 6.8 \\ 9t + 53.2 & \text{for} \quad 6.8 < t \leq 9.71. \end{cases}$$

Figure 3.17 gives a graph of the distance covered, d, as a function of time, t. Notice the three pieces.

---

[4]Data supplied by Susan Reid, Athletics Department, University of Arizona.

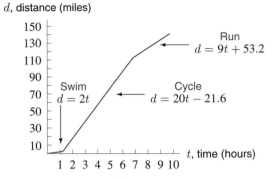

Run
$d = 9t + 53.2$

Swim
$d = 2t$

Cycle
$d = 20t - 21.6$

Figure 3.17: Ironman Triathlon: d as a function of t

Problems for Section 3.4

1. Consider the graph in Figure 3.18. An open circle represents a point which is not included.

    (a) Is y a function of x? Explain.
    (b) Is x a function of y? Explain.
    (c) The domain of $y = f(x)$ is $0 \leq x < 4$. What is the range of $y = f(x)$?

Figure 3.18

2. Many people believe that $\sqrt{x^2} = x$. We will investigate this claim graphically and numerically.
    (a) Graph the two functions x and $\sqrt{x^2}$ in the window $-5 \leq x \leq 5$, $-5 \leq y \leq 5$. Based on what you see, do you believe that $\sqrt{x^2} = x$? What function does the graph of $\sqrt{x^2}$ remind you of?
    (b) Complete Table 3.12. Based on this table, do you believe that $\sqrt{x^2} = x$? What function does the table for $\sqrt{x^2}$ remind you of? Is this the same function you found in part (a)?

    TABLE 3.12

x	-5	-4	-3	-2	-1	0	1	2	3	4	5
$\sqrt{x^2}$											

    (c) Explain how you know that $\sqrt{x^2}$ is the same as the function $|x|$.
    (d) Graph the function $\sqrt{x^2} - |x|$ in the window $-5 \leq x \leq 5$, $-5 \leq y \leq 5$. Explain what you see.

3. This problem deals with the function $u(x) = |x|/x$.
    (a) Graph $u(x)$ in the window $-5 \leq x \leq 5$, $-5 \leq y \leq 5$. Explain what you see.
    (b) Complete Table 3.13. Does this table agree with what you found in part (a)?

    TABLE 3.13

x	-5	-4	-3	-2	-1	0	1	2	3	4	5
$|x|/x$											

    (c) Identify the domain and range of $u(x)$.

    (d) Comment on the claim that $u(x)$ can be written as
    $$u(x) \begin{cases} -1 & \text{if } x < 0, \\ 0 & \text{if } x = 0, \\ 1 & \text{if } x > 0. \end{cases}$$

Graph the piecewise defined functions in Problems 4–7.

4. $f(x) = \begin{cases} -1, & -1 \le x < 0 \\ 0, & 0 \le x < 1 \\ 1, & 1 \le x < 2 \end{cases}$

5. $f(x) = \begin{cases} x + 4, & x \le -2 \\ 2, & -2 < x < 2 \\ 4 - x, & x \ge 2 \end{cases}$

6. $f(x) = \begin{cases} x^2, & x \le 0 \\ \sqrt{x}, & 0 < x < 4 \\ x/2, & x \ge 4 \end{cases}$

7. $f(x) = \begin{cases} x + 1, & -2 \le x < 0 \\ x - 1, & 0 \le x < 2 \\ x - 3, & 2 \le x < 4 \end{cases}$

8. The charge for a taxi ride is $1.50 for the first $1/8$ of a mile, and $0.25 for each additional $1/8$ of a mile (rounded up to the nearest $1/8$ mile).

   (a) Make a table showing the cost of a trip as a function of its length. Your table should start at zero and go up to one mile in $1/8$-mile intervals.
   (b) What is the cost for a $5/8$-mile ride?
   (c) How far can you go for $3.00?
   (d) Sketch a graph of the cost function in part (a).

9. A contractor purchases gravel one cubic yard at a time.

   (a) A gravel driveway L yards long and 6 yards wide is to be poured to a depth of 1 foot. Find a formula for $n(L)$, the number of cubic yards of gravel the contractor buys, assuming that he buys 10 more cubic yards of gravel than are needed (to be sure he'll have enough).
   (b) Assuming no driveway can be less than 5 yards long, state the domain and range of $n(L)$. Sketch a graph of $n(L)$ showing the domain and range.
   (c) If the function $n(L)$ did not represent an amount of gravel, but was a mathematical relationship defined by the formula in part (a), what is its domain and range?

10. A floor-refinishing company charges $1.83 per square foot to strip and refinish a tile floor for up to 1000 square feet. There is an additional charge of $350 for toxic waste disposal for any job which includes more than 150 square feet of tile.

    (a) Express the cost, y, of refinishing a floor as a function of the number of square feet, x, to be refinished.
    (b) Graph the function. Give the domain and range.

11. A private museum charges $40 for a group of 10 or fewer people. A group consisting of more than 10 people must, in addition to the $40, pay $2 per person for the number of people above 10. For example, a group of 12 pays $44 and a group of 15 pays $50. The maximum group size is 50.

    (a) Draw a graph that represents this situation.
    (b) What are the domain and range of the cost function?

12. At a supermarket checkout, a scanner records the prices of the foods you buy. In order to protect consumers, the state of Michigan passed a "scanning law" that says something similar to the following:

    > If there is a discrepancy between the price marked on the item and the price recorded by the scanner, the consumer is entitled to receive 10 times the difference between those prices; this amount given must be at least $1 and at most $5. Also, the consumer will be given the difference between the prices, in addition to the amount calculated above.

    For example: If the difference is 5¢, you should receive $1 (since 10 times the difference is only 50¢ and you are to receive at least $1), plus the difference of 5¢. Thus, the total you should receive is $1.00 + $0.05 = $1.05,
    If the difference is 25¢, you should receive 10 times the difference in addition to the difference, giving $(10)(0.25) + 0.25 = $2.75,$
    If the difference is 95¢, you should receive $5 (because $10(.95) = 9.50 is more than $5, the maximum penalty), plus 95¢, giving $5 + 0.95 = $5.95.$

    (a) What is the lowest possible refund?

(b) Suppose x is the difference between the price scanned and the price marked on the item, and y is the amount refunded to the customer. Write a formula for y in terms of x. (Hints: Look at the sample calculations.)

(c) What would the difference between the price scanned and the price marked have to be in order to obtain a $9.00 refund?

(d) Sketch a graph of y as a function of x.

13. Many printing presses are designed with large plates that print a fixed number of pages as a unit. Each unit is called a signature. Suppose a particular press prints signatures of 16 pages each. Suppose $C(p)$ is the cost of printing a book of p pages, assuming each signature printed costs $0.14.

(a) What is the cost of printing a book of 128 pages? 129 pages? p pages?

(b) What are the domain and range of C?

(c) Sketch a graph of $C(p)$ for $0 \leq p \leq 128$.

14. Gore Mountain is a ski resort in the Adirondack mountains in upstate New York. Table 3.14 shows the cost of a weekday ski-lift ticket for various ages and dates.

(a) Draw a graph of cost as a function of age for each time period given. (One graph will serve for times when rates are identical).

(b) For which age group does the date affect cost?

(c) Draw a graph of cost as a function of date for the age group mentioned in part (b).

(d) Why does the cost fluctuate as a function of date?

TABLE 3.14   *Ski-lift ticket prices at Gore Mountain, 1998-1999 season*[5]

Age	Opening-Dec 12	Dec 13-Dec 24	Dec 25-Jan 3	Jan 4-Jan 15	Jan 16-Jan 18	Jan 19-Feb 12	Feb 13-Feb 21	Feb 22-Mar 28	Mar 29-Closing
Up to 6	Free	Free	Free	Free	Free	Free	Free	Free	Free
7-12	$19	$19	$19	$19	$19	$19	$19	$19	$19
13-69	$29	$34	$39	$34	$39	$34	$39	$34	$29
70+	Free	Free	Free	Free	Free	Free	Free	Free	Free

15. Sometimes a relationship is defined by two independent variables. The wind chill index is one such function. The wind chill, denoted W, is a measure of how cold it *feels*; it takes into account both the temperature, T, and the wind velocity, V. The function can be called $W(T, V)$ to show that W depends on both T and V. One model for wind chill is the piecewise function shown below.

$$W(T, V) = \begin{cases} T & \text{for } 0 \leq V \leq 4 \\ 0.0817(5.81 + 3.71\sqrt{V} - 0.25V)(T - 91.4) + 91.4 & \text{for } 4 < V \leq 45 \\ 1.60T - 55 & \text{for } V > 45. \end{cases}$$

(a) Two of the three rules of the piecewise function are linear. Which are they?

(b) Beyond which velocity will the wind have no further effect on the perceived temperature?

(c) How hard must the wind blow so that a temperature of $40°$F feels like $9°$F?

---

[5]The Olympic Regional Development Authority.

REVIEW PROBLEMS FOR CHAPTER THREE

The graphs of $y = f(x)$ and $y = g(x)$ are in Figure 3.19. Show on a copy of the figure the coordinates of the points that you can obtain from the statements in Problems 1–4.

1. $f(0) = 2$

2. $f(-3) = f(3) = f(9) = 0$

3. $f(2) = g(2)$

4. $g(x) > f(x)$ for $x > 2$

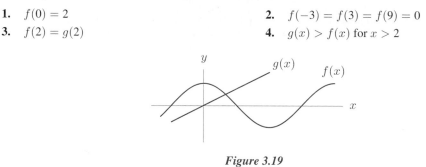

Figure 3.19

5. If $g(x) = x\sqrt{x} + 100x$, evaluate without a calculator

(a) $g(100)$ (b) $g(4/25)$ (c) $g(1.21 \cdot 10^4)$

6. Let $h(x) = x^2 + bx + c$. Evaluate and simplify:

(a) $h(1)$ (b) $h(b+1)$

Follow these steps for each of the functions in Problems 7–12.

(a) Complete the table of values. (If the function is undefined for a given x value, say so.)

x	-5	-4	-3	-2	-1	$-\frac{3}{4}$	$-\frac{1}{2}$	$-\frac{1}{4}$
y								

x	0	$\frac{1}{4}$	$\frac{1}{2}$	$\frac{3}{4}$	1	2	3	4	5
y									

(b) Sketch a graph of the function for $-5 \le x \le 5$. What is the range for the y-values you calculated for the table from part (a)?

(c) Give the complete domain and range of the function.

(d) For what values of x is the function increasing? Decreasing?

7. $y = x^2$ **8.** $y = x^3$ **9.** $y = \dfrac{1}{x}$ **10.** $y = \dfrac{1}{x^2}$ **11.** $y = \sqrt{x}$ **12.** $y = \sqrt[3]{x}$

13. (a) How can you tell from the graph of a function that a given x-value is not in the domain? Sketch an example.

(b) How can you tell from the formula for a function that a given x-value is not in the domain? Give an example.

State the domain and range of each of the functions in Problems 14–17

14. $f(x) = \sqrt{x - 4}$

15. $h(x) = x^2 + 8x$

16. $g(x) - \dfrac{1}{4 + x^2}$

17 $r(r) = \sqrt{4 - \sqrt{x - 4}}$

In Problems 18–20, if $f(x) = \dfrac{ax}{a + x}$, find and simplify

18. $f(a)$.

19. $f(1 - a)$.

20. $f\left(\dfrac{1}{1 - a}\right)$.

21. An epidemic of influenza spreads through a city. Figure 3.20 is the graph of $I = f(w)$, where I is the number of individuals (in thousands) infected w weeks after the epidemic begins.

    (a) Evaluate $f(2)$ and explain its meaning in terms of the epidemic.
    (b) Approximately how many people were infected at the height of the epidemic? When did that occur? Write your answer in the form $f(a) = b$.
    (c) Solve $f(w) = 4.5$ and explain what the solutions mean in terms of the epidemic.
    (d) The graph was obtained using the formula $f(w) = 6w(1.3)^{-w}$. Use the graph to estimate the solution of the inequality $6w(1.3)^{-w} \geq 6$. Explain what the solution means in terms of the epidemic.

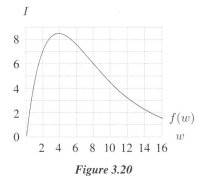

Figure 3.20

22. Let t be time in seconds and let $r(t)$ be the rate, in gallons/second, that water enters a reservoir:

    $$r(t) = 800 - 40t.$$

    (a) Evaluate the expressions $r(0), r(15), r(25)$, and explain their physical significance.
    (b) Graph $y = r(t)$ for $0 \leq t \leq 30$, labeling the intercepts. What is the physical significance of the slope and the intercepts?
    (c) For $0 \leq t \leq 30$, when does the reservoir have the most water? When does it have the least water?
    (d) What are the domain and range of $r(t)$?

23. Consider the functions f and g whose values are given in Table 3.15.

    (a) Evaluate $f(1)$ and $g(3)$.
    (b) Describe in full sentences the patterns you see in the values for each function.
    (c) Assuming that the patterns you observed in part (b) hold true for all values of x, calculate $f(5)$, $f(-2), g(5)$, and $g(-2)$.
    (d) Find possible formulas for $f(x)$ and $g(x)$.

    TABLE 3.15

x	-1	0	1	2	3	4
$f(x)$	-4	-1	2	5	8	11
$g(x)$	4	1	0	1	4	9

24. Let $k(x) = 6 - x^2$.

    (a) Find a point on the graph of $k(x)$ whose x-coordinate is -2.
    (b) Find two points on the graph whose y-coordinates are -2.
    (c) Sketch the graph and locate the points in parts (a) and (b).
    (d) Let $p = 2$. Calculate $k(p) - k(p - 1)$.

25. Let $f(a)$ be the cost in dollars of a kilograms of apples. What do the following statements tell you? What are the units of each of the numbers?

    (a) $f(2) = 1.50$ (b) $f(0.1) = 0.75$ (c) $f^{-1}(3) = 4$ (d) $f^{-1}(1.5) = 2$

26. A drug affects a patient's blood pressure. The patient's blood pressure, p, in millimeters of mercury (mm), soon after the drug is given is a function of the dose, q mg, so $p = f(q)$.

(a) What does the statement $f(200) = 135$ tell you about the patient and the drug?

(b) The patient has been taking a dose of q mg of the drug. Then there is a change, represented by the statements I–III. For each statement, choose the formula from (i)–(ix) which represents the blood pressure after the change.

    I   The blood pressure is decreased by 10%, while maintaining the same dose.

    II   The dose remains the same but the blood pressure is increased by 10 mm.

    III The dose is increased by 10 mg.

(i) $f(q) - 10$	(ii) $f(q) + 10$	(iii) $f(q - 10)$
(iv) $f(q + 10)$	(v) $f(q) - 0.1$	(vi) $f(q) + 0.1$
(vii) $0.9f(q)$	(viii) $1.1f(q)$	(ix) $0.1f(q)$

27. Suppose $w = j(x)$ is the average daily quantity of water (in gallons) required by an oak tree of height x feet.

(a) What does the expression $j(25)$ represent? What about $j^{-1}(25)$?

(b) What does the following equation tell you about v: $j(v) = 50$.
Rewrite this statement in terms of j^{-1}.

(c) Oak trees are on average z feet high and that a tree of average height requires p gallons of water. Represent this fact in terms of j and then in terms of j^{-1}.

(d) Using the definitions of z and p from part (c), what do the following expressions represent?

$$j(2z), \quad 2j(z), \quad j(z + 10), \quad j(z) + 10, \quad j^{-1}(2p), \quad j^{-1}(p + 10), \quad j^{-1}(p) + 10.$$

28. Let $t(x)$ be the time required, in seconds, to melt 1 gram of a certain compound at $x°$C.

(a) Express the following statement as an equation using $t(x)$: It takes 272 seconds to melt 1 gram of the compound at $400°$C.

(b) Explain the following equations in words:

    (i) $t(800) = 136$ (ii) $t^{-1}(68) = 1600$

(c) Above a certain temperature, doubling the temperature, x, halves the melting time. Express this fact with an equation involving $t(x)$.

29. Table 3.16 shows $N(s)$, the number of sections of Economics 101, as a function of s, the number of students in the course. If s is between two numbers listed in the table, then $N(s)$ is the higher number of sections.

TABLE 3.16

Number of students, s	50	75	100	125	150	175	200
Number of sections, $N(s)$	4	4	5	5	6	6	7

(a) Evaluate and interpret: (i) $N(150)$ (ii) $N(80)$ (iii) $N(55.5)$

(b) Solve for s and interpret: (i) $N(s) = 4$ (ii) $N(s) = N(125)$

30. The surface area of a cylindrical aluminum can is a measure of how much aluminum the can requires. If the can has radius r and height h, its surface area A and its volume V are given by the equations:

$$A = 2\pi r^2 + 2\pi r h \quad \text{and} \quad V = \pi r^2 h$$

(a) The volume, V, of a 12 oz cola can is 355 cm³. A cola can is approximately cylindrical. Express its surface area A as a function of its radius r, where r is measured in centimeters. [Hint: First solve for h in terms of r.]

(b)   Sketch a graph of $A = s(r)$, the surface area of a cola can whose volume is 355 cm^3, for $0 \leq r \leq 10$. Label your axes.

(c)   What is the domain of $s(r)$? Based on the sketch you made in (b), what, approximately, is the range of $s(r)$?

(d)   The manufacturers wish to use the least amount of aluminum (in cm^2) necessary to make a 12 oz cola can. Use your answer in (c) to find the minimum amount of aluminum needed. State the values of r and h that minimize the amount of aluminum used.

(e)   The radius of a real 12 oz cola can is about 3.25 cm. Show that real cola cans use more aluminum than necessary to hold 12 oz of cola. Why do you think real cola cans are made to this way?

31.   A subway train serves 19 stations, numbered 1 through 19. Although passengers can board the train at any of its 19 stops, most of its passengers board at the first station. The train has a capacity of 700 people. Define $p(n)$ to be the number of passengers on a given day who board the train at the first station and exit at station number n or lower.

(a)   State the domain and range of p.

(b)   Interpret the expression $p(1)$. Describe a scenario where $p(1) > 0$.

(c)   Is p increasing, decreasing, neither, or can't we tell? Explain.

CHAPTER FOUR

EXPONENTIAL AND LOGARITHMIC FUNCTIONS

Exponential functions represent quantities that increase or decrease at a constant percent rate. In contrast to linear functions, in which a constant amount is added per unit input, an exponential function involves multiplication by a constant factor for each unit increment in input value. Examples include the balance of a savings account, the size of some populations, and the quantity of a chemical that decays radioactively.

In this chapter, we study the family of exponential functions and their inverses, the logarithmic functions. We also compare exponential, linear, and logarithmic growth.

4.1 INTRODUCTION TO THE FAMILY OF EXPONENTIAL FUNCTIONS

Growing at a Constant Percent Rate

Linear functions represent quantities that change at a constant rate. In this section we introduce functions that change at a constant *percent* rate, the *exponential functions*.

Salary Raises

Example 1   After graduation from college, you will probably be looking for a job. Suppose you are offered a job at a starting salary of $40,000 per year. To strengthen the offer, the company promises annual raises of 6% per year for at least the first five years after you are hired. Let's compute your salary for the first few years.

If t represents the number of years since the beginning of your contract, then for $t = 0$, your salary is $40,000. At the end of the first year, when $t = 1$, your salary increases by 6% so

$$\text{Salary when } t = 1 = \text{Original salary } + 6\% \text{ of Original salary}$$
$$= 40000 + 0.06(40000)$$
$$= 42400.$$

After the second year, your salary again increases by 6%, so

$$\text{Salary when } t = 2 = \text{Former salary } + 6\% \text{ of Former salary}$$
$$= 42400 + 0.06(42400)$$
$$= 44944.$$

Notice that your raise is higher in the second year than in the first since the second 6% increase applies both to the original $40,000 salary and to the $2400 raise given in the first year.

Salary calculations for four years have been rounded and recorded in Table 4.1. At the end of the third and fourth years your salary again increases by 6%, and your raise is larger each year. Not only are you given the 6% increase on your original salary, but your raises earn raises as well.

TABLE 4.1 *Raise amounts and resulting salaries for a person earning 6% annual salary increases*

Year	Raise amount ($)	Salary ($)
0		40000.00
1	2400.00	42400.00
2	2544.00	44944.00
3	2696.64	47640.64
4	2858.44	50499.08

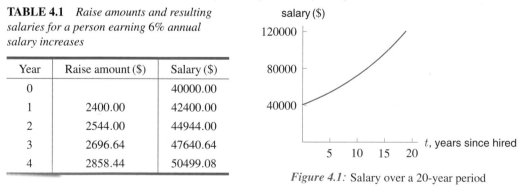

Figure 4.1: Salary over a 20-year period

Figure 4.1 gives a graph of salary over a 20-year period assuming that the annual increase remains 6%. Since the rate of change of your salary (in dollars per year) is not constant, the graph of this function is not a line. The salary increases at an increasing rate, giving the graph its upward curve.

Population Growth

Exponential functions provide a reasonable model for many growing populations.

Example 2 During the early 1980s, the population of Mexico increased at a constant annual percent rate of 2.6%. Since the population grew by the same percent each year, it can be modeled by an exponential function.

Let's calculate the population of Mexico for the first few years after 1980. In 1980, the population was 67.38 million. The population grew by 2.6%, so

$$\begin{aligned}\text{Population in 1981} &= \text{Population in 1980 } + 2.6\% \text{ of Population in 1980}\\ &= 67.38 + 0.026(67.38)\\ &\approx 67.38 + 1.75\\ &\approx 69.13.\end{aligned}$$

Similarly,

$$\begin{aligned}\text{Population in 1982} &= \text{Population in 1981 } + 2.6\% \text{ of Population in 1981}\\ &= 69.13 + 0.026(69.13)\\ &\approx 69.13 + 1.80\\ &\approx 70.93.\end{aligned}$$

The calculations for years 1980 through 1984 have been rounded and recorded in Table 4.2. The population of Mexico increased by slightly more each year than it did the year before, because each year the increase is 2.6% of a larger number.

TABLE 4.2 *Calculated values for the population of Mexico*

Year	ΔP, increase in population	P, population (millions)
1980	—	67.38
1981	1.75	69.13
1982	1.80	70.93
1983	1.84	72.77
1984	1.90	74.67

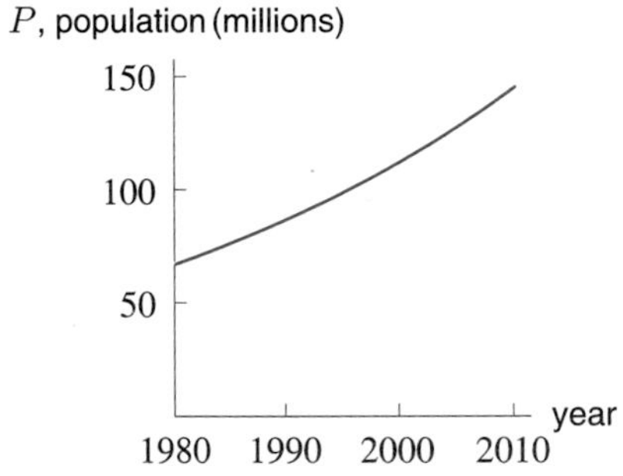

Figure 4.2: The projected population of Mexico, assuming 2.6% annual growth

Figure 4.2 gives a graph of the population of Mexico over a 30-year period, assuming a 2.6% annual growth rate. Notice that this graph curves upwards just like the graph in Figure 4.1.

Radioactive Decay

Exponential functions can also model decreasing quantities. A quantity which decreases at a constant percent rate is said to be decreasing exponentially.

Example 3 Carbon-14 is used to estimate the age of organic compounds. Over time, radioactive carbon-14 decays into a stable form. The decay rate is 11.4% every 1000 years. For example, if we begin with a 200 microgram (μg) sample of carbon-14 then

$$\begin{aligned}\begin{matrix}\text{Amount remaining}\\ \text{after 1000 years}\end{matrix} &= \text{Initial amount} - 11.4\% \text{ of Initial amount}\\ &= 200 - 0.114(200)\\ &= 177.2.\end{aligned}$$

Similarly,

$$\text{Amount remaining after 2000 years} = \text{Amount remaining after 1000 years} - 11.4\% \text{ of } \text{Amount remaining after 1000 years}$$

$$= 177.2 - 0.114(177.2) \approx 157,$$

and

$$\text{Amount remaining after 3000 years} = \text{Amount remaining after 2000 years} - 11.4\% \text{ of } \text{Amount remaining after 2000 years}$$

$$= 157 - 0.114(157) \approx 139.1.$$

These calculations are recorded in Table 4.3. During each 1000-year period, the amount of carbon-14 that decays is smaller than in the previous period. This is because we take 11.4% of a smaller quantity each time.

TABLE 4.3   *The amount of carbon-14 remaining over time*

Years elapsed	Amount decayed (μg)	Amount remaining (μg)
0	—	200.0
1000	22.8	177.2
2000	20.2	157.0
3000	17.9	139.1

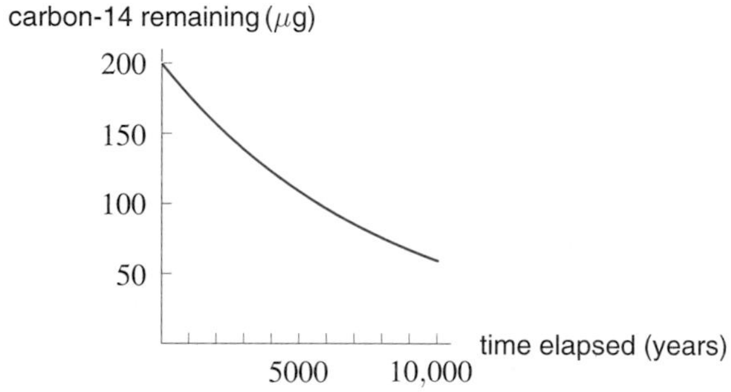

carbon-14 remaining (μg)

Figure 4.3: Amount of carbon-14 over 10,000 years

Figure 4.3 shows the amount of carbon-14 left from a 200 μg sample over 10,000 years. Because the amount decreases by a smaller amount over each successive time interval, the graph is not linear but bends upwards.

Growth Factors and Percent Growth Rates

The Growth Factor of an Increasing Exponential Function

The salary in Example 1 increases by 6% every year. We say that the annual percent growth rate is 6%. But there is another way to think about the growth of this salary. We know that each year,

New salary = Old salary + 6% of Old salary.

We can rewrite this as follows:

New salary = 100% of Old salary + 6% of Old salary.

So

New salary = 106% of Old salary.

Since 106% = 1.06, we have

New salary = 1.06 · Old salary.

We call the 1.06 the *annual growth factor*.

The Growth Factor of a Decreasing Exponential Function

In Example 3, the carbon-14 changes by −11.4% every 1000 years. The negative growth rate tells us that the quantity of carbon-14 decreases over time. We have

New amount = Old amount − 11.4% of Old amount,

which can be rewritten as

New amount = 100% of Old amount − 11.4% of Old amount.

So,
$$\text{New amount} = 88.6\% \text{ of Old amount.}$$

Since $88.6\% = 0.886$, we have

$$\text{New amount} = 0.886 \cdot \text{Old amount.}$$

Hence the growth factor is 0.886 per millenium. The fact that the growth factor is less than 1 indicates that the amount of carbon-14 is decreasing, since multiplying a quantity by a factor between 0 and 1 decreases the quantity.

Although it may sound strange to refer to the growth factor, rather than decay factor, of a decreasing quantity, we will use growth factor to describe both increasing and decreasing quantities.

A General Formula for the Family of Exponential Functions

Because it grows at a constant percentage rate each year, the salary, S, in Example 1 is an example of an exponential function. We want a formula for S in terms of t, the number of years since being hired. Since the annual growth factor is 1.06, we know that for each year,

$$\text{New salary} = \text{Previous salary} \cdot 1.06.$$

Thus, after one year, or when $t = 1$,

$$S = \underbrace{40{,}000}_{\text{Previous salary}} \cdot 1.06.$$

Similarly, when $t = 2$,

$$S = \underbrace{40{,}000(1.06)}_{\text{Previous salary}} \cdot 1.06 = 40{,}000(1.06)^2.$$

Here there are *two* factors of 1.06 because the salary has increased by 6% twice. When $t = 3$,

$$S = \underbrace{40{,}000(1.06)^2}_{\text{Previous salary}} \cdot 1.06 = 40{,}000(1.06)^3$$

and continues in this pattern so that after t years have elapsed,

$$S = 40{,}000 \underbrace{(1.06)(1.06)\ldots(1.06)}_{t \text{ factors of } 1.06} = 40{,}000(1.06)^t.$$

After t years the salary has increased by a factor of 1.06 a total of t times. Thus,

$$S = 40{,}000(1.06)^t.$$

These results, which are summarized in Table 4.4, are the same as in Table 4.1. Notice that in this formula we assume that t is an integer, $t \geq 0$, since the raises are given only once a year.

TABLE 4.4   *Salary after t years*

t (years)	S, salary ($)
0	40,000
1	$40{,}000(1.06) = 42{,}400.00$
2	$40{,}000(1.06)^2 = 44{,}944.00$
3	$40{,}000(1.06)^3 = 47{,}640.64$
t	$40{,}000(1.06)^t$

This salary formula can be written as

$$S = \text{Initial salary} \times (\text{Growth factor})^t.$$

In general, we have:[1]

An **exponential function** $Q = f(t)$ has the formula

$$f(t) = ab^t, \quad b > 0,$$

where the parameter a is the initial value of Q (at $t = 0$) and the parameter b is the growth factor: $b > 1$ gives exponential growth, $0 < b < 1$ gives exponential decay. The growth factor is given by

$$b = 1 + r$$

where r is the decimal representation of the percent rate of change.

Every function in the form $f(t) = ab^t$ with the input, t, in the exponent is an exponential function. Note that if $b = 1$, then $f(t) = a \cdot 1^t = a$ and $f(t)$ is a constant. Graphs showing exponential growth and decay are in Figures 4.4 and 4.5.

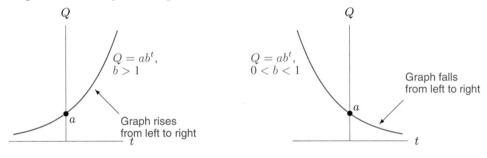

Figure 4.4: Exponential growth: $b > 1$

Figure 4.5: Exponential decay: $0 < b < 1$

Example 4 Use the formula $S = 40{,}000(1.06)^t$ to calculate your salary after 4 years, 12 years, and 40 years.

Solution After 4 years, $t = 4$, and we have

$$S = 40{,}000(1.06)^4 \approx 50{,}499.08.$$

Notice that this agrees with Table 4.1 on page 104. After 12 years, $t = 12$, and we have

$$S = 40{,}000(1.06)^{12} \approx 80{,}487.86.$$

So after 12 years, the salary has more than doubled from the initial salary of $40,000. When $t = 40$ we have

$$S = 40{,}000(1.06)^{40} \approx 411{,}428.72.$$

Thus if you work for 40 years and consistently earn 6% annual raises, your salary will be over $400,000 a year.

Example 5 Carbon-14 decays at a rate of 11.4% every 1000 years. Write a formula for the quantity, Q, of a 200 μg sample remaining as a function of time, t, in thousands of years.

Solution The growth factor of carbon-14 over 1000 years is $1 - 0.114 = 0.886$. Originally, there are 200 μg, so the quantity remaining after t thousand years is given by

$$Q = 200(0.886)^t.$$

---
[1] See page 122 for an explanation of why b is restricted to positive values.

Example 6   Using Example 2 on page 105, find a formula for P, the population of Mexico (in millions), in year t where $t = 0$ represents the year 1980.

Solution   In 1980, the population of Mexico was 67.38 million, and it was growing at a constant 2.6% annual rate. The growth factor is $b = 1 + 0.026 = 1.026$, and $a = 67.38$, so

$$P = 67.38(1.026)^t.$$

Because the growth factor may change eventually, this formula may not give accurate results for very large values of t.

Example 7   What does the formula $P = 67.38(1.026)^t$ predict when $t = 0$? When $t = -5$? What do these values tell you about the population of Mexico?

Solution   If $t = 0$, then , since $(1.026)^0 = 1$, we have

$$P = 67.38(1.026)^0 = 67.38.$$

This makes sense because $t = 0$ stands for 1980, and in 1980 the population was 67.38 million. When $t = -5$ we have

$$P = 67.38(1.026)^{-5} \approx 59.26.$$

To make sense of this figure, we must interpret the year $t = -5$ as five years before 1980; that is, as the year 1975. If the population of Mexico had been growing at a 2.6% annual rate from 1975 onward, then it was 59.26 million in 1975.

Example 8   On August 2, 1988, a US District Court judge imposed a fine on the city of Yonkers, New York, for defying a federal court order involving housing desegregation.[2] The fine started at \$100 for the first day and was to double daily until the city chose to obey the court order.

(a)   What was the daily percent growth rate of the fine?
(b)   Find a formula for the fine as a function of t, the number of days since August 2, 1988.
(c)   If Yonkers waited 30 days before obeying the court order, what would the fine have been?

Solution   (a)   Since the fine increased each day by a factor of 2, the fine grew exponentially with growth factor $b = 2$. To find the percent growth rate, we set $b = 1 + r = 2$, from which we find $r = 1$, or 100%. Thus the daily percent growth rate is 100%. This makes sense because when a quantity increases by 100%, it doubles in size.

(b)   If t is the number of days since August 2, the formula for the fine, P in dollars, is

$$P = 100 \cdot 2^t.$$

(c)   After 30 days, the fine is $P = 100 \cdot 2^{30} \approx 1.07 \cdot 10^{11}$ dollars, or over \$100 billion.

Problems for Section 4.1

1.   A quantity increases from 10 to 12. By what percent has it increased? Now suppose instead that it had increased from 100 to 102. What is the percent increase in this case?

2.   Explain why multiplying the quantity A by 1.15 has the effect of increasing the quantity of A by 15%.

3.   The value, V, of a \$100,000 investment that earns 3% annual interest is given by $V = f(t)$ where t is measured in years. How much is the investment worth in 3 years?

[2]*The Boston Globe*, August 27, 1988.

4.   In 1999, the population of a country was 70 million and growing at a rate of 1.9% per year. Assuming the percentage growth rate remains constant, express the population, P, of this country as a function of t, the number of years after 1999.

5.   In 1990 the number of people infected by a virus was P_0. Due to a new vaccine, the number of infected people has decreased by 20% each year since 1990. In other words, only 80% as many people are infected each year as were infected the year before. Find a formula for $P = f(n)$, the number of infected people n years after 1990. Sketch a graph of $f(n)$. Explain, in terms of the virus, why the graph has the shape it does.

6.   (a)   The annual inflation rate is 3.5% per year. If a movie ticket costs $7.50, find a formula for p, the price of the ticket t years from today, assuming that movie tickets keep up with inflation.
      (b)   According to your formula, how much will movie tickets cost in 20 years?

7.   A typical cup of coffee contains about 100 mg of caffeine and every hour approximately 16% of the amount of caffeine in the body is metabolized and eliminated.

      (a)   Let C represent the amount of caffeine in the body in mg and t represent the number of hours since a cup of coffee was consumed. Write C as a function of t.
      (b)   How much caffeine is in the body after 5 hours?

8.   Polluted water is passed through a series of filters. Each filter removes 85% of the remaining impurities. Initially, the untreated water contains impurities at a level of 420 parts per million (ppm). Find a formula for L, the remaining level of impurities, after the water has been passed through a series of n filters.

9.   An investment decreases by 5% per year for 4 years. By what total percent does it decrease?

10.   The *Home* section of many Sunday newspapers includes a mortgage table similar to Table 4.5. The table gives the monthly payment per $1000 borrowed for loans at various interest rates and time periods. Determine the monthly payment on a

      (a) $60,000 mortgage at 8% for fifteen years.
      (b) $60,000 mortgage at 8% for thirty years.
      (c) $60,000 mortgage at 10% for fifteen years.
      (d)   Over the life of the loan, how much money would be saved on a 15-year mortgage of $60,000 if the rate were 8% instead of 10%?
      (e)   Over the life of the loan, how much money would be saved on an 8% mortgage of $60,000 if the term of the loan was fifteen years rather than thirty years?

TABLE 4.5

Mortgage Rate	15-year loan	20-year loan	25-year loan	30-year loan
8.00	9.56	8.37	7.72	7.34
8.50	9.85	8.68	8.06	7.69
9.00	10.15	9.00	8.40	8.05
9.50	10.45	9.33	8.74	8.41
10.00	10.75	9.66	9.09	8.78
10.50	11.06	9.99	9.45	9.15
11.00	11.37	10.33	9.81	9.53
11.50	11.69	10.67	10.17	9.91

11.   You owe $2000 on a credit card. The card charges 1.5% monthly interest on your balance, and requires a minimum monthly payment of 2.5% of your balance. All transactions (payments and interest charges) are recorded at the end of the month. You make only the minimum required payment every month and you incur no additional debt.

      (a)   Complete Table 4.6 for a twelve-month period. (The first two rows have been filled in.)
      (b)   What is your unpaid balance after one year has passed? At that time, how much of your debt have you paid off? How much money in interest charges have you paid your creditors?

TABLE 4.6

Month	Balance	Interest	Minimum payment
0	$2000.00	$30.00	$50.00
1	$1980.00	$29.70	$49.50
2	$1960.20		
⋮			

12. The Gross National Product (GNP) of China increased from 1479.1 billion yuan in 1991 to 1690.4 billion yuan in 1992, and then to 1927.0 billion yuan in 1993. (A yuan is the Chinese unit of currency.)

    (a) In which interval, 1991-92 or 1992-93, did the Chinese GNP show the greatest increase? In which interval did it show the greatest percent increase?
    (b) Write the rates of change (in billions of yuan per year) over the three intervals 1991-92, 1992-93, 1991-93 in increasing order.

13. Figure 4.6 is the graph of $f(x) = 4 \cdot b^x$. Find the slope of the line segment PQ in terms of b.

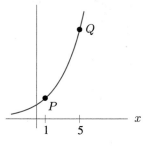

Figure 4.6

14. A one-celled animal is in a petri dish and reproduces (by division) once every minute, as do its offspring. In one hour the dish is full.   (a)   At what time was the dish half full?   (b)   One quarter full?

15. The amount (in milligrams) of a drug in the body t hours after taking a pill is given by $A(t) = 25(0.85)^t$.

    (a) What is the initial dose given?
    (b) What percent of the drug leaves the body each hour?
    (c) What is the amount of drug left after 10 hours?
    (d) After how many hours will there be less than 1 milligram left in the body?

16. In 1990 the population of the United States was about 248.7 million. Assume the population increases at a constant rate of 0.9% per year. In what year is the population projected to reach 300 million?

17. Every year, teams from 64 colleges qualify to compete in the NCAA basketball playoffs. For each round, every team is paired with an opponent. A team is eliminated from the tournament once it loses a round. So, at the end of a round, only one half the number of teams move on to the next round. Let $N(r)$ be the number of teams remaining in competition after r rounds of the tournament have been played.

    (a) Find a formula for $N(r)$ and sketch the graph of $y = N(r)$.
    (b) How many rounds will it take to determine the winner of the tournament?

18. A one-page letter is folded into thirds to go into an envelope. If it were possible to repeat this kind of tri-fold 20 times, how many miles thick would the letter be? (A stack of 150 pieces of stationery is one inch thick; 1 mile = 5280 feet.)

19. The population of a small town increases by a growth factor of 1.134 over a two-year period.

    (a) By what percent does the town increase in size during the two-year period?
    (b) If the town grows by the same percent each year, what is its annual percent growth rate?

20. Let $P(t) = 32(1.047)^t$ give the population of a town (in thousands) in year t. What is the town's annual growth rate? Monthly growth rate? Growth rate per decade?

4.2 COMPARING EXPONENTIAL AND LINEAR FUNCTIONS

The exponential function $Q = ab^t$ represents a quantity changing at a constant percent rate. In this section we compare exponential and linear models and we fit exponential models to data from tables and graphs.

Identifying Linear and Exponential Functions From a Table

Table 4.7 gives values of two functions, f and g. Notice that the value of x changes by equal steps of $\Delta x = 5$. The function f is linear because the difference between consecutive values of $f(x)$ is constant: $f(x)$ increases by 15 each time x increases by 5.

TABLE 4.7 *Two functions, one linear and one exponential*

x	20	25	30	35	40	45
$f(x)$	30	45	60	75	90	105
$g(x)$	1000	1200	1440	1728	2073.6	2488.32

On the other hand, the difference between consecutive values of $g(x)$ is *not* constant:

$$1200 - 1000 = 200$$
$$1440 - 1200 = 240$$
$$1728 - 1440 = 288.$$

Thus, g is not linear. However, the *ratio* of consecutive values of $g(x)$ is constant:

$$\frac{1200}{1000} = 1.2, \quad \frac{1440}{1200} = 1.2, \quad \frac{1728}{1440} = 1.2,$$

and so on. Note that $1200 = 1.2(1000)$, $1440 = 1.2(1200)$, $1728 = 1.2(1440)$. Thus, each time x increases by 5, the value of $g(x)$ increases by a factor of 1.2, so g is exponential. In general:

For a table of data that gives y as a function of x and in which Δx is constant:
- If the *difference* of consecutive y-values is constant, the table could represent a linear function.
- If the *ratio* of consecutive y-values is constant, the table could represent an exponential function.

Finding a Formula for an Exponential Function

To find a formula for the exponential function in Table 4.7, we must determine the values of a and b in the formula $g(x) = ab^x$. We do this by taking ratios. The table tells us that $g(20) = 1000$ and that $g(25) = 1200$. The ratio of $g(25)$ to $g(20)$ is

$$\frac{g(25)}{g(20)} = \frac{1200}{1000} = 1.2.$$

Now, substituting into the formula gives $g(25) = ab^{25}$ and $g(20) = ab^{20}$. Thus we can also write the ratio as

$$\frac{g(25)}{g(20)} = \frac{ab^{25}}{ab^{20}} = b^5.$$

Notice that the value of a cancels in this ratio. Equating these two expressions for the ratio $g(25)/g(20)$ gives

$$b^5 = 1.2.$$

We solve for b by raising each side to the $(1/5)^{\text{th}}$ power:

$$(b^5)^{1/5} = b = 1.2^{1/5} \approx 1.0371.$$

Now that we have the value of b, we can solve for a. Since $g(20) = ab^{20} = 1000$, we have

$$a(1.0371)^{20} = 1000$$

$$a = \frac{1000}{1.0371^{20}} \approx 482.6.$$

Thus, a formula for g is $g(x) = 482.6(1.0371)^x$. (Note: We could have used $g(25)$ or any other value from the table to find a.)

Modeling Linear and Exponential Growth Using Two Data Points

If we are given two data points, we can fit either a line or an exponential function to the points. The following example compares the predictions made by a linear model and an exponential model fitted to the same data.

Example 1 At time $t = 0$ years, a species of turtle is released into a wetland. When $t = 4$ years, a biologist estimates there are 300 turtles in the wetland. Three years later, the biologist estimates there are 450 turtles. Let P represent the size of the turtle population in year t.
  (a) Find a formula for $P = f(t)$ assuming linear growth. Interpret the slope and P-intercept of your formula in terms of the turtle population.
  (b) Now find a formula for $P = g(t)$ assuming exponential growth. Interpret the parameters of your formula in terms of the turtle population.
  (c) In year $t = 12$, the biologist estimates that there are 900 turtles in the wetland. What does this indicate about the two population models?

Solution (a) Assuming linear growth, we have $P = f(t) = b + mt$, and

$$m = \frac{\Delta P}{\Delta t} = \frac{450 - 300}{7 - 4} = \frac{150}{3} = 50.$$

Calculating b gives

$$300 = b + 50 \cdot 4$$

$$b = 100,$$

so $P = f(t) = 100 + 50t$. This formula tells us that 100 turtles were originally released into the wetland and that the number of turtles increases at the constant rate of 50 turtles per year.
  (b) Assuming exponential growth, we have $P = g(t) = ab^t$. The values of a and b are calculated from the ratio

$$\frac{g(7)}{g(4)} = \frac{ab^7}{ab^4} = b^3,$$

and also

$$\frac{g(7)}{g(4)} = \frac{450}{300} = 1.5.$$

Equating the two expressions for $g(7)/g(4)$ gives us

$$b^3 = 1.5$$

so

$$b = (1.5)^{1/3} \approx 1.145.$$

Using the fact that $g(4) = ab^4 = 300$ to find a gives

$$a(1.145)^4 = 300$$

$$a = \frac{300}{1.145^4} \approx 175, \quad \text{Rounding to the nearest whole turtle}$$

so $P = g(t) = 175(1.145)^t$. This formula tells us that 175 turtles were originally released into the wetland and the number increases at about 14.5% per year.

(c)   In year $t = 12$, there are approximately 900 turtles. The linear function from part (a) predicts

$$P = 100 + 50 \cdot 12 = 700 \text{ turtles.}$$

The exponential formula from part (b), however, predicts

$$P = 175(1.145)^{12} \approx 889 \text{ turtles.}$$

The fact that 889 is closer to the observed value of 900 turtles suggests that, during the first 12 years, exponential growth is a better model of the turtle population than linear growth. The two models are graphed in Figure 4.7.

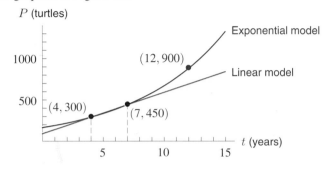

Figure 4.7: Comparison of the linear and exponential models of the turtle population

Similarities and Differences between Linear and Exponential Functions

In some ways the general formulas for linear and exponential functions are similar. If y is a linear function of x and x is an integer, we can write $y = b + mx$ as

$$y = b + \underbrace{m + m + m + \ldots + m}_{x \text{ times}}.$$

Similarly, if y is an exponential function of x, so that $y = a \cdot b^x$ and x is an integer, we can write

$$y = a \cdot \underbrace{b \cdot b \cdot b \cdot \ldots \cdot b}_{x \text{ times}}.$$

So linear functions involve repeated sums whereas exponential functions involve repeated products. In both cases, x determines the number of repetitions.

There are other similarities between the formulas for linear and exponential functions. The slope m of a linear function gives the rate of change of a physical quantity and the y-intercept gives the starting value. Similarly, in $y = a \cdot b^x$, the value of b gives the growth factor and a gives the starting value.

Exponential Growth Will Always Outpace Linear Growth in the Long Run

Figure 4.7 shows the graphs of the linear and exponential models for the turtle population from Example 1. The graphs highlight the fact that, although these two graphs remain fairly close for the first ten or so years, the exponential model predicts explosive growth later on.

It can be shown that an exponentially increasing quantity will, in the long run, always outpace a linearly increasing quantity. This fact led the 19th-century clergyman and economist, Thomas Malthus, to make some rather gloomy predictions, which are illustrated in the next example.

Example 2   The population of a country is initially 2 million people and is increasing at 4% per year. The country's annual food supply is initially adequate for 4 million people and is increasing at a constant rate adequate for an additional 0.5 million people per year.

(a) Based on these assumptions, in approximately what year will this country first experience shortages of food?

(b) If the country doubled its initial food supply, would shortages still occur? If so, when? (Assume the other conditions do not change).

(c) If the country doubled the rate at which its food supply increases, in addition to doubling its initial food supply, would shortages still occur? If so, when? (Again, assume the other conditions do not change.)

Solution   Let P represent the country's population (in millions) and N the number of people the country can feed (in millions). The population increases at a constant percent rate, so it can be modeled by an exponential function. The initial population is $a = 2$ million people and the annual growth factor is $b = 1 + 0.04 = 1.04$, so a formula for the population is

$$P = 2(1.04)^t.$$

In contrast, the food supply increases by a constant amount each year and is therefore modeled by a linear function. The initial food supply is adequate for $b = 4$ million people and the growth rate is $m = 0.5$ million per year, so the number of people that can be fed is

$$N = 4 + 0.5t.$$

(a) Figure 4.8(a) gives the graphs of P and N over a 105-year span. For many years, the food supply is far in excess of the country's needs. However, after about 78 years the population has begun to grow so rapidly that it catches up to the food supply and then outstrips it. After that time, the country will suffer from shortages.

(b) If the country can initially feed eight million people rather than four, the formula for N is

$$N = 8 + 0.5t.$$

However, as we see from Figure 4.8(b), this measure only buys the country three or four extra years with an adequate food supply. After 81 years, the population is growing so rapidly that the head start given to the food supply makes little difference.

(c) If the country doubles the rate at which its food supply increases, from 0.5 million per year to 1.0 million per year, the formula for N is

$$N = 8 + 1.0t.$$

Unfortunately the country still runs out of food eventually. Judging from Figure 4.8(c), this happens in about 102 years.

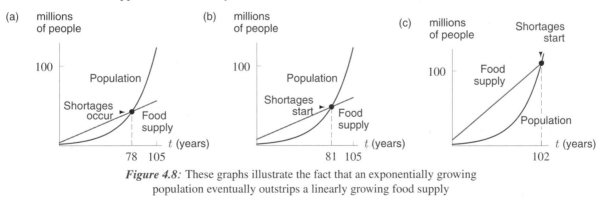

Figure 4.8: These graphs illustrate the fact that an exponentially growing population eventually outstrips a linearly growing food supply

Malthus believed that populations increase exponentially while food production increases linearly. The last example explains his gloomy predictions: Malthus believed that any population eventually outstrips its food supply, leading to famine and war.

Solving Exponential Equations Graphically

We are often interested in solving equations involving exponential functions. In the following example, we do this graphically. In Section 4.6, we will see how to solve equations using logarithms.

Example 3   In Example 8 on page 109, the fine, P, imposed on the city of Yonkers is given by $P = 100 \cdot 2^t$ where t is the number of days after August 2. In 1988, the annual budget of the city was \$337 million. If the city chose to disobey the court order, at what point would the fine have wiped out the entire annual budget?

Solution   We need to find the day on which the fine reaches \$337 million. That is, we must solve the equation
$$100 \cdot 2^t = 337,000,000.$$
Using a computer or graphing calculator we can graph $P = 100 \cdot 2^t$ to find the point at which the fine reaches 337 million. From Figure 4.9, we see that this occurs between $t = 21$ and $t = 22$.
At day $t = 21$, August 23, the fine is:
$$P = 100 \cdot 2^{21} = 209,715,200$$
or just over \$200 million. On day $t = 22$, the fine is
$$P = 100 \cdot 2^{22} = 419,430,400$$
or almost \$420 million – quite a bit more than the city's entire annual budget!

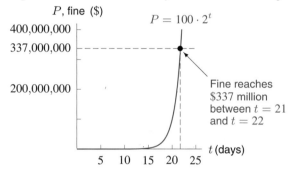

Figure 4.9: The fine imposed on Yonkers exceeds \$337 million after 22 days

Example 4   A 200 μg sample of carbon-14 decays according to the formula
$$Q = 200(0.886)^t$$
where t is in thousands of years. Estimate when there will be 25 μg of carbon-14 left.

Solution   We must solve the equation
$$200(0.886)^t = 25.$$
At the moment, we cannot find a formula for the solution to this equation. However, we can estimate the solution graphically. Figure 4.10 shows a graph of $Q = 200(0.886)^t$ and the line $Q = 25$. The amount of carbon-14 decays to 25 micrograms at $t \approx 17.2$. Since t is measured in thousands of years, this means in about 17,200 years.

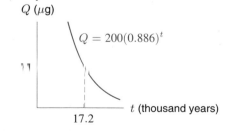

Figure 4.10: Solving the equation $200(0.886)^t = 25$

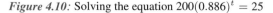

Problems for Section 4.2

1. Explain the difference between linear and exponential growth. That is, without writing down any formulas, describe how linear and exponential functions progress differently from one value to the next.

2. What is the value of the population at the end of 10 years, given each of the following assumptions? Graph each population against time.

    (a)   A population decreases linearly and the decrease is 10% in the first year.
    (b)   A population decreases exponentially at the rate of 10% a year.

3. Let $p(x) = 2 + x$ and $q(x) = 2^x$. Estimate the values of x such that $p(x) < q(x)$.

In Problems 4–7, find formulas for the exponential functions satisfying the given conditions.

4. $h(0) = 3$ and $h(1) = 15$ 5. $f(3) = -3/8$ and $f(-2) = -12$
6. $g(1/2) = 4$ and $g(1/4) = 2\sqrt{2}$ 7. $g(0) = 5$ and $g(-2) = 10$

8. Suppose $f(-3) = 5/8$ and $f(2) = 20$. Find a formula for f assuming it is:

    (a)   Linear   (b)   Exponential

For Problems 9–14, find possible formulas for the graphs of the exponential functions.

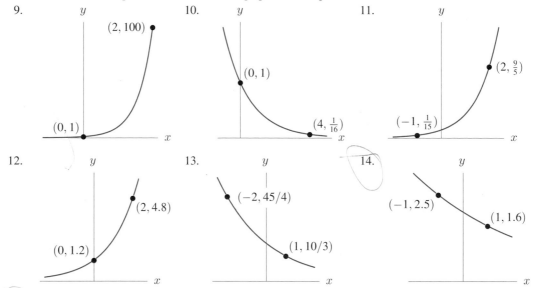

9. $(2, 100)$, $(0, 1)$

10. $(0, 1)$, $(4, \frac{1}{16})$

11. $(2, \frac{9}{5})$, $(-1, \frac{1}{15})$

12. $(2, 4.8)$, $(0, 1.2)$

13. $(-2, 45/4)$, $(1, 10/3)$

14. $(-1, 2.5)$, $(1, 1.6)$

15. Figure 4.11 gives the balance, P, of a bank account in year t.

    (a)   Find a possible formula for $P = f(t)$ assuming the balance grows exponentially.
    (b)   What was the initial balance?
    (c)   What annual interest rate does the account pay?

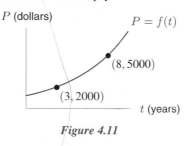

P (dollars) $P = f(t)$

$(8, 5000)$

$(3, 2000)$

t (years)

Figure 4.11

The tables in Problems 16–19 contain values from an exponential or a linear function. In each problem:

(a) Decide if the function is linear or exponential.

(b) Find a possible formula for each function and sketch its graph.

16.

x	$f(x)$
0	12.5
1	13.75
2	15.125
3	16.638
4	18.301

17.

x	$g(x)$
0	0
1	2
2	4
3	6
4	8

18.

x	$h(x)$
0	14
1	12.6
2	11.34
3	10.206
4	9.185

19.

x	$i(x)$
0	18
1	14
2	10
3	6
4	2

20. Determine which of the functions in Table 4.8 could be linear and which could be exponential. Write formulas for the linear and exponential functions.

TABLE 4.8

x	-2	-1	0	1	2
$f(x)$	0.43982	1.31947	2.19912	3.07877	3.95842
$g(x)$	1.02711	1.13609	1.25663	1.38996	1.53743
$h(x)$	0.95338	1.88152	2.72743	3.47375	4.02713

21. Determine whether the function whose values are in Table 4.9 could be exponential. Explain.

TABLE 4.9

x	1	2	4	5	8	9
$f(x)$	4096	1024	64	16	0.25	0.0625

22. The following formulas give the populations (in 1000s) of four different cities, A, B, C, and D. Which are changing exponentially? Describe in words how each of these populations is changing over time. Graph those that are exponential.

$$P_A = 200 + 1.3t, \quad P_B = 270(1.021)^t, \quad P_C = 150(1.045)^t, \quad P_D = 600(0.978)^t.$$

23. Five different stories are given in (a)-(e). Following the stories are five formulas in (i)-(v). Match each formula to the story it best models and state what the variables represent. Assume that the constants P_0, r, B, A are all positive.

(a) The percent of a lake's surface covered by algae, initially at 35%, was halved each year since the passage of anti-pollution laws.

(b) The amount of charge on a capacitor in an electric circuit decreases by 30% every second.

(c) Polluted water is passed through a series of filters. Each filter removes all but 30% of the remaining impurities from the water.

(d) In 1920, the population of a town was 3000 people. Over the course of the next 50 years, the town grew at a rate of 10% per decade.

(e) In 1920, the population of a town was 3000 people. Over the course of the next 50 years, the town grew at a rate of 230 people per year.

(i) $f(x) = P_0 + rx$ (ii) $g(x) = P_0(1 + r)^x$ (iii) $h(x) = B(0.7)^x$

(iv) $j(x) = B(0.3)^x$ (v) $k(x) = A(2)^{-x}$

24. By what year will the number of people infected by the virus described in Problem 5 on page 110 drop to 1% of its initial level?

25. Let $P = f(t) = 1000(1.04)^t$ be the population of a community in year t.

   (a) Evaluate $f(0)$ and $f(10)$. What do these expressions represent in terms of the population?
   (b) Using a calculator or a computer, find appropriate viewing windows on which to graph the population for the first 10 years and for the first 50 years. Give the viewing windows you used and sketch the resulting graphs.
   (c) If the percentage growth rate remains constant, approximately when will the population reach 2500 people?

26. Suppose y, the number of cases of a disease, is reduced by 10% each year.

   (a) If there are initially 10,000 cases, express y as a function of t, the number of years elapsed.
   (b) How many cases will there be 5 years from now?
   (c) How long does it take to reduce the number of cases to 1000?

27. The earth's atmospheric pressure, P, in terms of height above sea level is often modeled by an exponential decay function. Suppose the pressure at sea level is 1013 millibars and that the pressure decreases by 14% for every kilometer above sea level.

   (a) What is the atmospheric pressure at 50 km?
   (b) Estimate the altitude h at which the pressure equals 900 millibars.

28. Suppose the city of Yonkers is offered two alternative fines by the judge. (See Example 8 on page 109.)
   Penalty A: $1 million on August 2 and the fine increases by $10 million each day thereafter.
   Penalty B: 1¢ on August 2 and the fine doubles each day thereafter.

   (a) If the city of Yonkers plans to defy the court order until the end of the month (August 31), compare the fines incurred under Penalty A and Penalty B.
   (b) If t represents the number of days after August 2, express the fine incurred as a function of t under
      (i) Penalty A   (ii) Penalty B
   (c) Assuming your formulas in part (b) holds true for $t \geq 0$, is there a time such that the fines incurred under both penalties are equal? If so, estimate that time.

29. A 1999 Lexus costs $39,375 and the car depreciates a total of 46% during its first 7 years.

   (a) Suppose the depreciation is exponential. Find a formula for the value of the car at time t.
   (b) Suppose instead that the depreciation is linear. Find a formula for the value of the car at time t.
   (c) If this were your car and you were trading it in after 4 years, which depreciation model would you prefer (exponential or linear)?

30. On November 27, 1993, the *New York Times* reported that wildlife biologists have found a direct link between the increase in the human population in Florida and the decline of the local black bear population. From 1953 to 1993, the human population increased, on average, at a rate of 8% per year, while the black bear population decreased at a rate of 6% per year. In 1953 the black bear population was 11,000.

   (a) The 1993 human population of Florida was 13 million. What was the human population in 1953?
   (b) Find the black bear population for 1993.
   (c) If this trend continues, when will the black bear population number less than 100?

4.3 EXPONENTIAL GRAPHS AND CONCAVITY

As with linear functions, an understanding of the significance of the parameters a and b in the exponential model $Q = ab^t$ helps us analyze and compare exponential functions.

Graphs of the Exponential Family: The Effect of the Parameter a

In the formula $Q = ab^t$, the value of a tells us where the graph crosses the Q-axis, since a is the value of Q when $t = 0$. In Figure 4.12 each graph has the same value of b but different values of a and thus different vertical intercepts.

Graphs of the Exponential Family: The Effect of the Parameter b

The growth factor, b, is called the *base* of an exponential function. Provided a is positive, if $b > 1$, the graph climbs when read from left to right, and if $0 < b < 1$, the graph falls when read from left to right.

Figure 4.13 shows how the value of b affects the steepness of the graph of $Q = ab^t$. Each graph has a different value of b but the same value of a (and thus the same Q-intercept). For $b > 1$, the greater the value of b, the more rapidly the graph rises. For $0 < b < 1$, the smaller the value of b, the more rapidly the graph falls.

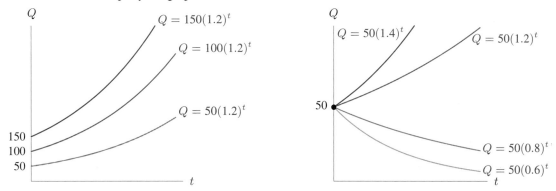

Figure 4.12: Graphs of $Q = a(1.2)^t$ for $a = 50, 100,$ and 150 *Figure 4.13:* Graphs of $Q = 50b^t$ for $b = 0.6, 0.8, 1.2$ and 1.4

The Number e

The irrational number $e = 2.71828\ldots$, introduced by Euler[3] in 1727, is often used for the base, b. Base e is so important that e is called the *natural base*. This may seem mysterious, as what could possibly be natural about using an irrational base such as e? The answer is that the formulas of calculus are much simpler if e is used as the base for exponentials. Since $2 < e < 3$, the graph of $Q = e^t$ lies between the graphs of $Q = 3^t$ and $Q = 2^t$. See Figure 4.14.

Some of the remarkable properties of the number e are introduced in Problems 45 and 46 on page 178.

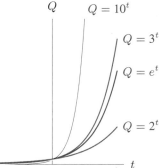

Figure 4.14: Graphs of exponential functions with various bases

Horizontal Asymptotes

The horizontal line $Q = 0$, the t-axis, is a *horizontal asymptote* for the graph of $Q = ab^t$, because Q approaches 0 as t gets large, either positively or negatively. For exponential decay, such as $Q = a(0.6)^t$ in Figure 4.13, the value of Q approaches to 0 as t gets large and positive. We write

$$Q \to 0 \quad \text{as} \quad t \to \infty.$$

---

[3]Leonhard Euler (1707-1783), a Swiss mathematician, introduced e, $f(x)$ notation, π, and i (for $\sqrt{-1}$).

This means that Q is as close to 0 as we like for all sufficiently large values of t. For exponential growth, the value of Q approaches zero as t grows more negative. See Figure 4.14. We write

$$Q \to 0 \quad \text{as} \quad t \to -\infty.$$

This means that Q is as close to 0 as we like for all sufficiently large negative values of t. We make the following definition:

> The horizontal line $y = k$ is a **horizontal asymptote** of a function, f, if the function values get arbitrarily close to k as x gets large (either positively or negatively or both). We describe this behavior using the notation
>
> $$f(x) \to k \quad \text{as} \quad x \to \infty$$
>
> or
>
> $$f(x) \to k \quad \text{as} \quad x \to -\infty.$$

Example 1   A capacitor is the part of an electrical circuit that stores electric charge. The quantity of charge stored decreases exponentially with time. Stereo amplifiers provide a familiar example: When an amplifier is turned off, the display lights fade slowly because it takes time for the capacitors to discharge. (This is why it can be unsafe to open a stereo or a computer immediately after it is turned off.)

If t is the number of seconds after the circuit is switched off, suppose that the quantity of stored charge (in micro-coulombs) is given by

$$Q = 200(0.9)^t, \quad t \geq 0,$$

(a) Describe in words how the stored charge changes over time.

(b) What quantity of charge remains after 10 seconds? 20 seconds? 30 seconds? 1 minute? 2 minutes? 3 minutes?

(c) Sketch a graph of the charge over the first minute. What does the horizontal asymptote of the graph tell you about the charge?

Solution   (a) The charge is initially 200 micro-coulombs. Since $b = 1 + r = 0.9$, we have $r = -0.10$, which means that the charge level decreases by 10% each second.

(b) Table 4.10 gives the value of Q at $t = 0, 10, 20, 30, 60, 120$, and 180. Notice that as t increases, Q gets closer and closer to, but doesn't quite reach, zero. The charge stored by the capacitor is getting smaller, but never completely vanishes.

(c) Figure 4.15 shows Q over a 60-second interval. The horizontal asymptote at $Q = 0$ corresponds to the fact that the charge gets very small as t increases. After 60 seconds, for all practical purposes, the charge is zero.

TABLE 4.10   *Charge (in micro-coulombs) stored by a capacitor over time*

t (seconds)	Q, charge level
0	200
10	69.7
20	24.3
30	8.48
60	0.359
120	0.000646
180	0.00000116

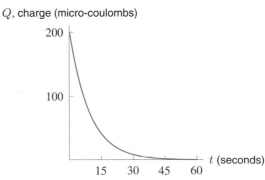

Figure 4.15: The charge stored by a capacitor over one minute

Why *b > 0* for Exponential Functions

Let's look at a specific example to see why the base b of an exponential function is restricted to positive values. For example, let $a = 1$ and $b = -2$, so $f(x) = (-2)^x$. At first, f seems like a reasonable function. For example,

$$f(1) = (-2)^1 = -2,$$
$$f(2) = (-2)^2 = +4,$$
$$f(3) = (-2)^3 = -8,$$

and so on. Granted, the value of f changes signs depending on whether x is even or odd, but that isn't the reason negative bases aren't allowed. We exclude negative bases because of the following even more unusual behavior:

$$f(x) = (-2)^x$$
$$f(1/2) = (-2)^{1/2} = \sqrt{-2} \qquad \text{Undefined}$$
$$f(1/3) = (-2)^{1/3} = \sqrt[3]{-2} \approx -1.260$$
$$f(1/4) = (-2)^{1/4} = \sqrt[4]{-2} \qquad \text{Undefined}$$
$$f(1/5) = (-2)^{1/5} = \sqrt[5]{-2} \approx -1.149$$

and so on. For some values of x, f is defined, but for many others it isn't. In fact, on *any* interval there are infinitely many values of x for which $f(x)$ is undefined! We avoid this awkward situation by requiring $b > 0$. In fact, many calculators and computer software packages are programmed not to evaluate some powers of negative numbers, even if the answer actually exists.

Concavity and Rates of Change

The graph of a linear function is a straight line because the average rate of change is a constant. The graphs of the exponential functions we have seen are not straight lines, but bend upward.

The salary figures in Table 4.11 are from Example 1 on page 104. Since the rate of change increases with time, the graph in Figure 4.16 bends upward. We say such graphs are *concave up*.

TABLE 4.11 *Salary: Increasing rate of change*

t (years)	S (\$1000s)	Rate of change $\Delta S/\Delta t$
0	40	
		3.2
10	72	
		5.6
20	128	
		10.2
30	230	
		18.1
40	411	

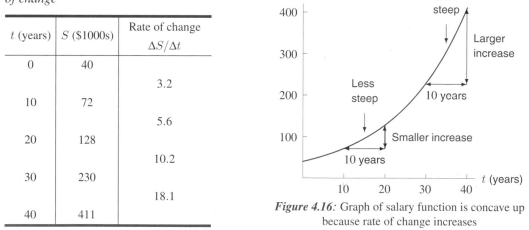

Figure 4.16: Graph of salary function is concave up because rate of change increases

The next example shows that a decreasing function can also be concave up.

Example 2   Table 4.12 shows $Q = 200(0.886)^t$, the quantity of carbon-14 (in μg) in a 200 μg sample remaining after t thousand years. (See Example 3 on page 105.) We see from Figure 4.17 that Q is a decreasing function of t, so its rate of change is always negative. What can we say about the concavity of the graph, and what does this mean about the rate of change of the function?

TABLE 4.12 *Carbon-14: Increasing rate of change*

t (thousand years)	Q (μg)	Rate of change $\Delta Q/\Delta t$
0	200	
		-18.2
5	109	
		-9.8
10	60	
		-5.4
15	33	

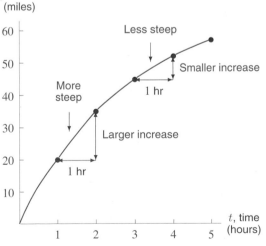

Figure 4.17: Graph of the quantity of carbon-14 is concave up

Solution   The graph bends upward, so it is concave up. Table 4.13 shows that the rate of change of the function is increasing, because the rate is becoming less negative. Figure 4.17 shows how the increasing rate of change can be visualized on the graph.

The exponential graphs we have seen have all been concave up. However, graphs can bend downward; we call such graphs *concave down*.

Example 3   On page 22, we considered the trip taken by a cyclist, Karim. Table 4.13 gives Karim's distance traveled as a function of time.[4] What is the concavity of the graph? Was Karim's speed (that is, the rate of change of distance with respect to time) increasing, constant, or decreasing?

Solution   Figure 4.18 bends downward so the graph is concave down. Table 4.13 shows Karim's speed was decreasing throughout the entire five hour trip. Figure 4.18 shows how the decreasing speed can be visualized on the graph.

TABLE 4.13   Karim's distance as a function of time, with the average speed for each hour

t, time (hours)	d, distance (miles)	Average speed, $\Delta d/\Delta t$ (mph)
0	0	
		20 mph
1	20	
		15 mph
2	35	
		10 mph
3	45	
		7 mph
4	52	
		5 mph
5	57	

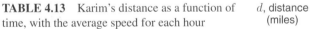

Figure 4.18: Karim's distance as a function of time

[4]This is not an exponential function.

Summary

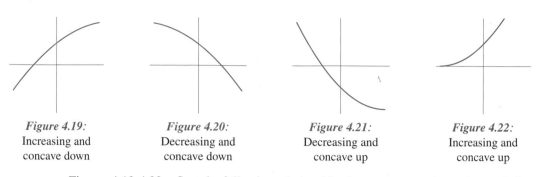

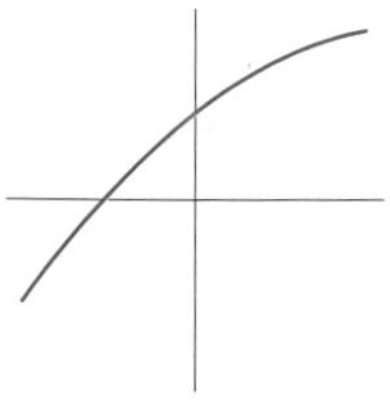

Figure 4.19:
Increasing and
concave down

Figure 4.20:
Decreasing and
concave down

Figure 4.21:
Decreasing and
concave up

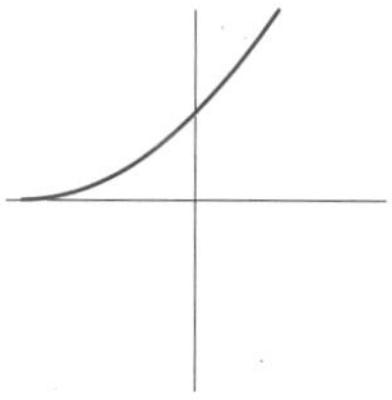

Figure 4.22:
Increasing and
concave up

Figures 4.19-4.22 reflect the following relationships between concavity and rate of change:

- If f is a function whose average rate of change increases (gets less and less negative or more and more positive as we move from left to right), then the graph of f is **concave up**. That is, the graph bends upward.
- If f is a function whose average rate of change decreases (gets less and less positive or more and more negative as we move from left to right), then the graph of f is **concave down**. That is, the graph bends downward.

If a function has a constant rate of change, its graph is a line and it is neither concave up nor concave down.

Problems for Section 4.3

1. Let $f(x) = 2^x$.
    (a) Make a table of values for f for $x = -3, -2, -1, 0, 1, 2, 3$.
    (b) Draw a graph of $f(x)$. Describe the graph in words.

2. Let $f(x) = \left(\dfrac{1}{2}\right)^x$.
    (a) Make a table of values for f for $x = -3, -2, -1, 0, 1, 2, 3$.
    (b) Draw a graph of $f(x)$. Describe the graph in words.

3. The three exponential functions $y = 2^x$, $y = 3^x$, and $y = e^x$ are shown in Figure 4.23. Which formula goes with which graph? Explain your reasoning.

4. The graphs of $f(x) = (1.1)^x$, $g(x) = (1.2)^x$, and $h(x) = (1.25)^x$ are given in Figure 4.24. Explain how you can match these formulas and the graphs without using a calculator.

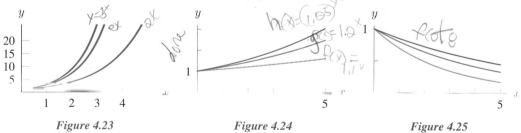

Figure 4.23 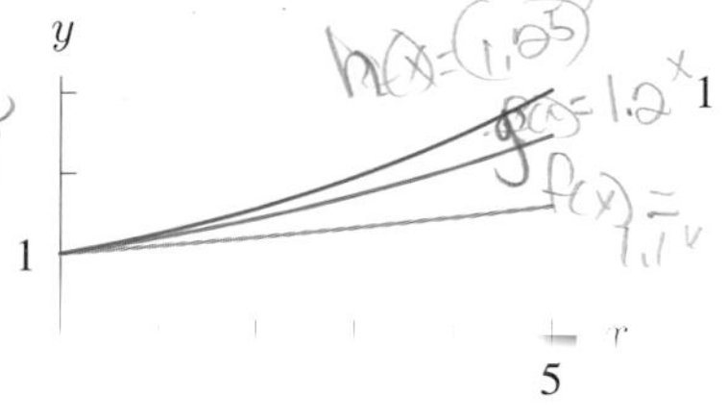   **Figure 4.24**   **Figure 4.25**

5. The graphs of $f(x) = (0.7)^x$, $g(x) = (0.8)^x$, and $h(x) = (0.85)^x$, are given in Figure 4.25. Explain how you can match these formulas and the graphs without using a calculator.

6. The graphs of $y = e^x$, $y = 2e^x$, and $y = 3e^x$ are given in Figure 4.26. Explain how you can match these formulas and graphs without using a calculator.

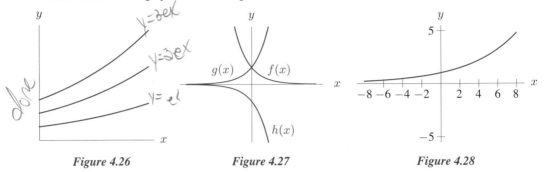

Figure 4.26   Figure 4.27   Figure 4.28

7. The graphs of $y = e^x$, $y = e^{-x}$, and $y = -e^x$ are given in Figure 4.27. Explain how you can match the formulas with the graphs without using a calculator.

8. Suppose you use your calculator to graph $y = 1.04^{5x}$. You correctly enter $y = 1.04\char`^(5x)$ and see the graph in Figure 4.28. A friend graphed the function by entering $y = 1.04\char`^5x$ and said, "The graph is a straight line, so I must have the wrong window." Explain why changing the window will not correct your friend's error.

9. Decide whether each of the following graphs is concave up, concave down, or neither.

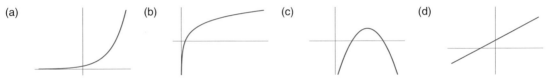

(a)   (b)   (c)   (d)

10. A function $y = f(x)$ is graphed in Figure 4.29. On which interval(s) is $f(x)$ increasing? Decreasing? Concave up? Concave down?

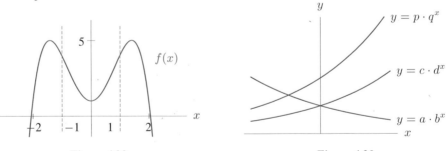

Figure 4.29   Figure 4.30

11. Consider the exponential functions graphed in Figure 4.30 and the six constants a, b, c, d, p, q.

   (a) Which of these constants are definitely positive?
   (b) Which of these constants are definitely between 0 and 1?
   (c) Which of these constants could be between 0 and 1?
   (d) Which two of these constants are definitely equal?
   (e) Which one of the following pairs of constants could be equal?

   a and p b and d b and q d and q

12. Set a window of $-4 \leq x \leq 4$, $-1 \leq y \leq 6$ and graph the following functions using several different values of a for each. Include some values of a with $a < 1$.

   (a) $y = a(2)^x$, $0 < a < 5$.   (b) $y = 2(a)^x$, $0 < a < 5$.

13. Write a paragraph that compares the function $f(x) = a^x$, where $a > 1$, and $g(x) = b^x$, where $0 < b < 1$. Include graphs in your answer.

14. For which value(s) of a and b is $y = ab^x$ an increasing function? A decreasing function? Concave up?

15. What are the domain and range of the exponential function $Q = ab^t$ where a and b are both positive constants?

16. The functions $f(x) = (\frac{1}{2})^x$ and $g(x) = 1/x$ are similar in that they both tend toward zero as x becomes large. Using your calculator, determine which function, f or g, approaches zero faster.

17. Three scientists, working independently of each other, arrive at the following formulas to model the spread of a species of mussel in a system of fresh water lakes:

$$f_1(x) = 3(1.2)^x, \qquad f_2(x) = 3(1.21)^x, \qquad f_3(x) = 3.01(1.2)^x,$$

where $f_n(x)$, $n = 1, 2, 3$, is the number of individual mussels (in 1000s) predicted by model number n to be living in the lake system after x months have elapsed.

(a) Graph these three functions for $0 \le x \le 60, 0 \le y \le 40{,}000$.

(b) The graphs of these three models do not seem all that different from each other. But do the three functions make significantly different predictions about the future mussel population? To answer this, graph the difference function, $f_2(x) - f_1(x)$, of the population sizes predicted by models 1 and 2, as well as the difference function, $f_3(x) - f_1(x)$, of the predictions made by models 1 and 3, and the function $f_3(x) - f_2(x)$. (Use the same window you used for part (a).)

(c) Based on the graphs you made in part (b), discuss the assertion that all three models are in good agreement as far as long-range predictions of mussel population are concerned. What conclusions can you draw about exponential functions in general?

18. Let f be a piecewise-defined function given by

$$f(x) = \begin{cases} 2^x, & x < 0 \\ 0, & x = 0 \\ 1 - \frac{1}{2}x, & x > 0 \end{cases}$$

(a) Graph f for $-3 \le x \le 4$.

(b) The domain of $f(x)$ is all real numbers. What is its range?

(c) What are the intercepts of f?

(d) Describe what happens to f as $x \to +\infty$ and $x \to -\infty$.

(e) Over what intervals is f increasing? Decreasing?

Find a formula for the exponential functions in Problems 19–20.

19. $f(0) = 5$ and $f(2) = 5e^2$

20. $g(0) = 4$ and $g(3) = 4e^3$

In Problems 21–23, find possible a formula for the exponential functions whose graphs are shown.

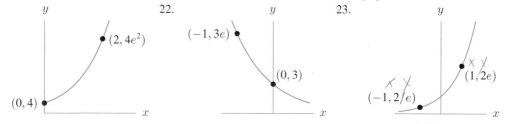

21.
$(2, 4e^2)$
$(0, 4)$

22.
$(-1, 3e)$
$(0, 3)$

23.
$(1, 2e)$
$(-1, 2/e)$

24. At time t in years, the value, V, of an investment of \$1000 is given by $V = 1000e^{0.02t}$. When is the investment worth \$3000?

25. From time $t = 0$, with t in years, a \$1200 deposit in a bank account grows according to the formula

$$B = 1200e^{0.03t}.$$

(a) What is the balance in the account at the end of 100 years?

(b) When does the balance first go over \$50,000?

26. How long does it take an investment to double if it grows according to the formula $V = 537e^{0.015t}$? Assume t is in years.

27. If $f(x)$ is increasing and concave down and $f(0) = 1$ and $f(10) = 7$, What can be said about $f(5)$?

28. Are each of the following functions increasing or decreasing? What do each of the following scenarios tell you about the concavity of the graph modeling them?

   (a) When money is deposited in the bank, the amount of money increases slowly at first. As the size of the account increases, the amount of money increases more rapidly, since the account is earning interest on the new interest, as well as on the original amount.

   (b) After a cup of hot chocolate is poured, the temperature cools off very rapidly at first, and then cools off more slowly, until the temperature of the hot chocolate eventually reaches room temperature.

   (c) When a rumor begins, the number of people who have heard the rumor increases slowly at first. As the rumor spreads, the rate of increase gets greater (as more people continue to tell their friends the rumor), and then slows down again (when almost everyone has heard the rumor).

   (d) When a drug is injected into a person's bloodstream, the amount of the drug present in the body increases rapidly at first. If the person receives daily injections, the body metabolizes the drug so that the amount of the drug present in the body continues to increase, but at a decreasing rate. Eventually, the quantity levels off at a saturation level.

   (e) When a new product is introduced, the number of people who use the product increases slowly at first, and then the rate of increase is faster (as more and more people learn about the product). Eventually, the rate of increase slows down again (when most people who are interested in the product are already using it).

29. Table 4.14 shows the number of US military personnel on active duty.[5]

   (a) Does the number of personnel appear to be an increasing or decreasing function of time?

   (b) Use the information in the table to analyze the concavity of the graph.

   (c) The point of inflection of a graph is the point where the concavity changes. Where does the point of inflection appear to be on this graph?

TABLE 4.14 *Military personnel on active duty*

Year	1990	1991	1992	1993	1994	1995	1996
Number of military personnel (thousands)	2044	1986	1807	1705	1611	1518	1472

30. Table 4.15 shows the population of Ireland[6] at various times between 1780 and 1910.

   (a) When was the population increasing? Decreasing?

   (b) For each successive time interval, construct a table showing the average rate of change of the population.

   (c) From the table you constructed in part (b), when is the graph of the population concave up? Concave down?

   (d) When was the average rate of change of the population the greatest? The least? How is this related to part (c)? What does this mean in human terms?

   (e) Graph the data in Table 4.15 and join the points by a curve to show the trend in the data. From this graph identify where the curve is increasing, decreasing, concave up and concave down. Compare your answers to those you got in parts (a) and (c). Identify the region you found in part (d).

   (f) Something catastrophic happened in Ireland between 1780 and 1910. When? What happened in Ireland at that time to cause this catastrophe?

TABLE 4.15 *The population of Ireland from 1780 to 1910, where 0 corresponds to 1780*

Years since 1780	0	20	40	60	70	90	110	130
Population (millions)	4.0	5.2	6.7	8.3	6.9	5.4	4.7	4.4

[5] *Statistical Abstract of the United States, 1998*, Table No. 582, p.366

[6] Adapted from D. N. Burghes and A. D. Wood, Ellis Horwood, *Mathematical Models in the Social, Management and Life Science*, p. 104 (Ellis Horwood, 1980).

4.4 LOGARITHMS AND THEIR PROPERTIES

In Example 3 on page 116, we solved the exponential equation $100 \cdot 2^t = 337{,}000{,}000$ graphically. In this section, we develop another technique for solving the equation. In the process, we define two logarithmic functions. In the language of Section 3.2, these new functions turn out to be inverses of exponential functions.

What is a Logarithm?

Suppose that a population grows according to the formula $P = 10^t$, where P is the colony size at any time t. When will the population exceed 2500? We want to find time t at which the population equals 2500. So, we solve the following equation for t:

$$10^t = 2500.$$

In Section 4.2, we introduced a graphical method which could be used to approximate t. This time, we introduce a function which returns precisely the exponent of 10 we need.

Since $10^3 = 1000$ and $10^4 = 10{,}000$, and $1000 < 2500 < 10{,}000$, we know that the exponent we are looking for is between 3 and 4. But how do we find the exponent exactly?

To answer this question, we introduce a new function, called the *common logarithm function*, or simply the *log function*, which is written $\log_{10} x$, or $\log x$. We define the log function as follows.

If x is a positive number,

$$\log x \text{ is the exponent of 10 that gives } x.$$

In other words, if

$$y = \log x \qquad \text{then} \qquad 10^y = x.$$

For example, $\log 100 = 2$, because 2 is the exponent of 10 that gives 100, or $10^2 = 100$.

To solve the equation $10^t = 2500$, we must find the power of 10 that gives 2500. Using the log button on a calculator, we can approximate this exponent. We find

$$\log 2500 \approx 3.398.$$

This means that $10^{3.398} \approx 2500$. Notice that this exponent is between 3 and 4 as predicted. The precise exponent is $\log 2500$; the approximate value is 3.398. Thus, it takes roughly 3.4 hours for the population to reach 2500.

Example 1   Rewrite the following statements using exponents instead of logs.

(a) $\log 100 = 2$ (b) $\log 0.01 = -2$ (c) $\log 30 = 1.477$

Solution   For each statement, we use the fact that if $y = \log x$ then $10^y = x$.
(a) $2 = \log 100$ means that $10^2 = 100$.
(b) $-2 = \log 0.01$ means that $10^{-2} = 0.01$.
(c) $1.477 = \log 30$ means that $10^{1.477} = 30$. (Actually, this is only an approximation. Using a calculator, we see that $10^{1.477} = 29.9916\ldots$ and that $\log 30 = 1.47712125\ldots$.)

Example 2   Rewrite the following statements using logs instead of exponents.

(a) $10^5 = 100{,}000$ (b) $10^{-4} = 0.0001$ (c) $10^{0.8} = 6.3096$.

Solution   For each statement, we use the fact that if $10^y = x$, then $y = \log x$.
(a) $10^5 = 100{,}000$ means that $\log 100{,}000 = 5$.
(b) $10^{-4} = 0.0001$ means that $\log 0.0001 = -4$.
(c) $10^{0.8} = 6.3096$ means that $\log 6.3096 = 0.8$. (This, too, is only an approximation because $10^{0.8}$ actually equals $6.30957344\ldots$.)

Logarithms Are Exponents

Note that logarithms are just exponents! Thinking in terms of exponents is often a good way to answer a logarithm problem.

Example 3   Without a calculator, evaluate the following, if possible:

(a) $\log 1$

(b) $\log 10$

(c) $\log 1{,}000{,}000$

(d) $\log 0.001$

(e) $\log \dfrac{1}{\sqrt{10}}$

(f) $\log(-100)$

Solution   (a)   We have $\log 1 = 0$, since $10^0 = 1$.

(b)   We have $\log 10 = 1$, since $10^1 = 10$.

(c)   Since $1{,}000{,}000 = 10^6$, the exponent of 10 that gives $1{,}000{,}000$ is 6. Thus, $\log 1{,}000{,}000 = 6$.

(d)   Since $0.001 = 10^{-3}$, the exponent of 10 that gives 0.001 is -3. Thus, $\log 0.001 = -3$.

(e)   Since $1/\sqrt{10} = 10^{-1/2}$, the exponent of 10 that gives $1/\sqrt{10}$ is $-\frac{1}{2}$. Thus $\log(1/\sqrt{10}) = -\frac{1}{2}$.

(f)   Since 10 to any power is positive, -100 cannot be written as a power of 10. Thus, $\log(-100)$ is undefined.

Example 4   Solve the equation $480(10)^{0.06x} = 1320$ using logarithms.

Solution   First, divide by 480 to isolate the power of 10 on one side of the equation:

$$480(10)^{0.06x} = 1320$$
$$10^{0.06x} = 2.75.$$

Thus, $0.06x$ is the power of 10 which gives 2.75. So, by the definition of log, we know that

$$0.06x = \log 2.75.$$

Evaluating $\log 2.75$ on a calculator, we solve for x, giving

$$x = \frac{\log 2.75}{0.06} \approx 7.3.$$

What is the Relationship Between the Log and Exponential Functions?

The operation of taking a logarithm "undoes" the exponential function. This means that the logarithm and the exponential are inverse functions. For example

$$\underbrace{10^6}_{\substack{\text{Raising} \\ 10 \text{ to } N = 6}} = \underbrace{1{,}000{,}000}_{\text{Result}}$$

and the definition of the log tells us that

$$\underbrace{\log(1{,}000{,}000)}_{\substack{\text{Taking log of the} \\ \text{result}}} = \underbrace{6.}_{\substack{\text{Returns} \\ \text{original } N = 6}}$$

Thus $\log(10^6) = 6$, or, for any N,

$$\boxed{\log 10^N = N.}$$

Similarly, the exponential function "undoes" the logarithm. For example,

$$\underbrace{\log 1000}_{\substack{\text{Taking log of} \\ N = 1000}} = \underbrace{3}_{\text{Result}} \qquad \text{and} \qquad \underbrace{10^3}_{\substack{\text{Raising 10} \\ \text{to result}}} = \underbrace{1000.}_{\substack{\text{Returns} \\ \text{original } N = 1000}}$$

Thus $10^{\log 1000} = 1000$, or, for any $N > 0$,

$$10^{\log N} = N.$$

Example 5 Evaluate without a calculator: (a) $\log\left(10^{8.5}\right)$ (b) $10^{\log(2.7)}$ (c) $10^{\log(x+3)}$

Solution Using $\log 10^N = N$ and $10^{\log N} = N$, we have:

(a) $\log\left(10^{8.5}\right) = 8.5$ (b) $10^{\log(2.7)} = 2.7$ (c) $10^{\log(x+3)} = x + 3$

You can check the first two results on a calculator.

Properties of Logarithms

The exponential equations we solved graphically, such as $100 \cdot 2^t = 337{,}000{,}000$ and $200(0.886)^t = 25$, did not have base 10. To use logarithms to solve these equations, we first need to know the properties of logarithms. A summary of these properties follows. The properties are justified on page 133.

Properties of the Common Logarithm

- By definition, $y = \log x$ means that $10^y = x$.

- Two important values of the log function are

$$\log 1 = 0 \quad \text{and} \quad \log 10 = 1.$$

- The relationship between the log and exponential functions means that

$$\log 10^x = x \qquad \text{for all } x,$$
$$10^{\log x} = x \qquad \text{for } x > 0.$$

- For a and b both positive and any value of t,

$$\log(ab) = \log a + \log b$$

$$\log\left(\frac{a}{b}\right) = \log a - \log b$$

$$\log b^t = t \cdot \log b.$$

We use these properties to solve the following equation.

Example 6 Solve $\log(2x + 1) + 3 = 0$.

Solution To solve for x, we use the exponential function to rewrite the equation without the log. Isolating the log expression gives

$$\log(2x + 1) = -3.$$

Thus, -3 is the power of 10 we need to get $(2x + 1)$. So, the definition of the log gives

$$2x + 1 = 10^{-3}$$

Solving for x, we get

$$2x = 10^{-3} - 1$$
$$x = \frac{10^{-3} - 1}{2}$$
$$x \approx -0.4995.$$

We can now use logarithms to solve the equations that we solved graphically in Section 4.2.

Example 7   Solve $100 \cdot 2^t = 337{,}000{,}000$ for t.

Solution   Dividing both sides of the equation by 100 gives

$$2^t = 3{,}370{,}000.$$

Taking logs of both sides gives

$$\log\left(2^t\right) = \log(3{,}370{,}000).$$

Since $\log(2^t) = t \cdot \log 2$, we have

$$t \log 2 = \log(3{,}370{,}000),$$

so, solving for t, we have

$$t = \frac{\log(3{,}370{,}000)}{\log 2} = 21.68.$$

In Example 3 on page 116, we solved the same equation graphically, getting between 21 and 22 years as the time for the Yonkers fine to exceed the city's annual budget.

Misconceptions and Calculator Errors Involving Logs

It is important to know how to use the properties of logarithms. It is equally important to recognize statements that are *not* true. Beware of the following:

- $\log(a + b)$ is not the same as $\log a + \log b$
- $\log(a - b)$ is not the same as $\log a - \log b$
- $\log(ab)$ is not the same as $(\log a)(\log b)$
- $\log\left(\dfrac{a}{b}\right)$ is not the same as $\dfrac{\log a}{\log b}$
- $\log\left(\dfrac{1}{a}\right)$ is not the same as $\dfrac{1}{\log a}$.

Notice that there are no formulas to simplify either $\log(a + b)$ or $\log(a - b)$. Another common mistake is to assume that the expression $\log 5x^2$ is the same as $2 \cdot \log 5x$. It is not, because the exponent, 2, applies only to the x and not to the 5. However, it is correct to write

$$\log 5x^2 = \log 5 + \log x^2 = \log 5 + 2 \log x.$$

We must also be very careful when using a calculator to evaluate expressions like $\log \frac{17}{3}$. On some calculators if you enter log 17/3, the result will be 0.410, which is incorrect. This is because the calculator assumes that you mean $(\log 17)/3$, which is not the same as $\log(17/3)$. Notice further that

$$\frac{\log 17}{\log 3} \approx \frac{1.230}{0.477} \approx 2.579,$$

which is not the same as either $(\log 17)/3$ or $\log(17/3)$. Thus, the following expressions are all different.

$$\log \frac{17}{3} \approx 0.753, \qquad \frac{\log 17}{3} \approx 0.410, \qquad \text{and} \qquad \frac{\log 17}{\log 3} \approx 2.579.$$

The Natural Logarithm

When e is used as the base for exponential functions, computations are easier with the use of another logarithm function, called log base e. The log base e is used so frequently that it has its own notation: $\ln x$, read as the *natural log of* x. We make the following definition:

> For $x > 0$,
>
> $$\ln x \text{ is the power of } e \text{ that gives } x$$
>
> or, in symbols,
>
> $$\ln x = y \quad \text{means} \quad e^y = x,$$
>
> and y is called the **natural logarithm** of x.

Example 8   The US population, P, is currently growing according to the formula

$$P = 263e^{0.009t},$$

where t is in years since 1995. When is the population predicted to reach 300 million?

Solution   We want to solve the following equation for t:

$$263e^{0.009t} = 300.$$

Dividing by 263 gives

$$e^{0.009t} = \frac{300}{263}$$

So $0.009t$ is the power of e which gives $300/263$. Thus, by the definition of the natural log,

$$0.009t = \ln\left(\frac{300}{263}\right).$$

Solving for t and evaluating $\ln(300/263)$ on a calculator gives

$$t = \frac{\ln(300/263)}{0.009} = 14.6 \text{ years.}$$

Just as the functions 10^x and $\log x$ "undo" each other, so do the functions e^x and $\ln x$. Therefore, the function $\ln x$ has the same properties as the common log function:

Properties of the Natural Logarithm

- By definition, $y = \ln x$ means $x = e^y$.

- In particular,

$$\ln 1 = 0 \quad \text{and} \quad \ln e = 1.$$

- The functions e^x and $\ln x$ "undo" each other, so

$$\ln e^x = x \qquad \text{for all } x$$
$$e^{\ln x} = x \qquad \text{for } x > 0.$$

- For a and b both positive and any value of t,

$$\ln(ab) = \ln a + \ln b$$
$$\ln\left(\frac{a}{b}\right) = \ln a - \ln b$$
$$\ln b^t = t \cdot \ln b.$$

Justification of the Properties of Logarithms

Justification of $\log(a \cdot b) = \log a + \log b$ and $\log(a/b) = \log a - \log b$. If a and b are both positive, we can write $a = 10^m$ and $b = 10^n$, so $\log a = m$ and $\log b = n$. Then, the product $a \cdot b$ can be written

$$a \cdot b = 10^m \cdot 10^n = 10^{m+n}.$$

Therefore $m + n$ is the power of 10 needed to give $a \cdot b$, so

$$\log(a \cdot b) = m + n,$$

which gives

$$\boxed{\log(a \cdot b) = \log a + \log b.}$$

Similarly, the quotient a/b can be written as

$$\frac{a}{b} = \frac{10^m}{10^n} = 10^{m-n}.$$

Therefore $m - n$ is the power of 10 needed to give a/b, so

$$\log\left(\frac{a}{b}\right) = m - n,$$

and thus

$$\boxed{\log\left(\frac{a}{b}\right) = \log a - \log b.}$$

Justification of $\log(b^t) = t \cdot \log b$. Suppose that b is positive, so we can write $b = 10^k$ for some value of k. Then

$$b^t = (10^k)^t.$$

Notice that we have rewritten the expression b^t so that the base is a power of 10. Using a property of exponents, we can write $(10^k)^t$ as 10^{kt}, so

$$b^t = (10^k)^t = 10^{kt}.$$

Therefore kt is the power of 10 which gives b^t, so

$$\log(b^t) = kt.$$

But since $b = 10^k$, we know $k = \log b$. This means

$$\log(b^t) = (\log b)t = t \cdot \log b.$$

Thus, for $b > 0$ we have

$$\boxed{\log(b^t) = t \cdot \log b.}$$

Problems for Section 4.4

1. Evaluate 10^n for $n = 3$, $n = 3.5$, $n = 3.48$, $n = 3.477$, and $n = 3.47712$. Based on your answer, estimate the value of $\log 3000$.

2. Evaluate the following expressions without using a calculator.
   - (a) $\log 1$
   - (b) $\log 0.1$
   - (c) $\log(10^0)$
   - (d) $\log \sqrt{10}$
   - (e) $\log(10^5)$
   - (f) $\log(10^2)$
   - (g) $\log\left(\frac{1}{\sqrt{10}}\right)$
   - (h) $10^{\log 100}$
   - (i) $10^{\log 1}$
   - (j) $10^{\log(0.01)}$

3. Evaluate the following expressions without using a calculator.

   (a) $\ln 1$ (b) $\ln e^0$ (c) $\ln e^5$ (d) $\ln \sqrt{e}$ (e) $e^{\ln 2}$ (f) $\ln \left(\dfrac{1}{\sqrt{e}}\right)$

4. Evaluate the following pairs of expressions without using a calculator.

   (a) $\log(10 \cdot 100)$ and $\log 10 + \log 100$ (b) $\log(100 \cdot 1000)$ and $\log 100 + \log 1000$

   (c) $\log\left(\dfrac{10}{100}\right)$ and $\log 10 - \log 100$ (d) $\log\left(\dfrac{100}{1000}\right)$ and $\log 100 - \log 1000$

   (e) $\log(10^2)$ and $2\log 10$ (f) $\log(10^3)$ and $3\log 10$

5. (a) Write general formulas based on what you observed in Problem 4.

   (b) Apply your formulas to rewrite $\log\left(\dfrac{AB}{C}\right)^p$ at least two different ways.

6. Evaluate the following expressions without using a calculator.

   (a) $\log(\log 10)$ (b) $\sqrt{\log 100} - \log \sqrt{100}$ (c) $\log(\sqrt{10}\sqrt[3]{10}\sqrt[5]{10})$

   (d) $1000^{\log 3}$ (e) $0.01^{\log 2}$ (f) $\dfrac{1}{\log(1/\log \sqrt[10]{10})}$

7. True or false?

   (a) $\log AB = \log A + \log B$ (b) $\dfrac{\log A}{\log B} = \log A - B$ (c) $\log A \log B = \log A + \log B$

   (d) $p \cdot \log A = \log A^p$ (e) $\log \sqrt{x} = \frac{1}{2}\log x$ (f) $\sqrt{\log x} = \log(x^{1/2})$

8. Suppose that $x = \log A$ and that $y = \log B$. Write the following expressions in terms of x and y.

   (a) $\log(AB)$ (b) $\log(A^3 \cdot \sqrt{B})$ (c) $\log(A - B)$

   (d) $\dfrac{\log A}{\log B}$ (e) $\log\dfrac{A}{B}$ (f) AB

9. Express the following in terms of x without using logs.

   (a) $\log 100^x$ (b) $1000^{\log x}$ (c) $\log 0.001^x$

10. Express the following in terms of x without using natural logs.

    (a) $\ln e^{2x}$ (b) $e^{\ln(3x+2)}$ (c) $\ln\left(\dfrac{1}{e^{5x}}\right)$ (d) $\ln \sqrt{e^x}$

11. Let $p = \log m$ and $q = \log n$. Write the following expressions in terms of p and/or q without using logs.

    (a) m (b) n^3 (c) $\log(mn^3)$ (d) $\log \sqrt{m}$

In Problems 12–23, solve the equations exactly for x.

12. $5(1.031)^x = 8$ 13. $4(1.171)^x = 7(1.088)^x$ 14. $3\log(2x+6) = 6$

15. $3 \cdot 2^x + 8 = 25$ 16. $e^{x+4} = 10$ 17. $e^{x+5} = 7 \cdot 2^x$

18. $\log(1-x) - \log(1+x) = 2$ 19. $\log(2x+5) \cdot \log(9x^2) = 0$ 20. $b^x = c$

21. $ab^x = c$ 22. $Pa^x = Qb^x$ 23. $Pe^{kx} = Q$

Solve exactly for t in the equations in Problems 24–32.

24. 25. $3(1.081)^t = 14$ 26. $5(1.014)^{3t} = 12$

27. $5(1.15)^t = 8(1.07)^t$ 28. $e^{0.044t} = 6$ 29. $121e^{} = 00$

30. $58e^{4t+1} = 30$ 31. $17e^{0.02t} = 18e^{0.03t}$ 32. $44e^{0.15t} = 50(1.2)^t$

33. Given that $10^{1.3} \approx 20$, approximate the value of $\log 200$ without using a calculator.

34.   Three students try to solve the equation

$$11 \cdot 3^x = 5 \cdot 7^x.$$

The first student finds that $x = \dfrac{\log(11/5)}{\log(7/3)}$. The second finds that $x = \dfrac{\log(5/11)}{\log(3/7)}$. The third finds that $x = \dfrac{\log 11 - \log 5}{\log 7 - \log 3}$. Which student (or students) is (are) correct? Explain.

4.5 THE LOGARITHMIC FUNCTION

The Graph, Domain, and Range of the Common Logarithm

In Section 4.4 we defined the log function (to base 10) for all positive numbers. In other words,

Domain of $\log x$ is all positive numbers.

By considering its graph in Figure 4.31, we determine the range of $y = \log x$. The log graph crosses the x-axis at $x = 1$, because $\log 1 = \log(10^0) = 0$. The graph climbs to $y = 1$ at $x = 10$, because $\log 10 = \log(10^1) = 1$. In order for the log graph to climb to $y = 2$, the value of x must reach 100, or 10^2, and in order for it to climb to $y = 3$, the value of x must be 10^3, or 1000. To reach the modest height of $y = 20$ requires x to equal 10^{20}, or 100 billion billion! The log function increases so slowly that it often serves as a benchmark for other slow-growing functions. Nonetheless, the graph of $y = \log x$ eventually climbs to any value we choose.

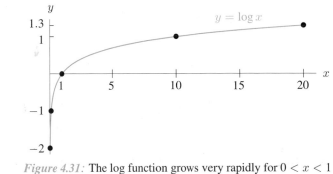

Figure 4.31: The log function grows very rapidly for $0 < x < 1$ and very slowly for $x > 1$. It has a vertical asymptote at $x = 0$

Although x cannot equal zero in the log function, we can choose $x > 0$ to be as small as we like. As x decreases toward zero, the values of $\log x$ get large and negative. For example,

$$\log 0.1 \;=\; \log 10^{-1} = -1,$$
$$\log 0.01 \;=\; \log 10^{-2} = -2,$$
$$\vdots \qquad\qquad \vdots$$
$$\log 0.0000001 \;=\; \log 10^{-7} = -7,$$

and so on. So, small positive values of x give exceedingly large negative values of y. Thus,

Range of $\log x$ is all real numbers.

The log function is increasing and its graph is concave down, because the rate of change is decreasing.

Graphs of the Functions $y = \log x$ and $y = 10^x$

The fact that $y = \log x$ and $y = 10^x$ are inverses means that their graphs are related. Looking at Tables 4.16 and 4.17, we see that the point $(0.01, -2)$ is on the graph of $y = \log x$ and the point $(-2, 0.01)$ is on the graph of $y = 10^x$. In general, if the point (a, b) is on the graph of $y = \log x$, the point (b, a) is on the graph of $y = 10^x$. Thus, the graph of $y = \log x$ is the graph of $y = 10^x$ with x and y-axes interchanged. This is equivalent to reflecting the graph of $y = 10^x$ across the diagonal line $y = x$. See Figure 4.32.

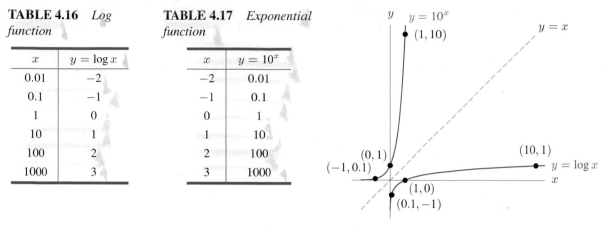

TABLE 4.16 Log function			**TABLE 4.17** Exponential function	
x	$y = \log x$		x	$y = 10^x$
0.01	-2		-2	0.01
0.1	-1		-1	0.1
1	0		0	1
10	1		1	10
100	2		2	100
1000	3		3	1000

Figure 4.32: The functions $y = \log x$ and $y = 10^x$ are inverse functions

Asymptotes

In Section 4.3 we saw that the graph of an exponential function has a horizontal asymptote. In Figure 4.32, we see that $y = 10^x$ has horizontal asymptote $y = 0$, because

$$\text{as } x \to -\infty, \quad 10^x \to 0.$$

Correspondingly, as x gets closer to zero, $y = \log x$ takes on larger and larger negative values. We write

$$\text{as } x \to 0^+, \quad f(x) \to -\infty.$$

The notation $x \to 0^+$ is read "x approaches zero from the right" and means that we are choosing smaller and smaller positive values of x—that is, we are sliding toward $x = 0$ through small positive values. We say the graph of the log function $y = \log x$ has a *vertical asymptote* of $x = 0$.

To describe vertical asymptotes in general, we use the notation

$$x \to a^+$$

meaning that x slides toward a from the right (that is, through values larger than a) and

$$x \to a^-$$

meaning that x slides toward a from the left (that is, through values smaller than a).

We summarize the information about both horizontal and vertical asymptotes:

Let $y = f(x)$ be a function and let a be a finite number.
- The graph of f has a **horizontal asymptote** of $y = a$ if

$$f(x) \to a \quad \text{as} \quad x \to \infty \quad \text{or} \quad x \to -\infty \quad \text{or both}$$

- The graph of f has a **vertical asymptote** of $x = a$ if

$$f(x) \to \infty \quad \text{or} \quad f(x) \to -\infty \quad \text{as} \quad x \to a \quad \text{(from the left or the right or both)}.$$

Notice that the process of finding a vertical asymptote is different from the process for finding a horizontal asymptote. Vertical asymptotes occur where the function values grow larger and larger, either positively or negatively, as x approaches a finite value (i.e. where $f(x) \to \infty$ or $f(x) \to -\infty$ as $x \to a$.) Horizontal asymptotes are determined by whether the function values approach a finite number as x takes on large positive or large negative values (i.e., as $x \to \infty$ or $x \to -\infty$).

Graph of Natural Logarithm

In addition to having similar algebraic properties, the natural log and the common log have similar graphs.

Example 1   Sketch the graph of $y = \ln x$ for $0 < x < 10$.

Solution   Values of $\ln x$ are in Table 4.18. Like the common log, the natural log is only defined for $x > 0$ and has a vertical asymptote at $x = 0$. The graph is slowly increasing and concave down.

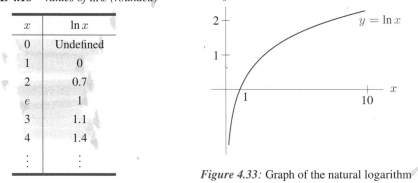

TABLE 4.18   *Values of* $\ln x$ *(rounded)*

x	$\ln x$
0	Undefined
1	0
2	0.7
e	1
3	1.1
4	1.4
$\vdots$	$\vdots$

Figure 4.33: Graph of the natural logarithm

The functions $y = \ln x$ and $y = e^x$ are inverses. Their graphs are reflections of one another across the line $y = x$. See Figure 4.34.

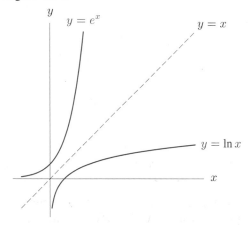

Figure 4.34: The functions $y = \ln x$ and $y = e^x$ are inverses

Problems for Section 4.5

1. What is the equation of the asymptote of the graph of $y = 10^x$? Of the graph of $y = 2^x$? Of the graph of $y = \log x$?

2. What is the equation for the asymptote of the graph of $y = e^x$? Of the graph of $y = e^{-x}$? Of the graph of $y = \ln x$?

3. What is the value (if any) of the following? (a) 10^{-x} as $x \to \infty$ (b) $\log x$ as $x \to 0^+$

4. What is the value (if any) of the following? (a) e^x as $x \to -\infty$ (b) $\ln x$ as $x \to 0^+$

Graph the functions in Problems 5–8. Label all asymptotes and intercepts.

5. $y = 2 \cdot 3^x + 1$ 6. $y = -e^{-x}$ 7. $y = \log(x - 4)$ 8. $y = \ln(x + 1)$

9. Each of the functions in (a) – (c) corresponds to one of the functions $r(x)$, $s(x)$, $t(x)$ whose values are given in one of the tables. Find the match and justify your answers.

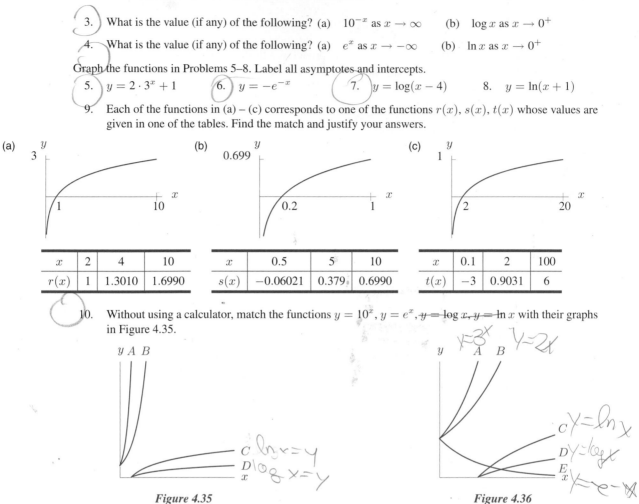

x	2	4	10
$r(x)$	1	1.3010	1.6990

x	0.5	5	10
$s(x)$	−0.06021	0.379	0.6990

x	0.1	2	100
$t(x)$	−3	0.9031	6

10. Without using a calculator, match the functions $y = 10^x$, $y = e^x$, $y = \log x$, $y = \ln x$ with their graphs in Figure 4.35.

Figure 4.35

Figure 4.36

11. Without using a calculator, match the functions $y = 2^x$, $y = e^{-x}$, $y = 3^x$, $y = \ln x$, $y = \log x$ with their graphs in Figure 4.36.

12. Immediately following the gold medal performance of the US women's gymnastic team in the 1996 Olympic Games, an NBC commentator, John Tesh, said of one of the team members: "Her confidence and performance have grown logarithmically." He clearly thought this was an enormous compliment. Is it a compliment? Is it realistic?

Find the domain of the functions in Problems 13–16.

13. $h(x) = \ln(x^2)$ 14. $g(x) = (\ln x)^2$ 15. $f(x) = \ln(\ln x)$ 16. $k(x) = \ln(x - 3)$

In Problems 17–22, find possible formulas for the functions using logs or exponentials.

17.   18.   19.

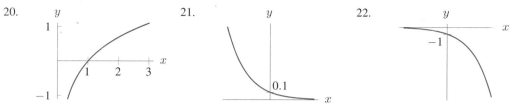

20. 21. 22.

23. (a) What is the domain and range of $f(x) = 10^x$? What is the asymptote of $f(x) = 10^x$?
  (b) What does your answer to part (a) tell you about the domain, range, and asymptotes of $g(x) = \log x$?

24. Since $e = 2.718\ldots$ we know that $2 < e < 3$, which means that $2^2 < e^2 < 3^2$. Without using a calculator, explain why

  (a) $1 < \ln 3 < 2$ (b) $1 < \ln 4 < 2$

4.6 LOGARITHMS AND EXPONENTIAL MODELS

The log function is often useful when answering questions about exponential models. Because logarithms "undo" the exponential functions, we use them to solve many exponential equations.

Example 1   In Example 4 on page 116 we solved the equation $200(0.886^t) = 25$ graphically, where t is in thousands of years. We found that a 200 microgram sample of carbon-14 decays to 25 micrograms in approximately 17,200 years. Now solve $200(0.886)^t = 25$ using logarithms.

Solution   First, isolate the power on one side of the equation

$$200(0.886^t) = 25$$
$$0.886^t = 0.125.$$

Take the log of both sides, and use the fact that $\log(0.886^t) = t \log 0.886$. Then

$$\log(0.886^t) = \log 0.125$$
$$t \log 0.886 = \log 0.125$$

so

$$t = \frac{\log 0.125}{\log 0.886} \approx 17.18 \text{ thousand years.}$$

This answer is close to the value we found from the graph, 17,200.

Example 2   The population of City A begins with 50,000 people and grows at a constant 3.5% annual rate. The population of City B begins with a larger population of 250,000 people but grows at the slower rate of 1.6% per year. Assuming that these growth rates hold constant, will the population of City A ever catch up to the population of City B? If so, when?

Solution   If t is time measured in years and P_A and P_B are the populations of these two cities, then

$$P_A = 50{,}000(1.035)^t \quad \text{and} \quad P_B = 250{,}000(1.016)^t.$$

We want to solve the equation

$$50{,}000(1.035)^t = 250{,}000(1.016)^t.$$

We first get the exponential terms together by dividing both sides of the equation by $50{,}000(1.016)^t$:

$$\frac{(1.035)^t}{(1.016)^t} = \frac{250{,}000}{50{,}000} = 5.$$

Since $\dfrac{a^t}{b^t} = \left(\dfrac{a}{b}\right)^t$, this gives

$$\left(\frac{1.035}{1.016}\right)^t = 5.$$

Taking logs of both sides and using $\log b^t = t \log b$, we have

$$\log \left(\frac{1.035}{1.016}\right)^t = \log 5$$

$$t \log \left(\frac{1.035}{1.016}\right) = \log 5$$

$$t = \frac{\log 5}{\log(1.035/1.016)} \approx 86.9.$$

Thus, the cities' populations will be equal in just under 87 years. To check this, notice that when $t = 86.9$,

$$P_A = 50{,}000(1.035)^{86.9} \approx 993{,}771$$

and

$$P_B = 250{,}000(1.016)^{86.9} \approx 993{,}123.$$

The answers aren't exactly equal because we rounded off the value of t. Rounding can introduce significant errors, especially when logs and exponentials are involved. Using $t = 86.86480867$, the values of P_A and P_B agree to three decimal places.

Doubling Time

Eventually, any exponentially growing quantity doubles, or increase by 100%. Since its percent growth rate is constant, the time it takes for it to grow by 100% is also a constant. This time period is called the *doubling time*.

Example 3   (a)   Find the time needed for the turtle population described by the function $P = 175(1.145)^t$ to double its initial size.

         (b)   How long does this population take to quadruple its initial size? To increase by a factor of 8?

Solution   (a)   The initial size is 175 turtles; doubling this gives 350 turtles. Thus, we need to solve the following equation for t:

$$175(1.145)^t = 350$$

$$1.145^t = 2$$

$$\log\left(1.145^t\right) = \log 2$$

$$t \cdot \log 1.145 = \log 2$$

$$t = \frac{\log 2}{\log 1.145} \approx 5.12 \text{ years.}$$

We check this by noting that

$$175(1.145)^{5.12} \approx 350,$$

which is double the initial population. In fact, at any time it takes the turtle population about 5.12 years to double in size.

         (b)   Since the population function is exponential, it increases by 100% every 5.12 years. Thus it doubles its initial size in the first 5.12 years, quadruples its initial size in two 5.12 year periods, or 10.24 years, and increases by a factor of 8 in three 5.12 year periods, or 15.36 years. We check this by noting that

$$175(1.145)^{10.24} \approx 700,$$

or 4 times the initial size, and that

$$175(1.145)^{15.36} \approx 1400,$$

or 8 times the initial size.

Half-Life

Just as an exponentially growing quantity doubles in a fixed amount of time, an exponentially decaying quantity decreases by a factor of 2 in a fixed amount of time. This time period is called the *half-life* of the quantity.

Example 4   Carbon-14 decays radioactively at a constant annual rate of 0.0121%. Show that the half-life of carbon-14 is about 5728 years.

Solution   Note that we are not given an initial amount of carbon-14, but we can solve this problem without that information. Let the symbol Q_0 represent the initial quantity of carbon-14 present. The growth rate is -0.000121 because carbon-14 is decaying. So the growth factor is $b = 1 - 0.000121 = 0.999879$. Thus, after t years the amount left will be

$$Q = Q_0(0.999879)^t.$$

We want to find how long it takes for the quantity to drop to half its initial level. Thus, we need to solve for t in the equation

$$\frac{1}{2}Q_0 = Q_0(0.999879)^t.$$

Dividing each side by Q_0, we have

$$\frac{1}{2} = 0.999879^t.$$

Taking logs

$$\log\frac{1}{2} = \log\left(0.999879^t\right)$$
$$\log 0.5 = t \cdot \log 0.999879$$
$$t = \frac{\log 0.5}{\log 0.999879} \approx 5728.14.$$

Thus, no matter how much carbon-14 there is initially, after about 5728 years, half will remain.

Example 5   The quantity, Q, of a substance decays according to the formula $Q = Q_0 e^{-kt}$, where t is in minutes. The half-life of the substance is 11 minutes. What is the value of k?

Solution   We know that after 11 minutes, $Q = \frac{1}{2}Q_0$. Thus, solving for k, we get

$$Q_0 e^{-k\cdot 11} = \frac{1}{2}Q_0$$
$$e^{-11k} = \frac{1}{2}$$
$$-11k = \ln\frac{1}{2}$$
$$k = \frac{\ln(1/2)}{-11} \approx 0.063,$$

so $k = 0.063$ per minute.

In Section 4.7, we see that k is called the *continuous decay rate*, so this substance decays at the continuous rate of 6.3% per minute.

Exponential Growth Problems That Cannot Be Solved By Logarithms

Some equations with the variable in the exponent cannot be solved using logarithms.

Example 6   With t in years, the population of a country (in millions) is given by $P = 2(1.02)^t$, while the food supply (in millions of people that can be fed) is given by $N = 4 + 0.5t$. Determine the year in which the country first experiences food shortages.

Solution   The country will start to experience shortages when the population equals the number of people that can be fed—that is, when $P = N$. We attempt to solve the equation $P = N$ by using logs:

$$2(1.02)^t = 4 + 0.5t$$
$$1.02^t = 2 + 0.25t \qquad \text{Dividing by 2}$$
$$\log 1.02^t = \log(2 + 0.25t)$$
$$t \log 1.02 = \log(2 + 0.25t).$$

Unfortunately, we cannot isolate t, so, this equation cannot be solved using logs. However, we can approximate the solution of the original equation numerically or graphically, as shown in Figure 4.37. The two functions, P and N, are equal when $t \approx 199.4$. Thus, it will be almost 200 years before shortages occur.

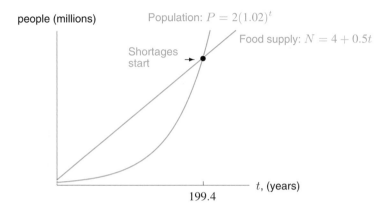

Figure 4.37: Finding the intersection of linear and exponential graphs

Chemical Acidity

Logarithms are useful in measuring quantities whose magnitudes vary widely, such as acidity (pH), sound (decibels), and earthquakes (the Richter scale). In chemistry, the acidity of a liquid is expressed using pH. The acidity depends on the hydrogen ion concentration in the liquid (in moles per liter); this concentration is written $[H^+]$. The greater the hydrogen ion concentration, the more acidic the solution. The pH is defined as: $pH = -\log[H^+]$.

Example 7   The hydrogen ion concentration of seawater is $[H^+] = 1.1 \cdot 10^{-8}$. Estimate the pH of seawater. Then check your answer with a calculator.

Solution   We want to estimate $pH = -\log(1.1 \cdot 10^{-8})$. Since $1.1 \cdot 10^{-8} \approx 10^{-8}$ and $\log 10^{-8} = -8$, we know that

$$pH = -\log(1.1 \cdot 10^{-8}) \approx -(-8) = 8.$$

Using a calculator, we have

$$pH = -\log(1.1 \cdot 10^{-8}) = 7.96.$$

Example 8   A vinegar solution has a pH of 3. Determine the hydrogen ion concentration.

Solution   Since $3 = -\log[H^+]$, we have $-3 = \log[H^+]$. This means that $10^{-3} = [H^+]$. So the hydrogen ion concentration is 10^{-3} moles per liter.

Logarithms and Orders of Magnitude

We often compare sizes or quantities by computing their ratios. If A is twice as tall as B, then

$$\frac{\text{Height of } A}{\text{Height of } B} = 2.$$

If one object is 10 times heavier than another, we say it is an *order of magnitude* heavier. If one quantity is two factors of 10 greater than another, we say it is two orders of magnitude greater, and so on. For example, the value of a dollar is two orders of magnitude greater than the value of a penny, because we have

$$\frac{\$1}{\$0.01} = 100 = 10^2.$$

The order of magnitude is the logarithm of their ratio.

Example 9   The sound intensity of a refrigerator motor is 10^{-11} watts/cm^2. A typical school cafeteria has sound intensity of 10^{-8} watts/cm^2. How many orders of magnitude more intense is the sound of the cafeteria?

Solution   To compare the two intensities, we compute their ratio:

$$\frac{\text{Sound intensity of cafeteria}}{\text{Sound intensity of refrigerator}} = \frac{10^{-8}}{10^{-11}} = 10^{-8-(-11)} = 10^3.$$

Thus, the sound intensity of the cafeteria is 1000 times greater than the sound intensity of the refrigerator. The log of this ratio is 3. We say that the sound intensity of the cafeteria is three orders of magnitude greater than the sound intensity of the refrigerator.

Decibels

The intensity of audible sound varies over an enormous range. The range is so enormous that we consider the logarithm of the sound intensity. This is the idea behind the *decibel* (abbreviated dB). To measure a sound in decibels, the sound's intensity, I, is compared to the intensity of a standard benchmark sound, I_0. The intensity of I_0 is defined to be 10^{-16} watts/cm^2, roughly the lowest intensity audible to humans. The comparison between a sound intensity I and the benchmark sound intensity I_0 is made as follows:

$$\boxed{\text{Noise level in decibels} = 10 \cdot \log\left(\frac{I}{I_0}\right).}$$

For instance, let's find the decibel rating of the refrigerator in Example 9. First, we find how many orders of magnitude more intense the refrigerator sound is than the benchmark sound:

$$\frac{I}{I_0} = \frac{\text{Sound intensity of refrigerator}}{\text{Benchmark sound intensity}} = \frac{10^{-11}}{10^{-16}} = 10^5.$$

Thus, the refrigerator's intensity is 5 orders of magnitude more than I_0, the benchmark intensity. We have

$$\text{Decibel rating of refrigerator noise} = 10 \times \underbrace{\text{Number of orders of magnitude}}_{5} = 50 \text{ dB}.$$

Note that 5, the number of orders of magnitude, is the log of the ratio I/I_0. We use the log function because it "counts" the number of powers of 10. Thus if N is the decibel rating, then

$$N = 10\log\left(\frac{I}{I_0}\right).$$

Example 10   (a)   If a sound doubles in intensity, by how many units does its decibel rating increase?

(b)   Loud music can measure 110 dB whereas normal conversation measures 50 dB. How many times more intense is loud music than normal conversation?

Solution   (a)   Let I be the sound's intensity before it doubles. Once doubled, the new intensity is $2I$. The decibel rating of the original sound is $10\log(I/I_0)$, and the decibel rating of the new sound is $10\log(2I/I_0)$. The difference in decibel ratings is given by

$$\text{Difference in decibel ratings} = 10\log\left(\frac{2I}{I_0}\right) - 10\log\left(\frac{I}{I_0}\right)$$

$$= 10\left(\log\left(\frac{2I}{I_0}\right) - \log\left(\frac{I}{I_0}\right)\right) \quad \text{Factoring out 10}$$

$$= 10\cdot\log\left(\frac{2I/I_0}{I/I_0}\right) \quad \text{Using the property } \log a - \log b = \log(a/b)$$

$$= 10\cdot\log 2 \quad \text{Canceling } I/I_0$$

$$\approx 3\text{ dB} \quad \text{Because } \log 2 \approx 0.3.$$

Thus, if the sound intensity is doubled, the decibel rating goes up by approximately 3 dB.

(b)   If I_M is the sound intensity of loud music, then

$$10\log\left(\frac{I_M}{I_0}\right) = 110\text{ dB}.$$

Similarly, if I_C is the sound intensity of conversation, then

$$10\log\left(\frac{I_C}{I_0}\right) = 50\text{ dB}.$$

Computing the difference of the decibel ratings gives

$$10\log\left(\frac{I_M}{I_0}\right) - 10\log\left(\frac{I_C}{I_0}\right) = 60.$$

Dividing by 10 gives

$$\log\left(\frac{I_M}{I_0}\right) - \log\left(\frac{I_C}{I_0}\right) = 6$$

$$\log\left(\frac{I_M/I_0}{I_C/I_0}\right) = 6 \quad \text{Using the property } \log b - \log a = \log(b/a)$$

$$\log\left(\frac{I_M}{I_C}\right) = 6 \quad \text{Canceling } I_0$$

$$\frac{I_M}{I_C} = 10^6 \quad \log x = 6 \text{ means that } x = 10^6.$$

So $I_M = 10^6 I_C$, which means that loud music is 10^6 times, or one million times, as intense as normal conversation.

Problems for Section 4.6

Solve the equations in Problems 1–6 if possible. Give an exact solution if there is one.

1. $1.7(2.1)^{3x} = 2(4.5)^x$
2. $3^{4\log x} = 5$
3. $5(1.044)^t = t + 10$
4. $12(1.221)^t = t + 3$
5. $10e^{3t} - e = 2e^{3t}$
6. $\log x + \log(x-1) = \log 2$

7. The populations (in thousands) of two different cities are given by

$$P_1 = 51(1.031)^t \quad \text{and} \quad P_2 = 63(1.052)^t,$$

where t is the number of years since 1980. When does the population of P_1 equal that of P_2?

8. The temperature, H, in °F, of a cup of coffee t hours after it is set out to cool is given by the equation:

$$H = 70 + 120(1/4)^t.$$

   (a) What is the coffee's temperature initially (that is, at time $t = 0$)? After 1 hour? 2 hours?
   (b) How long does it take the coffee to cool down to 90°F? 75°F?

9. A colony of bacteria grows exponentially. The colony begins with 3 bacteria, but 3 hours after the beginning of the experiment, it has grown to 100 bacteria.

   (a) Give a formula for the number of bacteria as a function of time.
   (b) How long does it take for the colony to triple in size?

10. In 1999, the population of the country Erehwon was 50 million people and has since been increasing by 2.9% every year. The population of the country Ecalpon, on other hand, was 45 million people and increasing by 3.2% every year.

   (a) For each country, write a formula expressing the population as a function of time t, where t is the number of years since 1999.
   (b) Find the value(s) of t, if any, when the two countries have the same population.
   (c) When is the population of Ecalpon double that of Erehwon?

11. Prices climb at a constant 3% annual rate.

   (a) By what percent will prices have climbed after 5 years?
   (b) How long will it take for prices to climb 25%?

12. A rubber ball is dropped onto a hard surface from a height of 6 feet, and it bounces up and down. At each bounce it rises to 90% of the height from which it fell.

   (a) Find a formula for $h(n)$, the height reached by the ball on bounce n.
   (b) How high will the ball bounce on the 12th bounce?
   (c) How many bounces before the ball rises no higher than an inch?

13. The growth of an animal population, P, is described by the function $P = 300 \cdot 2^{t/20}$.

   (a) How large is this population in year $t = 0$? $t = 20$?
   (b) When does this population reach 1000 members?

14. Due to inflation, the cost, C, in \$, of a week's groceries for a family of four increases over time according to the formula $C = 140(1.034)^t$, where t is the number of years since 1990.

   (a) How much did the family pay for a week's worth of groceries in 1990? 1991? 1995?
   (b) By what percent did the cost increase between 1990 and 1991? Between 1990 and 1995?
   (c) In what year will the family first pay at least \$500 for a week's worth of groceries?

15. The quantity remaining of a substance is diminishing exponentially, so that $Q(t) = 4.2(\frac{1}{2})^{t/11}$, where t is measured in years, and $Q(t)$ is measured in grams.

   (a) What is the half-life of this substance?
   (b) If $Q(t) = Q_0 b^t$, find b, and explain its physical significance.

16. The half-life of iodine-123 is about 13 hours. You begin with 50 grams of iodine-123.

   (a) Write an equation that gives the amount of iodine-123 remaining after t hours.
   (b) Determine the number of hours needed for your sample to decay to 10 grams.

17. Find the doubling time of a population growing according to $P = P_0 e^{0.2t}$.

18. A manager at Saks Fifth Avenue wants to estimate the number of customers to expect on the last shopping day before Christmas. She collects data from three previous years, and determines that the crowds follow the same general pattern. When the store opens at 10 am, 500 people enter, and the total number in the store doubles every 40 minutes. When the number of people in the store reaches 10,000, security guards need to be stationed at the entrances to control the crowds. At what time should the guards be commissioned?

19. Scientists observing owl and hawk populations collect the following data. Their initial count for the owl population is 245 owls, and the population grows by 3% per year. They initially observe 63 hawks, and this population doubles every 10 years.

    (a) Find a formula for the size of the population of owls in terms of time t.
    (b) Find a formula for the size of the population of hawks in terms of time t.
    (c) Use a graph to find how long it will take for these populations to be equal in number.

20. The size of a population, P, of toads t years after it is introduced into a wetland is given by
$$P = \frac{1000}{1 + 49(1/2)^t}.$$

    (a) How many toads are there in year $t = 0$? $t = 5$? $t = 10$?
    (b) How long does it take for the toad population to reach 500? 750?
    (c) What is the maximum number of toads that the wetland can support?

21. Figure 4.38 shows the graphs of the exponential functions f and g, and the linear function, h.
    (a) Find formulas for f, g, and h.
    (b) Find the exact value(s) of x such that $f(x) = g(x)$.
    (c) Estimate the value(s) of x such that $f(x) = h(x)$.

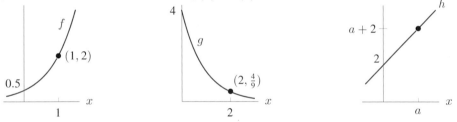

Figure 4.38

22. (a) Using the definition of pH on page 142, find the concentrations of hydrogen ions in solutions with
       (i) pH $= 2$ (ii) pH $= 4$ (iii) pH $= 7$
    (b) A high concentration of hydrogen ions corresponds to an acidic solution. From your answer to part (a), decide if solutions with high pHs are more or less acidic than solutions with low pHs.

23. (a) A 12 oz cup of coffee contains about $2.41 \cdot 10^{18}$ hydrogen ions. What is the concentration (moles/liter) of hydrogen ions in a 12 oz cup of coffee? [Hint: One liter equals 30.3 oz. One mole of hydrogen ions equals $6.02 \cdot 10^{23}$ hydrogen ions.]
    (b) Based on your answer to part (a) and the formula for pH, what is the pH of a 12 oz cup of coffee?

24. (a) The pH of lemon juice is about 2.3. What is the concentration of hydrogen ions in lemon juice?
    (b) A person squeezes 2 oz of lemon juice into a cup. Based on your answer to part (a), how many hydrogen ions does this juice contain?

25. (a) Let D_1 and D_2 represent the decibel ratings of sounds of intensity I_1 and I_2, respectively. Using log properties, find a simplified formula for the difference between the two ratings, $D_2 - D_1$, in terms of the two intensities, I_1 and I_2.
    (b) If a sound's intensity doubles, how many decibels louder does the sound become?

26. The magnitude of an earthquake is measured relative to the strength of a "standard" earthquake, whose seismic waves are of size W_0. The magnitude, M, of an earthquake whose seismic waves are of size W is defined to be
$$M = \log\left(\frac{W}{W_0}\right).$$

The value of M is called the *Richter scale* rating, the strength of an earthquake.

(a) Let M_1 and M_2 represent the magnitude of two earthquakes whose seismic waves are of sizes W_1 and W_2, respectively. Using log properties, find a simplified formula for the difference $M_2 - M_1$ in terms of W_1 and W_2.

(b) The 1989 earthquake in California had a rating of 7.1 on the Richter scale. How many times larger were the seismic waves in the 1906 earthquake in San Francisco which measured 8.4 on the Richter scale? Give your answer to the nearest integer.

4.7 CONTINUOUS GROWTH AND THE NUMBER e

Exponential Functions with Base e Represent Continuous Growth

Any positive base b can be written as a power of e:

$$b = e^k.$$

To calculate k, use the fact that $k = \ln b$. If $b > 1$, then k is positive; if $0 < b < 1$, then k is negative. Then the function $Q = ab^t$ can be rewritten in terms of e:

$$Q = ab^t = a\left(e^k\right)^t = ae^{kt}.$$

The constant k is called the *continuous percent growth rate*. The implications of a continuous growth rate are explored on page 149.

In general:

For the exponential function $Q = ab^t$, the **continuous percent growth rate**, k, is given by solving $e^k = b$. Then

$$Q = ae^{kt}.$$

If a is positive,

- If $k > 0$, then Q is increasing.
- If $k < 0$, then Q is decreasing.

The value of the continuous growth rate, k, may be given as a decimal or a percent. If t is in years, for example, then the units of k are per year; if t is in minutes, then k is per minute.

Example 1 Find the continuous growth rates of each of the following three functions and graph the functions.

$$P = 5e^{0.2t}, \qquad Q = 5e^{0.3t}, \qquad \text{and} \quad R = 5e^{-0.2t}.$$

Solution The function $P = 5e^{0.2t}$ is growing at a continuous growth rate of 20%, and $Q = 5e^{0.3t}$ is growing at a continuous 30% rate. The function $R = 5e^{-0.2t}$ has a continuous growth rate of -20%. The negative sign in the exponent of e tells us that R is decreasing instead of increasing.

Because $a = 5$ in all three formulas, all three populations start at 5. The graphs are shown in Figure 4.39. Note that the graphs of these functions have the same shape as the exponential functions in Section 4.5. They are concave up and have horizontal asymptotes of $y = 0$. (Note that $R \to 0$ and $Q \to 0$ as $t \to -\infty$, whereas $P \to 0$ as $t \to \infty$.)

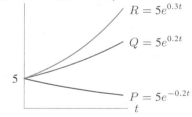

Figure 4.39: Exponential functions with different continuous growth rates

Example 2   A population increases from its initial level of 7.3 million at the continuous rate of 2.2% per year. When does the population reach 10 million?

Solution   We express the formula in base e since the continuous growth rate is given. If P is the population (in millions) in year t, then

$$P = 7.3e^{0.022t}.$$

Solving the equation $P = 10$ for t, we have

$$7.3e^{0.022t} = 10$$
$$e^{0.022t} = \frac{10}{7.3}.$$

Taking natural logs of both sides, we have

$$0.022t = \ln\left(\frac{10}{7.3}\right)$$
$$t = \frac{\ln(10/7.3)}{0.022} \approx 14.3.$$

Thus, it takes about 14.3 years for the population to reach 10 million.

Example 3   A substance decays at the continuous rate of 2.3% per minute. What is the half-life of the substance?

Solution   If Q is the quantity of the substance at time t, then $Q = Q_0 e^{-0.023t}$. Since Q_0 is the initial amount, $\frac{1}{2}Q_0$ is the amount remaining once half of the substance has decayed. Solving $Q = \frac{1}{2}Q_0$ for t gives

$$Q_0 e^{-0.023t} = \frac{1}{2}Q_0$$
$$e^{-0.023t} = \frac{1}{2}$$
$$-0.023t = \ln\left(\frac{1}{2}\right) \qquad \text{Taking ln of both sides}$$
$$t = \frac{\ln(1/2)}{-0.023} \approx 30.14,$$

and so the half-life is about 30.14 minutes, or a little more than half an hour.

Similarly, we can determine growth rate if given the half-life or doubling time.

Example 4   A population doubles in size every 20 years. What is its continuous growth rate?

Solution   We have $P = P_0 e^{kt}$. After 20 years, $P = 2P_0$, and so

$$P_0 e^{k \cdot 20} = 2P_0$$
$$e^{20k} = 2$$
$$20k = \ln 2 \qquad \text{Taking ln of both sides}$$
$$k = \frac{\ln 2}{20} \approx 0.0347.$$

Thus, the population grows at the continuous rate of 3.47% per year.

Converting Between $Q = ab^t$ and $Q = ae^{kt}$

Any exponential function can be written in either of the two forms:

$$Q = ab^t \qquad \text{or} \qquad Q = ae^{kt}.$$

If $b = e^k$, so $k = \ln b$, the two formulas represent the same function.

Example 5   Convert the exponential function $P = 175(1.145)^t$ to the form $P = ae^{kt}$.

Solution   Since the new formula represents the same function, we want $P = 175$ when $t = 0$. Thus, substituting $t = 0$, gives $175 = ae^{k(0)} = a$, so $a = 175$. The parameter a in both functions represents the initial population. For all t,

$$175(1.145)^t = 175(e^k)^t,$$

so we must find k such that

$$e^k = 1.145.$$

Therefore k is the power of e which gives 1.145. So, by the definition of ln, we have

$$k = \ln 1.145 \approx 0.1354.$$

Therefore, we have

$$P = 175e^{0.1354t}.$$

Example 6   Convert the formula $Q = 7e^{0.3t}$ to the form $Q = ab^t$.

Solution   Using the properties of exponents,

$$Q = 7e^{0.3t} = 7(e^{0.3})^t.$$

Using a calculator, we find $e^{0.3} \approx 1.3499$, so

$$Q = 7(1.3499)^t.$$

Example 7   Assuming t is in years, find the continuous and annual percent growth rates in Examples 5 and 6.

Solution   In Example 5, the annual percent growth rate is 14.5% and the continuous percent growth rate per year is 13.54%. In Example 6, the continuous percent growth rate is 30% and the annual percent growth rate is 34.99%.

Example 8   Find the continuous percent growth rate of $Q = 200(0.886)^t$, where t is in thousands of years.

Solution   Since this function describes exponential decay, we expect a negative value for k. We want

$$e^k = 0.886.$$

Solving for k gives

$$k = \ln(0.886) = -0.121.$$

So we have $Q = 200e^{-0.121t}$ and the continuous growth rate is -12.1% per thousand years.

Why k is Called the Continuous Growth Rate

Consider a \$1000 investment that earns 25% each year. The value of this investment in year t is given by $V = 1000(1.25)^t$. Because the value increases by 25% each year, the average rate of change of the value of the investment over an interval (in \$/year) is proportional to the value of the investment at the start of the interval:

Average rate of change of value (\$/year) = 25%/year · Initial value of investment.

For instance, during the first year,

Average rate of change of value = 25%/year · \$1000 = \$250/year.

In general:

For an exponential function $V = f(t)$, the average rate of change, $\Delta V / \Delta t$, is proportional to the value of V at the start of the interval:

$$\frac{\Delta V}{\Delta t} = k \cdot V.$$

The value of the constant of proportionality, k, depends on the length of the interval. The constant k is 25%/year for intervals of 1 year, but it is different for intervals of 1 month or 1 day. For instance, consider the average rate of change over the first quarter (three months) of the year. This interval runs from $t = 0$ to $t = 1/4$. The average rate of change is

$$\frac{\Delta V}{\Delta t} = \frac{1000(1.25)^{1/4} - 1000(1.25)^0}{1/4 - 0} = \frac{\$57.37}{1/4 \text{ year}} = \$229.49/\text{year}.$$

Notice that the average rate of change over the first three months (\$229.49/year) is less than the average rate of change over the entire year (\$250/year). This is because the investment grows by less during the first quarter than it does in the second, third, and fourth quarters.

To find the new constant of proportionality, k, for an interval of three months, we have

$$\frac{\Delta V}{\Delta t} = k \cdot V$$
$$\$229.49/\text{year} = k \cdot \$1000$$
$$k = \frac{\$229.49/\text{year}}{\$1000} = 0.22949/\text{year} = 22.949\%/\text{year}.$$

It turns out that the percentage growth rate is the same no matter which three-month period is used: it is 22.949% per year.

Table 4.19 shows the constant of proportionality, k, for other time intervals. For instance, during the first month of the year—that is, from time $t = 0$ to time $t = 1/12$—the investment grows from \$1000 to about \$1057.37. Therefore, $\Delta V/\Delta t = \$57.37/(1/12) = \$225.23/\text{year}$. For a one month period, $k = \$225.23/\$1000 = 0.22523 = 22.523\%/\text{year}$.

TABLE 4.19   *Constant of proportionality (% per year) for various time intervals*

Interval	Δt	V_{initial}	V_{final}	ΔV	$\Delta V/\Delta t$	k
year	1	1000	1250.0000	250.0000	\$250.00/year	25.000%/year
3 mo	1/4	1000	1057.3713	57.3713	\$229.49/year	22.949%/year
1 mo	1/12	1000	1018.7693	18.7693	\$225.23/year	22.523%/year
1 day	1/365	1000	1000.6115	0.6115	\$223.21/year	22.321%/year
1 hr	$1/(365 \cdot 24)$	1000	1000.0255	0.0255	\$223.15/year	22.315%/year
1 min	$1/(365 \cdot 24 \cdot 60)$	1000	1000.0004	0.0004	\$223.14/year	22.314%/year

Notice that as the interval gets shorter, from 1 month to 1 day to 1 hour to 1 minute, the value of k appears to level off to about 22.314%/year. This means that

$$\begin{matrix}\text{Average rate of change (\$/year)} \\ \text{over very short time interval}\end{matrix} \approx 22.314\%/\text{year} \times \begin{matrix}\text{Value at start} \\ \text{of interval}\end{matrix}.$$

This limiting value of k, namely 22.314%, is called the *continuous percent growth rate per year*. The connection between the continuous growth rate and the annual percent growth rate of 25% is provided by the number e:
$$e^{0.22314} = 1.25.$$

Thus, the formula $V = 1000(1.25)^t$ can be rewritten in the form $V = ae^{kt}$ as follows:

$$V = 1000(e^{0.22314})^t = 1000e^{0.22314t}.$$

In general, if k is the continuous growth rate, e^k gives the growth factor $b = 1 + r$, so $e^k = b$.

A more thorough explanation of continuous growth requires ideas from calculus. However, the important result is that, for the exponential function $V = ae^{kt}$,

$$\text{Instantaneous rate of change} = k \cdot \text{Value at that instant}.$$

The notion of *instantaneous rate of change* requires great care, but essentially the idea is to look at the rate of change of V on smaller and smaller intervals—much smaller than the intervals shown in Table 4.19. In effect, the idea is consider the rate of change of V at a particular instant in time.

Problems for Section 4.7

1. Rewrite $f(t) = 5(\frac{1}{2})^t$ as a function with base e. That is, if $f(t) = A_0 e^{-kt}$, find A_0 and k.

2. If $f(x) = 3(1.072)^x$ is rewritten as $f(x) = 3e^{kx}$, find k exactly.

3. The number of bacteria present in a culture after t hours is given by the formula $N = 1000e^{0.69t}$.

    (a) How many bacteria will there be after $1/2$ hour?
    (b) How long will it be before there are 1,000,000 bacteria?
    (c) What is the doubling time?

4. The US census projects the population of the state of Washington using the function $N(t) = 5.4e^{0.013t}$, where $N(t)$ is in millions and t is the number of years since 1995.

    (a) What is the population's continuous growth rate?
    (b) What is the population of Washington in year $t = 0$?
    (c) How many years will it be before the population triples?
    (d) In what year does this model indicate a population of only one person? Is this reasonable or unreasonable?

5. A population begins in year $t = 0$ with 25,000 people and grows at a continuous rate of 7.5% per year.

    (a) Find a formula for $P(t)$, the population in year t.
    (b) By what percent does the population actually increase each year? Explain why this is more than 7.5%.

6. A population grows from its initial level of 22,000 at a continuous growth rate of 7.1% per year.

    (a) Find a formula for $P(t)$, the population in year t.
    (b) By what percent does the population increase each year?

7. The voltage V across a charged capacitor is given by $V(t) = 5e^{-0.3t}$ where t is in seconds.

    (a) What is the voltage after 3 seconds?
    (b) When will the voltage be 1?
    (c) By what percent does the voltage decrease each second?

8. The population (in thousands) of a town in year t, where $t = 0$ represents 1990, is given by $P(t) = P_0 e^{kt}$. In 1995 the town's population was 18 thousand and in 1999 the town's population was 21 thousand.

    (a) Find the values of P_0 and k. Explain the significance of these quantities in terms of the town's population.
    (b) By what percent does the town's population increase during a given year?

9. A population grows exponentially. Let $P(t)$ be the population (in thousands) in year t, and suppose that $P(0) = 30$ and $P(10) = 45$.

    (a) If $P(t) = ab^t$, find a and b. What does b tell you about the population?
    (b) If $P(t) = P_0 e^{kt}$, find the approximate value of k. What does k tell you about the population?

10. A population increases exponentially. Let $P(t)$ be the population (in millions) in year t, and suppose $P(8) = 20$ and $P(15) = 28$.

    (a) Find a formula for $P(t)$ without using e.   (b) If $P(t) = ae^{kt}$, find k.

11. A population grows from 11000 to 13000 in three years. Assuming the growth is exponential, find the:

    (a) Annual growth rate   (b) Continuous annual growth rate
    (c) Why are your answers to parts (a) and (b) different?

12. A population doubles in size every 15 years. Assuming exponential growth, find the
    (a) Annual growth rate
    (b) Continuous growth rate

13. Oil leaks from a tank. At hour $t = 0$ there are 250 gallons of oil in the tank. Each hour after that, 4% of the oil leaks out.

    (a) What percent of the original 250 gallons has leaked out after 10 hours? Why is it less than $10 \cdot 4\% = 40\%$?
    (b) If $Q(t) = Q_0 e^{kt}$ is the quantity of oil remaining after t hours, find the value of k. What does k tell you about the leaking oil?

14. A population increases from 30,000 to 34,000 over a 5-year period at a constant annual percent growth rate.

    (a) By what percent did the population increase in total?
    (b) At what constant percent rate of growth did the population increase each year?
    (c) At what continuous annual growth rate did this population grow?

15. A population increases from 5.2 million at an annual rate of 3.1%. Find the continuous growth rate.

16. In 1991, the body of a man was found in melting snow in the Alps of Northern Italy. An examination of the tissue sample revealed that 46% of the carbon-14 present in his body at the time of his death had decayed. The half-life of carbon-14 is approximately 5,728 years. How long ago did this man die?

17. Radioactive carbon-14 decays according to the function $Q(t) = Q_0 e^{-0.000121t}$ where t is time in years, $Q(t)$ is the quantity remaining at time t, and Q_0 is the amount of carbon-14 present at time $t = 0$. Estimate the age of a skull if 23% of the original quantity of carbon-14 is still present.

18. What is the half-life of a radioactive substance that decays at a continuous rate of 11% per minute?

19. If 17% of a radioactive substance decays in 5 hours, what is the half-life of the substance?

20. In a country where inflation is a concern, prices have risen by 40% over a 5-year period.

    (a) By what percent do the prices rise each year?
    (b) How long does it take for prices to rise by 5%?
    (c) Use a continuous growth rate to model inflation by a function of the form $P = P_0 e^{rt}$.

21. A population begins at 5.2 million and grows at an annual rate of 3.1%. Will this population catch up to the population in Example 2 on page 148? If so, when?

22. Suppose 2 mg of a certain drug is injected into a person's bloodstream. As the drug is metabolized, the quantity diminishes at the continuous rate of 4% per hour.

    (a) Find a formula for $Q(t)$, the quantity of the drug remaining in the body after t hours.
    (b) By what percent does the drug level decrease during any given hour?
    (c) Suppose the person must receive an additional 2 mg of the drug whenever its level has diminished to 0.25 mg. When must the person receive the second injection?
    (d) When must the person receive the third injection?

23. In the early 1960s, radioactive strontium-90 was released during atmospheric testing of nuclear weapons and got into the bones of people alive at the time. If the half-life of strontium-90 is 29 years, what fraction of the strontium-90 absorbed in 1960 remained in people's bones in 1993?

24. A person's blood alcohol content (BAC) is a measure of how much alcohol is in the blood stream. When the person stops drinking, the BAC declines over time as the alcohol is metabolized. If Q is the amount of alcohol and Q_0 the initial amount, then $Q = Q_0 e^{-t/\tau}$, where τ is known as the *elimination time*. How long does it take for a person's BAC to drop from 0.10 to 0.04 if the elimination time is 2.5 hours?

25. The probability of a transistor failing within t months is given by $P(t) = 1 - e^{-0.016t}$,

    (a) What is the probability of failure within the first 6 months? Within the second six months?
    (b) Within how many months will the probability of failure be 99.99%?

26. (a) Use Table 4.19 on page 150 to find the average growth rate for a month. How does this compare to the average growth rate for the first quarter computed in the text? Explain.
    (b) Compare the average growth rates for smaller and smaller intervals. What trend do you observe?

4.8 COMPOUND INTEREST

Exponential Models of Investment

What is the difference between a bank account that pays 12% interest once per year and one that pays 1% interest every month? Imagine we deposit $1000 into the first account. Then, after 1 year, we have (assuming no other deposits or withdrawals)

$$\$1000(1.12) = \$1120.$$

But if we deposit $1000 into the second account, then after 1 year, or 12 months, we have

$$\$1000 \underbrace{(1.01)(1.01)\ldots(1.01)}_{\text{12 months of 1\% monthly interest}} = 1000(1.01)^{12} = \$1126.83.$$

Thus, we will earn $6.83 more in the second account than in the first. To see why this happens, notice that the 1% interest we earn in January will itself start earning interest at a rate of 1% per month. Similarly, the 1% interest we earn in February will start earning interest, and so will the interest earned in March, April, May, and so on. The extra $6.83 comes from interest earned on interest. This effect is known as *compounding*. We say that the first account earns 12% interest *compounded annually* and the second account earns 12% interest *compounded monthly*.

Nominal Versus Effective Rates

The expression 12% compounded monthly means that interest is added twelve times per year and that $12\%/12 = 1\%$ of the current balance is added each time. Banks refer to the 12% as the *annual percentage rate* or APR. We also call the 12% the *nominal rate* (nominal means "in name only"). When the interest is compounded more frequently than once a year, the account effectively earns more than the nominal rate. Thus, we distinguish between nominal rate and *effective annual yield*, or *effective rate*. The effective annual rate tells you how much interest the investment actually earns.

Example 1   What are the nominal and effective annual rates of an account paying 12% interest, compounded annually? Compounded monthly?

Solution   Since an account paying 12% annual interest, compounded annually, grows by exactly 12% in one year, we see that its nominal rate is the same as its effective rate: both are 12%.

The account paying 12% interest, compounded monthly, also has a nominal rate of 12%. On the other hand, since it pays 1% interest every month, after 12 months, its balance will increase by a factor of

$$\underbrace{(1.01)(1.01)\ldots(1.01)}_{\text{12 months of 1\% monthly growth}} = 1.01^{12} \approx 1.1268250.$$

Thus, effectively, the account earns 12.68% interest in a year.

Example 2   What is the effective annual rate of an account that pays interest at the nominal rate of 6% per year, compounded daily? Compounded hourly?

Solution   Since there are 365 days in a year, daily compounding pays interest at the rate of

$$\frac{6\%}{365} = 0.0164384\% \text{ per day.}$$

Thus, the daily growth factor is

$$1 + \frac{0.06}{365} = 1.000164384.$$

If at the beginning of the year the account balance is P, after 365 days the balance will be

$$\underbrace{P \cdot \left(1 + \frac{0.06}{365}\right)^{365}}_{\substack{\text{365 days of} \\ \text{0.0164384\% daily interest}}} = P \cdot (1.0618313).$$

Thus, this account earns interest at the effective annual rate of 6.18313%.

Notice that daily compounding results in a higher rate than yearly compounding (6.183% versus 6%), because with daily compounding the interest has the opportunity to earn interest.

If interest is compounded hourly, since there are 24×365 hours in a year, the balance at year's end would be

$$P \cdot \left(1 + \frac{0.06}{24 \cdot 365}\right)^{24 \cdot 365} = P \cdot (1.0618363).$$

The effective rate is now roughly 6.18363% instead of 6.18313% – that is, just slightly better than the rate of the account that compounds interest daily. The effective rate has increased with the frequency of compounding.

To summarize:

> If interest at an annual rate of r is compounded n times a year, then r/n times the current balance is added n times a year. Therefore, with an initial deposit of \$$P$, the balance t years later is
>
> $$B = P \cdot \left(1 + \frac{r}{n}\right)^{nt}.$$
>
> Note that r is the nominal rate; for example, $r = 0.05$ when the annual rate is 5%.

Continuous Compounding and the Number e

In Example 2 we calculated the effective rates for two accounts with a 6% nominal interest rate, but with different compounding periods. We see that the account with more frequent compounding earns a higher effective rate, though the increase is small.

This suggests compounding more and more frequently–say every minute or every second or even hundreds of times per second. However, there is a limit to how much more an account can earn by increasing the frequency of compounding.

TABLE 4.20   *Effect of increasing the frequency of compounding*

Compounding frequency	Annual growth factor	Effective annual rate
Annually	1.0600000	6%
Monthly	1.0616778	6.16778%
Daily	1.0618313	6.18313%
Hourly	1.0618363	6.18363%
$\vdots$	$\vdots$	$\vdots$
Continuously	$e^{0.06} \approx 1.0618365$	6.18365%

Table 4.20 shows several compounding periods with their annual growth factors and effective annual rates. As the compounding periods become shorter, the growth factor approaches e^k where

k is the nominal rate. Using the value

$$e = 2.718281828\cdots,$$

and a calculator, we check that

$$e^{0.06} \approx 1.0618365,$$

which is the final value for the annual growth factors in Table 4.20. If an account with a 6% nominal interest rate delivers this effective yield, we say that the interest has been *compounded continuously*.

Example 3   Find the effective annual rate if $1000 is deposited at 5% annual interest, compounded continuously.

Solution   The value of the deposit is given by

$$V = 1000e^{0.05t}.$$

To find the effective annual rate, we use the fact that $e^{0.05t} = (e^{0.05})^t$ to rewrite the function as

$$V = 1000(e^{0.05})^t.$$

Since $e^{0.05} = 1.05127$, we have

$$V = 1000(1.05127)^t.$$

This tells us that the effective annual rate is 5.127%.

To summarize:

> If interest on an initial deposit of P is *compounded continuously* at an annual rate r, the balance t years later can be calculated using the formula
>
> $$B = Pe^{rt}.$$
>
> Again, r is the nominal rate, and, for example, $r = 0.05$ when the annual rate is 5%.

It is important to realize that the formulas $V = 1000e^{0.05t}$ and $V = 1000(1.05127)^t$ both represent the *same* exponential function – they just describe it in different ways.[7]

Example 4   Which is better: An account that pays 8% annual interest compounded quarterly or an account that pays 7.95% annual interest compounded continuously?

Solution   The account that pays 8% interest compounded quarterly pays 2% interest 4 times a year. Thus, in one year the balance is

$$Q_0(1.02)^4 \approx Q_0(1.0824),$$

which means the effective annual rate is 8.24%.

The account that pays 7.95% interest compounded continuously has a year-end balance of

$$Q_0e^{0.0795} \approx Q_0(1.0827),$$

so, the effective annual rate is 8.27%. Thus, we see that 7.95% compounded continuously is better than 8% compounded quarterly.

[7]Actually, this isn't precisely true, because we rounded off when we found $b = 1.05127$. However, we can find b to as many digits as we want, and to this extent the two formulas are the same.

Problems for Section 4.8

1. Suppose $1000 is deposited into an account paying interest at a nominal rate of 8% per year. Find the balance 3 years later if the interest is compounded
   (a) Monthly   (b) Weekly   (c) Daily   (d) Continuously

2. A bank pays 6% annual interest, compounded daily. About how many days will it take for a deposit of $1000 to reach $1500?

3. For how many years must you invest $850 at an annual rate of 3.9%, compounded quarterly, in order to have $5,000 in the account?

4. You place $800 in an account that earns 4% annual interest, compounded annually. How long will it be until you have $2000?

5. One student deposits $500 into a savings account earning 4.5% annual interest compounded annually. Another deposits $800 into an account earning 3% annual interest compounded annually. When are the balances equal?

6. You deposit $4000 into an account that earns 6% annual interest, compounded annually. A friend deposits $3500 into an account that earns 5.95% annual interest, compounded continuously. Will your friend's balance ever equal yours? If so, when?

7. (a) Find the time required for an investment to triple in value if it earns 4% annual interest, compounded continuously.
   (b) Now find the time required assuming that the interest is compounded annually.

8. If you need $25,000 six years from now, what is the minimum amount of money you need to deposit into a bank account that pays 5% annual interest, compounded:

   (a) Annually   (b) Monthly   (c) Daily

   (d) Your answers get smaller as the number of times of compounding increases. Why is this so?

9. Suppose 1000 dollars earns 6% annual interest, compounded monthly.

   (a) Write a formula for the account balance, $V(t)$, as a function of t, the number of years elapsed.
   (b) What continuous annual interest rate would result in the same annual yield?

10. If the balance, M, at time t in years, of a bank account that compounds its interest payments monthly is given by
$$M = M_0(1.07763)^t.$$
   (a) What is the effective annual rate for this account?
   (b) What is the nominal annual rate?

11. Find the effective annual yield and the continuous growth rate if $Q = 5500\,e^{0.19\,t}$.

12. An investment grows by 3% per year for 10 years. By what percent does it increase over the 10-year period?

13. Suppose an account pays interest at a nominal rate of 8% per year. Find the effective annual yield if interest is compounded
    (a) Monthly   (b) Weekly   (c) Daily   (d) Continuously

14. An investment grows by 30% over a 5-year period. What is its effective annual percent growth rate?

15. An investment decreases by 60% over a 12-year period. At what effective annual percent rate does it decrease?

16. A sum of $850 is invested for 10 years and the interest is compounded quarterly. There is $1000 in the account at the end of 10 years. What is the nominal annual rate?

17. Three investments with different conditions are given in the following table.
    (a) Find the balance of each of the investments after the two-year period.
    (b) Rank them from best to worst in terms of rate of return. Explain your reasoning.

Investment A	Investment B	Investment C
$875 placed in an account giving 13.5% per year compounded daily for 2 years.	$1000 placed in an account giving 6.7% per year compounded continuously for 2 years.	$1050 placed in an account giving 4.5% per year compounded monthly for 2 years.

18. Rank the following three bank deposit options from best to worst.

Bank A	Bank B	Bank C
7% compounded daily	7.1% compounded monthly	7.05% compounded continuously

19. In the 1980s a northeastern bank experienced an unusual robbery. Each month an armored car delivered cash deposits from local branches to the main office, a trip requiring only one hour. One day, however, the delivery was six hours late. This delay turned out to be a scheme devised by an employee to defraud the bank. The armored car drivers had lent the money, a total of approximately $200,000,000, to arms merchants who then used it as collateral against the purchase of illegal weapons. The interest charged for this loan was 20% per year compounded continuously. How much was the fee for the six-hour period?

Problems 20–21 involve the Rule of 70, which gives quick estimates of the doubling time of an exponentially growing quantity. If $r\%$ is the annual growth rate of the quantity, then the Rule of 70 says

$$\text{Doubling time} \approx \frac{70}{r}.$$

20. Use the Rule of 70 by estimate how long it takes a $1000 investment to double if it grows at the following annual rates: 1%, 2%, 5%, 7%, 10%. Compare with the actual doubling times.

21. Using natural logs, solve for the doubling time for $Q = ae^{kt}$. Use your result to explain why the Rule of 70 works.

22. You want to borrow $25,000 to buy a Ford Explorer XL. The best available annual interest rate is 6.9%, compounded monthly. Determine how long it will take to pay off the loan if you can only afford monthly payments of $330. To do this, use the loan payment formula

$$P = \frac{Lr/12}{1 - (1 + (r/12))^{-m}},$$

where P is the monthly payment, L is the amount borrowed, r is the annual interest rate, and m is the number of months the loan is carried.

4.9 LOGARITHMIC SCALES

The Solar System and Beyond

Table 4.21 gives the distance from the sun to a number of different astronomical objects. Notice that the planet Mercury is 58,000,000 km from the sun, that earth is 149,000,000 km from the sun, and that Pluto is 5,900,000,000 km, or almost 6 billion kilometers from the sun. The table also gives the distance to Proxima Centauri, the star closest to the sun, and to the Andromeda Galaxy, the spiral galaxy closest to our own galaxy, the Milky Way.

TABLE 4.21   *Distance from the sun to various astronomical objects*

Object	Distance (million km)		
Mercury	58	Saturn	1426
Venus	108	Uranus	2869
Earth	149	Neptune	4495
Mars	228	Pluto	5900
Jupiter	778	Proxima Centauri	$4.1 \cdot 10^7$
		Andromeda Galaxy	$2.4 \cdot 10^{13}$

Linear Scales

We can represent the information in Table 4.21 graphically in order to get a better feel for the distances involved. Figure 4.40 shows the distance from the sun to the first five planets on a *linear scale*, which means that the evenly spaced units shown in the figure represent equal distances. In this case, each unit represents 100 million kilometers.

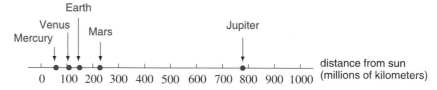

Figure 4.40: The distance from the sun of the first five planets (in millions of kilometers)

The drawback of Figure 4.40 is that the scale is too small to show all of the astronomical distances described by the table. For example, to show the distance to Pluto on this scale would require over six times as much space on the page. Even worse, assuming that each 100 million km unit on the scale measures half an inch on the printed page, we would need 3 miles of paper to show the distance to Proxima Centauri!

You might conclude that we could fix this problem by choosing a larger scale. In Figure 4.41 each unit on the scale is 1 billion kilometers. Notice that all five planets shown by Figure 4.40 are crowded into the first unit of Figure 4.41; even so, the distance to Pluto barely fits. The distances to the other objects certainly don't fit. For instance, to show the Andromeda Galaxy, Figure 4.41 would have to be almost 200,000 miles long. Choosing an even larger scale will not improve the situation.

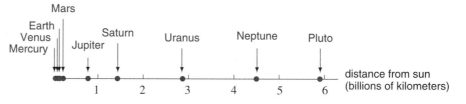

Figure 4.41: The distance to all nine planets (in billions of kilometers)

Logarithmic Scales

We conclude that the data in Table 4.21 cannot easily be represented on a linear scale. If the scale is too small, the more distant objects do not fit; if the scale is too large, the less distant objects are indistinguishable. The problem is not that the numbers are too big or too small; the problem is that the numbers vary too greatly in size.

We consider a different type of scale on which equal distances are not evenly spaced. All the objects from Table 4.21 are represented in Figure 4.42. The nine planets are still cramped, but it is possible to tell them apart. Each tick mark on the scale in Figure 4.42 represents a distance ten times larger than the one before it. This kind of scale is called *logarithmic*.

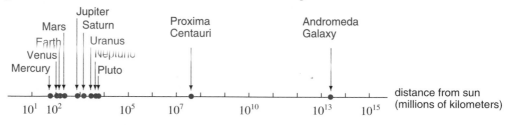

Figure 4.42: The distance from the sun (in millions of kilometers)

How Do We Plot Data on a Logarithmic Scale?

A logarithmic scale is marked with increasing powers of 10: 10^1, 10^2, 10^3, and so on. Notice that even though the distances in Figure 4.42 are not evenly spaced, the exponents are evenly spaced. Therefore the distances in Figure 4.42 are spaced according to their logarithms.

In order to plot the distance to Mercury, 58 million kilometers, we use the fact that

$$10 < 58 < 100,$$

so Mercury's distance is between 10^1 and 10^2, as shown in Figure 4.42. To plot Mercury's distance more precisely, calculate $\log 58 = 1.763$, so $10^{1.763} = 58$, and use 1.763 to represent Mercury's position. See Figure 4.43.

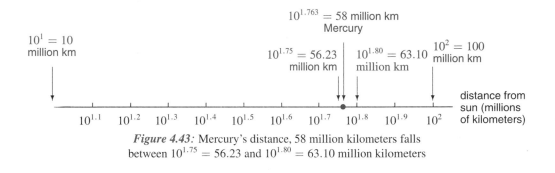

Figure 4.43: Mercury's distance, 58 million kilometers falls between $10^{1.75} = 56.23$ and $10^{1.80} = 63.10$ million kilometers

---

Example 1   Where should Saturn be on the logarithmic scale? What about the Andromeda Galaxy?

Solution   Saturn's distance is 1426 million kilometers, so we want the exponent of 10 that gives 1426, which is

$$\log 1426 \approx 3.154119526.$$

Thus $10^{3.154} \approx 1426$, so we use 3.154 to indicate Saturn's distance.

Similarly, the distance to the Andromeda Galaxy is $2.4 \cdot 10^{13}$ million kilometers, and since

$$\log(2.4 \cdot 10^{13}) \approx 13.38,$$

we use 13.38 to represent the galaxy's distance. See Figure 4.44.

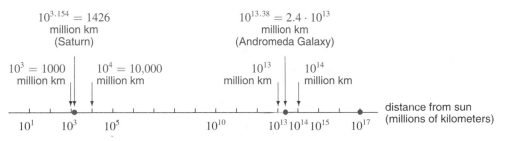

Figure 4.44: Saturn's distance is $10^{3.154}$ and the Andromeda Galaxy's distance is $10^{13.38}$

Logs of Small Numbers

The history of the world, like the distance to the stars and planets, involves numbers of vastly different sizes. Table 4.22 gives the ages of certain events[8] and the logarithms of their ages. The logarithms have been used to plot the events in Figure 4.45.

TABLE 4.22 *Ages of various events in earth's history and logarithms of the ages*

Event	Age (millions of years)	log (age)	Event	Age (millions of years)	log (age)
Man emerges	1	0	Rise of dinosaurs	245	2.39
Ape man fossils	5	0.70	Vertebrates appear	570	2.76
Rise of cats, dogs, pigs	37	1.57	First plants	2500	3.40
Demise of dinosaurs	67	1.83	Earth forms	4450	3.65

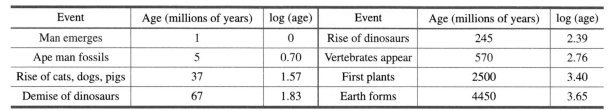

Figure 4.45: Logarithmic scale showing the ages of various events (in millions of years ago)

The events described by Table 4.22 all happened at least 1 million years ago. How do we indicate events which occurred less than 1 million years ago on the log scale?

Example 2 Where should the building of the pyramids be indicated on the log scale?

Solution The pyramids were built about 5000 years ago, or

$$\frac{5000}{1{,}000{,}000} = 0.005 \text{ million years ago.}$$

Notice that 0.005 is between 0.001 and 0.01, that is,

$$10^{-3} < 0.005 < 10^{-2}.$$

Since

$$\log 0.005 \approx -2.30,$$

we use -2.30 for the pyramids. See Figure 4.46.

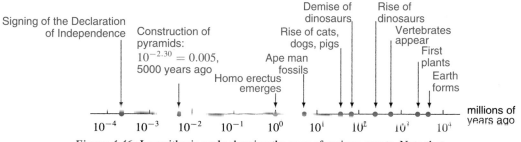

Figure 4.46: Logarithmic scale showing the ages of various events. Note that events that are less than 1 million years old are indicated by negative exponents

---

[8] *CRC Handbook, 75th ed.* 14-8.

Another Way to Label a Log Scale

In Figures 4.45 and 4.46, the log scale has been labeled so that exponents are evenly spaced. Another way to label a log scale is with the values themselves instead of the exponents. This has been done in Figure 4.47.

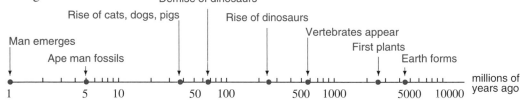

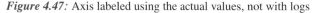

Figure 4.47: Axis labeled using the actual values, not with logs

Notice the characteristic way that the labels and tick marks "pile up" on each interval. The even spacing between exponents on log scales leads to uneven spacing in values. Although the values 10, 20, 30, 40, and 50 are evenly spaced, their corresponding exponents are not: $\log 10 = 1$, $\log 20 = 1.30$, $\log 30 = 1.48$, $\log 40 = 1.60$, and $\log 50 = 1.70$. Therefore, when we label an axis according to value on a scale that is spaced according to exponent, the labels get bunched up.

Log-Log Scales

Table 4.23 shows the average metabolic rate in kilocalories per day (kcal/day) for animals of different weights.[9] (A kilocalorie is the same as a standard nutritional calorie.) For instance, a 1-lb rat consumes about 35 kcal/day, whereas a 1750-lb horse consumes almost 9500 kcal/day.

TABLE 4.23   *The metabolic rate (in kcal/day) for animals of different weights*

Animal	Weight (lbs)	Rate (kcal/day)
Rat	1	35
Cat	8	166
Human	150	2000
Horse	1750	9470

It is not practical to plot these data on an ordinary set of axes. The values span too broad a range. However, we can plot the data using a log scales for both the horizontal (weight) axis and the vertical (rate) axes. See Figure 4.48. Figure 4.49 shows a close-up view of the data point for cats to make it easier to see how the labels work. Once again, notice the characteristic piling up of labels and gridlines. This happens for the same reason that it happened in Figure 4.47.

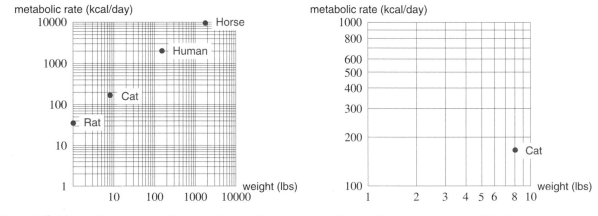

Figure 4.48: Metabolic rate (in kcal/hr) plotted against body weight   *Figure 4.49:* A close-up view of the Cat data point

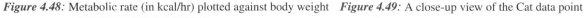

[9]*The New York Times*, January 11, 1999.

Problems for Section 4.9

1. The signing of the Declaration of Independence is marked on the log scale in Figure 4.46. To two decimal places, what is its position?

2. Use a calculator to fill in the following tables (round to 4 decimal digits).

n	1	2	3	4	5	6	7	8	9	10	20	30	40	50	60	70	80	90
$\log n$																		

n	100	200	300	400	500	600	700	800	900
$\log n$									

3. Using the results of Problem 2, plot the integer points 2 through 9 and the multiples of 10 from 20 to 90 on the log scaled axis shown in Figure 4.50.

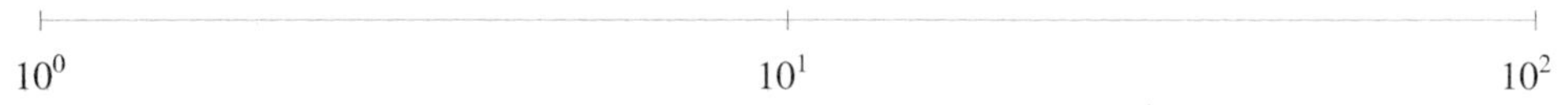

$$10^0 \qquad\qquad\qquad\qquad 10^1 \qquad\qquad\qquad\qquad 10^2$$

Figure 4.50

4. (a) Draw a line segment about 5 inches long. On it, choose an appropriate linear scale and mark points that represent the integral powers of two from zero to the sixth power. What is true about the location of the points as the exponents get larger?

   (b) Draw a second line segment. Repeat the process in (a) but this time use a logarithmic scale so that the units are now powers of ten. What do you notice about the location of these points?

5. Table 4.24 shows the typical body masses in kilograms for various animals.[10]

   (a) Find the log of the body mass of each animal to two decimal places.

   (b) Plot the body masses for each animal in Table 4.24 on a linear scale using A to identify the Blue Whale, B to identify the African Elephant, and so on down to L to identify the Hummingbird.

   (c) Plot and label the body masses for each animal in Table 4.24 on a logarithmic scale.

   (d) Which scale, (b) or (c), is more useful?

TABLE 4.24

Animal	Body mass (kg)	Animal	Body mass (kg)	Animal	Body mass (kg)
Blue Whale	91000	Black Rhinoceros	1170	Albatross	11
African Elephant	5450	Horse	700	Hawk	1
White Rhinoceros	3000	Lion	180	Robin	0.08
Hippopotamus	2520	Human	70	Hummingbird	0.003

6. Figure 4.51 shows the populations of eleven different localities, with the scale markings representing the logarithm of the population. Give the approximate populations of each locale.

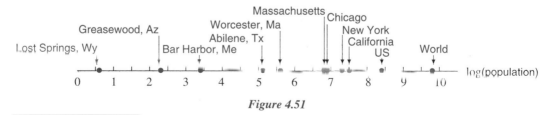

Figure 4.51

[10]R. McNiell Alexander, *Dynamics of Dinosaurs and Other Extinct Giants.* (New York: Columbia University Press, 1989) and H. Tennekes, *The Simple Science of Flight* (Cambridge: MIT Press, 1996).

7. The usual distances for track (running) events are 100 meters, 200 meters, 400 meters, 800 meters, 1500 meters, 3000 meters, 5000 meters, and 10,000 meters.

    (a) Plot the length of each track event on a linear scale.
    (b) Plot the length of each track event on a logarithmic scale.
    (c) Which scale, (a) or (b), is more useful to the runner?
    (d) On each figure identify the point corresponding to 50 meters.

8. Table 4.25 shows the dollar value of twelve items. Plot and label these values on a log scale.

    TABLE 4.25

Item	Dollar value	Item	Dollar value
Gum ball	0.25	Median family income	47,012
Big Mac	2.59	New house	125,000
Movie ticket	7.5	Lottery winnings	20 million
Plane ticket	500	Bill Gates' worth	40 billion
New computer	3000	National debt	5,200 billion
Year at private college	26,000	Gross domestic product	6,900 billion

9. Table 4.26 shows the sizes of various organisms. Plot and label these values on a log scale.

    TABLE 4.26

Animal	Size (cm)	Animal	Size (cm)
Virus	0.0000005	Domestic cat	60
Bacterium	0.0002	Wolf (with tail)	200
Human cell	0.002	Thresher shark	600
Ant	0.8	Giant squid (with tentacles)	2200
Hummingbird	12	Sequoia	7500

10. Figure 4.52 shows a graph of two data points from an ecological study of the relationship between the diameter, d, and height, h, of trees. This relationship is known to be of the form $h = kd^n$. (In practice, hundreds of data points might have been included instead of only two.) Notice that the graph has been drawn on a log-log scale.

    (a) It is traditional to show zero on axes scaled in the usual way. It is not possible to show zero on a log scaled axis. Why?
    (b) Find a formula for $h = f(d)$, the height as a function of diameter.

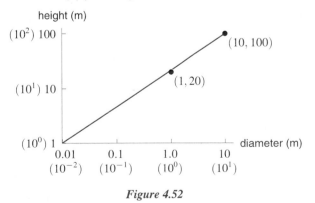

Figure 4.52

4.10 FITTING CURVES TO DATA

In Section 2.4 we used linear regression to find the equation for a line of best fit for a set of data. What if the data do not lie close to a straight line, but instead approximate the graph of some other function? In this section we see how logarithms help us fit data with an exponential function of the form $Q = a \cdot b^t$ or a power function of the form $Q = a \cdot t^p$.

Sales of Compact Discs

We look at data relating the fall in the sales of vinyl long-playing records (LPs) and the rise of compact discs (CDs). Table 4.27 gives the number of units (in millions) of CDs and LPs sold[11] for the years 1982 through 1993.

TABLE 4.27 *CD and LP sales*

t, years since 1982	c, CDs (millions)	l, LPs (millions)
0	0	244
1	0.8	210
2	5.8	205
3	23	167
4	53	125
5	107	107
6	150	72
7	207	35
8	287	12
9	333	4.8
10	408	2.3
11	495	1.2

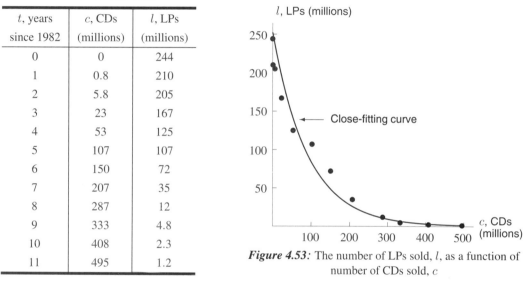

Figure 4.53: The number of LPs sold, l, as a function of number of CDs sold, c

From the table, we see that as CD sales rose dramatically during the 1980s and early 1990s, LP sales declined equally dramatically. Figure 4.53 shows the number of LPs sold in a given year as a function of the number of CDs sold that year.

Using a Log Scale to Linearize Data

In Section 4.9, we saw that a log scale allows us to compare values that vary over a wide range. Let's see what happens when we use a log scale to plot the data shown in Figure 4.53. Table 4.28 shows values $\log l$, where l is LP sales. These are plotted against c, CD sales, in Figure 4.54. Notice that plotting the data in this way tends to *linearize* the graph – that is, make it look more like a line. A line has been drawn in to emphasize the trend in the data.

Finding a Formula for the Curve

We say that the data in the third column of Table 4.28 have been *transformed*. A calculator or computer gives a regression line for the transformed data:[12]

$$y = 5.52 - 0.011c.$$

---

[11] Data from Recording Industry Association of America, Inc., 1998
[12] The values obtained by a computer or another calculator may vary slightly from the ones given.

TABLE 4.28   *Values of* $y = \ln l$ *and* c.

c, CDs	l, LPs	$y = \ln l$
0	244	5.50
0.8	210	5.35
5.8	205	5.32
23	167	5.12
53	125	4.83
102	107	4.67
150	72	4.28
207	35	3.56
287	12	2.48
333	4.8	1.57
408	2.3	0.83
495	1.2	0.18

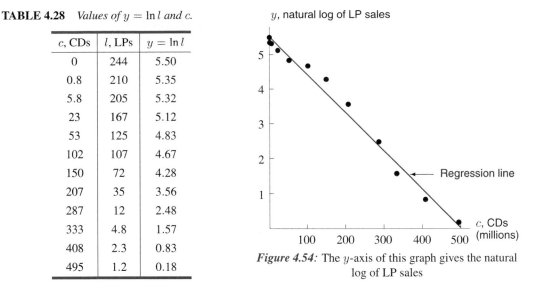

Figure 4.54: The y-axis of this graph gives the natural log of LP sales

Notice that this equation gives y in terms of c. To transform the equation back to our original variables, l and c, we substitute $\ln l$ for y, giving

$$\ln l = 5.52 - 0.011c.$$

We solve for l by raising e to both sides:

$$e^{\ln l} = e^{5.52 - 0.011c}$$
$$= (e^{5.52})(e^{-0.011c}) \qquad \text{Using an exponent rule.}$$

Since $e^{\ln l} = l$ and $e^{5.52} \approx 250$, we have

$$l = 250e^{-0.011c}.$$

This is the equation of the curve in Figure 4.53.

Fitting An Exponential Function To Data

In general, to fit an exponential formula, $N = ae^{kt}$, to a set of data of the form (t, N), we use three steps. First, we transform the data by taking the natural log of both sides and making the substitution $y = \ln N$. This leads to the equation

$$y = \ln N = \ln \left(ae^{kt}\right)$$
$$= \ln a + \ln e^{kt}$$
$$= \ln a + kt.$$

Setting $b = \ln a$ gives a linear equation with k as the slope and b as the y-intercept.

$$y = b + kt.$$

Secondly we can now use linear regression on the variables t and y. (Remember that $y = \ln N$.) Finally, as step three, we transform the linear regression equation back into our original variables by substituting $\ln N$ for y and solving for N.

The Spread of AIDS

The data in Table 4.29 give the total number of deaths in the US from AIDS since the start of 1981. Figure 4.55 suggests that a linear function may not give the best possible fit for these data.

Fitting an Exponential

We first fit an exponential function to the data[13] in Table 4.29

$$N = ae^{kt},$$

where N is the total number of deaths t years after 1980. We begin by linearizing the data, using logs. Table 4.30 shows the transformed data, where $y = \ln N$. Figure 4.56 shows that the transformed data lie close to a straight line.

TABLE 4.29   *Domestic deaths from AIDS, 1981–96*

t	N	t	N
1	159	9	90039
2	622	10	121577
3	2130	11	158193
4	5635	12	199287
5	12607	13	243923
6	24717	14	292586
7	41129	15	340957
8	62248	16	375904

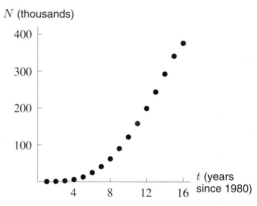

Figure 4.55: Domestic deaths from AIDS, 1981–96

TABLE 4.30

Transformed data from Table 4.29

t	$y = \ln N$	t	$y = \ln N$
1	5.069	9	11.408
2	6.433	10	11.708
3	7.664	11	11.972
4	8.637	12	12.203
5	9.442	13	12.405
6	10.115	14	12.587
7	10.624	15	12.740
8	11.039	16	12.837

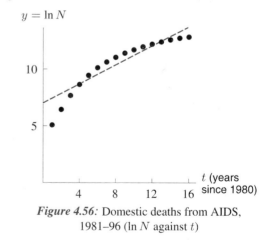

Figure 4.56: Domestic deaths from AIDS, 1981–96 ($\ln N$ against t)

We now use linear regression to estimate a line to fit the data points $(t, \ln N)$. The formula provided by a calculator (and rounded) is

$$y = 6.445 + 0.47t.$$

To find the formula for N in terms of t, we substitute $\ln N$ for y and solve for N:

$$\ln N = 6.445 + 0.47t$$
$$N = e^{6.445+0.47t}$$
$$= \left(e^{6.445}\right)\left(e^{0.47t}\right),$$

---

[13]*HIV/AIDS Surveillance Report*, Year-end Edition, Vol 9, No 2, Table 13, US Department of Health and Human Services, Centers for Disease Control and Prevention, Atlanta. Data does not include 450 people whose dates of death are unknown.

and since $e^{6.445} \approx 630$, we have
$$N \approx 630e^{0.47t}.$$

Figure 4.58 on page 168 shows how the graph of this formula fits the data points.

Fitting a Power Function

Now we fit the AIDS data with a power function of the form
$$N = at^p,$$

where a and p are constants. Some scientists have suggested that a power function may be a better model for the growth of AIDS than an exponential function.[14] We take the log of both sides and make the substitution $y = \ln N$ to obtain

$$\begin{aligned} y = \ln N = \ln(at^p) \\ = \ln a + \ln t^p \\ = \ln a + p \ln t. \end{aligned}$$

As before, we make the substitution $b = \ln a$, so the equation becomes
$$y = b + p \ln t.$$

Notice that this substitution does not result in y being a linear function of t. However, if we make a second substitution, $x = \ln t$, we have
$$y = b + px.$$

So y *is* a linear function of x.

We can now use linear regression to find a formula for y in terms of x. From there we can determine a formula for N in terms of t. Let's carry out this procedure for the AIDS data. We first transform the data in Table 4.29, using the substitutions $x = \ln t$ and $y = \ln N$. The result is in Table 4.31.

TABLE 4.31

$x = \ln t$	$y = \ln N$	$x = \ln t$	$y = \ln N$
0	5.069	2.20	11.408
0.69	6.433	2.30	11.708
1.10	7.664	2.40	11.972
1.39	8.637	2.48	12.203
1.61	9.442	2.56	12.405
1.79	10.115	2.64	12.587
1.95	10.624	2.71	12.740
2.08	11.039	2.77	12.837

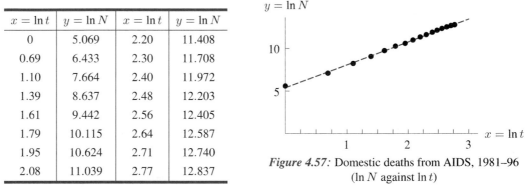

Figure 4.57: Domestic deaths from AIDS, 1981–96 ($\ln N$ against $\ln t$)

The transformed data from Table 4.31 are plotted in Figure 4.57. The points show a more linear pattern than the points in Figure 4.56. Using regression to fit a line to these data gives
$$y = 4.670 + 3.005x.$$

---

[14] "Risk behavior-based model of the cubic growth of acquired immunodeficiency syndrome in the United States", by Stirling A. Colgate, E. Ann Stanley, James M. Hyman, Scott P. Layne, and Alifford Qualls, in *Proc. Natl. Acad. Sci. USA*, Vol 86, June 1989, Population Biology.

Now transform the equation back to the original variables, t and N. Since our original substitutions were $y = \ln N$ and $x = \ln t$, this equation becomes

$$\ln N = 4.670 + 3.005 \ln t.$$

Raise e to the power of both sides

$$e^{\ln N} = e^{4.670+3.005 \ln t}$$

$$N = (e^{4.670})(e^{3.005 \ln t}).$$

Since $e^{4.670} \approx 107$ and $e^{3.005 \ln t} = \left(e^{\ln t}\right)^{3.005} = t^{3.005}$, we have

$$N \approx 107t^{3.005},$$

which is the formula of a power function. Figure 4.58 shows the graph of this formula with the data.

Which Function Best Fits the Data?

Both the exponential function

$$N = 630e^{0.47t}$$

and the power function

$$N = 107t^{3.005}$$

fit the AIDS data reasonably well. By visual inspection alone, the power function arguably provides the better fit. If we fit a linear function to the original data we get

$$N = -97311 + 25946t.$$

Even this linear function gives a possible fit for $t \geq 4$, that is, for 1984 to 1996. (See Figure 4.58.)

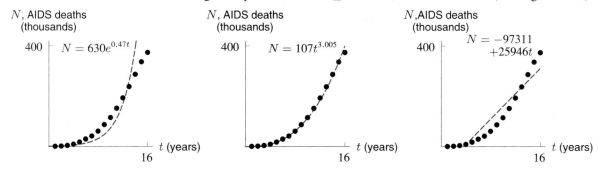

Figure 4.58: The AIDS data since 1980 together with an exponential model, a power-function model, and a linear model

Despite the fact that all three functions fit the data reasonably well up to 1996, it's important to realize that they give wildly different predictions for the future. If we use each model to estimate the total number of AIDS deaths by the year 2010 (when $t = 30$), the exponential model gives

$$N = 630e^{(0.47)(30)} \approx 837{,}322{,}467, \quad \text{over triple the current US population;}$$

the power model gives

$$N = 107(30)^{3.005} \approx 2{,}938{,}550, \quad \text{or about 1\% of the current population;}$$

and the linear model gives

$$N = -97311 + 25946 \cdot 30 \approx 681{,}069, \quad \text{or about 0.25\% of the current population.}$$

Which function is the best predictor for the future? This is an extremely hard question and one which cannot be answered by mathematics alone. An understanding of the processes leading to the data, in this case the nature of the disease and the behavior of the US population, is necessary in order to tackle such a question.

Problems for Section 4.10

1. (a) Complete the Table 4.32 with values of $y = 3^x$.
    (b) Complete Table 4.33 with values for $y = \log(3^x)$. What kind of function is this?
    (c) Complete tables for $f(x) = 2 \cdot 5^x$ and $g(x) = \log(2 \cdot 5^x)$. What kinds of functions are these?
    (d) What seems to be true about a function which is the logarithm of an exponential function? Is this true in general?

TABLE 4.32

x	0	1	2	3	4	5
$y = 3^x$						

TABLE 4.33

x	0	1	2	3	4	5
$y = \log(3^x)$						

2. Repeat part (b) and (c) of Problem 1 using the natural log function. Is your answer to part (d) the same?

3. (a) Plot the data in Table 4.34.
    (b) What kind of function might the data from part (a) represent?
    (c) Now plot $\log y$ versus x instead of y versus x. What do you notice?

TABLE 4.34

x	0.2	1.3	2.1	2.8	3.4	4.5
y	5.7	12.3	21.4	34.8	52.8	113.1

For the tables in Problem 4–6,

(a) Use linear regression to find a linear function $y = b + mx$ that fits the data. Record the correlation coefficient.

(b) Use linear regression on the values x and $\ln y$ to fit a function of the form $\ln y = b + mx$. Record the correlation coefficient. Convert to an exponential function $y = ae^{kx}$.

(c) Use linear regression on the values $\ln x$ and $\ln y$ to fit a function of the form $\ln y = b + m \ln x$. Record the correlation coefficient. Convert to a power function $y = ax^p$.

(d) Compare the correlation coefficients. Sketch of a graph of the data and the three functions to assess which function fits best.

4.

x	y
30	70
85	120
122	145
157	175
255	250
312	300

5.

x	y
8	23
17	150
23	496
26	860
32	2720
37	8051

6.

x	y
3.2	35
4.7	100
5.1	100
5.5	150
6.8	200
7.6	300

7. Table 4.35 shows the value, y, of US imports from China with x in years since 1985.

    (a) Find a formula for a linear function $y = b + mx$ that approximates the data.
    (b) Find $\ln y$ for each y value, and use the x and $\ln y$ values to find a formula for a linear function $\ln y = b + mx$ that approximates the data.
    (c) Use the equation in part (b) to find an exponential function of the form $y = ae^{kx}$ that fits the data.

TABLE 4.35 *Value of US imports from China in millions of dollars*

Year	1985	1986	1987	1988	1989	1990	1991
x, years since 1985	0	1	2	3	4	5	6
y, value of imports	3862	4771	6293	8511	11,990	15,237	18,976

8. Table 4.36 shows the newspapers' share of the national expenditure on advertising. Using the method of Problem 7, fit an exponential function of the form $y = ae^{kx}$ to the data, where y is percent share and x is the number of years since 1950.

TABLE 4.36 *Newspapers' share of all spending by national advertisers*

Year	1950	1960	1970	1980	1990	1992
x, years since 1950	0	10	20	30	40	42
y, percent share	16.0	10.8	8.0	6.7	5.8	5.0

9. In this problem, we will determine whether or not the compact disc data from Table 4.27 on page 164 can be well modeled using a power function of the form $l = kc^p$, where l and c give the number of LPs and CDs (in millions) respectively, and where k and p are constant.

(a) Based on the plot of the data given in Figure 4.53, what do you expect to be true about the sign of the power p?
(b) Let $y = \ln l$ and $x = \ln c$. Find a linear formula for y in terms of x by making substitutions in the equation $l = kc^p$.
(c) Transform the data in Table 4.27 to create a table comparing $x = \ln c$ and $y = \ln l$. What data point must necessarily be omitted?
(d) Plot your transformed data from part (c). Based on your plot, is there a power function that gives a good fit to the data from Table 4.27? If so, find its formula; if not, explain why not.

10. An analog radio dial can be measured in millimeters from left to right. Although the scale of the dial can be different from radio to radio, Table 4.37 gives typical measurements.

(a) Which radio band data appear linear? Graph and connect the data points for each band.
(b) Which radio band data appear exponential?
(c) Find a possible formula for the FM station number in terms of x.
(d) Find a possible formula for the AM station number in terms of x.

TABLE 4.37

x, millimeters	5	15	25	35	45	55
FM (mhz)	88	92	96	100	104	108
AM (khz/10)	53	65	80	100	130	160

11. To study how recognition memory decreases with time, the following experiment was conducted. The subject read a list of 20 words slowly aloud, and later, at different time intervals, was shown a list of 40 words containing the 20 words that he or she had read. The percentage, P, of words recognized was recorded as a function of t, the time elapsed in minutes. Table 4.38 shows the averages for 5 different subjects.[15] This is modeled by $P = a \ln t + b$.

(a) Estimate a and b.
(b) Graph the data points and regression line on a coordinate system of P against $\ln t$.
(c) When does this model predict that the subjects will recognize no words? All words?
(d) Graph the data points and curve $P = a \ln t + b$ on a coordinate system with P against t, with $0 \le t \le 10{,}500$.

TABLE 4.38

t, min	5	15	30	60	120	240	480	720	1440	2880	5760	10,080
$P\%$	73.0	61.7	58.3	55.7	50.3	46.7	40.3	38.3	29.0	24.0	18.7	10.3

---

[15] Adapted from D. Lewis, *Quantitative Methods in Psychology.* (New York: McGraw-Hill, 1960).

12. On a piece of paper draw a straight line through the origin with slope 2 in the first quadrant. What is the equation of the function represented by your straight line in each of the following cases?

    (a) The horizontal axis is labeled x and the vertical axis is labeled y.
    (b) The horizontal axis is labeled $\ln x$ and the vertical axis is labeled $\ln y$.
    (c) The horizontal axis is labeled x and the vertical axis is labeled $\ln y$.

13. In this problem you will fit a quartic polynomial to the AIDS data.

    (a) With N as the total number of AIDS deaths t years after 1980, use a calculator or computer to fit the data in Table 4.29 on page 166 with a polynomial of the form

    $$N = at^4 + bt^3 + ct^2 + dt + e.$$

    (b) Graph the data and your quartic for $0 \leq t \leq 16$. Comment on the fit.
    (c) Graph the data and your quartic for $0 \leq t \leq 30$. Comment on the predictions made by this model.

14. (a) Find a linear function that fits the data in Table 4.39. How good is the fit?
    (b) The data in the table was generated using the power function $y = 5x^3$. Explain why (in this case) a linear function gives such a good fit to a power function. Does the fit remain good for other values of x?

TABLE 4.39

x	2.00	2.01	2.02	2.03	2.04	2.05
y	40.000	40.603	41.212	41.827	42.448	43.076

REVIEW PROBLEMS FOR CHAPTER FOUR

1. The balance B (in \$) of an investment after t years is given by the formula $B = 5000(1.12)^t$.

   (a) What is the investment's value after 5 years? 10 years?
   (b) When will the investment be worth \$10,000? \$20,000?

2. Accion is a non-profit microlending organization which makes small loans to entrepeneurs who do not qualify for bank loans.[16] A New York woman who sells clothes from a cart has the choice of a \$1000 loan from Accion to be repaid by \$1160 a year later and a \$1000 loan from a loan shark with an annual interest rate of 22%, compounded annually.

   (a) What is the annual interest rate charged by Accion?
   (b) To pay off the loan shark for a year's loan of \$1000, how much would the woman have to pay?
   (c) Which loan is a better deal for the woman? Why?

3. Two reductions are available on a copy machine: 70% and 85%. A page can be further reduced by copying an already reduced copy.

   (a) If a 70% reduction is followed by an 85% reduction, what is the overall percent reduction?
   (b) Write a formula for the percent reduction if a page is copied n times in succession at 70%.
   (c) Estimate the number of times a page has to be copied at 70% reduction before it is less than 10% of its original size.

---

[16]From *Hemispheres*, December 1998 (United Airlines).

4. Without a calculator, match each of the following formulas to one of the graphs in Figure 4.59.

(a) $y = 0.8^t$ (b) $y = 5(3)^t$ (c) $y = -6(1.03)^t$

(d) $y = 15(3)^{-t}$ (e) $y = -4(0.98)^t$ (f) $y = 82(0.8)^{-t}$

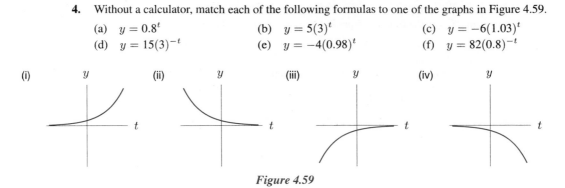

Figure 4.59

5. Without a calculator, match each of the following formulas to one of the graphs in Figure 4.59.

(a) $y = 8.3e^{-t}$ (b) $y = 2.5e^t$ (c) $y = -4e^{-t}$

6. Find a formula for $f(x)$, an exponential function such that $f(2) = 1/27$ and $f(-1) = 27$.

7. Find the equation of an exponential curve through the points $(-1, 2)$, $(1, 0.3)$.

For Problems 8–10, find a possible formula for each exponential function graphed.

8.   **9.**   **10.**

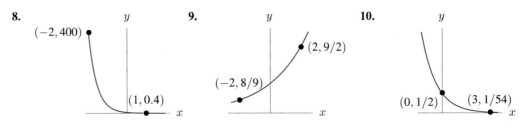

In Problems 11–13, find an exact solution for t if possible.

11. $16.3(1.072)^t = 18.5$ **12.** $13e^{0.081t} = 25e^{0.032t}$ **13.** $87e^{0.066t} = 3t + 7$

In Problems 14–16, simplify fully.

14. $\log\left(100^{x+1}\right)$ **15.** $\ln\left(e \cdot e^{2+M}\right)$ **16.** $\ln(A + B) - \ln(A^{-1} + B^{-1})$

17. One of the functions in Table 4.40 is linear and the other is exponential. Find formulas for these functions.

TABLE 4.40

x	0.21	0.37	0.41	0.62	0.68
$f(x)$	0.03193	0.04681	0.05053	0.07006	0.07564
$g(x)$	3.324896	3.423316	3.448373	3.582963	3.622373

18. Decide whether the following functions are approximately linear, approximately exponential, or neither. For those that are nearly linear or nearly exponential, find a possible formula.

(a)

t	3	10	14
$Q(t)$	7.51	8.7	9.39

(b)

t	5	9	15
$R(t)$	2.32	2.61	3.12

(c)

t	5	12	16
$S(t)$	4.35	6.72	10.02

19. If $f(x) = 12 + 20x$ and $g(x) = \frac{1}{2} \cdot 3^x$, for what values of x is $g(x) < f(x)$?

20. In 1980, the population of a town was 18,500 and it grew by 250 people by the end of the year. By 1990, its population had reached 22,500.

    (a) Can this town be best described by a linear or an exponential model, or neither? Explain.

    (b) If possible, find a formula for $P(t)$, this town's population t years after 1980.

21. In 1980, the population of a town was 20,000, and it grew by 4.14% that year. By 1990, the town's population had reached 30,000.

    (a) Can this town be best described by a linear or an exponential model, or neither? Explain.

    (b) If possible, find a formula for $P(t)$, this town's population t years after 1980.

22. Figure 4.60 gives the voltage, $V(t)$, across a circuit element at time t seconds. For $t < 0$, the voltage is a constant 80 volts; for $t \geq 0$, the voltage decays exponentially.

    (a) Find a piecewise formula for $V(t)$.

    (b) At what value of t will the voltage reach 0.1?

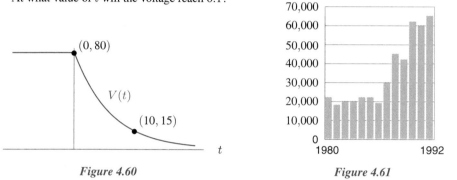

Figure 4.60   *Figure 4.61*

23. Hong Kong shifted from British to Chinese rule in 1997. Figure 4.61 shows[17] the number of people who emigrated from Hong Kong during each of the years from 1980 to 1992.

    (a) Find an exponential function that approximates the data.

    (b) What does the model predict about the number of emigrants in 1996?

    (c) Write a short paragraph explaining why this model is or is not useful to predict emigration in the year 2000.

24. The annual inflation rate, r, for a five-year period is given in Table 4.41.

    (a) By what total percent did prices rise between the start of 1980 and the end of 1984?

    (b) What is the average annual inflation rate for this time period?

    (c) At the beginning of 1980, a shower curtain costs $20. Make a prediction for the good's cost at the beginning of 1990, using the average inflation rate found in part (b).

TABLE 4.41

t	1980	1981	1982	1983	1984
r	5.1%	6.2%	3.1%	4.7%	3.3%

TABLE 4.42

t	1990	1991	1992	1993	1994
Growth	27%	36%	19%	44%	57%

25. A high-risk investment reports the annual yields in Table 4.42. What is the average annual yield of this investment over the five-year period shown?

26. Suppose that $u = \log 2$ and $v = \log 5$.

    (a) Find possible formulas for the following expressions in terms of u and/or v. Your answers should not involve logs.

        (i) $\log(0.4)$ (ii) $\log 0.25$ (iii) $\log 40$ (iv) $\log \sqrt{10}$

    (b) Justify the statement: $\log(7) \approx \frac{1}{2}(u + 2v)$.

---

[17]Adapted from the *New York Times* July 5, 1995.

27. Solve the following equations. Give approximate solutions if exact ones can't be found.

(a) $e^{x+3} = 8$ (b) $4(1.12^x) = 5$ (c) $e^{-0.13x} = 4$

(d) $\log(x - 5) = 2$ (e) $2\ln(3x) + 5 = 8$ (f) $\ln x - \ln(x - 1) = 1/2$

(g) $e^x = 3x + 5$ (h) $3^x = x^3$ (i) $\ln x = -x^2$

28. Solve for x exactly.

(a) $\dfrac{3^x}{5^{(x-1)}} = 2^{(x-1)}$ (b) $-3 + e^{x+1} = 2 + e^{x-2}$

(c) $\ln(2x - 2) - \ln(x - 1) = \ln x$ (d) $9^x - 7 \cdot 3^x = -6$

(e) $\ln\left(\dfrac{e^{4x} + 3}{e}\right) = 1$ (f) $\dfrac{\ln(8x) - 2\ln(2x)}{\ln x} = 1$

29. The population of Botswana[18] from 1975 to 1990 is shown in Table 4.43.

(a) Fit an exponential growth model, $P = ab^t$, to this data set, where P is the population in millions and t measures the years since 1975 in 5-year intervals — so $t = 1$ corresponds to 1980. Estimate the values of a and b. Plot the data set and $P = ab^t$ on the same graph.

(b) Starting from 1975, how long does it take for the population of Botswana to double? When is the population of Botswana projected to exceed 214 million, the 1975 population of the US?

TABLE 4.43

Year	1975	1980	1985	1990
Population (millions)	0.755	0.901	1.078	1.285

30. The following is excerpted from an article that appeared in the January 8, 1990 *Boston Globe*.

> Men lose roughly 2 percent of their bone mass per year in the same type of loss that can severely affect women after menopause, a study indicates. "There is a problem with osteoporosis in men that hasn't been appreciated. It's a problem that needs to be recognized and addressed," said Dr. Eric Orwoll, who led the study by the Oregon Health Sciences University. The bone loss was detected at all ages and the 2 percent rate did not appear to vary, Orwoll said.

(a) Assume that the average man starts losing bone mass at age 30. Let M_0 be the average man's bone mass at this age. Express the amount of remaining bone mass as a function of the man's age, a.

(b) At what age will the average man have lost half his bone mass?

31. The following populations, $P(t)$, are given in millions in year t. Describe the growth of each population in words. Give both the percent annual growth rate and the continuous growth rate per year.

(a) $P(t) = 51(1.03)^t$ (b) $P(t) = 15e^{0.03t}$ (c) $P(t) = 7.5(0.94)^t$

(d) $P(t) = 16e^{-0.051t}$ (e) $P(t) = 25(2)^{t/18}$ (f) $P(t) = 10(\frac{1}{2})^{t/25}$

32. Suppose \$300 was deposited into one of five bank accounts. Each of the following formulas gives the balance of an account in dollars, as a function of the time, t, in years. Following the formulas are verbal descriptions of five situations. For each situation, state which formula(s) could represent it.

(a) $B = 300(1.2)^t$ (b) $B = 300(1.12)^t$ (c) $B = 300(1.06)^{2t}$

(d) $B = 300(1.06)^{t/2}$ (e) $B = 300(1.03)^{4t}$

(i) This investment earned 12% annually, compounded annually.

(ii) This investment earned, on average, more than 1% each month.

(iii) This investment earned 12% annually, compounded semi-annually.

(iv) This investment earned, on average, less than 3% each quarter.

(v) This investment earned, on average, more than 6% every 6 months.

---

[18]*World Population Growth and Aging* by N. Keyfitz, University of Chicago Press, 1990.

33. Rewrite the following formulas as indicated.

   (a) If $f(x) = 5(1.121)^x$, find a and k such that $f(x) = ae^{kx}$.
   (b) If $g(x) = 17e^{0.094x}$, find a and b such that $f(x) = ab^x$.
   (c) If $h(x) = 22(2)^{x/15}$, find a and b such that $h(x) = ab^x$, and a and k such that $h(x) = ae^{kx}$.

34. (a) Let $B = 5000(1.06)^t$ give the balance of a bank account after t years. If the formula for B is written $B = 5000e^{kt}$, estimate the value of k correct to four decimal places. What is the financial meaning of k?

   (b) The balance of a bank account after t years is given by the formula $B = 7500e^{0.072t}$. If the formula for B is written $B = 7500b^t$, find b exactly, and give the value of b correct to four decimal places. What is the financial meaning of b?

35. Forty percent of a radioactive substance decays in five years. By what percent does the substance decay each year?

36. Find the annual growth rates of a quantity which:

   (a) Doubles in size every 7 years   (b) Triples in size every 11 years
   (c) Grows by 3% per month   (d) Grows by 18% every 5 months

37. The Richter scale is a measure of the ground motion that occurs during an earthquake. The intensity, R, of an earthquake as measured on the Richter scale is given by

$$R = \log\left(\frac{a}{T}\right) + B$$

where a is the amplitude (in microns) of vertical ground motion, T is the period (in seconds) of the seismic wave, and B is a constant. Let $B = 4.250$ and $T = 2.5$. Find a if

   (a) $R = 6.1$ (b) $R = 7.1$

   (a) Compare the values of R in parts (a) and (b). How do the corresponding values of a compare?

38. If y and x are given by the following graphs, find equations for y in terms of x.

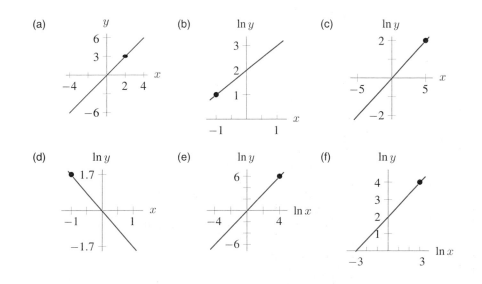

39. It is a well-documented fact that the earning power of men is higher than that of women.[19] Table 4.44 gives the median income of year-round full-time workers in the US. Salaries are given in thousands of dollars.

(a) Plot the points with domain 1980 to 1990 and connect the points.
(b) Let t be the year. Construct two functions of the form $W(t) = ae^{b(t-1980)}$, one each for the men's and women's earning power data.
(c) Graph the two functions on the interval 1980 to 2000.
(d) Graph the two functions on the interval 1980 to 2030.
(e) Does the graph in part (d) predict women's salaries will catch up with mens?
(f) Do you think it likely that the median salaries of women will ultimately be higher than mens?

TABLE 4.44

Year	1980	1985	1988	1989	1990
Female	11.591	16.252	18.545	19.643	20.586
Male	19.173	24.999	27.342	28.605	29.172

40. Table 4.45 shows the US minimum wage[20] from 1950 to 1997. In this problem you will fit the data with an exponential function of the form $f(t) = ae^{b(t-1950)}$.

(a) Substitute $t = 1950$ to find the coefficient a.
(b) By trial and error, find an acceptable b.
(c) What does this function predict for the minimum wage in the year 2000? In 2010?
(d) Sketch a graph of the data as a step function. Why is this realistic?

TABLE 4.45

Year, t	1950	1956	1961	1963	1967	1968	1974	1975	1976
Minimum wage, $	0.75	1.00	1.15	1.25	1.40	1.60	2.00	2.10	2.30
Year, t	1978	1979	1980	1981	1990	1991	1996	1997	
Minimum wage, $	2.65	2.90	3.10	3.35	3.80	4.25	4.75	5.15	

41. There have been frequent debates concerning the inequity of health care for children in the US. Figure 4.62 shows infant mortality rate, or number of deaths in the first year of life for every 1000 live births, for African-American and Caucasian children.[21]

(a) Use average rate of change to decide whether the infant mortality rate declined faster for African-American or Caucasian infants from 1950 to 1992.
(b) Compare the ratio of African-American to Caucasian infant mortality in 1950 to that in 1992. What does this ratio suggest about which mortality rate declined faster?
(c) Construct two tables for the data in Figure 4.62, one for African-American infants and one for Caucasian infants (use 5-year intervals).
(d) Using the tables constructed in part (c), determine if either Caucasian or African-American infant mortality rate declined exponentially.

---

[19] Source: The American Almanac, 1992–1993, Table 710
[20] Data from R. Famighetti, ed. *The World Almanac and Book of Facts: 1999*. (New Jersey: Funk and Wagnalls, 1998).
[21] *New York Times*. July 10, 1995, B9.

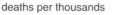

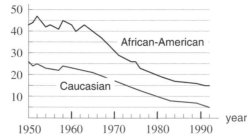

Figure 4.62: Infant mortality rates

42. A computer virus is a program that can damage a computer's operating system. According to one expert, in March 1991 there were just over 700 known computer viruses. By June 1993, there were more than 2000, with about 110 new viruses being discovered each month.

    (a) What was the monthly rate of change of the number of computer viruses since June 1993? Assuming that the rate of change remains constant, how many viruses are predicted for the year 2000?

    (b) What was the monthly percent increase of the number of computer viruses since June 1993? Assuming that this percent increase remains constant, how many viruses are predicted for the year 2000?

    (c) Define $N = f(t)$ as the number of computer viruses as a function of t, the number of months since March 1991. Show that $N = f(t)$ is not a linear function.

    (d) Show that $N = f(t)$ is not an exponential function either.

43. According to a letter to the *New York Times* on April 10, 1993, "... the probability of [a driver's] involvement in a single-car accident increases exponentially with increasing levels of blood alcohol." The letter goes on to state that when a driver's blood-alcohol content (BAC) is 0.15, the risk of such an accident is about 25 times greater than for a nondrinker.

    (a) Let p_0 be a nondrinker's probability of being involved in a single-car accident. Let $f(x)$ be the probability of an accident for a driver whose blood alcohol level is x. Find a formula for $f(x)$. (This only makes sense for some values of x.)

    (b) In many states, the legal definition of intoxication is having a BAC of 0.1 or higher. According to your formula for $f(x)$, how many times more likely is a driver at the legal limit to be involved in a single-car accident than a nondrinker?

    (c) Suppose that new legislation were proposed to change the definition of legal intoxication. The new definition would hold that a person was legally intoxicated when their likelihood of involvement in a single-car accident is three times that of a non-drinker. To what BAC would the new definition of legal intoxication correspond?

44. The relationship between the swimming speed U (in cm/sec) of a salmon to the length l of the salmon (in cm) is given by the function[22]
    $$U = 19.5\sqrt{l}.$$

    (a) If one salmon is 4 times the length of another salmon, how are their swimming speeds related?

    (b) Graph the function $U = 19.5\sqrt{l}$. Describe the graph using words such as increasing, decreasing, concave up, concave down.

    (c) Using a property that you described in part (b), answer the question "Do larger salmon swim faster than smaller ones?"

    (d) Using a property that you described in part (b), answer the question "Imagine four salmon—two small and two large. The smaller salmon differ in length by 1 cm, as do the two larger. Is the difference in speed between the two smaller fish, greater than, equal to, or smaller than the difference in speed between the two larger fish?"

---

[22]From K. Schmidt-Nielsen, *Scaling–Why is animal size so important?* (London: Cambridge University Press, 1984).

45. (a) Using a computer or a graphing calculator, sketch a graph of $f(x) = 2^x$.

(b) Find the slope of the line tangent to f at $x = 0$ to an accuracy of two decimals. [Hint: Zoom in on the graph until it is indistinguishable from a line and estimate the slope using two points on the graph.]

(c) Find the slope of the line tangent to $g(x) = 3^x$ at $x = 0$ to an accuracy of two decimals.

(d) Find a base b (to two decimals) such that the line tangent to the function $h(x) = b^x$ at $x = 0$ has slope 1.

46. It can be shown that $e = 1 + \dfrac{1}{1} + \dfrac{1}{1 \cdot 2} + \dfrac{1}{1 \cdot 2 \cdot 3} + \dfrac{1}{1 \cdot 2 \cdot 3 \cdot 4} + \cdots$, where the approximation improves as more terms are included.

(a) Use your calculator to find the sum of the five terms shown.

(b) Find the sum of the first seven terms.

(c) Compare your sums with the calculator's displayed value for e (which you can find by entering $e^{\wedge}1$) and state the number of correct digits in the five and seven term sum.

(d) How many terms of the sum are needed in order to give a nine decimal digit approximation equal to the calculator's displayed value for e?

CHAPTER FIVE

TRANSFORMATIONS OF FUNCTIONS AND THEIR GRAPHS

We have introduced the families of linear, exponential, and logarithmic functions and we will study several other families in the later chapters. Before going on to the next family, we introduce some tools that are useful for analyzing every family of functions. These tools allow us to transform members of a family into one another by shifting, flipping, and stretching their graphs. In the process, we construct another family, the family of quadratic functions.

Throughout the chapter, we consider the relationship between changes made to the formula of a function and changes made to its graph.

5.1 VERTICAL AND HORIZONTAL SHIFTS

Suppose we shift the graph of some function vertically or horizontally. It is then the graph of a new function. In this section we investigate the relationship between the formulas for the original function and the new function.

Vertical and Horizontal Shift: The Heating Schedule For an Office Building

We start with an example of a vertical shift in the context of the heating schedule for a building.

Example 1   To save money, an office building is kept warm only during business hours. Figure 5.1 shows the temperature, H, in °F, as a function of time, t, in hours after midnight. At midnight ($t = 0$), the building's temperature is 50°F. This temperature is maintained until 4 am. Then the building begins to warm up so that by 8 am the temperature is 70°F. At 4 pm the building begins to cool. By 8 pm, the temperature is again 50°F.

Suppose that the building's superintendent decides to keep the building 5°F warmer than before. Sketch a graph of the resulting function.

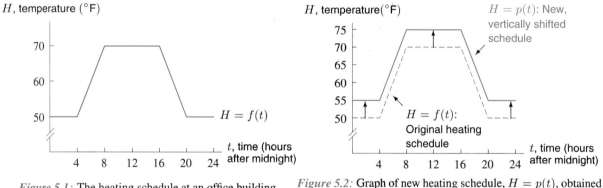

Figure 5.1: The heating schedule at an office building

Figure 5.2: Graph of new heating schedule, $H = p(t)$, obtained by shifting original graph, $H = f(t)$, upward by 5 units

Solution   The graph of f, the heating schedule function of Figure 5.1, is shifted upward by 5 units. The new heating schedule, $H = p(t)$, is graphed in Figure 5.2. The building's overnight temperature is now 55°F instead of 50°F and its daytime temperature is 75°F instead of 70°F. The 5°F increase in temperature corresponds to the 5-unit vertical shift in the graph.

The next example involves shifting a graph horizontally.

Example 2   The superintendent then changes the original heating schedule to start two hours earlier. The building now begins to warm at 2 am instead of 4 am, reaches 70°F at 6 am instead of 8 am, begins cooling off at 2 pm instead of 4 pm, and returns to 50°F at 6 pm instead of 8 pm. How are these changes reflected in the graph of the heating schedule?

Solution   Figure 5.3 gives a graph of $H = q(t)$, the new heating schedule, which is obtained by shifting the graph of the original heating schedule, $H = f(t)$, two units to the left.

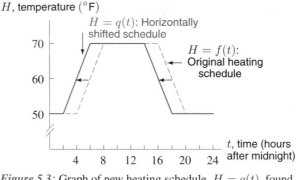

Figure 5.3: Graph of new heating schedule, $H = q(t)$, found by shifting, f, the original graph 2 units to the left

Notice that the upward shift in Example 1 results in a warmer temperature, whereas the leftward shift in Example 2 results in an earlier schedule.

Formulas for a Vertical or Horizontal Shift

How does a horizontal or vertical shift of a function's graph affect its formula?

Example 3   In Example 1, the graph of the original heating schedule, $H = f(t)$, was shifted upward by 5 units; the result was the warmer schedule $H = p(t)$. How are the formulas for $f(t)$ and $p(t)$ related?

Solution   The temperature under the new schedule, $p(t)$, is always 5°F warmer than the temperature under the old schedule, $f(t)$. Thus,

$$\begin{array}{ccc} \text{New temperature} & = & \text{Old temperature} \\ \text{at time } t & & \text{at time } t \end{array} + 5.$$

Writing this algebraically:

$$\underbrace{p(t)}_{\substack{\text{New temperature} \\ \text{at time } t}} = \underbrace{f(t)}_{\substack{\text{Old temperature} \\ \text{at time } t}} + 5.$$

The relationship between the formulas for p and f is given by the equation $p(t) = f(t) + 5$.

We can get information from the relationship $p(t) = f(t) + 5$, although we do not have an explicit formula for f or p.

Suppose we need to know the temperature at 6 am under the schedule $p(t)$. The graph of $f(t)$ shows that under the old schedule $f(6) = 60$. Substituting $t = 6$ into the equation relating f and p gives $p(6)$:

$$p(6) = f(6) + 5 = 60 + 5 = 65.$$

Thus, at 6 am the temperature under the new schedule is 65°.

Example 4   In Example 2 the heating schedule was changed to 2 hours earlier, shifting the graph horizontally 2 units to the left. Find a formula for q, this new schedule, in terms of f, the original schedule.

Solution   The old schedule always reaches a given temperature 2 hours after the new schedule. For example, at 4 am the temperature under the new schedule reaches 60°. The temperature under the old schedule reaches 60° at 6 am, 2 hours later. The temperature reaches 65° at 5 am under the new schedule, but not until 7 am, under the old schedule. In general, we see that

$$\begin{array}{c} \text{Temperature under new schedule} \\ \text{at time } t \end{array} = \begin{array}{c} \text{Temperature under old schedule} \\ \text{at time } (t+2), \text{ two hours later.} \end{array}$$

Algebraically, we have

$$q(t) = f(t+2).$$

This is a formula for q in terms of f.

Let's check the formula from Example 4 by using it to calculate $q(14)$, the temperature under the new schedule at 2 pm. The formula gives

$$q(14) = f(14+2) = f(16).$$

Figure 5.1 shows that $f(16) = 70$. Thus, $q(14) = 70$. This agrees with Figure 5.3.

Translations of a Function and Its Graph

In the heating schedule example, the function representing a warmer schedule,

$$p(t) = f(t) + 5,$$

has a graph which is a vertically shifted version of the graph of f. On the other hand, the earlier schedule is represented by

$$q(t) = f(t+2)$$

and its graph is a horizontally shifted version of the graph of f. Adding 5 to the temperature, or output value, $f(t)$, shifted its graph *up* five units. Adding 2 to the time, or input value, t, shifted its graph to the *left* two units. Generalizing these observations to any function g:

If $y = g(x)$ is a function and k is a constant, then the graph of
- $y = g(x) + k$ is the graph of $y = g(x)$ shifted vertically $|k|$ units. If k is positive, the shift is up; if k is negative, the shift is down.
- $y = g(x + k)$ is the graph of $y = g(x)$ shifted horizontally $|k|$ units. If k is positive, the shift is to the left; if k is negative, the shift is to the right.

A vertical or horizontal shift of the graph of a function is called a *translation* because it does not change the shape of the graph, but simply translates it to another position in the plane. Shifts or translations are the simplest examples of *transformations* of a function. We will see others in later sections of Chapter 5.

Inside and Outside Changes: Horizontal and Vertical Changes

Since $y = g(x + k)$ involves a change to the input value, x, it is an *inside change* to g. Similarly, since $y = g(x) + k$ involves a change to the output value, $g(x)$, it is an *outside change*. In general, as with horizontal and vertical shifts, an inside change in a function results in a horizontal change in its graph, whereas an outside change results in a vertical change.

Combining Horizontal and Vertical Shifts

We have seen what happens when we shift a function's graph either horizontally or vertically. What happens if we shift it both horizontally and vertically?

Example 5   Let r be the transformation of the heating schedule function, $H = f(t)$, defined by the equation

$$r(t) = f(t - 2) - 5.$$

(a)   Sketch the graph of $H = r(t)$.

(b)   Describe in words the heating schedule determined by r.

Solution   (a)   To graph r, we break this transformation into two steps. First, we sketch a graph of $H = f(t-2)$. This is an inside change to the function f and it results in the graph of f being shifted 2 units to the right. Next, we sketch a graph of $H = f(t - 2) - 5$. This graph can be found by shifting our sketch of $H = f(t - 2)$ down 5 units. The resulting graph is shown in Figure 5.4. The graph of r is the graph of f shifted 2 units to the right and 5 units down.

(b)   The function r represents a schedule that is both 2 hours later and 5 degrees cooler than the original schedule.

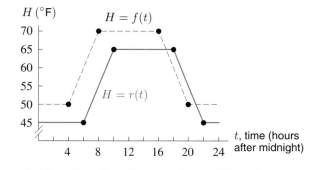

Figure 5.4: Graph of $r(t) = f(t - 2) - 5$ is graph of $H = f(t)$ shifted right by 2 and down by 5

We can use transformations to understand an unfamiliar function by relating it to a function we already know.

Example 6   A graph of $f(x) = x^2$ is in Figure 5.5. Define g by shifting the graph of f to the right 2 units and down 1 unit; see Figure 5.6. Find a formula for g in terms of f. Find a formula for g in terms of x.

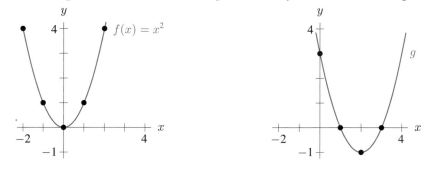

Figure 5.5: The graph of $f(x) = x^2$ *Figure 5.6:* The graph of g, a transformation of f

Solution   The graph of g is the graph of f shifted to the right 2 units and down 1 unit, so a formula for g is $g(x) = f(x - 2) - 1$. Since $f(x) = x^2$, we have $f(x - 2) = (x - 2)^2$. Therefore,

$$g(x) = (x - 2)^2 - 1.$$

It is a good idea to check by graphing $g(x) = (x - 2)^2 - 1$ and comparing the graph with Figure 5.6.

Problems for Section 5.1

1. The values for the function $f(x)$ are given in Table 5.1. Complete the following tables, where:

   (a) $g(x) = f(x - 1)$ (b) $h(x) = f(x + 1)$ (c) $k(x) = f(x) + 3$ (d) $m(x) = f(x-1)+3$

   Explain how the graph of each function relates to the graph of $f(x)$.

   TABLE 5.1

x	-2	-1	0	1	2
$f(x)$	-3	0	2	1	-1

x	-1	0	1	2	3
$g(x)$					

x	-3	-2	-1	0	1
$h(x)$					

x	-2	-1	0	1	2
$k(x)$					

x	-1	0	1	2	3
$m(x)$					

2. The function $H = f(t)$, graphed in Figure 5.1 on page 180, gives the heating schedule of an office building during winter, where H is the building's temperature in $°F$ at t hours after midnight.

   (a) Graph the function $H = f(t) - 2$. If the company decides to schedule its heating according to this function, what has it decided to do?

   (b) Graph the function $H = f(t - 2)$. If the company decides to schedule its heating according to this function, what has it decided to do?

   (c) At 8 am, is the building warmer under the $f(t)$ schedule, the $f(t) - 2$ schedule, or the $f(t - 2)$ schedule? What is the temperature be under that schedule?

   (d) Which schedule saves the company most on heating costs, assuming that the cost of heating depends only on the thermostat setting?

3. Complete the following tables using $f(p) = p^2 + 2p - 3$, and $g(p) = f(p + 2)$, and $h(p) = f(p - 2)$. Sketch graphs of the three functions. Explain how the graphs of g and h are related to the graph of f.

p	-3	-2	-1	0	1	2	3
$f(p)$							

p	-3	-2	-1	0	1	2	3
$g(p)$							

p	-3	-2	-1	0	1	2	3
$h(p)$							

4. Graph $m(n) = \frac{1}{2}n^2$. Write the formulas and sketch the graphs for the following transformations of m.

   (a) $y = m(n) + 1$ (b) $y = m(n + 1)$ (c) $y = m(n) - 3.7$

   (d) $y = m(n - 3.1)$ (e) $y = m(n) + \sqrt{13}$ (f) $y = m(n + 2\sqrt{2})$

   (g) $y = m(n + 3) + 7$ (h) $y = m(n - 17) - 159$

5. Let $k(w) = 3^w$, write a formula and sketch a graph for the following transformations of k.

   (a) $y = k(w) - 3$ (b) $y = k(w - 3)$ (c) $y = k(w) + 1.8$

   (d) $y = k(w + \sqrt{5})$ (e) $y = k(w + 2.1) - 1.3$ (f) $y = k(w - 1.5) - 0.9$

6. Let $f(x) = \left(\dfrac{1}{3}\right)^x$, $g(x) = \left(\dfrac{1}{3}\right)^{x+4}$, and $h(x) = \left(\dfrac{1}{3}\right)^{x-2}$. How do the graphs of $g(x)$ and $h(x)$ compare to the graph of $f(x)$?

7. Match each graph in parts (a)–(f) with the corresponding formula from (i)–(vi).

    (i) $y = |x|$
    (iv) $y = |x| + 2.5$

    (ii) $y = |x| - 1.2$
    (v) $y = |x + 3.4|$

    (iii) $y = |x - 1.2|$
    (vi) $y = |x - 3| + 2.7$

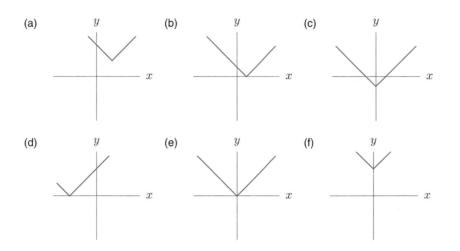

8. The graph in Figure 5.7 defines the function $s = c(t)$. Graph the following transformations of c.

    (a) $s = c(t) + 3$
    (d) $s = c(t - 1.5)$

    (b) $s = c(t + 3)$
    (e) $s = c(t - 2) + 2$

    (c) $s = c(t) - 1.5$
    (f) $s = c(t + 0.5) - 3$

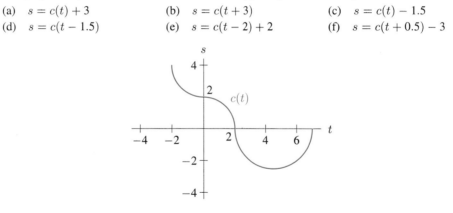

Figure 5.7

9. Sketch graphs of the following functions by treating them as transformations of $y = |x|$.

    (a) $g(x) = |x| + 1$
    (b) $h(x) = |x + 1|$
    (c) $j(x) = |x - 2| + 3$

10. Let $f(x) = 4^x$, $g(x) = 4^x + 2$, and $h(x) = 4^x - 3$. What is the relationship between the graph of $f(x)$ and the graphs of $h(x)$ and $g(x)$?

11. Suppose $S(d)$ gives the height of high tide in Seattle on a specific day, d, of the year. Use shifts of the function $S(d)$ to find formulas for each of the following functions:

    (a) $T(d)$, the height of high tide in Tacoma on day d, given that high tide in Tacoma is always one foot higher than high tide in Seattle.
    (b) $P(d)$, the height of high tide in Portland on day d, given that high tide in Portland is the same height as the previous day's high tide in Seattle.

12. The graph of $y = m(r)$ is given in Figure 5.8. The graph of each function in parts (a) – (d) resulted from translations of $y = m(r)$. Give a formula for each of these functions in terms of m.

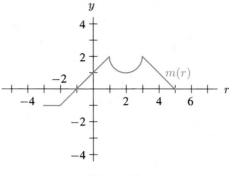

Figure 5.8

(a)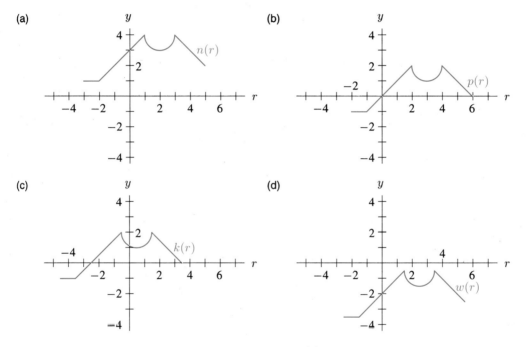

(b)

(c)

(d)

13. The data in Table 5.2 are from the function $f(x)$. Each function in parts (a) – (c) is a translation of $f(x)$. Find a possible formula for each of these functions in terms of f. For example, if you were given the data in Table 5.3, you could say that $k(x) = f(x) + 1$.

TABLE 5.2

x	0	1	2	3	4	5	6	7
$f(x)$	0	0.5	2	4.5	8	12.5	18	24.5

TABLE 5.3

x	0	1	2	3	4	5	6	7
$k(x)$	1	1.5	3	5.5	9	13.5	19	25.5

(a)

x	0	1	2	3	4	5	6	7
$h(x)$	-2	-1.5	0	2.5	6	10.5	16	22.5

(b)

x	0	1	2	3	4	5	6	7
$g(x)$	0.5	2	4.5	8	12.5	18	24.5	32

(c)

x	0	1	2	3	4	5	6	7
$i(x)$	-1.5	0	2.5	6	10.5	16	22.5	30

14. The data in Table 5.4 are from the function $g(t)$. Each function in parts (a) – (e) is a translation of $g(t)$. Find a possible formula for each of these functions in terms of g.

TABLE 5.4

t	-2	-1.5	-1	-0.5	0	0.5	1	1.5	2
$g(t)$	-0.48	0	0.48	0.84	1.0	0.91	0.60	0.14	-0.35

(a)

t	-2	-1.5	-1	-0.5	0	0.5	1	1.5	2
$a(t)$	0.02	0.5	0.98	1.34	1.50	1.41	1.10	0.64	0.15

(b)

t	-2	-1.5	-1	-0.5	0	0.5	1	1.5	2
$b(t)$	0.84	1.0	0.91	0.60	0.14	-0.35	-0.76	-0.98	-0.96

(c)

t	-2	-1.5	-1	-0.5	0	0.5	1	1.5	2
$c(t)$	0.54	0.7	0.61	0.30	-0.16	-0.65	-1.06	-1.28	-1.26

(d)

t	-2	-1.5	-1	-0.5	0	.5	1	1.5	2
$d(t)$	1.32	-0.48	0	0.48	0.84	1	0.91	0.6	0.14

(e)

t	-2	-1.5	-1	-0.5	0	0.5	1	1.5	2
$e(t)$	2.52	0.72	1.2	1.68	2.04	2.20	2.11	1.80	1.34

15. (a) What is a reasonable window to use to graph $y = x$ on a calculator? (Many answers are possible.)
    (b) Find a window in which you can see the graph of $y = x + 100$. Sketch a copy of the graph on paper, recording the window used.

16. (a) Find a calculator window in which you can see the graph of $y = \dfrac{1}{(x^2 + 1)}$. Sketch the graph and record the window.
    (b) Find a window in which you can see the graph of $y = \dfrac{1}{(x - 20)^2 + 1}$. Sketch the graph and record the window.

17. Graph $y = \log x$, $y = \log(10x)$, and $y = \log(100x)$. How do the graphs compare? Use a property of logs to show that the graphs are vertical shifts of one another.

For Problems 18–19, find a formula for the family of functions obtained from $f(x)$ by: (a) Shifting horizontally (b) Shifting vertically (c) Shifting horizontally and vertically. In each case, discuss the similarities and differences between the graph of a typical member of the family and the graph of $f(x)$. Include intercepts and asymptotes in your discussion. Are the three families the same?

18. $f(x) = 2x$ 19. $f(x) = 2^x$

20. Suppose $H(t)$ gives the temperature of a cup of coffee in degrees Fahrenheit, t minutes after it is brought to class. We are told that $H(t) = 68 + 93(0.91)^t$ for $t \geq 0$.

   (a) Find formulas for $H(t + 15)$ and $H(t) + 15$.
   (b) Sketch graphs of $H(t)$, $H(t + 15)$, and $H(t) + 15$ using a calculator or computer.
   (c) Describe in practical terms what situation might be modeled by the function $H(t + 15)$. What about $H(t) + 15$?
   (d) Which function, $H(t + 15)$ or $H(t) + 15$, approaches the same final temperature as the function $H(t)$? What is that temperature?

21. Suppose $T(d)$ gives the average temperature in your hometown on the dth day of last year (where $d = 1$ is January 1st, and so on).

   (a) Sketch a possible graph of $T(d)$ for $1 \leq d \leq 365$.
   (b) Give a possible value for each of the following: $T(6)$; $T(100)$; $T(215)$; $T(371)$.
   (c) What is the relationship between $T(d)$ and $T(d + 365)$? Explain.
   (d) If you were to graph $w(d) = T(d + 365)$ on the same axes as $T(d)$, how would the two graphs compare?
   (e) Do you think the function $T(d) + 365$ has any practical significance? Explain.

22. A carpenter remodeling your kitchen quotes you a price based on the cost of labor and materials. Materials are subject to a sales tax of 8.2%. Labor is not taxed.

   (a) One option is based on a fixed labor cost of $800 and the variable cost of materials. The total cost is given by $C(x) = 800 + x + 0.082x$ where x is the cost of materials in dollars. Later the carpenter says the job will actually cost $C(x) - 50$. Find a formula for $C(x) - 50$ and describe in practical terms what might have changed.
   (b) Under another option, you pay a fixed amount of $1000 for the materials (including tax) plus an hourly labor rate. This time, the total cost is given by $D(x) = 1000 + 15x$ where x is the number of hours to complete the job. Find a formula for $D(x) + 250$ and explain what this might mean in terms of the job.
   (c) Using the option from part (b), find a formula for $D(x - 8)$ and explain what this might suggest about the job.

23. At a jazz club, the cost of an evening is based on a cover charge of $5 plus a beverage charge of $3 per drink.

   (a) Find a formula for $t(x)$, the total cost for an evening in which x drinks are consumed.
   (b) If the price of the cover charge is raised by $1, express the new total cost function, $n(x)$, as a transformation of $t(x)$.
   (c) The management decides to increase the cover charge to $10, leave the price of a drink at $3, but include the first two drinks for free. For $x \geq 2$, express $p(x)$, the new total cost, as a transformation of $t(x)$.

24. A hot brick is removed from a kiln and set on the floor to cool. According to a law attributed to Isaac Newton, the difference between the brick's temperature and room temperature decays exponentially

over time. The brick's temperature is initially 350°F and room temperature is 70°F. The temperature difference, $D(t)$, decays exponentially at the constant rate of 3% per minute. The function $H(t)$, the brick's temperature t minutes after being removed from the kiln, is a transformation of $D(t)$. Find a formula for $H(t)$. Compare the graphs of $D(t)$ and $H(t)$, paying attention to the asymptotes.

25. If your total taxable income for one year is d dollars, let $I(d)$ be the federal income tax that you owe. For 1993, the value of $I(d)$ was given (approximately) by:

$$I(d) = \begin{cases} 0.15d & \text{if } 0 \le d \le 20{,}000 \\ 3000 + 0.28(d - 20{,}000) & \text{if } 20{,}000 < d \le 49{,}000 \\ 11{,}120 + 0.31(d - 49{,}000) & \text{if } d > 49{,}000. \end{cases}$$

(a) Sketch a graph of $I(d)$.
(b) Suppose the Internal Revenue Service (IRS) suddenly declared that your new tax obligation was $I(d) + 200$. Sketch a graph of this new function. Explain how this changes your tax obligation.
(c) Suppose the IRS suddenly declared that your new tax obligation was $I(d + 1000)$. Sketch a graph of this new function. Explain how this changes your tax obligation.
(d) If your taxable income were $15,000, would you prefer the $I(d) + 200$ system or the $I(d + 1000)$ system? Explain your reasoning.
(e) Would your answer to part (d) be different if your taxable income were $30,000?
(f) At what income level do the $I(d) + 200$ system and the $I(d + 1000)$ system produce the same tax obligation?

5.2 REFLECTIONS AND SYMMETRY

In Section 5.1 we saw that a horizontal shift of the graph of a function results from a change to the input of the function. (Specifically, adding or subtracting a constant inside the function's parentheses.) On the other hand, a vertical shift corresponds to an outside change.

In this section we consider the effect on a function's formula of reflecting its graph over the x or y-axis. A reflection over the x-axis is a vertical reflection and corresponds to an outside change. A reflection over the y-axis is a horizontal reflection and corresponds to an inside change.

A Formula for a Reflection

Figure 5.9 shows the graph of a function $y = f(x)$ and Table 5.5 gives a corresponding table of values. Note that we do not need an explicit formula for f.

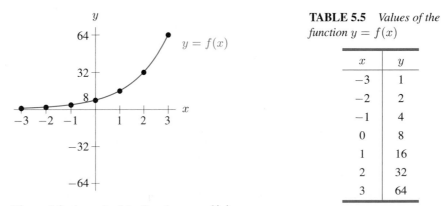

Figure 5.9: A graph of the function $y = f(x)$

TABLE 5.5 *Values of the function* $y = f(x)$

x	y
-3	1
-2	2
-1	4
0	8
1	16
2	32
3	64

Figure 5.10 shows a graph of a function $y = g(x)$, resulting from a vertical reflection of the graph of f across the x-axis. Figure 5.11 is a graph of a function $y = h(x)$, resulting from a horizontal reflection of the graph of f across the y-axis. Figure 5.12 is a graph of a function $y = k(x)$, resulting from a horizontal reflection of the graph of f across the y-axis followed by a vertical reflection across the x-axis.

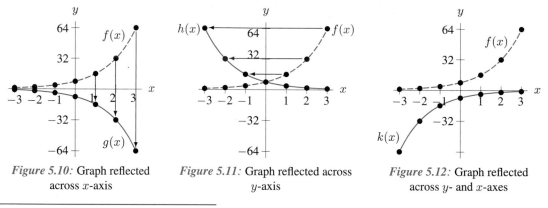

Figure 5.10: Graph reflected across x-axis   *Figure 5.11:* Graph reflected across y-axis   *Figure 5.12:* Graph reflected across y- and x-axes

Example 1   Find a formula in terms of f for   (a) $y = g(x)$ (b) $y = h(x)$ (c) $y = k(x)$

Solution   (a)   The graph of $y = g(x)$ is obtained by reflecting the graph of f vertically across the x-axis. For example, the point $(3, 64)$ on the graph of f reflects to become the point $(3, -64)$ on the graph of g. The point $(2, 32)$ on the graph of f becomes $(2, -32)$ on the graph of g. See Table 5.6.

TABLE 5.6   *Values of the functions $g(x)$ and $f(x)$ graphed in Figure 5.10*

x	-3	-2	-1	0	1	2	3
$g(x)$	-1	-2	-4	-8	-16	-32	-64
$f(x)$	1	2	4	8	16	32	64

Notice that when a point is reflected vertically over the x-axis, the x-value stays fixed, while the y-value changes sign. That is, for a given x-value,

y-value of g is the negative of y-value of f.

Algebraically, this means
$$g(x) = -f(x).$$

(b)   The graph of $y = h(x)$ is obtained by reflecting the graph of $y = f(x)$ horizontally across the y-axis. In part (a), a vertical reflection corresponded to an outside change in the formula, specifically, multiplying by -1. Thus, you might guess that a horizontal reflection of the graph corresponds to an inside change in the formula. This is correct. To see why, consider Table 5.7.

TABLE 5.7   *Values of the functions $h(x)$ and $f(x)$ graphed in Figure 5.11*

x	-3	-2	-1	0	1	2	3
$h(x)$	64	32	16	8	4	2	1
$f(x)$	1	2	4	8	16	32	64

Notice that when a point is reflected horizontally across the y-axis, the y-value remains fixed, while the x-value changes sign. For example, since $f(-3) = 1$ and $h(3) = 1$, we have $h(3) = f(-3)$. Since $f(-1) = 4$ and $h(1) = 4$, we have $h(1) = f(-1)$. In general,

$$h(x) = f(-x).$$

(c) The graph of the function $y = k(x)$ results from a horizontal reflection of the graph of f across the y-axis, followed by a vertical reflection across the x-axis. Since a horizontal reflection corresponds to multiplying the inputs by -1 and a vertical reflection corresponds to multiplying the outputs by -1, we have

Vertical reflection over the x-axis
$$k(x) = -f(-x).$$
Horizontal reflection over the y-axis

Let's check a point. If $x = 1$, then the formula $k(x) = -f(-x)$ gives:

$$k(1) = -f(-1) = -4 \qquad \text{since } f(-1) = 4.$$

This result is consistent with the graph, since $(1, -4)$ is on the graph of $k(x)$.

For a function f:
- The graph of $y = -f(x)$ is a reflection of the graph of $y = f(x)$ across the x-axis.
- The graph of $y = f(-x)$ is a reflection of the graph of $y = f(x)$ across the y-axis.

Symmetry About the y-Axis

The graph of $p(x) = x^2$ in Figure 5.13 is *symmetric* about the y-axis. In other words, the part of the graph to the left of the y-axis is the mirror image of the part to the right of the y-axis. Reflecting the graph of $p(x)$ across the y-axis gives the graph of $p(x)$ again.

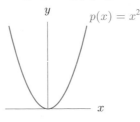

Figure 5.13: Reflecting the graph of $p(x) = x^2$ across the y-axis does not change its appearance

Symmetry about the y-axis is called *even symmetry*, because power functions with even exponents, such as $y = x^2$, $y = x^4$, $y = x^6$, $\cdots$ have this property.

Since $y = p(-x)$ is a reflection of the graph of p across the y-axis and $p(x)$ has even symmetry, we have

$$p(-x) = p(x).$$

To check this relationship, let $x = 2$. Then $p(2) = 2^2 = 4$, and $p(-2) = (-2)^2 = 4$, so $p(-2) = p(2)$. This means that the point $(2, 4)$ and its reflection across the y-axis, $(-2, 4)$, are both on the graph of $p(x)$.

Example 2   For the function $p(x) = x^2$, check algebraically that $p(-x) = p(x)$ for all x.

Solution   Substitute $-x$ into the formula for $p(x)$ giving

$$p(-x) = (-x)^2 = (-x) \cdot (-x)$$
$$= x^2$$
$$= p(x).$$

Thus, $p(-x) = p(x)$.

In general,

> If f is a function, then f is called an **even function** if, for all values of x in the domain of f,
>
> $$f(-x) = f(x).$$
>
> The graph of f is symmetric across the y-axis.

Symmetry About the Origin

Figures 5.14 and 5.15 show the graph of $q(x) = x^3$. Reflecting the graph of q first across the y-axis and then across the x-axis (or vice-versa) gives the graph of q again. This kind of symmetry is called symmetry about the origin, or *odd symmetry*.

In Example 1, we saw that $y = -f(-x)$ is a reflection of the graph of $y = f(x)$ across both the y-axis and the x-axis. Since $q(x) = x^3$ is symmetric about the origin, q is the same function as this double reflection. That is,

$$q(x) = -q(-x) \quad \text{which means that} \quad q(-x) = -q(x).$$

To check this relationship, let $x = 2$. Then $q(2) = 2^3 = 8$, and $q(-2) = (-2)^3 = -8$, so $q(-2) = -q(2)$. This means the point $(2, 8)$ and its reflection across the origin, $(-2, -8)$, are both on the graph of q.

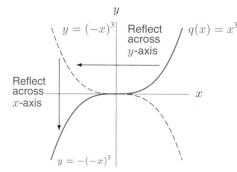

Figure 5.14: If the graph is reflected across the y-axis and then across the x-axis, it does not change

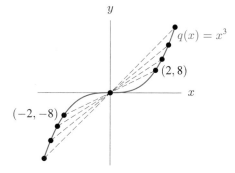

Figure 5.15: If every point on this graph is reflected about the origin, the graph is unchanged

Example 3   For the function $q(x) = x^3$, check algebraically that $q(-x) = -q(x)$ for all x.

Solution   We evaluate $q(-x)$ giving

$$q(-x) = (-x)^3 = (-x) \cdot (-x) \cdot (-x)$$
$$= -x^3$$
$$= -q(x).$$

Thus, $q(-x) = -q(x)$.

In general,

> If f is a function, then f is called an **odd function** if, for all values of x in the domain of f,
>
> $$f(-x) = -f(x).$$
>
> The graph of f is symmetric about the origin.

Example 4   Determine whether the following functions are symmetric across the y-axis, the origin, or neither.
(a) $f(x) = |x|$ (b) $g(x) = 1/x$ (c) $h(x) = -x^3 - 3x^2 + 2$

Solution   The graphs of the functions in Figures 5.16, 5.17, and 5.18 can be helpful in identifying symmetry.

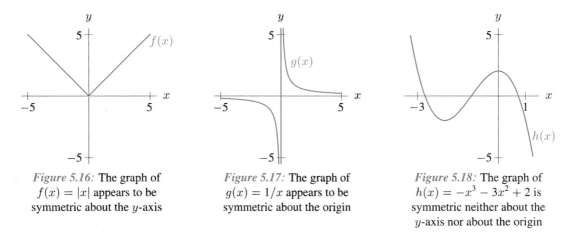

Figure 5.16: The graph of $f(x) = |x|$ appears to be symmetric about the y-axis

Figure 5.17: The graph of $g(x) = 1/x$ appears to be symmetric about the origin

Figure 5.18: The graph of $h(x) = -x^3 - 3x^2 + 2$ is symmetric neither about the y-axis nor about the origin

From the graphs it appears that f is symmetric about the y-axis (even symmetry), g is symmetric about the origin (odd symmetry), and h has neither type of symmetry. However, how can we be sure that $f(x)$ and $g(x)$ are really symmetric? We check algebraically.

If $f(-x) = f(x)$, then f has even symmetry. We check by substituting $-x$ in for x:

$$f(-x) = |-x|$$
$$= |x|$$
$$= f(x).$$

Thus, f does have even symmetry.

If $g(-x) = -g(x)$, then g is symmetric about the origin. We check by substituting $-x$ for x:

$$g(-x) = \frac{1}{-x}$$
$$= -\frac{1}{x}$$
$$= -g(x).$$

Thus, g is symmetric about the origin.

The graph of h does not exhibit odd or even symmetry. To confirm, look at an example, say $x = 1$:

$$h(1) = -1^3 - 3 \cdot 1^2 + 2 = -2.$$

Now substitute $x = -1$, giving

$$h(-1) = -(-1)^3 - 3 \cdot (-1)^2 + 2 = 0.$$

Thus $h(1) \neq h(-1)$, so the function is not symmetric about the y-axis. Also, $h(-1) \neq -h(1)$, so the function is not symmetric about the origin.

Combining Shifts and Reflections

We can combine the horizontal and vertical shifts from Section 5.1 with the horizontal and vertical reflections of this section to make more complex transformations of functions.

Example 5   A cold yam is placed in a hot oven. Assuming that Newton's Law of Heating applies, the difference between the oven's temperature and the yam's temperature is modeled by an exponential decay function. The yam's temperature is initially $0°$F, the oven's temperature is $300°$F, and the temperature difference decreases by 3% per minute. Find a formula for $Y(t)$, the yam's temperature at time t.

Solution   Let $D(t)$ be the difference between the oven's temperature and the yam's temperature, which is given by an exponential function $D(t) = ab^t$. The initial temperature difference is $300°$F $- 0°$F $= 300°$F, so $a = 300$. The temperature difference decreases by 3% per minute, so $b = 1 - 0.03 = 0.97$. Thus,

$$D(t) = 300(0.97)^t.$$

If the yam's temperature is represented by $Y(t)$, then the temperature difference is given by

$$D(t) = 300 - Y(t),$$

so, solving for $Y(t)$, we have

$$Y(t) = 300 - D(t),$$

giving

$$Y(t) = 300 - 300(0.97)^t.$$

Writing $Y(t)$ in the form

$$Y(t) = \underbrace{-D(t)}_{\text{Reflect}} + \underbrace{300}_{\text{Shift}}$$

shows that the graph of Y is obtained by reflecting the graph of D across the t-axis and then shifting it vertically up 300 units. Notice that the horizontal asymptote of D, which is on the t-axis, is also shifted upward, resulting in a horizontal asymptote at $300°$F for Y.

Figures 5.19 and 5.20 give the graphs of D and Y. Figure 5.20 shows that the yam heats up rapidly at first and then its temperature levels off toward $300°\,F$, the oven temperature.

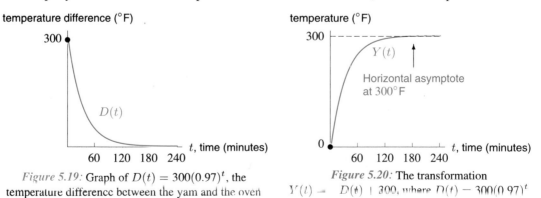

Figure 5.19: Graph of $D(t) = 300(0.97)^t$, the temperature difference between the yam and the oven

Figure 5.20: The transformation $Y(t) = -D(t) + 300$, where $D(t) = 300(0.97)^t$

Note that the temperature difference, D, is a decreasing function, so its average rate of change is negative. However, Y, the yam's temperature, is an increasing function, so its average rate of change is positive. Reflecting the graph of D over the t-axis to obtain the graph of Y changed the sign of the average rate of change.

Problems for Section 5.2

1. Complete the following tables using $f(p) = p^2 + 2p - 3$, and $g(p) = f(-p)$, and $h(p) = -f(p)$. Sketch graphs of the three functions. Explain how the graphs of g and h are related to the graph of f.

p	-3	-2	-1	0	1	2	3
$f(p)$							

p	-3	-2	-1	0	1	2	3
$g(p)$							

p	-3	-2	-1	0	1	2	3
$h(p)$							

2. Graph $m(n) = n^2 - 4n + 5$. Give a formula and graph for each of the following transformations of m.
   (a) $y = m(-n)$ (b) $y = -m(n)$ (c) $y = -m(-n)$ (d) $y = -m(n+2)$
   (e) $y = m(-n) - 4$ (f) $y = -m(-n) + 3$ (g) $y = 1 - m(n)$

3. Sketch the graphs of $y = f(x) = 4^x$ and $y = f(-x)$ on the same set of axes. How are these graphs related? Give an explicit formula for $y = f(-x)$.

4. Sketch the graphs of $y = g(x) = \left(\frac{1}{3}\right)^x$ and $y = -g(x)$ on the same set of axes. How are these graphs related? Give an explicit formula for $y = -g(x)$.

5. Sketch a graph of $k(w) = 3^w$. Give a formula and graph for each of the following transformations of k.
   (a) $y = k(-w)$ (b) $y = -k(w)$ (c) $y = -k(-w)$ (d) $y = -k(w-2)$
   (e) $y = k(-w) + 4$ (f) $y = -k(-w) - 1$ (g) $y = -3 - k(w)$

6. If the graph of a line $y = b + mx$ is reflected across the y-axis, what are the slope and intercepts of the resulting line?

7. Graph $y = \log(1/x)$ and $y = \log x$ on the same axes. How are the two graphs related? Use the properties of logarithms to explain the relationship algebraically.

8. Figure 5.21 shows a function $y = f(x)$. Match each formula with a graph from (a)–(f).
   (i) $y = f(-x)$ (ii) $y = -f(x)$ (iii) $y = f(-x) + 3$
   (iv) $y = -f(x-1)$ (v) $y = -f(-x)$ (vi) $y = -2 - f(x)$

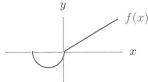

Figure 5.21

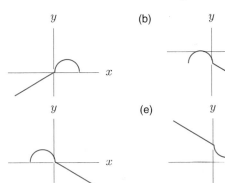

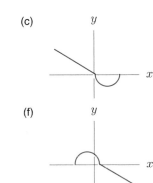

9. Given the graph of $y = f(x)$ in Figure 5.22, graph the following transformations of f on separate axes.

    (a) $y = f(x) - 2$ (b) $y = f(x - 2)$ (c) $y = -f(x)$ (d) $y = f(-x)$

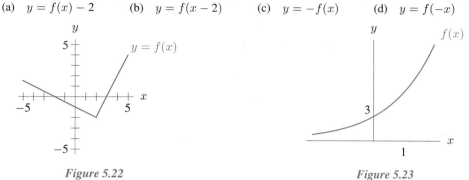

Figure 5.22   Figure 5.23

10. Figure 5.23 shows the graph of $y = f(x) = 3 \cdot 2^x$. Without using a calculator, sketch a graph of:

    (a) $y = f(-x)$ (b) $y = -f(x)$ (c) $y = 4 - f(-x)$

11. (a) If $f(x) = \sqrt{4 - x^2}$, find a formula for $f(-x)$.
     (b) Graph $y = f(x) = \sqrt{4 - x^2}$, $y = f(-x)$ and $y = -f(x)$ on the same axes.
     (c) Is the function $f(x) = \sqrt{4 - x^2}$ even, odd, or neither?

12. (a) If $g(x) = \sqrt[3]{x}$, find a formula for $g(-x)$.
     (b) Graph $y = g(x) = \sqrt[3]{x}$, $y = g(-x)$, and $y = -g(x)$ on the same axes.
     (c) Is $g(x) = \sqrt[3]{x}$ even, odd, or neither?

13. Let $g(x) = 3 \cdot 2^x$. Write a paragraph that compares the graph of $y = g(x)$ to the graphs of $y = -g(x)$, $y = g(-x)$, and $y = -g(-x)$.

14. Let f be defined by the graph in Figure 5.24. Find formulas (in terms of f) for the following transformations of f and sketch a graph of each. Show that these transformations lead to different outcomes.

    (a) First shift the graph of f upward by 3 units, then reflect it across the x-axis.
    (b) First reflect the graph of f across the x-axis. Then, shift it upward by 3 units.

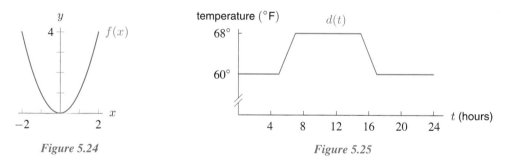

Figure 5.24   Figure 5.25

15. The function $d(t)$ graphed in Figure 5.25 gives the winter temperature in °F at a high school, t hours after midnight.

    (a) Describe in words the heating schedule for this building during the winter months.
    (b) Sketch a graph of $c(t) = 142 - d(t)$.
    (c) Explain why c might describe the cooling schedule for summer months.

16. Let $f(x) = x^3$.

    (a) Sketch the graph of the function obtained from f by first reflecting about the x-axis, then translating up two units. Write a formula for the resulting function.
    (b) Sketch the graph of the function obtained from f by first translating up two units, then reflecting about x-axis. Write a formula for the resulting function.
    (c) Are the functions you found in parts (a) and (b) the same?

17. Let $g(x) = 2^x$.

    (a) Sketch the graph of the function obtained from g by first reflecting about the y-axis, then translating down three units. Write a formula for the resulting function.

    (b) Sketch the graph of the function obtained from g by first translating down three units, then reflecting about y-axis. Write a formula for the resulting function.

    (c) Are the functions you found in parts (a) and (b) the same?

18. Table 5.8 gives values for a function $y = f(x)$. Fill in as many y-values as you can if you know that f is (a) An even function   (b) An odd function.

    TABLE 5.8

x	-3	-2	-1	0	1	2	3
y	5		-4			-8	

19. Are the following functions even, odd, or neither?

    (a) $m(x) = \dfrac{1}{x^2}$ (b) $n(x) = x^3 + x$ (c) $p(x) = x^2 + 2x$ (d) $q(x) = 2^{x+1}$

20. Are the following functions even, odd, or neither?

    (a) $a(x) = \dfrac{1}{x}$ (b) $b(x) = |x|$ (c) $e(x) = x + 3$ (d) $f(x) = \dfrac{-1}{(x-2)^2} + 1$

21. For each table, decide whether the function could be symmetric about the y-axis, across the origin, or neither.

    (a)

x	-4	-3	-2	-1	0	1	2	3	4
$f(x)$	13	6	1	-2	-3	-2	1	6	13

    (b)

x	-4	-3	-2	-1	0	1	2	3	4
$g(x)$	-19.2	-8.1	-2.4	-0.3	0	0.3	2.4	8.1	19.2

    (c)

x	-4	-3	-2	-1	0	1	2	3	4
$h(x) = f(x) + g(x)$	-6.2	-2.1	-1.4	-2.3	-3	-1.7	3.4	14.1	32.2

    (d)

x	-4	-3	-2	-1	0	1	2	3	4
$j(x) = f(x + 1)$	6	1	-2	-3	-2	1	6	13	22

22. A function is called symmetric across the line $y = x$ if interchanging x and y gives the same graph. The simplest example is the function $y = x$. Sketch the graph of another straight line that is symmetric across the line $y = x$ and give its equation.

23. Figure 5.26 shows the graph of a function f in the second quadrant. In each of the following cases, sketch a graph of $y = f(x)$, given that f is symmetric across

    (a) The y-axis.   (b) The origin.   (c) The line $y = x$.

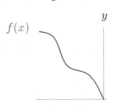

Figure 5.26

24. Show that the function $g(x) = (x - x^3)^2$ is an even function.

25. Show that the graph of the function h is symmetric across the origin, given that

$$h(x) = \frac{1 + x^2}{x - x^3}.$$

26. Comment on the following justification that the function $f(x) = x^3 - x^2 + 1$ is an odd function: For a function to be odd we need $f(-x) = -f(x)$. With $x = 1$ we find $f(1) = 1^3 - 1^2 + 1 = 1$. With $x = -1$ we find $f(-1) = (-1)^3 - (-1)^2 + 1 = -1$. So $f(-1) = -f(1)$ and the function is odd.

27. Comment on the following justification that the function $f(x) = x^3 - x^2 + 1$ is an even function: Because $f(0) = 1 \neq -f(0)$, we know that $f(x)$ is not odd. If a function is not odd, it must be even.

28. Is it possible for an odd function whose domain is all real numbers to be strictly concave up?

29. If f is an odd function and defined at $x = 0$, what is the value of $f(0)$? Explain how you can use the result of part (a) to show that the following functions are not odd: (a) $c(x) = x + 1$ (b) $d(x) = 2^x$

30. In the first quadrant an even function is increasing and concave down. What can you say about the function's behavior in the second quadrant?

31. Show that the power function $f(x) = x^{1/3}$ is an odd function. Give a counterexample to the statement that all power functions of the form $f(x) = x^p$ are odd.

32. Graph the functions $s(x) = 2^x + (\frac{1}{2})^x$, $c(x) = 2^x - (\frac{1}{2})^x$, and $n(x) = 2^x - (\frac{1}{2})^{x-1}$. State whether you think these functions are even, odd or neither. Show that your statements are true using algebra. That is, prove or disprove statements such as $s(-x) = s(x)$.

33. Consider $f(x) = b + mx$.

    (a)   Can $f(x)$ be even? How?   (b)   Can $f(x)$ be odd? How?
    (c)   Can $f(x)$ be both odd and even? How?

34. There are functions which are *neither* even nor odd. Is there a function that is *both* even and odd?

35. Some functions are symmetric across the y-axis. Is it possible for a function to be symmetric across the x-axis?

36. The force exerted by an electric charge Q on a moving charge q is given by

$$F = qQr^{-2},$$

where r is the distance between the charges q and Q. The fixed charge $Q = +1$ is located at $x = 0$ on a number line.

    (a)   A charge $q = +1$ moves along the positive half of the number line, so its position is given by x, where $x > 0$. Write a formula for the force F as a function of x, that is, $F = R(x)$. Sketch a graph of $R(x)$. [Note that the distance between q and Q is x.]
    (b)   A charge $q = +1$ moves along the negative half of the number line, so its position is given by x, where $x < 0$. Write a formula for the force F as a function of x, that is, $F = L(x)$. Sketch a graph of $L(x)$.
    (c)   Compare the graphs of $R(x)$ and $L(x)$. Express $L(x)$ as a transformation of $R(x)$.

5.3 VERTICAL STRETCHES AND COMPRESSIONS

We have studied translations and reflections of graphs. In this section, we consider vertical stretches and compressions of graphs. As with a vertical translation, a vertical stretch or compression of a function is represented by an outside change to its formula.

Vertical Stretch: A Stereo Amplifier

A stereo amplifier takes a weak signal from a cassette-tape deck, compact disc player, or radio tuner, and transforms it into a stronger signal to power a set of speakers.

Figure 5.27 shows a graph of a typical radio signal (in volts) as a function of time, t, both before and after amplification. In this illustration, the amplifier has boosted the strength of the signal by a factor of 3. (The amount of amplification, or *gain*, of most stereos is considerably greater than this.)

Notice that the wave crests of the amplified signal are 3 times as high as those of the original signal; similarly, the amplified wave troughs are 3 times deeper than the original wave troughs. If f is the original signal function and V is the amplified signal function, then

$$\underbrace{\text{Amplified signal strength at time } t}_{V(t)} = 3 \cdot \underbrace{\text{Original signal strength at time } t,}_{f(t)}$$

so we have

$$V(t) = 3 \cdot f(t).$$

This formula tells us that values of the amplified signal function are 3 times the values of the original signal. The graph of V is the graph of f stretched vertically by a factor of 3. As expected, a vertical stretch of the graph of $f(t)$ corresponds to an outside change in the formula.

Notice that the t-intercepts remain fixed under a vertical stretch, because the f-value of these points is 0, which is unchanged when multiplied by 3.

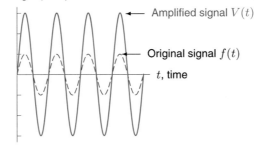

Figure 5.27: A stereo amplifier transforms a weak signal into a signal 3 times as strong

Negative Stretch Factor

What happens if we multiply a function by a negative stretch factor? Figure 5.28 gives a graph of a function $y = f(x)$, together with a graph of $y = -2 \cdot f(x)$. The stretch factor of f is $k = -2$. We think of $y = -2f(x)$ as a combination of two separate transformations of $y = f(x)$. First, the graph is stretched by a factor of 2, then it is reflected across the x-axis.

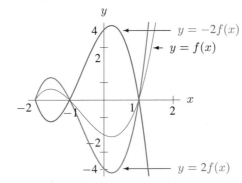

Figure 5.28: The graph of $y = -2f(x)$ is a vertically stretched version of the graph of $y = f(x)$ that has been reflected across the x-axis

Formula for Vertical Stretch or Compression

Generalizing the examples gives the following result:

> If f is a function and k is a constant, then the graph of $y = k \cdot f(x)$ is the graph of $y = f(x)$
> - Vertically stretched by a factor of k, if $k > 1$.
> - Vertically compressed by a factor of k, if $0 < k < 1$.
> - Vertically stretched or compressed by a factor $|k|$ and reflected across x-axis, if $k < 0$.

Example 1   A yam is placed in a 300°F oven. Suppose that Table 5.9 gives values of $H = r(t)$, the yam's temperature t minutes after being placed in the oven. Figure 5.29 shows these data points with a curve drawn in to emphasize the trend.

TABLE 5.9   *Temperature of a yam*

t, time (min)	$r(t)$, temperature (°F)
0	0
10	150
20	225
30	263
40	281
50	291
60	295

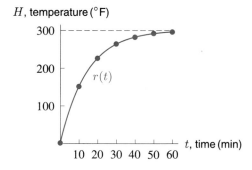

Figure 5.29: The temperature of a yam at time t

(a) Describe the function r in words. What do the data indicate about the yam's temperature?

(b) Make a table of values for $q(t) = 1.5r(t)$. Sketch a graph of this new function. Under what condition might q describe the yam's temperature?

Solution   (a) The function r is increasing and concave down. The yam starts out at 0°F and warms up quickly, reaching 150°F after 10 minutes. It continues to heat up, although more slowly, climbing by 75°F in the next 10 minutes and by 38°F in the 10 minutes after that. The temperature levels off around 300°F, the oven's temperature, represented by a horizontal asymptote.

(b) We calculate values of $q(t)$ from values of r. For example, Table 5.9 gives $r(0) = 0$ and $r(10) = 150$. Thus,

$$q(0) = 1.5r(0)$$
$$= 1.5 \cdot 0$$
$$= 0.$$

Similarly, $q(10) = 1.5(150) = 225$, and so on. The values for $q(t)$ are 1.5 times as large as the corresponding values for $r(t)$.

The data in Table 5.10 are plotted in Figure 5.30. The graph of q is a vertically stretched version of the graph of r, because the stretch factor of $k = 1.5$ is larger than 1. The horizontal asymptote of r was $H = 300$, so the horizontal asymptote of q is $H = 1.5 \cdot 300 = 450$. This suggests that the yam has been placed in a 450°F oven instead of a 300°F oven.

TABLE 5.10 *Values of $q(t) = 1.5r(t)$*

t, time (min)	$q(t)$, temperature (°F)
0	0
10	225
20	337.5
30	394.5
40	421.5
50	436.5
60	442.5

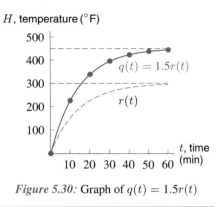

Figure 5.30: Graph of $q(t) = 1.5r(t)$

Stretch Factors and Average Rates of Change

Consider again the graph of the radio signal and its amplification in Figure 5.27. Notice that the amplified signal, V, is increasing on the same intervals as the original signal, f. Similarly, both functions decrease on the same intervals.

Stretching or compressing a function vertically does not change the intervals on which the function increases or decreases. However, the average rate of change of a function, visible in the steepness of the graph, is altered by a vertical stretch or compression.

Example 2   In Example 1, the function $H = r(t)$ gives the temperature (in °F) of a yam placed in a 300°F oven. The function $q(t) = 1.5r(t)$ gives the temperature of the yam placed in a 450°F oven.

(a) Calculate the average rate of change of r over 10-minute intervals. What does this tell you about the yam's temperature?

(b) Now calculate the average rate of change of q over 10-minute intervals. What does this tell you about the yam's temperature?

Solution   (a) On the first interval, from $t = 0$ to $t = 10$, we have

$$\begin{array}{l} \text{Average rate of change} \\ \text{of temperature, } r \end{array} = \frac{\Delta H}{\Delta t} = \frac{r(10) - r(0)}{10 - 0}$$

$$= \frac{150 - 0}{10} \quad \text{(referring to Table 5.9)}$$

$$= 15°\text{F/min.}$$

Thus, during the first 10 minute interval, the yam's temperature increased at an average rate of 15°F per minute.

On the second time interval from $t = 10$ to $t = 20$,

$$\begin{array}{l} \text{Average rate of change} \\ \text{of temperature} \end{array} = \frac{\Delta H}{\Delta t} = \frac{225 - 150}{10} = 7.5°\text{F/min,}$$

and on the third interval from $t = 20$ to $t = 30$,

$$\begin{array}{l} \text{Average rate of change} \\ \text{of temperature} \end{array} = \frac{\Delta H}{\Delta t} = \frac{263 - 225}{10} = 3.8°\text{F/min.}$$

See Table 5.11.

TABLE 5.11   *The average rate of change of yam's temperature, $r(t)$*

Time interval (min)	0 − 10	10 − 20	20 − 30	30 − 40	40 − 50	50 − 60
Average rate of change of r (°F/min)	15	7.5	3.8	1.8	1.0	0.4

(b)   The data in Table 5.10 was used to calculate the average rate of change of $q(t) = 1.5r(t)$ in Table 5.12.

TABLE 5.12   *The average rate of change of yam's temperature, $q(t)$*

Time interval (min)	$0 - 10$	$10 - 20$	$20 - 30$	$30 - 40$	$40 - 50$	$50 - 60$
Average rate of change of q (°F/min)	22.5	11.25	5.7	2.69	1.51	0.6

Comparing the average rates of change on each 10-minute interval, we see that q's average rate of change is 1.5 times r's. Thus, q depicts a yam whose temperature increases more quickly than r does.

In the last example, multiplying a function by a stretch factor k has the effect of multiplying the function's average rate of change on each interval by the same factor. We check this statement algebraically for the function $g(x) = k \cdot f(x)$. On the interval from a to b,

$$\begin{array}{c} \text{Average rate of change} \\ \text{of } y = g(x) \end{array} = \frac{\Delta y}{\Delta x} = \frac{g(b) - g(a)}{b - a}.$$

But $g(b) = k \cdot f(b)$ and $g(a) = k \cdot f(a)$. Thus,

$$\begin{array}{c} \text{Average rate of change} \\ \text{of } y = g(x) \end{array} = \frac{\Delta y}{\Delta x} = \frac{k \cdot f(b) - k \cdot f(a)}{b - a}$$

$$= k \cdot \frac{f(b) - f(a)}{b - a} \quad \text{(factoring out } k\text{)}$$

$$= k \cdot \left(\begin{array}{c} \text{Average rate of change} \\ \text{of } f \end{array} \right).$$

In general,

> If $g(x) = k \cdot f(x)$, then on any interval,
>
> Average rate of change of $g = k \cdot$ (Average rate of change of f).

Combining Transformations

Any transformations of functions can be combined.

Example 3   The function $y = f(x)$ is graphed in Figure 5.31. Graph the function $g(x) = -\dfrac{1}{2}f(x + 3) - 1$.

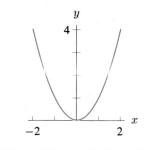

Figure 5.31: Graph of $y = f(x)$

Solution   To combine several transformations, always work from inside the parentheses outward as in Figure 5.32. The graphs corresponding to each step are shown in Figure 5.33. Note that we did not need a formula for f to graph g.

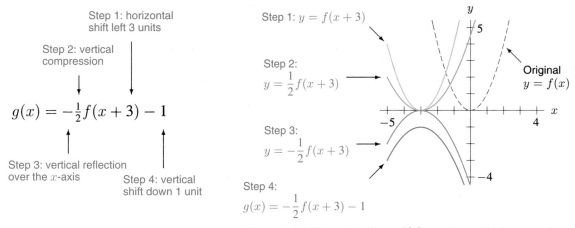

Step 1: horizontal
shift left 3 units

Step 2: vertical
compression

$$g(x) = -\tfrac{1}{2}f(x+3) - 1$$

Step 3: vertical reflection
over the x-axis

Step 4: vertical
shift down 1 unit

Figure 5.32

Step 1: $y = f(x+3)$

Step 2:
$y = \dfrac{1}{2}f(x+3)$

Step 3:
$y = -\dfrac{1}{2}f(x+3)$

Step 4:
$g(x) = -\dfrac{1}{2}f(x+3) - 1$

Figure 5.33: **The graph of** $y = f(x)$ **transformed in four steps into** $g(x) = -(1/2)f(x+3) - 1$

Problems for Section 5.3

1. Let $f(x)$ be defined by Table 5.13. Make tables for the following transformations of f using an appropriate domain.

   (a) $\tfrac{1}{2}f(x)$ (b) $-2f(x+1)$ (c) $f(x) + 5$
   (d) $f(x-2)$ (e) $f(-x)$ (f) $-f(x)$

   TABLE 5.13

x	-3	-2	-1	0	1	2	3
$f(x)$	2	3	7	-1	-3	4	8

2. Table 5.14 gives values for a function f. Fill in all the blanks for which you have sufficient information.

   TABLE 5.14

x	-3	-2	-1	0	1	2	3
$f(x)$	-4	-1	2	3	0	-3	-6
$f(-x)$							
$-f(x)$							
$f(x) - 2$							
$f(x - 2)$							
$f(x) + 2$							
$f(x + 2)$							
$2f(x)$							
$-f(x)/3$							

3. Using the values of f in Table 5.15, create a table of values for
    (a) $f(-x)$.   (b) $-f(x)$.   (c) $3f(x)$.
    (d) Which of these tables from parts (a), (b), and (c) represents an even function?

    TABLE 5.15

x	-4	-3	-2	-1	0	1	2	3	4
$f(x)$	13	6	1	-2	-3	-2	1	6	13

4. Figure 5.34 is a graph of $y = x^{3/2}$. Match the following functions with the graphs in Figure 5.35.
    (a) $y = x^{3/2} - 1$ (b) $y = (x-1)^{3/2}$ (c) $y = 1 - x^{3/2}$ (d) $y = \frac{3}{2}x^{3/2}$

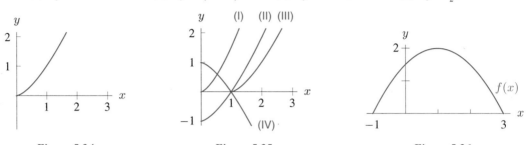

Figure 5.34   Figure 5.35   Figure 5.36

5. Graph the following functions using the graph of f in Figure 5.36.
    (a) $y = -f(x) + 2$ (b) $y = 2f(x)$ (c) $y = f(x-3)$ (d) $y = -\frac{1}{2}f(x+1) - 3$

6. Sketch graphs of the following transformations. Label at least three points on each graph.
    (a) $y = f(x+3)$ if $f(x) = |x|$ (b) $y = f(x) + 3$ if $f(x) = |x|$
    (c) $y = -g(x)$ if $g(x) = x^2$ (d) $y = g(-x)$ if $g(x) = x^2$
    (e) $y = 3h(x)$ if $h(x) = 2^x$ (f) $y = 0.5h(x)$ if $h(x) = 2^x$

7. Graph the transformations (a) – (f) of the function f given in Figure 5.37. Label the new positions of the points A and B and the horizontal asymptote.
    (a) $y = 2f(x)$ (b) $y = f(-x)$ (c) $y = -f(x)$
    (d) $y = f(x+3)$ (e) $y = f(x) + 3$ (f) $y = \frac{1}{2}f(x)$

Figure 5.37

8. Find formulas, in terms of f, for the following transformations of the graph of $f(x)$ in Figure 5.37.

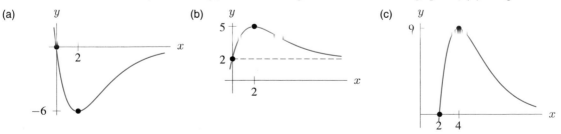

9. Let $y = f(x)$ be given by the graph in Figure 5.38. For each of the following functions, choose the letter (a) – (i) corresponding to the graph.

(i) $y = 2f(x)$ (ii) $y = \frac{1}{3}f(x)$ (iii) $y = -f(x) + 1$
(iv) $y = f(x + 2) + 1$ (v) $y = f(-x)$

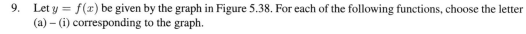

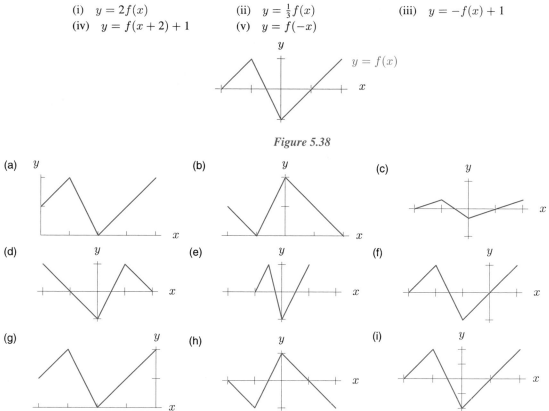

Figure 5.38

10. Let $f(x) = 2^x$. Find possible formulas in terms of f and then in terms of x for the transformations of f in (a) – (d). *Example:* The graph in Figure 5.39 appears to be f flipped across the y-axis. Because the horizontal asymptote is at $y = -3$ instead of $y = 0$, it appears that f is shifted downward by 3 units. Therefore, a formula is $y = f(-x) - 3 = 2^{-x} - 3$.

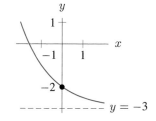

Figure 5.39

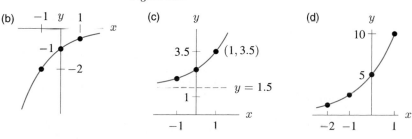

11. For each of the following formulas, find a graph in Figure 5.40, which could represent the function. (There maybe no answer or several answers.) Explain your reasoning.

(a) $y = 3 \cdot 2^x$

(b) $y = 5^{-x}$

(c) $y = -5^x$

(d) $y = 2 - 2^{-x}$

(e) $y = 1 - \left(\frac{1}{2}\right)^x$

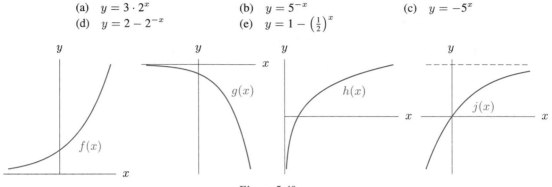

Figure 5.40

12. The graph of $y = f(x)$ is in Figure 5.41. Graph each of the following functions on separate sets of axes, together with the graph of the original function. Label intercepts and asymptotes.

(a) $y = 3f(x)$

(b) $y = f(x - 1)$

(c) $y = f(x) - 1$

(d) $y = -2f(x)$

(e) $y = \frac{1}{2}f(x + 2) - 1$

(f) $y = -f(-x)$

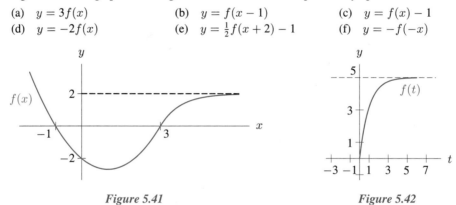

Figure 5.41   Figure 5.42

13. The graph of $y = f(t)$ is given in Figure 5.42. Find formulas in terms of f for the horizontal and vertical shifts of the graph of f shown in parts (a) – (c). What is the equation of the asymptote in each case?

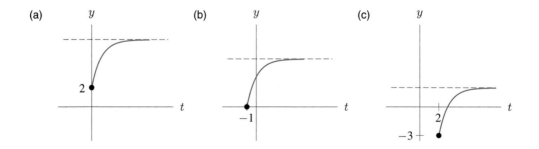

For Problems 14–15, find a formula for the family of functions obtained from $f(x)$ by: (a) Stretching vertically (b) Stretching vertically and shifting horizontally. In each case, discuss the similarities and differences between the graph of a typical member of the family and the graph of $f(x)$. Include intercepts and asymptotes in your discussion. Are the two families the same?

14. $f(x) = x$

15. $f(x) = 2^x$

16. Figure 5.43 shows the graphs of $y = x^2$, $y = ax^2$, and $y = bx^2$. Estimate the values of a and b.

17. Figure 5.44 gives a graph of $y = f(x)$. Consider the transformations $y = \frac{1}{2}f(x)$ and $y = 2f(x)$. Which points on the graph of $y = f(x)$ stay fixed under these transformations? Compare the intervals on which all three functions are increasing and decreasing.

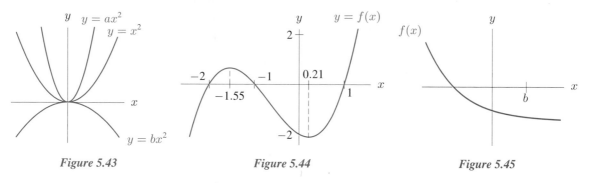

Figure 5.43   Figure 5.44   Figure 5.45

18. In Figure 5.45, the point b is labeled on the x-axis. On the y-axis, locate and label the output values:

(a) $f(b)$ (b) $-2f(b)$ (c) $-2f(-b)$

5.4 HORIZONTAL STRETCHES AND COMPRESSIONS

In Section 5.3, we observed that a vertical stretch of a function's graph corresponds to an outside change in its formula, specifically, multiplication by a stretch factor. Since horizontal changes generally correspond to inside changes, we expect that a horizontal stretch will correspond to a constant multiple of the inputs. This turns out to be the case.

Horizontal Stretch: A Lighthouse Beacon

The beacon in a lighthouse turns once per minute, and its beam sweeps across a beach house. Figure 5.46 gives a graph of $L(t)$, the intensity, or brightness, of the light striking the beach house as a function of time.

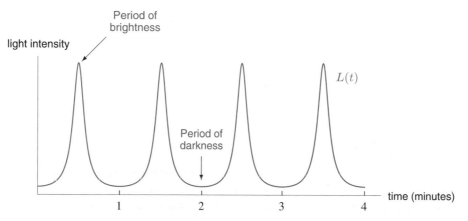

Figure 5.46: Light intensity or brightness, $L(t)$, as a function of time

Now suppose the lighthouse beacon turns twice as fast as before, so that its beam sweeps past the beach house twice instead of once each minute. The periods of brightness now occur twice as

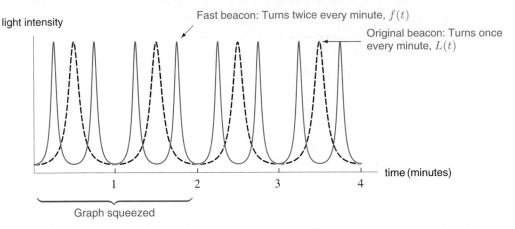

Figure 5.47: Comparing light intensity from the fast beacon, $f(t)$, to light intensity from the original beacon, $L(t)$

often. See Figure 5.47. The graph of $f(t)$, the intensity of light from this faster beacon, is a horizontal squeezing or compression of the original graph of $L(t)$.

   If the lighthouse beacon turns at half its original rate, so that its beam sweeps past the beach house once every two minutes instead of once every minute, the periods of brightness occur half as often as originally. Slowing the beacon's speed results in a horizontal stretch of the original graph, illustrated by the graph of $s(t)$, the light intensity of the slow beacon, in Figure 5.48.

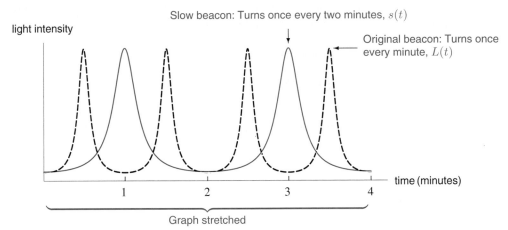

Figure 5.48: Comparing light intensity from the slow beacon, $s(t)$, to light intensity from the original beacon, $L(t)$

Formula for Horizontal Stretch or Compression

How are the formulas for the three light functions related? We expect that multiplying the function's input by a constant will horizontally stretch or compress its graph. The fast beacon corresponds to speeding up by a factor of 2, or multiplying the input times by 2. Thus

$$f(t) = L(2t).$$

Similarly for the slow beacon, the input times are multiplied by $1/2$, so

$$s(t) = L(\tfrac{1}{2}t).$$

Generalizing the lighthouse example gives the following result:

If f is a function and k a positive constant, then the graph of $y = f(kx)$ is the graph of f
- Horizontally compressed by a factor of $1/k$ if $k > 1$,
- Horizontally stretched by a factor of $1/k$ if $k < 1$.

If $k < 0$, then the graph of $y = f(kx)$ also involves a horizontal reflection across the y-axis.

Example 1   Values of the function $f(x)$ are in Table 5.16 and its graph is in Figure 5.49. Make a table and a graph of the function $g(x) = f(\frac{1}{2}x)$.

TABLE 5.16   *Values of $f(x)$*

x	$f(x)$
-3	0
-2	2
-1	0
0	-1
1	0
2	-1
3	1

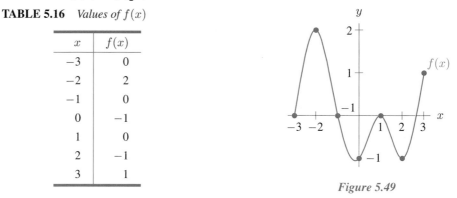

Figure 5.49

Solution   To make a table for $g(x) = f(\frac{1}{2}x)$, we substitute values for x. For example, if $x = 4$, then
$$g(4) = f(\tfrac{1}{2} \cdot 4) = f(2).$$
Table 5.16 shows that $f(2) = -1$, so
$$g(4) = f(2) = -1.$$
This result is recorded in Table 5.17. If $x = 6$, since Table 5.16 gives $f(3) = 1$, we have
$$g(6) = f(\tfrac{1}{2} \cdot 6) = f(3) = 1.$$
In Figure 5.50, we see that the graph of g is the graph of f stretched horizontally away from the y-axis. Substituting $x = 0$, gives
$$g(0) = f(\tfrac{1}{2} \cdot 0) = f(0) = -1,$$
so the y-intercept remains fixed (at -1) under a horizontal stretch.

TABLE 5.17   *Values of $g(x) = f(\frac{1}{2}x)$*

x	$g(x)$
-6	0
-4	2
-2	0
0	-1
2	0
4	-1
6	1

Figure 5.50: The graph of $g(x) = f(\frac{1}{2}x)$ is the graph of $y = f(x)$ stretched away from the y-axis by a factor of 2

Example 1 shows the effect of an inside multiple of $1/2$. The next example shows the effect on the graph of an inside multiple of 2.

Example 2   Let $f(x)$ be the function in Example 1. Make a table and a graph for the function $h(x) = f(2x)$.

Solution   We use Table 5.16 and the formula $h(x) = f(2x)$ to evaluate $h(x)$ at several values of x. For example, if $x = 1$, then

$$h(1) = f(2 \cdot 1) = f(2).$$

Table 5.16 shows that $f(2) = -1$, so $h(1) = -1$. These values are recorded in Table 5.18. Similarly, substituting $x = 1.5$, gives

$$h(1.5) = f(2 \cdot 1.5) = f(3) = 1.$$

Since $h(0) = f(2 \cdot 0) = f(0)$, the y-intercept remains fixed (at -1). In Figure 5.51 we see that the graph of h is the graph of f compressed by a factor of 2 horizontally toward the y-axis.

TABLE 5.18   *Values of* $h(x) = f(2x)$

x	$h(x)$
-1.5	0
-1.0	2
-0.5	0
0.0	-1
0.5	0
1.0	-1
1.5	1

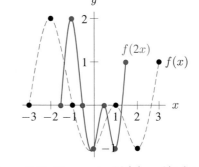

Figure 5.51: The graph of $h(x) = f(2x)$ is the graph of $y = f(x)$ compressed horizontally by a factor of 2

In Chapter 4, we used the function $P = 263e^{0.009t}$ to model the US population in millions. This function is a transformation of the basic exponential function $f(t) = e^t$, since we can write

$$P = 263e^{0.009t} = 263f(0.009t).$$

The US population is the basic exponential function $f(t) = e^t$ stretched vertically by a factor of 263 and stretched horizontally by a factor of $1/0.009 \approx 111$.

Example 3   Match the functions $f(t) = e^t$, $g(t) = e^{0.5t}$, $h(t) = e^{0.8t}$, $j(t) = e^{2t}$ with the graphs in Figure 5.52.

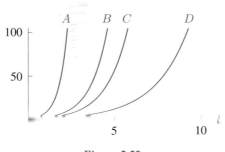

Figure 5.52

Solution   Since the function $j(t) = e^{2t}$ climbs fastest of the four and $g(t) = e^{0.5t}$ climbs slowest, graph A must be j and graph D must be g. Similarly, graph B is f and graph C is h.

Problems for Section 5.4

1. Let $f(x)$ be defined by Table 5.19. Make a table of values for $f(\frac{1}{2}x)$ using an appropriate domain.

    TABLE 5.19

x	-3	-2	-1	0	1	2	3
$f(x)$	2	3	7	-1	-3	4	8

2. Table 5.20 gives values for a function f. Fill in all the blanks for which you have sufficient information.

    TABLE 5.20

x	-3	-2	-1	0	1	2	3
$f(x)$	-4	-1	2	3	0	-3	-6
$f(\frac{1}{2}x)$							
$f(2x)$							

3. (a) Make a table of values (rounded to four decimal places) for the functions $f(x) = e^{0.04x}$ and $g(x) = e^{0.08x}$, with $x = 0, 0.1, 0.2, \ldots, 1.0$.
    (b) Do the first five values of $g(x)$ appear among the $f(x)$ values? Explain.
    (c) Do the first five values of $f(x)$ appear among the $g(x)$ values? Explain.

4. Let $y = f(x)$ be given by the graph in Figure 5.53. Choose the letter (a) – (i) corresponding to the graph (if any) that represents the given function:

    (i) $y = f(2x)$ (ii) $y = 2f(2x)$ (iii) $y = f(\frac{1}{2}x)$

Figure 5.53

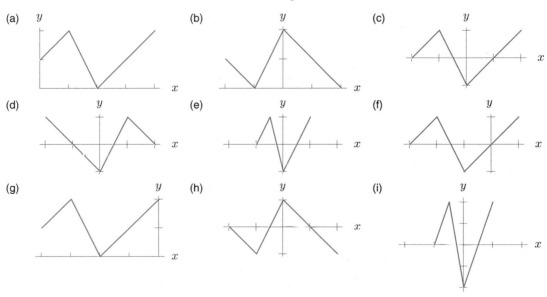

5. Graph $m(x) = e^x$, $n(x) = e^{2x}$, and $p(x) = 2e^x$ on the same axes and describe how the graphs of $n(x)$ and $p(x)$ compare with that of $m(x)$.

6. Figure 5.54 shows the graph of $f(x) = e^x$ in the window $-3 \leq x \leq 3$ and $-1 \leq y \leq 5$. For what window does Figure 5.55 represent the function: (a) $g(x) = e^{3x}$? (b) $h(x) = e^{0.4x}$?

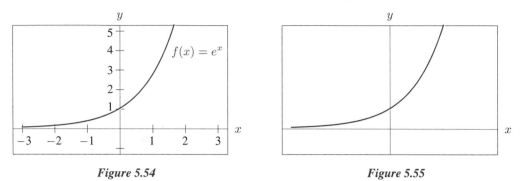

Figure 5.54   Figure 5.55

7. (a) Graph the functions $f(x) = e^x$ and $g(x) = f(4x) = e^{4x}$ in the window $-2 \leq x \leq 2$ and $-5 \leq y \leq 10$.

   (b) Using new x-values and the original y-values, graph g so that it looks like the original graph of f. What are the new x-values?

   (c) With the same y-values, find another set of new x-values so that the graph of f looks like the original graph of g.

8. Sketch a graph of $y = h(3x)$ if $h(x) = 2^x$.

9. Graph the following transformations of the function f given in Figure 5.56. Relabel points A and B as well as the horizontal asymptote: (a) $y = f(2x)$ (b) $y = f\left(-\dfrac{x}{3}\right)$

Figure 5.56   Figure 5.57

10. The function in Figure 5.57 is a transformation of the function, f, graphed in Figure 5.56. Find the formula for the function in Figure 5.57 in terms of f.

11. The graph of $y = f(x)$ is given in Figure 5.58. Graph the following functions on separate sets of axes, together with the graph of the original function. Label any intercepts or special points.

    (a) $y = f(3x)$ (b) $y = f(-2x)$ (c) $y = f(\tfrac{1}{2}x)$

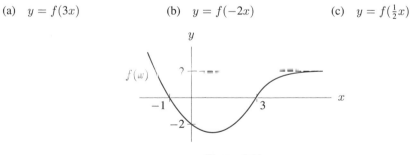

Figure 5.58

12. This problem investigates the effect of a horizontal stretch on the zeros of a function.

    (a) Graph $f(x) = 4 - x^2$. Mark the zeros of f on the graph.
    (b) Graph and find a formula for $g(x) = f(0.5x)$. What are the zeros of $g(x)$?
    (c) Graph and find a formula for $h(x) = f(2x)$. What are the zeros of $h(x)$?
    (d) Without graphing, what are the zeros of $f(10x)$?

13. The log function has the property that the graph resulting from a horizontal stretch can also be obtained by a vertical shift.

    (a) Graph $f(x) = \log x$ and $g(x) = \log(10x)$ and determine the vertical shift.
    (b) Explain how you could have predicted the answer to part (a) from the properties of logarithms.
    (c) If $h(x) = \log(ax)$, what is the vertical shift k making $h(x) = \log(x) + k$?

14. In Figure 5.59, the point c is labeled on the x-axis. On the y-axis, locate and label output values:
    (a) $g(c)$ (b) $2g(c)$ (c) $g(2c)$

15. Figure 5.60 shows the graphs of two functions, $f(x)$ and $g(x)$. It is claimed that $f(x)$ is a horizontal stretch of $g(x)$. If that could be true, find the stretch. If that could not be true, explain.

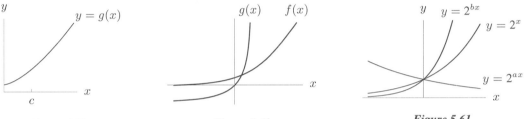

Figure 5.59   **Figure 5.60**   **Figure 5.61**

16. Figure 5.61 shows three functions: $y = 2^x$, $y = 2^{ax}$, and $y = 2^{bx}$. Estimate the values of a and b.

17. A company projects a total profit, $P(t)$ dollars, in year t. Explain the economic meaning of $r(t) = 0.5P(t)$ and $s(t) = P(0.5t)$.

18. You are a banker with a table showing year-end values of \$1 invested at an interest rate of 1% per year, compounded annually, for a period of fifty years.

    (a) Can this table be used to show 1% monthly interest charges on a credit card? Explain.
    (b) Can this table be used to show values of an annual interest rate of 5%? Explain.

5.5 THE FAMILY OF QUADRATIC FUNCTIONS

A baseball is "popped" straight up by a batter. The path of the ball is straight up and down. The ball goes up fast at first and then more slowly because of gravity. The height of the ball above the ground is modeled by the function $y = f(t) = -16t^2 + 64t + 3$, where t is time in seconds after the ball leaves the bat and y is in feet. The function is graphed in Figure 5.62.

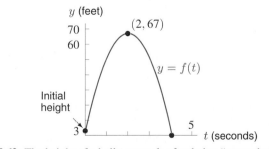

Figure 5.62: The height of a ball t seconds after being "popped up". (Note: This graph does not represent the ball's path)

A natural question to ask is when the ball hits the ground. The graph suggests that $y = 0$ when $t \approx 4$. We can phrase the question symbolically: For what value of t does $f(t) = 0$? Input values of t which make the output $f(t) = 0$ are called *zeros* of f.

We may also ask for the maximum height reached by the baseball. The point on the graph with the largest y-value appears to be $(2, 67)$. This means that the baseball reaches its maximum height of 67 feet 2 seconds after being hit. For this graph, the maximum point $(2, 67)$ is called the *vertex*.

The baseball height function is an example of a *quadratic function*, whose general form is $y = ax^2 + bx + c$. The graph of a quadratic is called a *parabola*; its maximum (or minimum, if the parabola opens upward) is the vertex. In this section, we see how to find the vertex and zeros of a quadratic function algebraically.

The Vertex of a Parabola

The graph of the function $y = x^2$ is a parabola with vertex at the origin. All other functions in the quadratic family turn out to be transformations of this function. Let's first graph a quadratic function of the form $y = a(x - h)^2 + k$ and locate its vertex.

Example 1   Let $f(x) = x^2$ and $g(x) = -2(x + 1)^2 + 3$.

    (a)   Express the function g in terms of the function f.
    (b)   Sketch a graph of f. Transform the graph of f into the graph of g.
    (c)   Multiply out and simplify the formula for g.
    (d)   Explain how the formula for g can be used to obtain the vertex of the graph of g.

Solution   (a)   Since $f(x + 1) = (x + 1)^2$, we have

$$g(x) = -2f(x + 1) + 3.$$

    (b)   The graph of $f(x) = x^2$ is shown at the left in Figure 5.63. The graph of g is obtained from the graph of f in four steps, as shown in Figure 5.63.

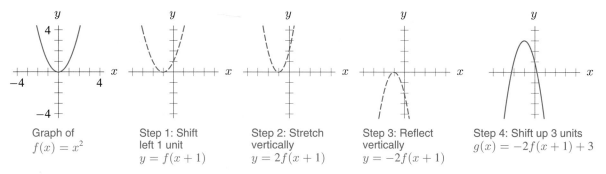

| Graph of $f(x) = x^2$ | Step 1: Shift left 1 unit $y = f(x + 1)$ | Step 2: Stretch vertically $y = 2f(x + 1)$ | Step 3: Reflect vertically $y = -2f(x + 1)$ | Step 4: Shift up 3 units $g(x) = -2f(x + 1) + 3$ |

Figure 5.63: The graph of $f(x) = x^2$, on the left, is transformed in four steps into the graph of $g(x) = -2(x + 1)^2 + 3$, on the right

    (c)   Multiplying out gives $g(x) = -2(x^2 + 2x + 1) + 3 = -2x^2 - 4x + 1$
    (d)   The vertex of the graph of f is $(0, 0)$. In Step 1 the vertex shifts 1 unit to the left (because of the $(x + 1)$ in the formula), and in Step 4 the vertex shifts 3 units up (because of the $+3$ in the formula). Thus, the vertex of the graph of g is at $(-1, 3)$.

In general, the graph of $g(x) = a(x - h)^2 + k$ is obtained from the graph of $f(x) = x^2$ by shifting horizontally $|h|$ units, stretching vertically by a factor of a (and reflecting vertically over the x-axis if $a < 0$), and shifting vertically $|k|$ units. In the process, the vertex is shifted from $(0, 0)$ to the point (h, k). The graph of the function is symmetrical about a vertical line through the vertex, called the *axis of symmetry*.

Formulas for Quadratic Functions

The function g in Example 1 can be written in two ways:

$$g(x) = -2(x + 1)^2 + 3$$

and

$$g(x) = -2x^2 - 4x + 1.$$

The first version is helpful for understanding the graph of the quadratic function and finding its vertex. In general, we have the following:

The **standard form** for a **quadratic function** is

$$y = ax^2 + bx + c, \quad \text{where } a, \ b, \ c \text{ are constants, } a \neq 0.$$

The **vertex form** is

$$y = a(x - h)^2 + k, \quad \text{where } a, \ h, \ k \text{ are constants, } a \neq 0.$$

The graph of a quadratic function is called a **parabola**. The parabola
- Has vertex (h, k)
- Has axis of symmetry $x = h$
- Opens upward if $a > 0$ or downward if $a < 0$

Thus, any quadratic function can be expressed in both standard form and vertex form. To convert from vertex form to standard form, we multiply out the squared term. To convert from standard form to vertex form, we *complete the square*.

Example 2   Put these quadratic functions into vertex form by completing the square and then graph them.

(a) $s(x) = x^2 - 6x + 8$ (b) $t(x) = -4x^2 - 12x - 8$

Solution   (a) To complete the square,[1] find the square of half of the coefficient of the x-term, $(-6/2)^2 = 9$. Add and subtract this number after the x-term:

$$s(x) = \underbrace{x^2 - 6x + 9}_{\text{Perfect square}} - 9 + 8,$$

so

$$s(x) = (x - 3)^2 - 1.$$

---

[1] A more detailed explanation of this method is in the Appendix on page 520.

The vertex of s is $(3, -1)$ and the axis of symmetry is the vertical line $x = 3$. There is no vertical stretch since $a = 1$, and the parabola opens upward. See Figure 5.64.

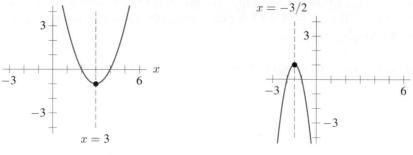

Figure 5.64: $s(x) = x^2 - 6x + 8$ Figure 5.65: $t(x) = -4x^2 - 12x - 8$

(b)   To complete the square, first factor out -4, the coefficient of x^2, giving

$$t(x) = -4(x^2 + 3x + 2).$$

Now add and subtract the square of half the coefficient of the x-term, $(3/2)^2 = 9/4$, inside the parentheses. This gives

$$t(x) = -4\left(\underbrace{x^2 + 3x + \frac{9}{4}}_{\text{Perfect square}} - \frac{9}{4} + 2 \right)$$

$$t(x) = -4\left(\left(x + \frac{3}{2} \right)^2 - \frac{1}{4} \right)$$

$$t(x) = -4\left(x + \frac{3}{2} \right)^2 + 1.$$

The vertex of t is $(-3/2, 1)$, the axis of symmetry is $x = -3/2$, the vertical stretch factor is 4, and the parabola opens downward. See Figure 5.65.

Finding a Formula for a Parabola

If we know the vertex of a quadratic function and one other point, we can use the vertex form to find its formula.

Example 3   Find the formula for the quadratic function graphed in Figure 5.66.

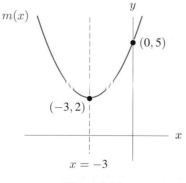

Figure 5.66

Solution   Since the vertex is given, we use the form $m(x) = a(x - h)^2 + k$ and find a, h, and k. The vertex is $(-3, 2)$, so $h = -3$ and $k = 2$. Thus,

$$m(x) = a(x - (-3))^2 + 2,$$

so

$$m(x) = a(x + 3)^2 + 2.$$

To find a, use the y-intercept $(0, 5)$. Substitute $x = 0$ and $y = m(0) = 5$ into the formula for $m(x)$ and solve for a:

$$5 = a(0 + 3)^2 + 2$$
$$3 = 9a$$
$$a = \frac{1}{3}.$$

Thus, the formula is

$$m(x) = \frac{1}{3}(x + 3)^2 + 2.$$

If we want the formula in standard form, we multiply out:

$$m(x) = \frac{1}{3}x^2 + 2x + 5.$$

If a parabola has three intercepts and we know them all, we can use the standard form to find its formula.

Example 4   Find a formula for the parabola in Figure 5.67.

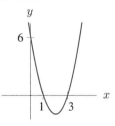

Figure 5.67

Solution   In standard form the parabola has the equation

$$y = ax^2 + bx + c.$$

Since the y-intercept is 6, we know the point $x = 0, y = 6$ satisfies the equation. Substituting gives

$$6 = a(0)^2 + b(0) + c$$
$$c = 6$$

so $y = ax^2 + bx + 6$.

The parabola goes through the points $x = 1, y = 0$ and $x = 3, y = 0$ so we have

$$0 = a(1)^2 + b(1) + 6$$
$$0 = a(3)^2 + b(3) + 6$$

giving

$$a + b = -6$$
$$9a + 3b = -6.$$

To solve these simultaneous equations, first divide both sides of the second equation by 3 to get

$$a + b = -6$$
$$3a + b = -2.$$

Then subtract the first equation from the second to get

$$2a = 4,$$

so

$$a = 2.$$

Now solve for b, using $a + b = -6$:

$$2 + b = -6$$
$$b = -8.$$

Thus we have found a, b, and c, and the equation of the parabola is

$$y = 2x^2 - 8x + 6.$$

___

If we know three points on a parabola, but not the intercepts, it is still possible to find a, b, c for the standard form. However, the algebra may be harder.

Finding the Zeros of a Quadratic Function

The zeros of a function f are values of x for which $f(x) = 0$. It is easy to find the zeros of a quadratic function if its formula can be factored.

___

Example 5   Find the zeros of $f(x) = x^2 - x - 6$.

Solution   To find the zeros, set $f(x) = 0$ and solve for x:

$$x^2 - x - 6 = 0$$
$$(x - 3)(x + 2) = 0.$$

Thus the zeros are $x = 3$ and $x = -2$.

___

Any quadratic function can be expressed both in standard form and in vertex form. Some quadratic functions can also be expressed in *factored form*,

$$q(x) = a(x - r)(x - s),$$

where a, r, and s are constants, $a \neq 0$. Note that r and s are zeros of the function q. The factored form of the function f in Example 5 is $f(x) = (x - 3)(x + 2)$.

___

Example 6   Find the equation of the parabola in Example 4 using the factored form.

Solution   Since the parabola has x-intercepts at $x = 1$ and $x = 3$, its formula is

$$y = a(x - 1)(x - 3)$$

Substituting $x = 0, y = 6$ gives

$$6 = a(3)$$
$$a = 2.$$

Thus, the equation is

$$y = 2(x - 1)(x - 3).$$

Multiplying out gives $y = 2x^2 - 8x + 6$, the same result as before.

Many quadratic functions, however, cannot be easily factored.

Example 7   Find the zeros of $g(x) = x^2 - \dfrac{8}{3}x - 1$.

Solution   To find the zeros of g, set $g(x) = 0$:

$$x^2 - \frac{8}{3}x - 1 = 0.$$

Since the formula for g cannot be factored easily, we complete the square. We take $(-8/3)/2 = -4/3$, and $(-4/3)^2 = 16/9$. Add and subtract 16/9 to make a perfect square:

$$\underbrace{x^2 - \frac{8}{3}x + \frac{16}{9}}_{\text{Perfect square}} \underbrace{- \frac{16}{9} - 1}_{-\frac{16}{9} - 1 = -\frac{25}{9}} = 0$$

$$\left(x - \frac{4}{3}\right)^2 - \frac{25}{9} = 0$$

$$\left(x - \frac{4}{3}\right)^2 = \frac{25}{9}$$

$$x - \frac{4}{3} = \pm\sqrt{\frac{25}{9}} = \pm\frac{5}{3} \qquad \text{(Taking the square root)}$$

$$x = \frac{4}{3} \pm \frac{5}{3} \qquad \text{(Adding 4/3 to both sides).}$$

So the zeros are $x = 3$ and $x = -1/3$.

Alternatively, we can find the zeros of a quadratic function using the quadratic formula, which is derived by completing the square for $y = ax^2 + bx + c$. See the Appendix, page 520. You should check that the quadratic formula gives the zeros of the function in Example 7.

The solutions to the equation $ax^2 + bx + c = 0$ are given by the **quadratic formula**:

$$x = \frac{-b \pm \sqrt{b^2 - 4ac}}{2a}.$$

The zeros of a function occur at the x-intercepts of its graph. Not every quadratic function has x-intercepts, as we see in the next example.

Example 8 Figure 5.68 shows a graph of $h(x) = -\dfrac{1}{2}x^2 - 2$. What happens if we try to use algebra to find its zeros?

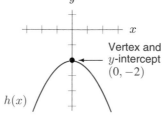

Vertex and
y-intercept
$(0, -2)$

$h(x)$

Figure 5.68

Solution To find the zeros, we solve the equation

$$-\frac{1}{2}x^2 - 2 = 0$$
$$-\frac{1}{2}x^2 = 2$$
$$x^2 = -4$$
$$x = \pm\sqrt{-4}.$$

Since $\sqrt{-4}$ is not a real number, there are no real solutions, so h has no real zeros. This corresponds to the fact that the graph of h in Figure 5.68 does not cross the x-axis.

An Application of Quadratic Functions

In applications, it is often useful to find the maximum or minimum value of a quadratic function. First, we return to the baseball example which started this section.

Example 9 For t in seconds, the height of a baseball in feet is given by the formula

$$y = f(t) = -16t^2 + 64t + 3.$$

Using algebra, find the maximum height reached by the baseball and the time at which the ball reaches the ground.

Solution To find the maximum height, complete the square to find the vertex:

$$\begin{aligned}
y = f(t) &= -16(t^2 - 4t) + 3 \\
&= -16(t^2 - 4t + 4 - 4) + 3 \\
&= -16(t^2 - 4t + 4) - 16(-4) + 3 \\
&= -16(t - 2)^2 + 16 \cdot 4 + 3 \\
&= -16(t - 2)^2 + 67.
\end{aligned}$$

Thus, the vertex is at the point $(2, 67)$. This means that the ball reaches it maximum height of 67 feet at $t = 2$ seconds.

The time at which the ball hits the ground is found by solving $f(t) = 0$. We have

$$-16(t - 2)^2 + 67 = 0$$
$$(t - 2)^2 = \frac{67}{16}$$
$$t - 2 = \pm\sqrt{\frac{67}{16}} \approx \pm 2.05.$$

The solutions are $t \approx -0.05$ and $t \approx 4.05$. Since the ball was thrown at $t = 0$, we want $t \geq 0$. Thus, the ball hits the ground approximately 4.05 seconds after being hit.

Example 10   A city decides to make a park by fencing off a section of riverfront property. Funds are allotted to provide 80 meters of fence. The area enclosed will be a rectangle, but only three sides will be enclosed by fence—the other side will be bound by the river. What is the maximum area that can be enclosed in this way?

Solution   Two sides are perpendicular to the bank of the river and have equal length, which we call h. The other side is parallel to the bank of the river. Call its length b. See Figure 5.69. Since the fence is 80 meters long,

$$2h + b = 80$$
$$b = 80 - 2h.$$

The area of the park, A, is the product of the lengths of two adjacent sides, so

$$A = bh = (80 - 2h)h$$
$$= -2h^2 + 80h.$$

The function $A = -2h^2 + 80h$ is quadratic. Since the coefficient of h^2 is negative, the parabola opens downward and we have a maximum at the vertex. The zeros of this quadratic function are $h = 0$ and $h = 40$, so the axis of symmetry, which is midway between the zeros, is $h = 20$. The vertex of a parabola occurs on its axis of symmetry. Thus, substituting $h = 20$ gives the maximum area:

$$A = (80 - 2(20))(20) = (80 - 40)(20) = (40)(20) = 800 \text{ meter}^2.$$

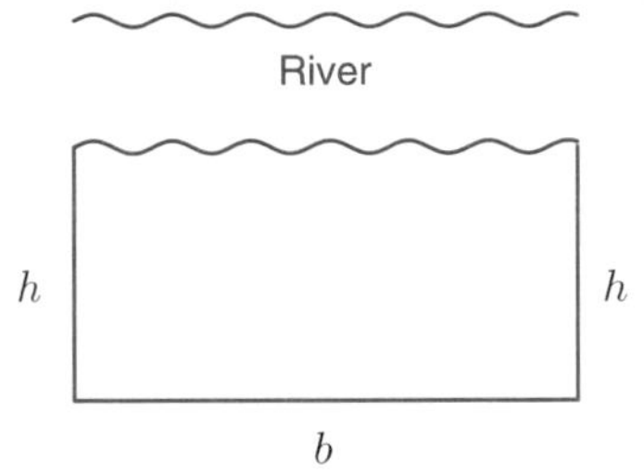

River

h h

b

Figure 5.69

Problems for Section 5.5

1. The vertex of the parabola in Example 3 on page 216 is $(-3, 2)$ and the y-intercept is $(0, 5)$. We used the vertex form of a quadratic to find its formula. Now use the standard form to obtain the same result.

2. The intercepts of the parabola in Example 4 on page 217 are $x = 1$, $x = 3$, and $y = 6$. We used the standard form of a quadratic to find its equation. Now use the vertex form to obtain the same result.

3. Sketch the following quadratic functions given in standard form. Identify the values of the parameters a, b, and c. Label the zeros, axis of symmetry, vertex, and y-intercept.
   (a) $f(x) = -2x^2 + 4x + 16$ (b) $g(x) = x^2 + 3$

4. Let $f(x) = x^2$ and let $g(x) = (x - 3)^2 + 2$.
   (a) Give the formula for g in terms of f, and describe the relationship between f and g in words.
   (b) Is g a quadratic function? If so, find its standard form and the parameters a, b, and c.
   (c) Sketch a graph of g, labeling all important features.

5. (a) Sketch a graph of the quadratic function $h(x) = -2x^2 - 8x - 8$.
   (b) Compare the graphs of $h(x)$ and $f(x) = x^2$. How are these two graphs related? Be specific.

6. Let f represent a quadratic function whose graph is a concave up parabola with a vertex at $(1, -1)$, and a zero at the origin.

    (a) Sketch a graph of $y = f(x)$.   (b) Determine a formula for $f(x)$.

    (c) Determine the range of f.   (d) Find any other zeros.

For Problems 7–12, give a possible formula for the function graphed.

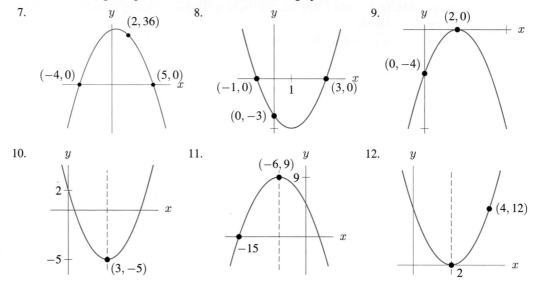

7.

8.

9.

10.

11.

12.

13. Graph $y = x^2 - 10x + 25$ and $y = x^2$. Use a shift transformation to explain the relationship between the two graphs.

14. Find a formula for the quadratic function f with zeros at $x = 1$ and $x = 2$ and vertex at $(3/2, -1)$.

15. If we know a quadratic function f has a zero at $x = -1$ and vertex at $(1, 4)$, do we have enough information to find a formula for this function? If your answer is yes, find it; if not, give your reasons.

16. Without a calculator, graph the function $y = 3x^2 - 16x - 12$ by factoring and plotting zeros.

17. Without a calculator, graph the following function by factoring and plotting zeros:

$$y = -4cx + x^2 + 4c^2 \quad \text{for} \quad c > 0.$$

18. Find two quadratic functions with zeros $x = 1$, $x = 2$.

19. Is there a quadratic function with zeros $x = 1$, $x = 2$ and $x = 3$?

20. Put the equation in vertex form by completing the square and then graph it without a calculator:

$$y - 12x = 2x^2 + 19.$$

21. Complete the square in the function $r(x) = x^2 - 12x + 28$. Find the vertex and axis of symmetry.

22. Find the vertex and axis of symmetry of the graph of $v(t) = t^2 + 11t - 4$.

23. Find the vertex and axis of symmetry of the graph of $w(x) = -3x^2 - 30x + 31$.

24. By completing the square, find the zeros of $y = x^2 + 8x + 5$ exactly.

25. Show that the function $y = -x^2 + 7x - 13$ has no real zeros.

26. Solve for x using the quadratic formula and demonstrate your solution graphically:

    (a) $6x - \frac{1}{3} = 3x^2$ (b) $2x^2 + 7.2 = 5.1x$

27. Suppose $f(x) = ax^2 + bx + c$, with $a \neq 0$. Without any calculation, explain briefly why

$$f\left(\frac{-b + \sqrt{b^2 - 4ac}}{2a}\right) = f\left(\frac{-b - \sqrt{b^2 - 4ac}}{2a}\right) = 0.$$

28. Let $f(x) = x^2$ and $g(x) = x^2 + 2x - 8$.

(a) Sketch graphs f and g in the window $-10 \le x \le 10$, $-10 \le y \le 10$. How are the two graphs similar? How are they different?

(b) Graph f and g in the window $-10 \le x \le 10$, $-10 \le y \le 100$. Why do the two graphs appear more similar on this window than on the window from part (a)?

(c) Graph f and g in the window $-20 \le x \le 20$, $-10 \le y \le 400$, the window $-50 \le x \le 50$, $-10 \le y \le 2500$, and the window $-500 \le x \le 500$, $-2500 \le y \le 250{,}000$. Describe the change in appearance of f and g on these three successive windows.

29. Gwendolyn, a common parabola, was taking a peaceful nap when her dream turned into a nightmare: she dreamt that a low-flying pterodactyl was swooping towards her. Startled, she flipped over the horizontal axis, darted up (vertically) by three units, and to the left (horizontally) by two units. Finally she woke up and realized that her equation was $y = (x-1)^2 + 3$. What was her equation before she had the bad dream?

30. Table 5.21 shows the percentage US households engaged in gardening[2] each year from 1990 to 1993.

(a) Show that this data set can be modeled by the quadratic function $q(x) = -\frac{1}{2}x^2 - \frac{3}{2}x + 80$, where x is in years since 1990.

(b) According to this model, when was the percentage highest? What was that percentage?

(c) When is the percentage predicted to be 50%?

(d) What does this model predict for the year 2000? 2005?

(e) Is this a good model?

TABLE 5.21

Year	1990	1991	1992	1993
Percentage	80	78	75	71

31. The percentage of schools with interactive videodisc players[3] each year from 1992 to 1996 is shown in Table 5.22. If x is in years since 1992, show that this data set can be approximated by the quadratic function $p(x) = -0.8x^2 + 8.8x + 7.2$. What does this model predict for the year 2004? How good is this model for predicting the future?

TABLE 5.22

Year	1992	1993	1994	1995	1996
Percentage	8	14	21	29.1	29.3

32. A ball is thrown into the air. Its height (in feet) t seconds later is given by $h(t) = 80t - 16t^2$.

(a) Evaluate and interpret $h(2)$.

(b) Solve the equation $h(t) = 80$. Interpret your solutions and illustrate them on a graph of $h(t)$.

33. (a) Consider the data in Table 5.23. Using the first three data points, fit a quadratic function to the data. Use the fifth point as a check.

(b) Find the formula for a linear function that passes through the second two data points.

(c) Compare the value of the linear function at $x = 3$ to the value of the quadratic at $x = 3$.

(d) Compare the values of the linear and quadratic functions at $x = 50$.

(e) For approximately what x values do the quadratic function values and the linear function values differ from each other by less than 0.05? Using a calculator or computer, graph both functions on the same axes and estimate an answer.

TABLE 5.23

x	0	1	2	3	50
y	1.0	3.01	5.04	7.09	126.0

---

[2] Data from *The American Almanac: 1995 – 1996.* (Texas: The Reference Press, 1995).
[3] Data from R. Famighetti, ed. *The World Almanac and Book of Facts: 1999.* (New Jersey: Funk and Wagnalls, 1998).

34.   A tomato is thrown vertically into the air at time $t = 0$. Its height, $d(t)$ (in feet), above the ground at time t (in seconds) is given by

$$d(t) = -16t^2 + 48t.$$

(a)   Sketch a graph of $d(t)$.
(b)   Find t when $d(t) = 0$. What is happening to the tomato the first time $d(t) = 0$? The second time?
(c)   When does the tomato reach its maximum height?
(d)   What is the maximum height that the tomato reaches?

35.   An espresso stand finds that the weekly profit for their business is a function of the price they charge per cup. If x is the price (in dollars) of one cup, the weekly profit is $P(x) = -2900x^2 + 7250x - 2900$.

(a)   Approximate the maximum profit and the price per cup that produces that profit.
(b)   Which function, $P(x - 2)$ or $P(x) - 2$, gives a function that has the same maximum profit? What price per cup produces that maximum profit?
(c)   Which function, $P(x + 50)$ or $P(x) + 50$, gives a function where the price per cup that produces the maximum profit remains unchanged? What is the maximum profit?

36.   If you have a string of length 50 cm, what are the dimensions of the rectangle of maximum area that you can enclose with your string? Explain your reasoning. What about a string of length k cm?

37.   Plot the functions $y = 2x$ and $y = x^2 + 1$. Is one graph always above the other? Make a conjecture involving an inequality between these two functions. Show algebraically why your conjecture is true.

38.   A relief package is dropped from an airplane moving at a speed of 500 km/hr. Since the package is initially released with this forward horizontal velocity, it follows a trajectory of projectile motion (instead of dropping straight down). The graph in Figure 5.70 shows the height of the package, h, as a function of the horizontal distance, d, it has traveled since it was dropped.

(a)   From what height was the package released?
(b)   How far away from the spot above which it was released does the package hit the ground?
(c)   Write a formula for $h(d)$. [Hint: The path is a parabola. Since it is not thrown upward, the package starts falling at the vertex of the parabola].

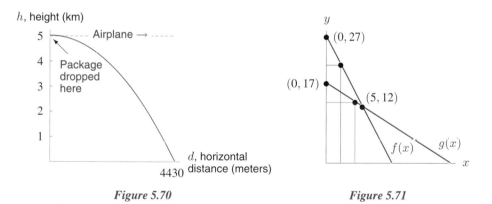

Figure 5.70   Figure 5.71

39.   Figure 5.71 shows two lines, given by $f(x)$ and $g(x)$, with a rectangle inscribed under each one. For each line, we define a function giving the area of the rectangle of base x under that line:

$$R(x) = x \cdot f(x) \qquad S(x) = x \cdot g(x)$$

(a)   Give formulas for $f(x)$ and $g(x)$.
(b)   Solve $f(x) - g(x) = 4$ for x.
(c)   For what value of x will the area of the rectangle under $f(x)$ be 40?
(d)   Define a new function $D(x) = R(x) - S(x)$. What information is given by $D(x)$?
(e)   Find the x-value for which $D(x)$ is a maximum. What does this tell you?

40. If a football player kicks a ball at an angle of $37°$ above the ground with an initial speed of 20 meters/second, then the height, h, as a function of the horizontal distance traveled, d, is given by:

$$h = 0.75d - 0.0192d^2.$$

 (a) Sketch a graph of the path the ball follows.
 (b) When the ball hits the ground, how far is it from the spot where the football player kicked it?
 (c) What is the maximum height the ball reached during its flight?
 (d) What is the horizontal distance the ball has traveled when it reaches its maximum height?[4]

41. A ballet dancer jumps in the air. The height, $h(t)$, in feet, of the dancer at time t, in seconds since the start of the jump, is given by[5]
$$h(t) = -16t^2 + 16Tt,$$

 where T is the total time in seconds that the ballet dancer is in the air.

 (a) Why does this model make sense only for $0 \leq t \leq T$?
 (b) When, in terms of T, does the maximum height of the jump occur?
 (c) Show that the time, T, that the dancer is in the air is related to H, the maximum height of the jump, by the equation
$$H = 4T^2.$$

REVIEW PROBLEMS FOR CHAPTER FIVE

1. Suppose $x = 2$. Determine the value of the input of the function f in each of the following expressions:

 (a) $f(2x)$ (b) $f(\frac{1}{2}x)$ (c) $f(x + 3)$ (d) $f(-x)$

2. Determine the value of x in each of the following expressions which leads to an input of 2 to the function f:

 (a) $f(2x)$ (b) $f(\frac{1}{2}x)$ (c) $f(x + 3)$ (d) $f(-x)$

3. The point $(2, 5)$ is on the graph of $y = f(x)$. Give the coordinates of one point on the graph of each of the following functions.

 (a) $y = f(x - 4)$ (b) $y = f(x) - 4$ (c) $y = f(4x)$ (d) $y = 4f(x)$

4. The point $(-3, 4)$ is on the graph of $y = g(x)$. Give the coordinates of one point on the graph of each of the following functions.

 (a) $y = g(\frac{1}{3}x)$ (b) $y = \frac{1}{3}g(x)$ (c) $y = g(-3x)$ (d) $y = -g(3x)$

5. On a calculator, a student entered the expression $e\hat{\ }x + 1$, which may be interpreted as $e^x + 1$ or as e^{x+1}.

 (a) Are they the same function?
 (b) Which form does your calculator think $e\hat{\ }x + 1$ is?

---

[4]Adapted from R. Halliday, D. Resnick, and K. Krane, *Physics*. (New York: Wiley, 1992), p.58.
[5]K. Laws, *The Physics of Dance*. (Schirmer, 1984).

6.   Without a calculator, match each of the following functions with one of the graphs (I) – (VI).

(a) $y = e^x$

(b) $y = e^{5x}$

(c) $y = 5e^x$

(d) $y = e^{x+5}$

(e) $y = e^{-x}$

(f) $y = e^x + 5$

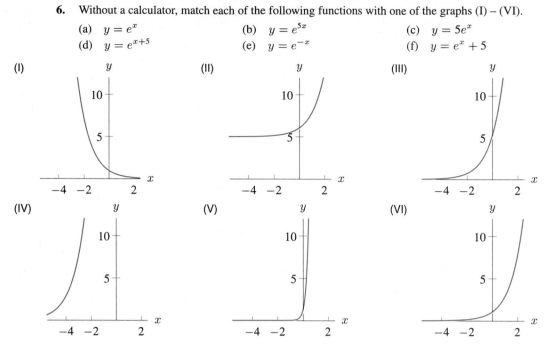

7.   The functions graphed in parts (a)–(c) are transformations of some basic function. Give a possible formula for each one.

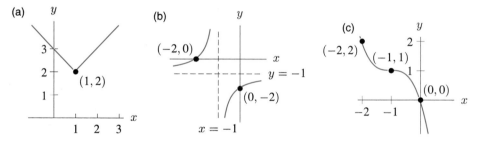

8.   On the graph in Figure 5.72, the value c is labeled on the x-axis. Locate the following quantities on the y-axis: (a) $f(c)$ (b) $f(-c)$ (c) $-f(c)$

9.   On the graph in Figure 5.73, the value of d is labeled on the x-axis. Sketch the graph, then locate the following quantities on the y-axis:

(a) $g(d)$

(b) $g(-d)$

(c) $-g(-d)$

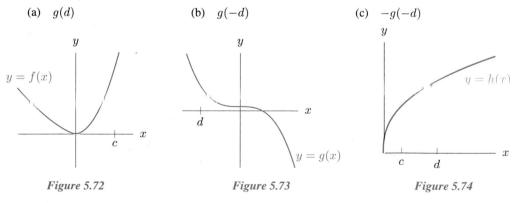

Figure 5.72   Figure 5.73   Figure 5.74

10. On the graph in Figure 5.74, the values c and d are labeled on the x-axis. On the y-axis, locate and label the following quantities:

   (a) $h(c)$ (b) $h(d)$ (c) $h(c+d)$ (d) $h(c) + h(d)$

11. The graph of $f(x) = |x|$ is in Figure 5.75. Without using a calculator, graph the following:

   (a) $y = -f(x)$ (b) $y = f(x+3)$ (c) $y = f(x) - 4$
   (d) $y = 5f(x)$ (e) $y = f(\frac{1}{4}x)$

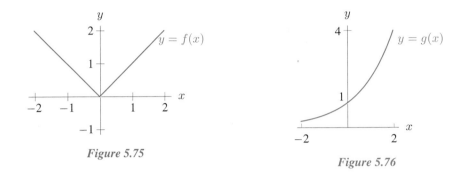

Figure 5.75

Figure 5.76

12. Figure 5.76 shows the graph of $g(x) = 2^x$. Graph the following pairs of functions:

   (a) $y = -g(x)$ and $g(x)$ (b) $y = g(-x)$ and $g(x)$ (c) $y = -g(-x)$ and $g(x)$
   (d) $y = g(x) + 3$ and $g(x)$ (e) $y = 5 - g(x)$ and $g(x)$

13. The signum function, $\text{sign}(x)$, is defined as follows:

$$\text{sign}(x) = 1 \qquad \text{if} \quad x > 0,$$
$$\text{sign}(x) = 0 \qquad \text{if} \quad x = 0,$$
$$\text{sign}(x) = -1 \quad \text{if} \quad x < 0.$$

Sketch graphs of the following transformations of the signum function.

   (a) $y = \text{sign}(2x)$ (b) $y = \text{sign}(-2x)$ (c) $y = 2\text{sign}(x)$ (d) $y = 2 - \text{sign}(x)$

14. Graph and describe the differences between $f(x) = 2^x$, $g(x) = x^2$, and $h(x) = 2x$.

15. Graph $f(x) = 3^x$, $g(x) = 2(3^x)$, and $h(x) = 3^{2x}$. Describe the relationships between the three functions.

16. Graph $f(x) = \sqrt{1 - x^2}$, $g(x) = 2\sqrt{1 - x^2}$, and $h(x) = \sqrt{1 - (2x)^2}$. Describe the relationships between the three graphs.

17. Let $f(x) = ax^2 + bx + c$, with $a \neq 0$. Simplify: $f\left(\frac{-b + \sqrt{b^2 - 4ac}}{2a}\right)$.

18. Suppose $f(x) = (x + a_1)^2$ and $g(x) = (x + a_2)^2$ where a_1 and a_2 are constants. In addition, suppose there is a number $c > 0$ such that $a_1 = -a_2 = 2c$.

   (a) Solve $f(x) = g(x)$. Your answer may involve c but no other constants.
   (b) Sketch $f(x)$ and $g(x)$, labeling the two graphs, the intercepts and the point of intersection you found in part (a).
   (c) As c increases, what happens to the x and y coordinates of the point of intersection?

19. Suppose $f(x) = c(x + c)^2$ and $g(x) = (x + c^2)^2$, for $c > 1$.

   (a) Solve $f(x) = g(x)$. Your answer may contain c.
   (b) Sketch $f(x)$ and $g(x)$ on the same axes. Label the two graphs, the intercepts, and the point(s) of intersection.

Problems 20–24 use Table 5.24 which gives the total cost, C, for a carpenter to build n wooden chairs. We write $C = f(n)$ to represent the cost, C, of building n chairs.

TABLE 5.24

n	0	10	20	30	40	50
$f(n)$	5000	6000	6800	7450	8000	8500

20. Evaluate the following expressions. Explain in everyday terms what they mean.

(a) $f(10)$ (b) $f(x)$ if $x = 30$ (c) z if $f(z) = 8000$ (d) $f(0)$

21. Find approximate values for p and q if $f(p) = 6400$ and $q = f(26)$.

22. Let $d_1 = f(30) - f(20)$, $d_2 = f(40) - f(30)$, and $d_3 = f(50) - f(40)$.

(a)   Evaluate d_1, d_2 and d_3.

(b)   What do these numbers tell you about the carpenter's cost of building chairs?

23. Sketch a graph of $f(n)$. Label the quantities you found in Problems 20–22 on your graph.

24. Suppose the carpenter currently builds k chairs per week.

(a)   What do the following expressions represent?

(i) $f(k + 10)$ (ii) $f(k) + 10$ (iii) $f(2k)$ (iv) $2f(k)$

(b)   If the carpenter sells his chairs at 80% above cost, plus an additional 5% sales tax, write an expression for his gross income (including sales tax) each week.

25. Figure 5.77 gives the graph of $y = h(x)$. Each of the graphs in parts (a) – (c) is a transformation of the function $y = h(x)$. Find a formula for each of these graphs in terms of $h(x)$.

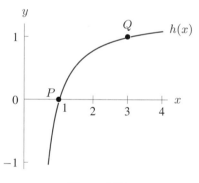

Figure 5.77

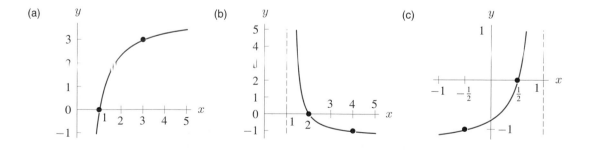

26. Without using a calculator, match each of the following formulas to one of the graphs (i) – (v).

(a) $y = \log(x - a), a > 0$ (b) $y = \log(x/a), a > 1$ (c) $y = \dfrac{1}{\log x}$

(d) $y = \log(x + a), a > 0$ (e) $y = a \log x, a > 0$

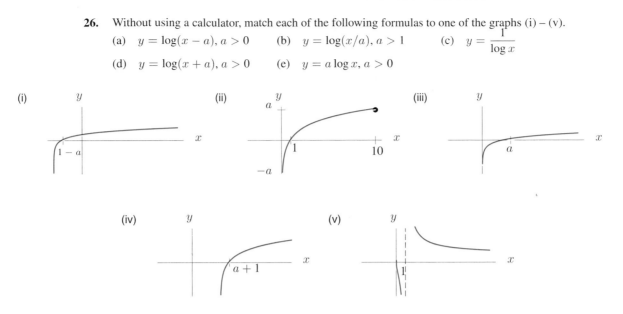

27. The graph of $h = L(d)$ shown in Figure 5.78 gives the number of hours of daylight in Charlotte, North Carolina on a given day d of the year, where $d = 0$ means January 1. Sketch a graph of the number of hours of daylight in Buenos Aires, Argentina, which is as far south of the equator as Charlotte is north. [Hint: When it is summer in the Northern Hemisphere, it is winter in the Southern Hemisphere.]

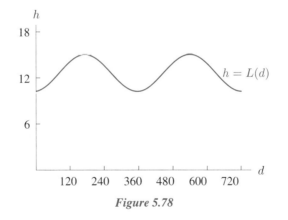

Figure 5.78

28. $T(d)$ is the low temperature in degrees Fahrenheit on the dth day of last year (where $d = 1$ is January 1st, and so on).

(a)   Sketch a possible graph of T for your home town for $1 \le d \le 365$.

(b)   Suppose n is a new function that reports the low temperature on day, d, in terms of degrees above (or below) freezing. (For example, if the low temperature on the 100th day of the year is 42°F, then $n(100) = 42 - 32 = 10$.) Write an expression for $n(d)$ as a shift of T. Graph n. How does the graph of n relate to the graph of T?

(c)   This year, something different has happened. The temperatures for each day are exactly the same as they were last year, except each temperature occurs a week earlier than it did last year. Suppose $p(d)$ gives the low temperature on the dth day of this year. Write an expression for $p(d)$ in terms of $T(d)$. Graph p. How does the graph of p relate to the graph of T?

Filling in the Graph of the Ferris Wheel Function

It is tempting to connect the points in Figure 6.4 with straight lines, but this would be incorrect. Consider the first five minutes of your ride, starting at the 6 o'clock position and ending at the 3 o'clock position. (See Figure 6.5). Halfway through this part of the ride, the wheel has turned halfway from the 6 o'clock to the 3 o'clock position. However, as is clear from Figure 6.5, your seat rises less than half the vertical distance from $y = 0$ to $y = 250$. At the same time, the seat glides more than half the horizontal distance. If the points in Figure 6.4 were connected with straight lines, $f(2.5)$ would be halfway between $f(0)$ and $f(5)$, which is incorrect.

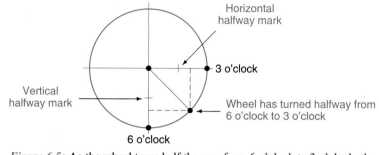

Figure 6.5: As the wheel turns half the way from 6 o'clock to 3 o'clock, the seat rises less than half the vertical distance but glides more than half the horizontal distance

The graph of $f(t)$ in Figure 6.6 is a smooth, wave-shaped curve. The curve repeats itself: It looks the same from $t = 0$ to $t = 20$ as it does from $t = 20$ to $t = 40$, or from $t = 40$ to $t = 60$, or from $t = 60$ to $t = 80$.

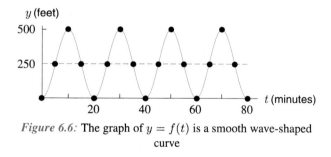

Figure 6.6: The graph of $y = f(t)$ is a smooth wave-shaped curve

Periodic Functions: Period, Midline, and Amplitude

The ferris wheel function, f, is an example of a *periodic function*. The length of the smallest time interval over which a function completes one full cycle is called its *period*. The period of the ferris wheel function, f, is 20 minutes and is represented as a horizontal distance in Figure 6.7.

We can think about the period in terms of horizontal shifts. If the graph of f is shifted to the left by 20 units, the resulting graph looks exactly the same. That is,

$$\underbrace{\text{Graph of } f \text{ shifted}}_{f(t+20)} \text{left by 20 units} \quad \text{is the same as} \quad \underbrace{\text{Original graph,}}_{f(t)}$$

and so

$$f(t + 20) = f(t).$$

This relationship holds for all values of t. In general:

A function f is **periodic** if its values repeat at regular intervals. Graphically, this means that if the graph of f is shifted horizontally by c units, the new graph is identical to the original. In function notation, periodic means

$$f(t + c) = f(t)$$

for all t in the domain of f. The smallest positive constant c for which this relationship holds for all values of t is called the **period** of f.

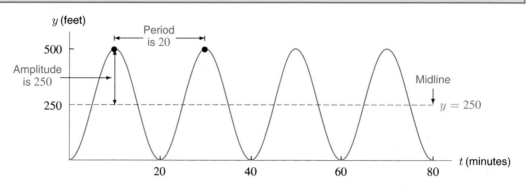

Figure 6.7: The graph of $y = f(t)$ showing the amplitude, period, and midline

In Figure 6.7, the dashed horizontal line is called the *midline* of the graph of f. The midline of a periodic function is the horizontal line on its graph halfway between the function's maximum and minimum values. For the ferris wheel function, the midline is $y = 250$.

The *amplitude* of a wave-like periodic function is the distance between its maximum and the midline (or the distance between the midline and the minimum). Thus the amplitude of f is 250 because the ferris wheel's maximum height is 500 feet and its midline is at 250 feet. The amplitude is represented graphically as a vertical distance. (See Figure 6.7.) In general:

The **midline** of a periodic function is the horizontal line midway between the function's maximum and minimum values. The **amplitude** is the vertical distance between the function's maximum (or minimum) value and the midline.

Problems for Section 6.1

You board the London ferris wheel described in this section. For each of the stories in Problems 1–3, sketch a graph of $h = f(t)$, your height in feet above the ground t minutes after the wheel begins to turn. Label the period, the amplitude, and the midline of each graph, as well as both axes. In each problem, first determine an appropriate interval for t, with $t \geq 0$.

1. The London ferris wheel has increased its rotation speed. The wheel completes one full revolution every ten minutes. You get off when you reach the ground after having made two complete revolutions.

2. Everything is the same as Problem 1 (including the rotation speed) except the developers decide to build the wheel even taller, with a 600 foot diameter.

3. The London ferris wheel is rotating at twice the speed as the wheel in Problem 1.

Problems 4–6 involve different ferris wheels. For each of them, draw a graph of $h = f(t)$. Label the period, the amplitude, and the midline for each graph. In each problem, first determine an appropriate interval for t, $t \geq 0$.

4. A ferris wheel is 50 meters in diameter and must be boarded from a platform that is 5 meters above the ground. Assume the six o'clock position on the ferris wheel is level with the loading platform. The wheel completes one full revolution every 8 minutes. You make two complete revolutions on the wheel, starting at $t = 0$.

5. A ferris wheel is 20 meters in diameter and must be boarded from a platform that is 4 meters above the ground. Assume the six o'clock position on the ferris wheel is level with the loading platform. The wheel completes one full revolution every 2 minutes. At $t = 0$ you are in the twelve o'clock position. You then make two complete revolutions and any additional part of a revolution needed to return to the boarding platform.

6. A ferris wheel is 35 meters in diameter and can be boarded at ground level. The wheel completes one full revolution every 5 minutes. At $t = 0$ you are in the three o'clock position and ascending. You then make two complete revolutions and return to the boarding platform.

The graphs in Problems 7–10 describe your height, $h = f(t)$, above the ground on different ferris wheels, where h is in meters and t is time in minutes. You boarded the wheel before $t = 0$. For each graph, determine the following: your position and direction at $t = 0$, how long it takes the wheel to complete one full revolution, the diameter of the wheel, at what height above the ground you board the wheel, and the length of time the graph shows you riding the wheel. The boarding platform is level with the bottom of the wheel.

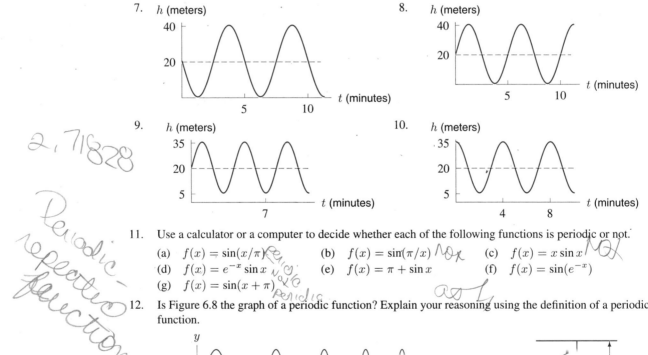

7. 8.

9. 10.

11. Use a calculator or a computer to decide whether each of the following functions is periodic or not.
   (a) $f(x) = \sin(x/\pi)$
   (b) $f(x) = \sin(\pi/x)$
   (c) $f(x) = x \sin x$
   (d) $f(x) = e^{-x} \sin x$
   (e) $f(x) = \pi + \sin x$
   (f) $f(x) = \sin(e^{-x})$
   (g) $f(x) = \sin(x + \pi)$

12. Is Figure 6.8 the graph of a periodic function? Explain your reasoning using the definition of a periodic function.

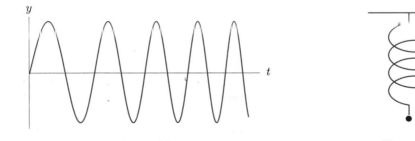

Figure 6.8   *Figure 6.9*

Problems 13–16 concern a weight suspended from the ceiling by a spring. (See Figure 6.9.) Let d be the distance in centimeters from the ceiling to the weight. When the weight is motionless, $d = 10$. If the weight is disturbed, it begins to bob up and down, or *oscillate*. Then d is a periodic function of t, time in seconds, so $d = f(t)$.

13.   Determine the midline, period, amplitude, and the minimum and maximum values of f from the graph in Figure 6.10. Interpret these quantities physically; that is, use them to describe the motion of the weight.

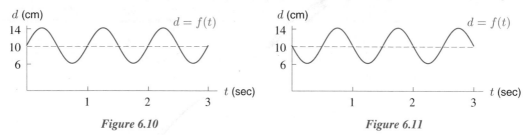

Figure 6.10   *Figure 6.11*

14.   A new experiment with the same weight and spring is represented by Figure 6.11. Compare Figure 6.11 to Figure 6.10. How do the oscillations differ? For both figures, the weight was disturbed at time $t = -0.25$ and then left to move naturally; determine the nature of the initial disturbances.

15.   The weight in Figure 6.9 is gently pulled down to a distance of 14 cm from the ceiling and released at time $t = 0$. Sketch a graph of the motion of the weight for $0 \le t \le 3$.

16.   Figures 6.12 and 6.13 describe the motion of two different weights, A and B, attached to two different springs. Based on these graphs, answer the following questions.

   (a)   Which weight, when not in motion, is closest to the ceiling?
   (b)   Which weight makes the largest oscillations?
   (c)   Which weight makes the fastest oscillations?

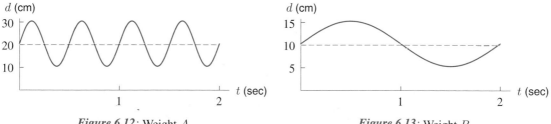

Figure 6.12: Weight A *Figure 6.13:* Weight B

17.   The temperature of a chemical reaction oscillates between a low of $30°C$ and a high of $110°C$. The temperature is at its lowest point when $t = 0$ and completes one cycle over a five-hour period.

   (a)   Sketch a graph of the temperature, T, against the elapsed time, t, over a ten-hour period.
   (b)   Find the period, the amplitude, and the midline of the graph you drew in part (a).

18.   In the US, household electricity is in the form of *alternating current* (AC) at 155.6 volts and 60 hertz. This means that the voltage cycles from -155.6 volts to $+155.6$ volts and back to -155.6 volts, and that 60 cycles occur each second. Suppose that at $t = 0$ the voltage at a given outlet is at 0 volts.

   (a)   Sketch a graph of $V = f(t)$, the voltage as a function of time, for the first 0.1 seconds.
   (b)   State the period, the amplitude, and the midline of the graph you made in part (a). Describe the physical significance of these quantities.

19.   Table 6.2 gives data from a vibrating string experiment, with time, t, in seconds, and height, h, in centimeters. Find the midline, amplitude and period of the function $h = f(t)$ given in the table.

TABLE 6.2

t	0	0.1	0.2	0.3	0.4	0.5	0.6	0.7	0.8	0.9	1
$h = f(t)$	2	2.588	2.951	2.951	2.588	2	1.412	1.049	1.049	1.412	2

20. Table 6.3 gives the height in feet of a weight on a spring where t is the time in seconds. Find the midline, amplitude and period of the function $h = f(t)$ given in the table.

TABLE 6.3

t	0	1	2	3	4	5	6	7
$h = f(t)$	4.0	5.25	6.165	6.5	6.167	5.25	4.0	2.75
t	8	9	10	11	12	13	14	15
$h = f(t)$	1.835	1.5	1.8348	2.75	4.0	5.22	6.16	6.5

21. US imports of petroleum, measured in quadrillion BTUs (a unit of energy), are shown in Table 6.4.

   (a) Graph this data for the period shown.
   (b) What trends in US oil imports do you notice?
   (c) Determine an approximate midline, amplitude and period.

TABLE 6.4

Year	BTU (quadrillion)	Year	BTU (quadrillion)	Year	BTU (quadrillion)	Year	BTU (quadrillion)
1973	14.2	1978	17	1983	13	1988	17
1974	14.5	1979	17	1984	15	1989	18
1975	14	1980	16	1985	12	1990	17
1976	16.5	1981	15	1986	14	1991	16
1977	18	1982	14	1987	16	1992	16

22. The number of white blood cells in a patient with chronic myelogenous leukemia with nearly periodic relapses is given in Table 6.5. Plot these data and estimate the midline, amplitude and period.

TABLE 6.5

Day	0	10	40	50	60	70	75	80	90
WBC ($\times 10^4$/ml)	0.9	1.2	10	9.2	7.0	3.0	0.9	0.8	0.4
Day	100	110	115	120	130	140	145	150	160
WBC ($\times 10^4$/ml)	1.5	2.0	5.7	10.7	9.5	5.0	2.7	0.6	1.0
Day	170	175	185	195	210	225	230	240	255
WBC ($\times 10^4$/ml)	2.0	6.0	9.5	8.2	4.5	1.8	2.5	6.0	10.0

6.2 THE SINE AND COSINE FUNCTIONS

In this section we construct a formula for the ferris wheel function. To do this, we first represent positions on the ferris wheel using angles.

Using Angles to Describe Position on the Ferris Wheel

Imagine the ferris wheel superimposed on a coordinate system with the origin at the center of the ferris wheel and the positive x-axis extending horizontally to the right. Wherever your seat is on the wheel, we can measure the angle that the line from the center of the wheel to your seat makes with this axis. By convention, angles are measured in the counterclockwise direction from the x-axis. Figure 6.14 shows the seat in two different positions on the wheel. The first position is at 1 o'clock, which corresponds to a 60° angle. The second position is at 10 o'clock, and it corresponds to a 150°

angle. A point at the 3 o'clock position corresponds to an angle of $0°$.

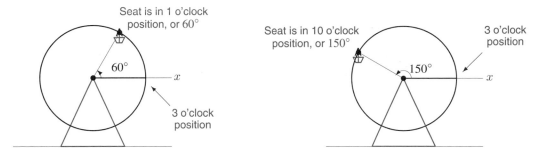

Figure 6.14: The 1 o'clock position forms a $60°$ angle with the positive x-axis, and the 10 o'clock position forms a $150°$ angle

Using Angles to Describe Rotations

It is useful to think of angles as rotations, since then we can make sense of angles of over $360°$ and of negative angles. For instance, as the ferris wheel turns through an angle of $360°$, it carries you once around the circle and brings you back to where you started. If it turns through an angle of $720°$, it completes two full revolutions; it completes three revolutions as it turns through an angle of $1080°$. Thus, the angles $0°$, $360°$, $720°$, $1080°$ all leave you at the same position (the positive x-axis), but each represents a different rotation. See Figure 6.15. Angles larger than $360°$ are drawn so that they wrap around the circle more than once. (Notice that in the figure we assume your ride starts on the x-axis in the 3 o'clock position.)

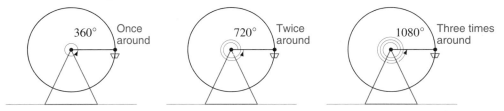

Figure 6.15: The angles $0°$, $360°$, $720°$ and $1080°$ all specify the same position on the ferris wheel, but when thought of as rotations they are different

What about negative angles? For instance, what does $-90°$ mean? Negative angles represent rotations in the clockwise (opposite) direction. Since $90°$ is one-quarter of a turn from $0°$ in the *counterclockwise* direction, $-90°$ is one-quarter of a turn in the *clockwise* direction. Figure 6.16 shows that $-90°$ and $270°$ leave you at the same point (the 6 o'clock position), even though the two angles represent different rotations. (A turn through $270°$ is a three-quarter turn in the counterclockwise direction.)

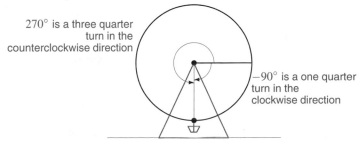

Figure 6.16: The angles $-90°$ and $270°$

Example 1   A revolving door has three glass panels. Each panel is one meter in width and separates the door into three compartments of the same size. See Figure 6.17. The points A, B, C, D, E, and F are equally spaced on a circle of radius one meter.

(a)   What is the angle θ?

(b)   What is the angle swept out as a panel moves from C to D in Figure 6.17 in a counterclockwise direction? A clockwise direction?

(c)   A child playing in the hotel lobby pushes the door from C through an angle of 3900° before his father stops him. Describe the motion of the door in words.

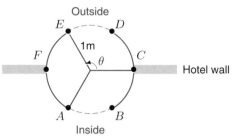

Figure 6.17: A three panel revolving door

Solution   (a)   Since there are 360° in a circle and since the angles between each panel are equal,

$$\theta = \frac{360°}{3} = 120°.$$

(b)   The angle θ between C and E is 120°. The counterclockwise angle from C to D is one half of this, or 60°. Turning clockwise, the angle is −300°.

(c)   The door completes ten full rotations (3600°), and then turns an additional 300°. Thus the panel the child was pushing comes to rest at B.

The Unit Circle

The *unit circle* is the circle of radius one that is centered at the origin. (See Figure 6.18.) Since the distance from the point P with coordinates (x, y) to the origin is 1, we have

$$\sqrt{x^2 + y^2} = 1,$$

so squaring both sides gives the equation of the circle:

$$x^2 + y^2 = 1.$$

Angles can be used to locate points on the unit circle as we did on the ferris wheel. On the unit circle, angles are measured counterclockwise from the positive x-axis.

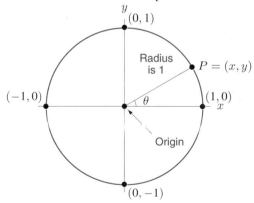

Figure 6.18: The unit circle, with equation $x^2 + y^2 = 1$

Example 2   Figure 6.19 shows the point P determined by $\theta = 90°$. The coordinates of P are $(0, 1)$. Figure 6.20 shows the point Q corresponding to $\alpha = 180°$. It has coordinates $(-1, 0)$. Figure 6.21 shows the point R determined by the angle[2] $\phi = 210°$. Shortly, we will see how to find the coordinates of point R.

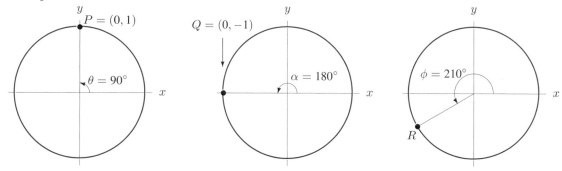

Figure 6.19: The angle $\theta = 90°$ specifies the point $P = (0, 1)$ on the unit circle

Figure 6.20: The angle $\alpha = 180°$ specifies the point Q on the unit circle

Figure 6.21: The angle $\phi = 210°$ specifies the point R on the unit circle

Definition of Sine and Cosine

In Example 2, angles were used to designate points on the unit circle. The trigonometric functions *sine* and *cosine* give the coordinates of a point in terms of its angle.

> Suppose $P = (x, y)$ is the point on the unit circle specified by the angle θ. We define the functions, **cosine** of θ, or $\cos \theta$, and the **sine** of θ, or $\sin \theta$, by the formulas
>
> $$\cos \theta = x \qquad \text{and} \qquad \sin \theta = y.$$
>
> In other words, $\cos \theta$ is the x-coordinate of the point on the unit circle specified by the angle θ and $\sin \theta$ is the y-coordinate. (See Figure 6.22.)

Notice that we often omit the parentheses around the independent variable, writing $\cos \theta$ instead of $\cos(\theta)$. The independent variable here is θ, not x. In fact, x and y are both functions of θ.

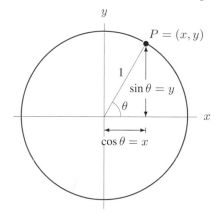

Figure 6.22: For point P, $\cos \theta$ is the x-coordinate and $\sin \theta$ is the y-coordinate

---

[2]The Greek letters θ, pronounced theta, α, pronounced alpha, and ϕ, pronounced phi, are often used for the names of angles. Other Greek letters are also used.

Values of the Sine and Cosine Functions

In principle, we can find values of $\sin\theta$ and $\cos\theta$ for any value of θ. However, there are no easy algebraic formulas for calculating the sine and cosine in terms of θ. So in practice we rely on tables of values, or on calculators and computers. However, there are some values of $\sin\theta$ and $\cos\theta$ we can find on our own.

Example 3   Find the values of $\cos 90°$, $\sin 90°$, $\cos 180°$, $\sin 180°$, $\cos 210°$, $\sin 210°$.

Solution   See Figure 6.19 on page 241. The point P has coordinates $(0, 1)$, so

$$\cos 90° = 0 \qquad \text{and} \qquad \sin 90° = 1.$$

In Figure 6.20, point Q has coordinates $(-1, 0)$, so

$$\cos 180° = -1 \qquad \text{and} \qquad \sin 180° = 0.$$

In Figure 6.21 on page 241, we do not know the coordinates of point R. To find the values of $\cos 210°$ and $\sin 210°$, we use a calculator, which tells us that, approximately

$$\cos 210° = -0.866 \qquad \text{and} \qquad \sin 210° = -0.500.$$

Notice that $\cos 210°$ and $\sin 210°$ are both negative because the point R is in the third quadrant.

Example 4   The point Q on the unit circle is designated by the angle $130°$, as shown in Figure 6.23.
(a)   Find the coordinates of point Q.
(b)   Find the lengths of the line segments labeled m and n in Figure 6.23.

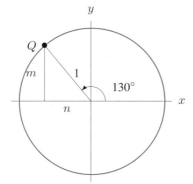

Figure 6.23: The point Q designated by $130°$ on the unit circle

Solution   (a)   The coordinates of the point Q are $(\cos 130°, \sin 130°)$. A calculator gives, approximately,

$$\cos 130° = -0.643 \qquad \text{and} \qquad \sin 130° = 0.766.$$

(b)   The length of line segment m is the same as the y-coordinate of point Q, so $m = \sin 130° = 0.766$. Note that the x-coordinate of Q is negative because Q is in the second quadrant. The length of line segment n has the same magnitude as the x-coordinate of Q, but it has opposite sign, because lengths are always positive. Thus, the length of $n = -\cos 130° = 0.643$.

Coordinates of a Point on a Circle of Radius *r*

Using the sine and cosine, we can find the coordinates of points on circles of any size. Figure 6.24 shows two concentric circles: the inner circle has radius 1 and the outer circle has radius r. The angle θ designates point P on the unit circle and point Q on the larger circle. We know that the coordinates of point $P = (\cos\theta, \sin\theta)$. We want to find (x, y), the coordinates of Q.

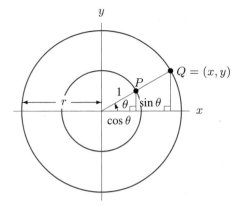

Figure 6.24: Points P and Q on circles of different radii specify the same angle θ

The coordinates of P and Q are the lengths of the sides of the two right triangles in Figure 6.24. Since both right triangles include the angle θ, they are similar and their sides are proportional. This means that the larger triangle is a "magnification" of the smaller triangle. Since the radius of the large circle is r times the radius of the small, we have

$$\frac{x}{\cos\theta} = \frac{r}{1} \qquad \text{and} \qquad \frac{y}{\sin\theta} = \frac{r}{1}.$$

Solving for x and y gives us the following result:

The coordinates (x, y) of the point Q in Figure 6.24 are given by

$$x = r\cos\theta \qquad \text{and} \qquad y = r\sin\theta.$$

Example 5 Find the coordinates of the points A, B, and C in Figure 6.25 to two decimal places.

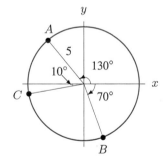

Figure 6.25: Finding coordinates of points on a circle of radius $r = 5$

Solution Since the circle has radius 5, the coordinates of point A are given by

$$x = 5\cos 130° = 5(-0.643) \approx -3.2,$$
$$y = 5\sin 130° = 5(0.766) \approx 3.8.$$

Point B corresponds to an angle of $-70°$, (because the angle is measured clockwise), so B has coordinates

$$x = 5\cos(-70°) = 5(0.342) \approx 1.7,$$
$$y = 5\sin(-70°) = 5(-0.940) \approx -4.7.$$

For point C, we must first calculate the corresponding angle, since the $10°$ is not measured from the positive x-axis. The angle we want is $180° + 10° = 190°$, so

$$x = 5\cos(190°) = 5(-0.985) \approx -4.9,$$
$$y = 5\sin(190°) = 5(-0.174) \approx -0.9.$$

Example 6   In Figure 6.26, write the height of the point P above the x-axis as a function of the angle θ.

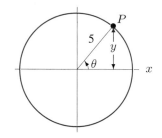

Figure 6.26: Height of point P above x-axis

Solution   We want y as a function of the angle θ. Since the radius of the circle is 5, we have

$$y = 5\sin\theta.$$

Height on the Ferris Wheel as a Function of Angle

We can now find a formula for the height on the London ferris wheel as a function of the angle θ.

Example 7   The ferris wheel described in Section 6.1 has a radius of 250 feet. Find your height above the ground as a function of the angle θ measured from the 3 o'clock position. What is your height when the angle is $60°$?

Solution   Think of the ferris wheel as a circle centered at the origin, with your position at point P as shown in Figure 6.27. Since $r = 250$, the y-coordinate of point P is given by

$$y = r\sin\theta = 250\sin\theta.$$

Your height above the ground is given by $250 + y$, so the formula for your height in terms of θ is

$$\text{Height} = 250 + y = 250 + 250\sin\theta.$$

When $\theta = 60°$, we have

$$\text{Height} = 250 + 250\sin 60°$$
$$= 250 + 250(0.866) \approx 466.5 \text{ feet.}$$

So you are approximately 466.5 feet above the ground.

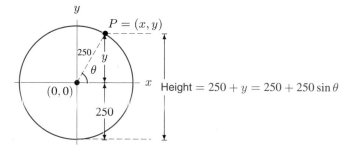

Figure 6.27: Height above the 3 o'clock position is given by the value of y; height above the ground is given by $250 + y$

Problems for Section 6.2

1. Mark the following angles on a unit circle and give the coordinates of the point determined by each angle.

    (a) 100°   (b) 200°   (c) −200°   (d) −45°   (e) 1000°   (f) −720°

For Problems 2–5, sketch the position of the point corresponding to each angle on the unit circle and find the coordinates of each point.

2. S is at 225°, T is at 270°, and U is at 330°   3. A is at 390°, B is at 495°, and C is at 690°

4. D is at −90°, E is at −135°, and F is at −225°   5. P is at 540°, Q is at −180°, and R is at 450°

6. Suppose the angles in Problem 2 are on a circle of radius 5, evaluate the coordinates of S, T and U.

7. Suppose the angles in Problem 3 are on a circle of radius 3, evaluate the coordinates of A, B and C.

8. Describe the angles $\phi = 420°$ and $\theta = -150°$ in terms of a displacement on the ferris wheel, assuming the car of the ferris wheel starts in the 3 o'clock position. What position on the wheel do these angles indicate?

9. (a) Given that $P \approx (0.71, 0.71)$ is a point on the unit circle with angle 45°, estimate sin 135° and cos 135°.

    (b) Given that $Q \approx (0.26, 0.97)$ is a point on the unit circle with angle 75°, estimate sin 285° and cos 285°.

10. Find an angle θ, with $0° < \theta < 360°$, that has the same (a) Cosine as 240° (b) Sine as 240°

11. Find an angle ϕ, with $0° < \phi < 360°$, that has the same (a) Cosine as 53° (b) Sine as 53°

12. Find approximations to two decimal places for the coordinates of point Z in Figure 6.28.

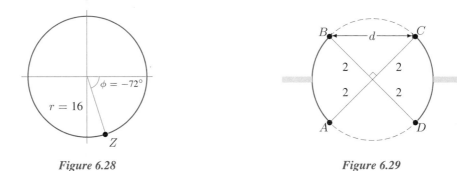

Figure 6.28   **Figure 6.29**

13. A four-panel revolving door is shown in Figure 6.29. What is the width of the opening if each panel is two meters wide?

14. The revolving door in Figure 6.30 rotates counterclockwise and has four equally spaced panels.

    (a) What is the angle between two adjacent panels?
    (b) What is the angle created by a panel rotating from B to A?
    (c) When the door is as shown in Figure 6.30, a person going outside rotates the door from D to B. What is this angle of rotation?
    (d) If the door is initially as shown in Figure 6.30, a person coming inside rotates the door from B to D. What is this angle of rotation?

(e)   The door starts in the position shown in Figure 6.30. Where is the panel at A after three people enter and five people exit? Assume that people going in and going out do so in the manner described in parts (d) and (c), and that each person goes completely through the door before the next enters.

<table>
<tr><td>Figure 6.30</td><td>Figure 6.31</td></tr>
</table>

15.   A revolving door (which rotates counterclockwise in Figure 6.31) was designed with five equally spaced panels for the entrance to the Pentagon. The arcs BC and AD have equal length.

(a)   What is the angle between two adjacent panels?
(b)   A Four Star General enters by pushing on the panel at point B, and leaves the panel at point D. What is the angle of rotation?
(c)   With the door in the position shown in Figure 6.31, an Admiral leaves the Pentagon by pushing the panel between A and D to point B. What is the angle of rotation?

16.   For the angle ϕ shown in Figure 6.32, sketch each of the following angles.

(a) $180 + \phi$ (b) $180 - \phi$ (c) $90 - \phi$ (d) $360 - \phi$

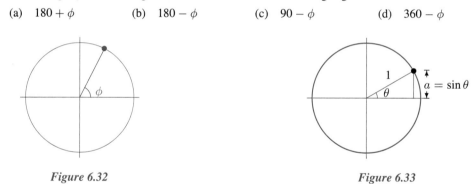

Figure 6.32   Figure 6.33

17.   Let θ be an angle in the first quadrant, and suppose $\sin \theta = a$. Evaluate the following expressions in terms of a. (See Figure 6.33.)

(a) $\sin(\theta + 360°)$ (b) $\sin(\theta + 180°)$ (c) $\cos(90° - \theta)$
(d) $\sin(180° - \theta)$ (e) $\sin(360° - \theta)$ (f) $\cos(270° - \theta)$

18.   Explain in your own words the definition of $\sin \theta$ on the unit circle (θ in degrees).

19.   You have been riding on the London ferris wheel (see Section 6.1) for 11 minutes and 40 seconds. What is your height above the ground?

20.   A ferris wheel is 20 meters in diameter and makes one revolution every 4 minutes. For how many minutes of any revolution will your seat be above 15 meters?

21.   A compact disc is 120 millimeters across with a center hole of diameter 15 millimeters. The center of the disc is at the origin. What are the coordinates of the points at which the inner and outer edge intersect the positive x-axis? What are the coordinates of the points at which the inner and outer edges cut a line making an angle θ with the positive x-axis?

6.3 RADIANS

So far we have measured angles in degrees. There is another way to measure an angle, which involves comparing the length of arc that the angle cuts off on a circle to the radius of the circle. This is the idea behind radians; it turns out to be very helpful in calculus.

Definition of a Radian

The arc length *spanned*, or cut off, by an angle is shown in Figure 6.34.

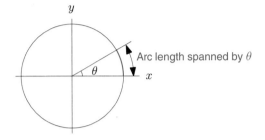

Figure 6.34: Arc length spanned by the angle θ

If the radius of a circle is fixed, (say the radius is 1), the arc length is completely determined by the angle θ. Following Figure 6.35, we make the following definition:

> An angle of **1 radian** is defined to be the angle, in the counterclockwise direction, at the center of a unit circle which spans an arc of length 1.

The radius and arc length must be measured in the same units.

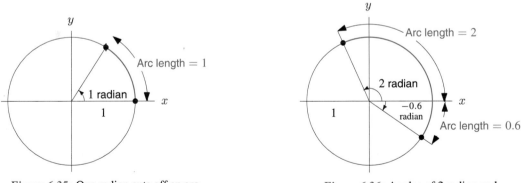

Figure 6.35: One radian cuts off an arc length of one in a unit circle

Figure 6.36: Angles of 2 radian and -0.6 radian

An angle of 2 radians cuts off an arc of length 2 in a unit circle; an angle of -0.6 radian is measured clockwise and cuts off an arc of length 0.6. See Figure 6.36. In general:

> The radian measure of a positive angle is the length of the arc spanned by the angle in a unit circle. For a negative angle, the radian measure is the negative of the arc length.

Relationship Between Radians and Degrees

The circumference, C, of a circle of radius r is given by

$$C = 2\pi r.$$

In a unit circle, $r = 1$, so $C = 2\pi$. This means that the arc length spanned by a complete revolution of $360°$ is 2π, so

$$360° = 2\pi \text{ radians}.$$

Dividing by 2π gives

$$1 \text{ radian} = \frac{360°}{2\pi} \approx 57.3°.$$

Thus, one radian is approximately $57.3°$. One-quarter revolution, or $90°$, is equal to $\frac{1}{4}(2\pi)$ or $\pi/2$ radians. Figure 6.37 shows several positions on the unit circle described in radian measure. Since $\pi \approx 3.14$, one complete revolution is about 6.28 radians and one-quarter revolution is about 1.57 radians.

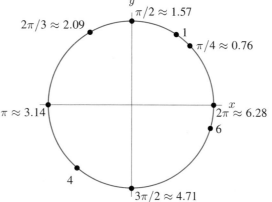

Figure 6.37: Angles in radian measure marked on circle

Example 1   In which quadrant is an angle of 2 radians? An angle of 5 radians?

Solution   Refer to Figure 6.37. The second quadrant includes angles between $\pi/2$ and π, (that is, between 1.57 and 3.14 radians), so 2 radians lies in the second quadrant. An angle of 5 radians is between 4.71 and 6.28, that is, between $3\pi/2$ and 2π radians, so 5 radians lies in the fourth quadrant.

Converting Between Degrees and Radians

To convert degrees to radians, or vice versa, we use the fact that 2π radians $= 360°$. So

$$1 \text{ radian} = \frac{180°}{\pi} \approx 57.3°.$$

Similarly

$$1° = \frac{\pi}{180} \approx 0.01745 \text{ radians}.$$

Thus, to convert from radians to degrees, multiply the radian measure by $180°/\pi$ radians. To convert from degrees to radians, multiply the degree measure by π radians$/180°$.

Example 2   (a)   Convert 3 radians to degrees.   (b)   Convert 3 degrees to radians.

Solution   (a) $3 \text{ radians} \times \dfrac{180°}{\pi \text{ radians}} = \dfrac{540°}{\pi \text{ radians}} \approx 171.9°$.

   (b) $3° \times \dfrac{\pi \text{ radians}}{180°} = \dfrac{\pi \text{ radians}}{60} \approx 0.05 \text{ radians}$.

The word radians is often dropped, so if an angle or rotation is referred to without units, it is understood to be in radians. We can write, for instance, $90° = \pi/2$ and $\pi = 180°$.

Example 3   Find the arc length spanned by an angle of 30° in a circle of radius 1 meter.

Solution   Convert 30° to radians, giving

$$30° = 30° \cdot \frac{\pi}{180°} = \frac{\pi}{6}.$$

The definition of a radian tells us that the arc length spanned by an angle of $\pi/6$ in a circle of radius 1 meter is $\pi/6 \approx 0.52$ meter. See Figure 6.38.

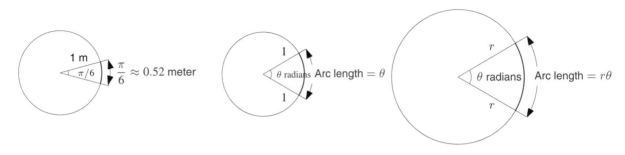

Figure 6.38: Arc length spanned by $30° = \pi/6$ radian in a unit circle

Figure 6.39: Arc length spanned by angle θ is proportional to the radius of the circle

Arc Length

We defined a radian using arc length in a unit circle. However, radians can be used to calculate arc length in a circle of any size. An angle of θ radians spans an arc of length θ in a unit circle. An angle of θ radians spans an arc of length $r\theta$ in a circle of radius r. Figure 6.39 reflects the following result:

> The **arc length**, s, spanned in a circle of radius r by an angle of θ radians, $0 \leq \theta \leq 2\pi$, is given by
> $$s = r\theta.$$

Thus if the size of an angle is fixed, the arc length it spans is proportional to the radius of the circle. (See Figure 6.40). Note that θ must be in radians in this arc length formula.

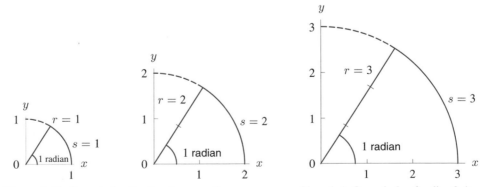

Figure 6.40: On a circle of radius 1, one radian spans an arc of length 1. On a circle of radius 2, it spans an arc of length 2; on a circle of radius 3, it spans an arc of length 3

Example 4   What length of arc is cut off by an angle of 120° on a circle of radius 12 cm?

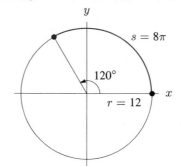

Figure 6.41: An arc cut off by 120°

Solution   Converting 120° to radians gives

$$\text{Angle} = 120° \cdot \frac{\pi}{180°} = \frac{2}{3}\pi \text{ radians},$$

so

$$\text{Arc length} = 12 \cdot \frac{2}{3}\pi = 8\pi \text{ cm.}$$

Example 5   You walk 4 miles around a circular lake. Give an angle in radians which represents your final position relative to your starting point if the radius of the lake is: (a)   1 mile   (b)   3 miles

Solution   (a)   The path around the lake is a unit circle, so you have traversed an angle in radians equal to the arc length traveled, 4 miles. An angle of 4 radians is in the middle of the third quadrant relative to your starting point. (See Figure 6.42.)

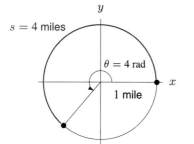

Figure 6.42: Arc length 4 and radius 1, so angle is 4 radians

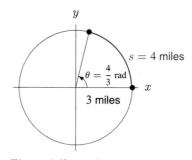

Figure 6.43: Arc length 4 and radius 3, so angle is 4/3 radians

(b)   See Figure 6.43. To find θ, use $s = r\theta$ giving

$$\theta = \frac{s}{r}.$$

The angle you have traversed in

$$\frac{s}{r} = \frac{4}{3} \approx 1.33 \text{ radians.}$$

One-quarter revolution is $\pi/4 \approx 1.57$ radians, so you are nearly one-quarter of the way around the lake.

Problems for Section 6.3

Determine the radian measure of the angles in Problems 1–4.

1. 45° $\pi/4 = .785$ 2. 60° $\frac{\pi}{3} = 1.047$ 3. 17° $\frac{17\pi}{180} = .28706$ 4. 100° $\frac{5}{9}\pi = 1.74533$

5. Convert each of the following angles to radians in two forms: as a multiple of π and as a decimal approximation rounded to two decimal places.

    (a) 30° $\frac{\pi}{6} = .52$ (b) 120° 2.09   (c) 200° $\frac{10\pi}{9} = 3.49$ (d) 315° $\frac{7\pi}{4} = 5.50$

6. An angle with radian measure $\pi/4$ corresponds to a point on the unit circle in quadrant I. Give the quadrants corresponding to angles with the following radian measures:

    (a) 1 radian   (b) 2 radians   (c) 3 radians   (d) 4 radians   (e) 5 radians
    (f) 6 radians   (g) 7 radians   (h) 8 radians   (i) 9 radians   (j) 10 radians

7. A radian can be defined to be the angle at the center of a circle which cuts off an arc of length equal to the radius of the circle. This is because the arc length formula in radians, $s = r \cdot \theta$, becomes just $s = r$ when $\theta = 1$. Figure 6.44 shows circles of radii 2 cm and 3 cm.

    (a) Does the angle 1 radian appear to be the same in both circles?
    (b) Estimate the number of arcs of length 2 cm that fit in the circumference of the circle of radius 2 cm.

Figure 6.44

8. Without using a calculator, give the sign of each of the following numbers:

    (a) cos 3   (b) sin 4   (c) sin(−4)   (d) cos 7

9. Without using a calculator, rank the following in order from smallest to largest.

    (a) The angles: $\dfrac{2\pi}{3}, 2.3, \dfrac{2}{3}, -\dfrac{2\pi}{3}$

    (b) The numbers: $\cos\left(\dfrac{2\pi}{3}\right), \cos(2.3), \cos\left(\dfrac{2}{3}\right), \cos\left(-\dfrac{2\pi}{3}\right)$

10. Explain in your own words how to determine the radian measure of an angle given in degrees, and why this method works.

11. What is the angle determined by an arc of length 2π meters on a circle of radius 18 meters? $\frac{\pi}{9} = .34906 = 20°$

12. What is the length of an arc cut off by an angle of 2 radians on a circle of radius 8 inches? $= 16$

13. What is the length of an arc which is cut off by an angle of 225° in a circle of radius 4 feet?

14. What is the radius of a circle in which an angle of 3 radians cuts off an arc of 30 cm?

Evaluate $\sin\theta$ and $\cos\theta$ for the angle θ on the unit circle in Problems 15–16.

15.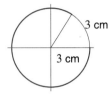
(−0.8, 0.6)

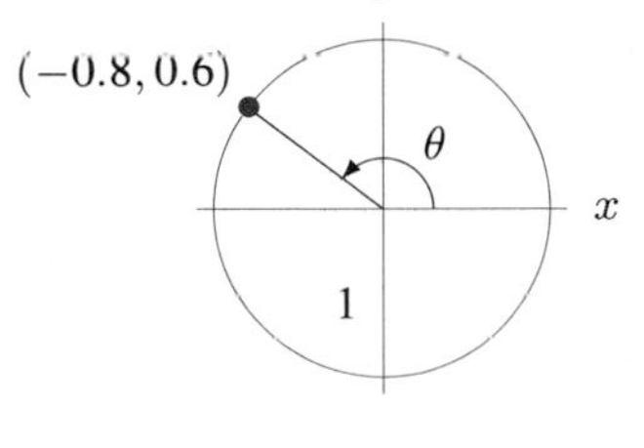

16.
(−0.6, 0.8)

8) $\frac{s}{3} =$

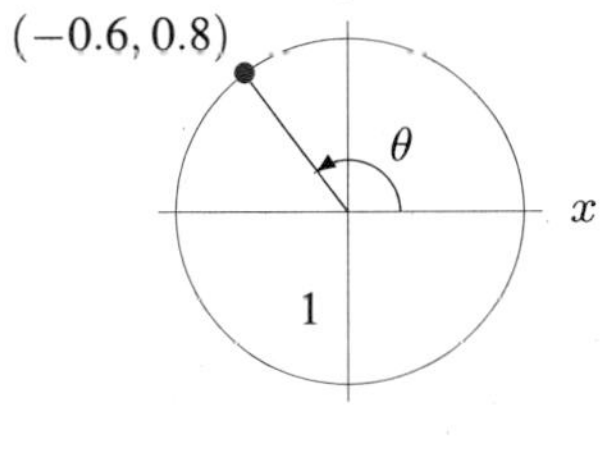

17. An ant starts at the point $(1, 0)$ on the unit circle and walks counterclockwise a distance of 3 units around the circle. Find the x and y coordinates (accurate to 2 decimal places) of the final location of the ant.

18. Estimate the radian measure of the angle that designates each point in Figure 6.45. Your answer should be an angle between 0 and 2π, measured counterclockwise from $(1, 0)$.

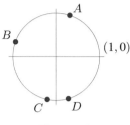

Figure 6.45

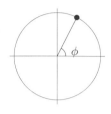

Figure 6.46

19. For the angle ϕ shown in Figure 6.46, sketch each of the following angles.

    (a) $\pi + \phi$ (b) $\pi - \phi$ (c) $\pi/2 - \phi$ (d) $2\pi - \phi$

20. An art student wants to make a string collage by connecting equally spaced points on the circumference of a circle to its center. What would be the radian and degree measure of the angle between two adjacent pieces of string if there were

    (a) 6 points   (b) 12 points   (c) 24 points   (d) 48 points

21. (a) What is the length, s, of the arc on a circle of diameter 38 cm cut off by an angle of $\theta = 3.83$ radians?
    (b) Find the angle, φ, in radians which cuts off an arc of length 3.83 cm on the circumference of a circle of diameter 38 cm.

22. If a weight hanging on a string of length 3 feet swings through $5°$ on either side of the vertical, how long is the arc through which the weight moves from one high point to the next high point?

23. How far does the tip of the minute hand of a clock move in 35 minutes if the hand is 6 inches long?

24. Using a weight on a string called a plumb bob, it is possible to erect a pole that is exactly vertical, which means that the pole points directly toward the center of the earth. Two such poles are erected one hundred miles apart. If the poles were extended they would meet at the center of the earth at an angle of $1.4333°$. Compute the radius of the earth.

25. A weather satellite orbits the earth in a circular orbit 500 miles above the earth's surface. What is the radian measure of the angle (measured at the center of the earth) through which the satellite moves in traveling 600 miles along its orbit? (The radius of the earth is 3960 miles.)

26. If you have the wrong degree/radian mode set on your calculator, you can get unexpected results.

    (a) If you intend to enter an angle of 30 degrees but you are in radian mode, the angle value would be considered 30 radians. How many degrees is a 30 radian angle?
    (b) If you intend to enter an angle of $\pi/6$ radians but you are in degree mode, the angle value would be considered $\pi/6$ degrees. How many radians is a $\pi/6$ degree angle?

27. Using a protractor and a ruler, construct a right triangle with hypotenuse 4 inches and one angle 0.4 radian. Measure the two legs and the remaining angle. The three angles (in radians) should add to what value? Do they? Do the three sides satisfy the Pythagorean theorem?

28. Without a calculator, decide which is bigger, t or $\sin t$, for $0 < t < \pi/2$. Illustrate your answer with a sketch.

29. Do you think there is a value of t for which $\cos t = t$? If so, estimate the value of t. If not, explain why not.

6.4 GRAPHS OF THE SINE AND COSINE

Before graphing, we calculate values of the sine and cosine. Some values can be found exactly. In Example 3 on page 242, we found $\cos 90° = 0$, $\sin 90° = 1$ and $\cos 180° = -1$, $\sin 180° = 0$.

Exact Values of the Sine and Cosine

Example 1   Evaluate $\sin \theta$ and $\cos \theta$ for $\theta = 0°, 270°$, and $360°$.

Solution   Figure 6.47 gives the coordinates of the points on the unit circle specified by $0°, 270°$, and $360°$. We use these coordinates to evaluate the sines and cosines of these angles.

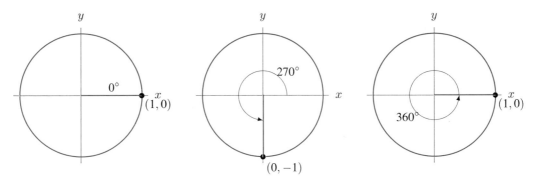

Figure 6.47: The coordinates of the points on the unit circle specified by the angles $0°, 270°$, and $360°$

The definition of the cosine function tells us that $\cos 0°$ is the x-coordinate of the point on the unit circle specified by the angle $0°$. Since the x-coordinate of this point is $x = 1$, we have

$$\cos 0° = 1.$$

Similarly, the y-coordinate of the point $(1, 0)$ is $y = 0$ which gives

$$\sin 0° = 0.$$

From Figure 6.47 we see that the coordinates of the remaining points are $(0, -1)$, and $(1, 0)$, so

$$\cos 270° = 0, \qquad \sin 270° = -1, \qquad \text{and} \qquad \cos 360° = 1, \qquad \sin 360° = 0.$$

In Section 7.1, triangles are used to compute the exact values of the sine and cosine of $30° = \pi/6$, $45° = \pi/4$, $60° = \pi/3$. See page 289 and Problem 16 at the end of this section. The results are:

$$\cos 30° = \cos \frac{\pi}{6} = \frac{\sqrt{3}}{2} \qquad \cos 45° = \cos \frac{\pi}{4} = \frac{1}{\sqrt{2}} \qquad \cos 60° = \cos \frac{\pi}{3} = \frac{1}{2}$$

$$\sin 30° = \sin \frac{\pi}{6} = \frac{1}{2} \qquad \sin 45° = \sin \frac{\pi}{4} = \frac{1}{\sqrt{2}} \qquad \sin 60° = \sin \frac{\pi}{3} = \frac{\sqrt{3}}{2}$$

Graphs of the Sine and Cosine Functions

Table 6.6 records values of the sine and cosine functions, using $1/\sqrt{2} \approx 0.7$. The graphs are plotted in Figure 6.48. To model periodic phenomena, we usually work in radians. Thus, the axes are labeled in radians and degrees. Notice that both the sine and cosine functions are periodic with a period of $2\pi = 360°$ and an amplitude of 1.

TABLE 6.6   *Table of values (rounded) for* $\sin\theta$ *and* $\cos\theta$, *with* θ *in degrees and radians*

θ (degrees)	0°	45°	90°	135°	180°	225°	270°	315°	360°	450°	540°	630°	720°	⋯
θ (radians)	0	$\pi/4$	$\pi/2$	$3\pi/4$	π	$5\pi/4$	$3\pi/2$	$7\pi/4$	2π	$5\pi/2$	3π	$7\pi/2$	4π	⋯
$\sin\theta$	0	0.7	1	0.7	0	−0.7	−1	−0.7	0	1	0	−1	0	⋯
$\cos\theta$	1	0.7	0	−0.7	−1	−0.7	0	0.7	1	0	−1	0	1	⋯

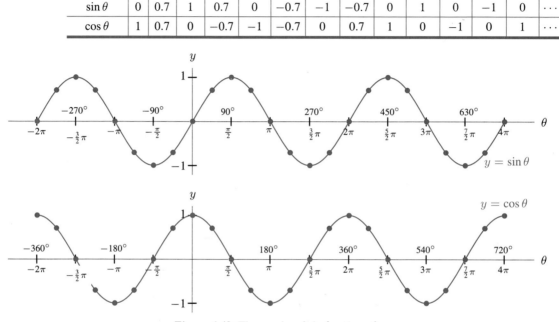

Figure 6.48: The graphs of $\sin\theta$ and $\cos\theta$

Several properties of the sine and cosine functions are illustrated by their graphs. Since the outputs of the sine and cosine functions are the coordinates of points on the unit circle, they lie between −1 and 1. So the range of the sine and cosine are

$$-1 \le \sin\theta \le 1 \qquad \text{and} \qquad -1 \le \cos\theta \le 1.$$

The domain of both of these functions is all real numbers, since any angle, positive or negative, specifies a point on the unit circle. Both functions are periodic with period 2π, because adding a multiple of 2π (or 360°) to an angle does not change the position of the point that it designates on the unit circle. The midline of both graphs is $y = 0$ and their amplitude is 1.

Amplitude

We can determine the amplitude of a trigonometric function from its formula. Recall that the amplitude is the distance between the midline and the highest or lowest point on the graph.

Example 2   Compare the graph of $y = \sin t$ to the graphs of $y = 2\sin t$ and $y = -0.5\sin t$, for $0 \le t \le 2\pi$. How are these graphs similar? How are they different? What are their amplitudes?

Solution   The graphs are in Figure 6.49. The amplitude of $y = \sin t$ is 1, the amplitude of $y = 2\sin t$ is 2 and the amplitude of $y = -0.5\sin t$ is 0.5. The graph of $y = -0.5\sin t$ is "upside-down" relative to $y = \sin t$. These observations are consistent with the fact that the constant A in the equation

$$y = A\sin t$$

stretches or shrinks the graph vertically, and reflects it about the t-axis if A is negative.

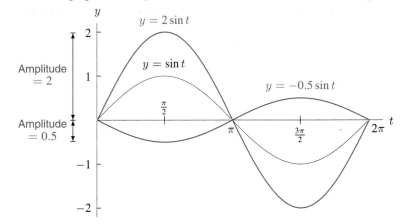

Figure 6.49: The graphs of $y = \sin t$, $y = 2\sin t$, and $y = -0.5\sin t$ all have different amplitudes

Midline

Recall that the graph of $y = f(t) + k$ is the graph of $y = f(t)$ shifted vertically by k units. For example, the graph of $y = \cos t + 2$ is the graph of $y = \cos t$ shifted up by 2 units, as shown in Figure 6.50. Notice that the midline of the new graph is the line $y = 2$; it has been shifted up 2 units from its old position at $y = 0$.

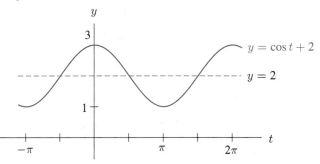

Figure 6.50: The graph of $y = \cos t + 2$ and its midline $y = 2$

Generalizing, we conclude that the graphs of $y = \sin t + k$ and $y = \cos t + k$ have midlines $y = k$. Notice that the expression $\sin t + k$ could also be written as $k + \sin t$; it is *not* the same as $\sin(t + k)$.

Example 3   Graph the ferris wheel function giving your height, $h = f(\theta)$, in feet, above ground as a function of the angle θ:

$$f(\theta) = 250 + 250\sin\theta.$$

(See Example 7 on page 244.) What are the period, midline, and amplitude?

Solution   Using a calculator, we get the graph in Figure 6.51. The period of this function is 360°, because 360° is one full rotation, so the function repeats every 360°. The midline is $h = 250$ feet, since the values of h oscillate about this value. The amplitude is also 250 feet, since the maximum value of h is 500 feet.

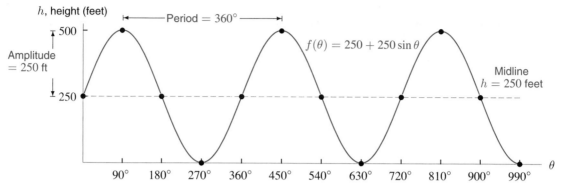

Figure 6.51: On the ferris wheel: Height, h, above ground as a function of the angle, θ

Problems for Section 6.4

1. (a)   Draw an x-axis and scale it with equally spaced units from 0 to 8.
   (b)   On your x-axis, mark the points $\pi/2$, π, $3\pi/2$, and 2π.
   (c)   Now add a y-axis to your figure and sketch a graph of $y = \sin x$.

2. Sketch a graph of $y = \sin t$ on the interval $-\pi \le t \le 3\pi$.
   (a)   Indicate the interval(s) on which the function is
       (i)   Positive   (ii)   Increasing   (iii)   Concave up
   (b)   Estimate the t values at which the function is increasing most rapidly.

3. Match each of the letters A–G in Figure 6.52 to one of the following values of x (in radians): $1, 2, 4, 5, \pi/2, \pi,$ and $3\pi/2$.

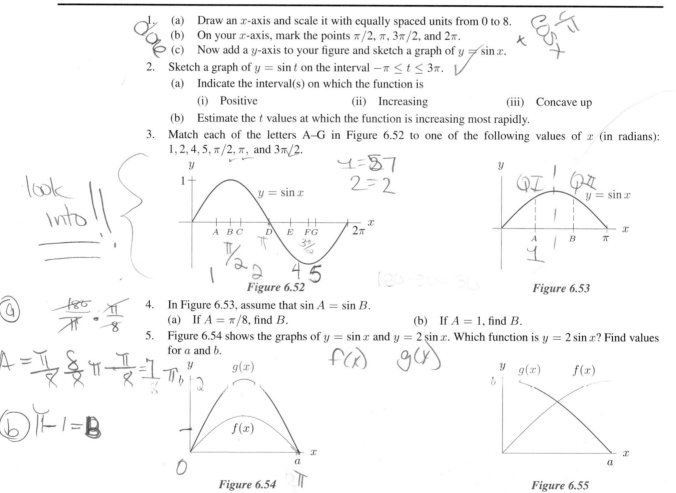

Figure 6.52

Figure 6.53

4. In Figure 6.53, assume that $\sin A = \sin B$.
   (a)   If $A = \pi/8$, find B.   (b)   If $A = 1$, find B.

5. Figure 6.54 shows the graphs of $y = \sin x$ and $y = 2\sin x$. Which function is $y = 2\sin x$? Find values for a and b.

Figure 6.54

Figure 6.55

6. Figure 6.55 on page 256 shows graphs of $y = \sin x$ and $y = \cos x$ starting at $x = 0$. Which is $y = \cos x$? Find values for a and b.

7. Figure 6.56 shows the graphs of $y = (\sin x) - 1$ and $y = \sin x + 1$. Identify which is which.

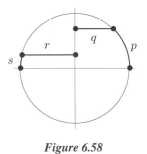

Figure 6.56   Figure 6.57

8. Figure 6.57 shows graphs of $y = \sin(x - \frac{\pi}{2})$ and $y = \sin(x + \frac{\pi}{2})$ starting at $x = 0$. Identify which is which.

9. Compare the graph of $y = \sin \theta$ to the graphs of $y = 0.5 \sin \theta$ and $y = -2 \sin \theta$ for $0 \le \theta \le 2\pi$. How are these graphs similar? How are they different?

10. Compare the values of $\sqrt{1/2}$ and $\sqrt{2}/2$ using a calculator. Explain your observations.

11. Compare the values of $\sqrt{3/4}$ and $\sqrt{3}/2$ using your calculator. Explain your observations.

12. The graph of $y = \sin \theta$ never goes higher than 1. Explain why this is true by using a sketch of the unit circle and the definition of the $\sin \theta$ function.

13. (a) Match the lengths p, q, r, s marked on the unit circle in Figure 6.58 with the following values:

    (i) $t = 0.8$ (ii) $t = \pi - 2.9$ (iii) $\cos(0.8)$ (iv) $-\cos(2.9)$

    (b) On a graph of $y = \cos t$, sketch segments corresponding to each of the values above.

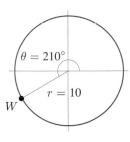

Figure 6.58   Figure 6.59

14. Find exact values for the coordinates of point W in Figure 6.59.

15. Find exact values for each of the following:

    (a) $\sin\left(\dfrac{3\pi}{4}\right)$ (b) $\cos\left(\dfrac{5\pi}{3}\right)$ (c) $\cos\left(\dfrac{7\pi}{6}\right)$ (d) $\sin\left(\dfrac{11\pi}{6}\right)$ (e) $\sin\left(\dfrac{9\pi}{4}\right)$

16. Calculate $\sin 45°$ and $\cos 45°$ exactly. Use the fact that the point P corresponding to $45°$ on the unit circle, $x^2 + y^2 = 1$, lies on the line $y = x$.

17. (a) For each of the following angles, sketch an arc of length θ on a unit circle and a line segment of length $\cos \theta$: (i) $\theta = 1.1$ (ii) $\theta = 5.2$

    (b) Now sketch $y = \cos t$. Mark on your graph of $y = \cos t$ line segments of length θ and $\cos \theta$ for the same values of θ that you used in part (a).

18. Using your knowledge of the absolute value function, explain in a few sentences the relationship between the graph of $y = |\sin x|$ and the graph of $y = \sin x$.

19. How many complete cycles of $y = \cos x$ appear on the interval $0 \le x \le 10$? On the interval $0 \le x \le 20$?

20.   (a)   Write an expression for the slope of the line segment joining P and Q in Figure 6.60.
      (b)   Evaluate your expression for $a = \pi/4$, $b = 4\pi/3$. Give an exact value for your answer.

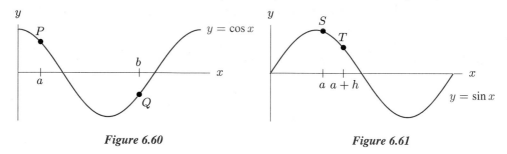

| Figure 6.60 | Figure 6.61 |

21.   (a)   Write an expression for the slope of the line segment joining S and T in Figure 6.61.
      (b)   Evaluate your expression for $a = 1.7$, $h = 0.05$. Round your answer to two decimal places.

22.   Sketch graphs of the following functions for $-2\pi \leq x \leq 2\pi$.

      (a) $f(x) = \sin x$ (b) $g(x) = |\sin x|$ (c) $h(x) = \sin |x|$ (d) $i(x) = |\sin |x||$

      (e)   Do any two of these functions have identical graphs? If so, explain why this makes sense.

23.   (a)   Using what you know about the graph of $y = \sin x$, make predictions about the graph of $y = (\sin x)^2$. Compare your predictions to the actual graph.
      (b)   Using a computer or a graphing calculator, graph $g(x) = (\sin x)^2$, $h(x) = (\cos x)^2$ and $f(x) = g(x) + h(x)$, on the interval $-2\pi \leq x \leq 2\pi$. Given your result from part (a), does the graph of $f(x)$ make sense? Why or why not?

24.   A circle of radius 5 is centered at the point $(-6, 7)$. Define $f(\theta)$ to be the x-coordinate of the point $P = (x, y)$ on the circle specified by the angle θ in Figure 6.62. Find a formula for $f(\theta)$.

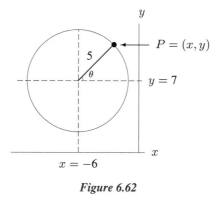

Figure 6.62

6.5   SINUSOIDAL FUNCTIONS

Section 6.1 introduced $f(t)$, your height above the ground while riding a ferris wheel. Comparing the graph of f to the graph of $y = \sin \theta$, we see that f looks like a transformation of $y = \sin \theta$. Transformations of the sine and cosine are called *sinusoidal* functions.

We consider functions which can be expressed in the form

$$y = A \sin B(t - h) + k \qquad \text{and} \qquad y = A \cos B(t - h) + k,$$

where A, B, h, and k are constants. Their graphs resemble the graphs of sine and cosine, but may also be shifted, flipped, or stretched. These transformations may change the period, amplitude, and midline of the function as well as its value at $t = 0$.

From Section 6.4 we know the following:

> The functions of $y = A\sin t$ and $y = A\cos t$ have **amplitude** $|A|$. If A is negative, the graph is reflected across the t-axis.

> The **midline** of the functions $y = \sin t + k$ and $y = \cos t + k$ is the horizontal line $y = k$.

Period

Next, we consider the effect of the constant B. We usually have $B > 0$.

Example 1   Graph $y = \sin t$ and $y = \sin 2t$ for $0 \le t \le 2\pi$. Describe any similarities and differences. What are their periods?

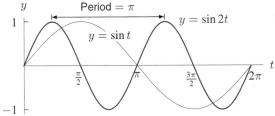

Figure 6.63: The functions $y = \sin t$ and $y = \sin 2t$ have different periods

Solution   The graphs are in Figure 6.63. The two functions have the same amplitude and midline, but their periods are different. The period of $y = \sin t$ is 2π, but the period of $y = \sin 2t$ is π. This is because the factor of 2 causes a horizontal compression, squeezing the graph twice as close to the y-axis.

If $B > 0$ the function $y = \sin(Bt)$ resembles the function $y = \sin t$ except that it is stretched or compressed horizontally. The constant B determines how many cycles the function completes on an interval of length 2π. For example, we see from Figure 6.63 that the function $y = \sin 2t$ completes two cycles on the interval $0 \le t \le 2\pi$. The constant B is called the *angular frequency* of the function.

Since, for $B > 0$, the graph of $y = \sin(Bt)$ completes B cycles on the interval $0 \le t \le 2\pi$, each cycle has length $2\pi/B$. The period is thus $2\pi/B$. In general, for B of any sign, we have:

> The functions of $y = \sin(Bt)$ and $y = \cos(Bt)$ have **period** $P = 2\pi/|B|$.

Example 2   Find possible formulas for the functions f and g shown in Figures 6.64 and 6.65.

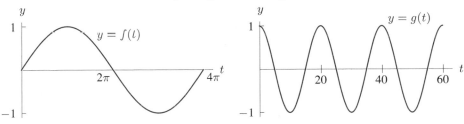

Figure 6.64: This function has period 4π **Figure 6.65:** This function has period 20

Solution   The graph of f resembles the graph of $y = \sin t$ except that its period is $P = 4\pi$. Using $P = 2\pi/B$ gives

$$4\pi = \frac{2\pi}{B} \qquad \text{so} \qquad B = \frac{1}{2}.$$

Thus, $f(t) = \sin\left(\frac{1}{2}t\right)$.

The function g resembles the function $y = \cos t$ except that its period is $P = 20$. This gives

$$20 = \frac{2\pi}{B} \qquad \text{so} \qquad B = \frac{\pi}{10}.$$

Thus, $g(t) = \cos\left(\frac{\pi}{10}t\right)$.

Example 3   Household electrical power in the US is provided in the form of alternating current. Typically the voltage cycles smoothly between $+155.6$ volts and -155.6 volts 60 times per second.[3] Use a cosine function to model the alternating voltage.

Solution   If V is the voltage at time, t, in seconds, then V begins at $+155.6$ volts, drops to -155.6 volts, and then climbs back to $+155.6$ volts, repeating this process 60 times per second. We use a cosine with amplitude $A = 155.6$. Since the function alternates 60 times in one second, the period is $1/60$ of a second. We know that $P = 2\pi/B = 1/60$, so $B = 120\pi$. We have $V = 155.6\cos(120\pi t)$.

Example 4   Describe in words the function $y = 300\cos(0.2\pi t) + 600$ and sketch its graph.

Solution   This function resembles $y = \cos t$ except that it has an amplitude of 300, a midline of $y = 600$, and a period of $P = 2\pi/(0.2\pi) = 10$. See Figure 6.66.

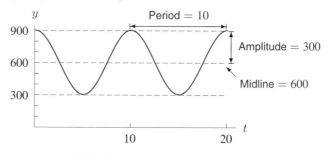

Figure 6.66: The function $y = 300\cos(0.2\pi t) + 600$

Horizontal Shift

Figure 6.67 shows the graphs of two trigonometric functions, f and g, with period $P = 12$. The graph of f resembles a sine function, so a possible formula for f is $f(t) = \sin Bt$. Since the period of f is 12, we have $12 = 2\pi/B$, so $B = 2\pi/12$, so $f(t) = \sin(\pi t/6)$.

The graph of g looks like the graph of f shifted to the right by 2 units. Thus a possible formula for g is

$$g(t) = f(t - 2),$$

or

$$g(t) = \sin\left(\frac{\pi}{6}(t - 2)\right).$$

---

[3]A voltage cycling between $+155.6$ volts and -155.6 volts has an average magnitude, over time, of 110 volts.

Notice that we can also write the formula for $g(t)$ as

$$g(t) = \sin\left(\frac{\pi}{6}t - \frac{\pi}{3}\right),$$

but $\pi/3$ is *not* the horizontal shift in the graph! To pick out the horizontal shift from the formula, we must write the formula in factored form, that is, as $\sin B(t - h)$.

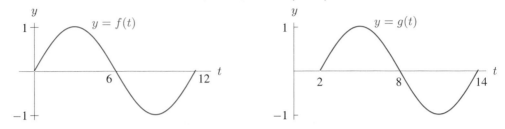

Figure 6.67: The graphs of two trigonometric functions f and g, related by a horizontal shift

> The graphs of $y = \sin B(t - h)$ and $y = \cos B(t - h)$ are the graphs of $y = \sin Bt$ and $y = \cos Bt$ **shifted horizontally** by h units.

Example 5   Describe in words the graph of the function $g(t) = \cos\left(3t - \pi/4\right)$.

Solution   Write the formula for g in the form $\cos B(t - h)$ by factoring 3 out from the expression $3t - \pi/4$ to get $g(t) = \cos(3(t - \pi/12))$. The period of g is $2\pi/3$ and the graph is the graph of $f = \cos 3t$ shifted $\pi/12$ units to the right, as shown in Figure 6.68.

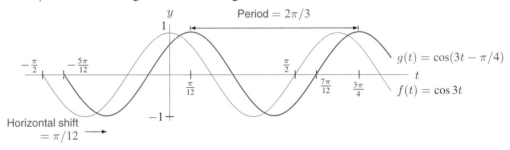

Figure 6.68: The graphs of $g(t) = \cos\left(3t - \frac{\pi}{4}\right)$ and $f(t) = \cos 3t$

Summary of Transformations

The parameters A, B, h, and k describe completely the graph of a transformed sine or cosine function.

> For the **sinusoidal** functions
>
> $$y = A \sin B(t - h) + k \qquad \text{and} \qquad y = A \cos B(t - h) + k,$$
>
> - $|A|$ is the amplitude
> - $2\pi/|B|$ is the period
> - h is the horizontal shift
> - $y = k$ is the midline
> - $|B|$ is the angular frequency; that is, the number of cycles completed in $0 \leq t \leq 2\pi$.

Example 6   The temperature, T, in °C, of the surface water in a pond varies according to the graph in Figure 6.69. If t is the number of hours since sunrise at 6 am, find a possible formula for $T = f(t)$.

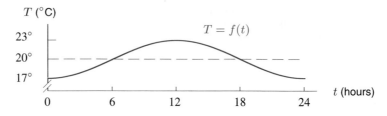

Figure 6.69: Surface water temperature of a pond since sunrise

Solution   The graph of $T = f(t)$ resembles a cosine function with amplitude $|A| = 3$, period $= 24$ and midline $k = 20$. Compared to $y = \cos t$, there is no horizontal shift, but the graph has been reflected across the midline, so A is negative, $A = -3$. We have

$$24 = \frac{2\pi}{B} \qquad \text{so} \qquad B = \frac{2\pi}{24} = \frac{\pi}{12}.$$

So a possible formula for f is

$$T = f(t) = -3\cos\left(\frac{\pi}{12}t\right) + 20.$$

Phase Shift

In Example 5, we factored $(3t - \pi/4)$ to write the function as $g(t) = \cos(3(t - \pi/12))$. This allowed us to recognize the horizontal shift, $\pi/12$. However, in most physical applications, the quantity $\pi/4$, known as the *phase shift*, is more important than the horizontal shift. The phase shift enables us to calculate the fraction of a full period that the curve has been shifted. For instance, in Example 5, the wave has been shifted

$$\frac{\text{Phase shift}}{2\pi} = \frac{\pi/4}{2\pi} = \frac{1}{8} \text{ of a full period,}$$

and the graph of g in Figure 6.70 is the graph of $f(t) = \cos 3t$ shifted $1/8$ of its period to the right.

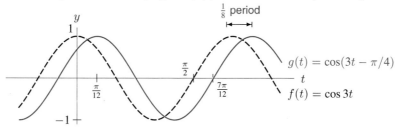

Figure 6.70: The graph of $g(t) = \cos(3t - \frac{\pi}{4})$ has phase shift $\frac{\pi}{4}$ relative to $f(t) = \cos 3t$

Phase shift is significant because in many applications, such as optical interference, we want to know if two waves reinforce or cancel each other. For two waves of the same period, a phase shift of 0 or 2π tells us that the two waves reinforce each other; a phase shift of π tells us that the two waves cancel. Thus, the phase shift tells us the relative positions of two waves of the same period.[4]

Using the Transformed Sine and Cosine Functions

Sinusoidal functions are used to model oscillating quantities. Starting with $y = A\sin B(t - h) + k$ or $y = A\cos B(t - h) + k$, we calculate values of the parameters A, B, h, k to fit the particular case.

---

[4]The phase shift which tells us that two waves cancel is independent of their period. The horizontal shift that gives the same information is not independent of period.

Example 7   A rabbit population in a national park rises and falls each year. It is at its minimum of 5000 rabbits in January. By July, as the weather warms up and food grows more abundant, the population triples in size. By the following January, the population again falls to 5000 rabbits, completing the annual cycle. Use a trigonometric function to find a possible formula for $R = f(t)$, where R is the size of the rabbit population as a function of t, the number of months since January.

Solution   Notice that January is month 0, so July is month 6. The five points in Table 6.7 have been plotted in Figure 6.71 and a curve drawn in. This curve has midline $k = 10{,}000$, amplitude $|A| = 5000$, and period $= 12$ so $B = 2\pi/12 = \pi/6$. It resembles a cosine function reflected across its midline. Thus, a possible formula for this curve is

$$R = f(t) = -5000\cos\left(\frac{\pi}{6}t\right) + 10{,}000.$$

There are other possible formulas.

TABLE 6.7 *Rabbit population over time*

t (month)	R
0	5000
6	15,000
12	5000
18	15,000
24	5000

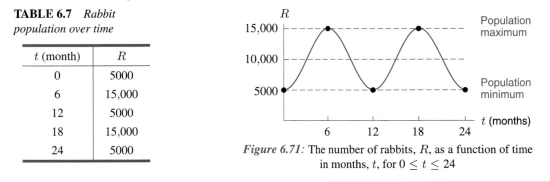

Figure 6.71: The number of rabbits, R, as a function of time in months, t, for $0 \le t \le 24$

Now let's return to the ferris wheel example in Section 6.1.

Example 8   Use the sinusoidal function $f(t) = A\sin B(t - h) + k$ to represent your height above ground while riding the ferris wheel.

Solution   The diameter of the ferris wheel is 500 feet, so the midline is $k = 250$ and the amplitude, A, is also 250. The period of the ferris wheel is 20 minutes, so

$$B = \frac{2\pi}{20} = \frac{\pi}{10}.$$

Figure 6.72 shows a sine graph shifted 5 minutes to the right because we reach $y = 250$ (the 3 o'clock position) when $t = 5$. Thus, the horizontal shift is $h = 5$, so

$$f(t) = 250\sin\frac{\pi}{10}(t - 5) + 250.$$

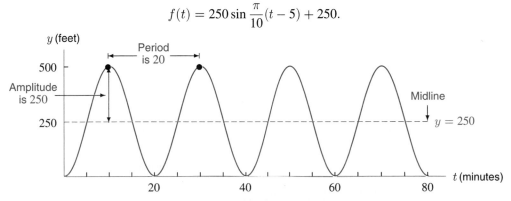

Figure 6.72: Graph of the ferris wheel height function $f(t) = 250\sin\frac{\pi}{10}(t - 5) + 250$

This formula can be checked at some particular values of t. For example, when $t = 0$, we have the correct starting height:

$$f(0) = 250 + 250 \sin \frac{\pi}{10}(0 - 5) = 250 + 250 \sin\left(-\frac{\pi}{2}\right) = 250 - 250 = 0.$$

There are other possible formulas for the function graphed in Figure 6.72. For example, we could have used a cosine reflected about its midline, as in Example 7. (See Problem 30.)

Problems for Section 6.5

1. Which of the following functions are periodic? Justify your answers. State the periods of those that are periodic.

   (a) $y = \sin(-t)$ (b) $y = 4\cos(\pi t)$ (c) $y = \sin(t) + t$ (d) $y = \sin(t/2) + 1$

Without using a calculator, graph one full period of each of the functions in Problems 2–5.

2. $y = \sin(\frac{1}{2}t)$ 3. $y = 4\cos(t + \frac{\pi}{4})$ 4. $y = 5 - \sin t$ 5. $y = \cos(2t) + 4$

State the amplitude, period, phase shift, and horizontal shifts for each of the functions in Problems 6–11. Without using a calculator, graph each of them on the given interval.

6. $y = -4\sin t, \quad -2\pi \leq t \leq 2\pi$

7. $y = -20\cos(4\pi t), \quad -\frac{3}{4} \leq t \leq 1$

8. $y = \cos\left(2t + \frac{\pi}{2}\right), \quad -\pi \leq t \leq 2\pi$

9. $y = \cos\left(\frac{t}{4} - \frac{\pi}{4}\right), \quad 0 \leq t \leq 2\pi$

10. $y = 3\sin(4\pi t + 6\pi), \quad -\frac{3}{2} \leq t \leq \frac{1}{2}$

11. $y = 10\sin(0.1t + \pi), \quad -20\pi \leq t \leq 20\pi$

Find possible formulas for the trigonometric functions in Problems 12–20.

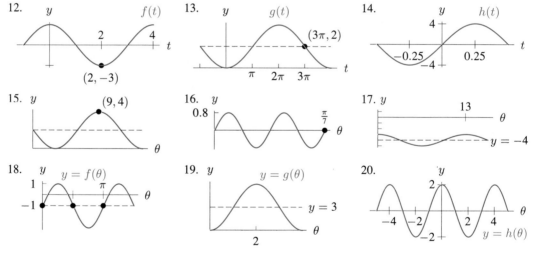

21. Figure 6.73 shows the graphs of $y = \sin x$ and $y = \sin 2x$. Which graph is $y = \sin x$? Identify the points a to e.

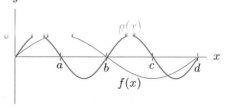

Figure 6.73

22. Describe in words how you can obtain the graph of the function $y = \cos\left(5t + \dfrac{\pi}{4}\right)$ from the graph of $y = \cos(5t)$.

23. A person's blood pressure, P, (in millimeters of mercury, abbreviated mm Hg) is given by

$$P = 100 - 20\cos\left(\frac{8\pi}{3}t\right),$$

where t is the time in seconds. Sketch the graph of this function. State the period and amplitude and explain the practical significance of these quantities.

Problems 24–26 involve ferris wheels. Find a formula, using the sine function, for your height above ground after t minutes on the ferris wheel. Graph the function to check that it is correct.

24. A ferris wheel is 35 meters in diameter and can be boarded at ground level. The wheel completes one full revolution every 5 minutes. At $t = 0$ you are in the three o'clock position and ascending.

25. A ferris wheel is 20 meters in diameter and must be boarded from a platform that is 4 meters above the ground. The six o'clock position on the ferris wheel is level with the loading platform. The wheel completes one full revolution every 2 minutes. At $t = 0$ you are in the twelve o'clock position.

26. A ferris wheel is 50 meters in diameter and must be boarded from a platform that is 5 meters above the ground. The six o'clock position on the ferris wheel is level with the loading platform. The wheel completes one full revolution every 8 minutes. At $t = 0$ you are at the loading platform.

The graphs in Problems 27–29 show your height h meters above ground after t minutes on various ferris wheels. Using the sine function, find a formula for h as a function of t. Graph your function to check that it is correct.

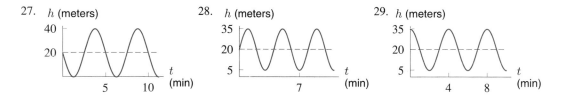

27. h (meters)

28. h (meters)

29. h (meters)

30. Find formula of the form $y = A\cos B(t - h) + k$ for the graph in Figure 6.72 on page 263.

31. The London ferris wheel has diameter 500 feet and one complete revolution takes 20 minutes.

    (a) Find the rate (in degrees per minute) that the London ferris wheel is rotating.
    (b) Let t, the time in minutes, be 0 when you are in the 6 o'clock position. Write θ, measured from the 3 o'clock position, as a function of t.
    (c) Find a formula for the ferris wheel function, $h = f(t)$, giving your height in feet above the ground.
    (d) Graph $h = f(t)$. What are the period, midline, and amplitude?

32. The following formulas give animal populations as functions of time, t, in years. Describe the growth of each population in words.

    (a) $P = 1500 + 200t$ (b) $P = 2700 - 80t$ (c) $P = 1800(1.03)^t$

    (d) $P = 800e^{-0.04t}$ (e) $P = 230\sin\left(\dfrac{2\pi}{7}t\right) + 3800$

33. Graph $y = (\sin x)^2$. Use the graph to find values of A, B, h, and k so that $y = A\cos B(x - h) + k$.

34. A population of animals varies sinusoidally over a year between a low of 700 on January 1 and a high of 900 on July 1. It returns to a low of 700 on the following January 1.

    (a) Graph the population as a function of time for a period of one year.
    (b) Find a formula for the population as a function of time, t, in months since the start of the year.

35. A population of animals oscillates between a low of 1300 on January 1 ($t = 0$), to a high of 2200 on July 1 ($t = 6$), and back to a low of 1300 the following January 1 ($t = 12$).

(a) Find a formula for the population, P, in terms of the time, t, in months.
(b) Interpret the amplitude, period, and midline of the function $P = f(t)$.
(c) Use a graph to estimate when $P = 1500$ during the year.

36. Find a possible formula for the trigonometric function whose values are in the following table.

x	0	0.1	0.2	0.3	0.4	0.5	0.6	0.7	0.8	0.9	1
$g(x)$	2	2.588	2.951	2.951	2.588	2	1.412	1.049	1.049	1.412	2

37. Let $f(x) = \sin(2\pi x)$ and $g(x) = \cos(2\pi x)$. State the periods, amplitudes, and midlines of f and g.

For Problems 38–43, let $f(x) = \sin(2\pi x)$ and $g(x) = \cos(2\pi x)$. Use what you know about graph transformations to sketch graphs of the following functions on the interval $0 \le x \le 3$ without a calculator.

38. $y = 2f(x)$
39. $y = -\frac{1}{2}g(x)$
40. $y = 3 + 2f(x)$
41. $y = g(2x)$
42. $y = f(\frac{1}{3}x)$
43. $y = 2f(3x)$

For Problems 44–47, let $f(x) = \sin(2\pi x)$ and $g(x) = \cos(2\pi x)$. Find possible formulas in terms of f or g for each of the graphs given.

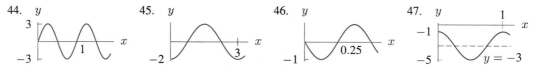

44. 45. 46. 47.

48. A company sells $S(t)$ thousand electric blankets in month t (with $t = 0$ being January). An approximation of this function is given by

$$S(t) \approx 72.25 + 41.5 \sin\left(\frac{\pi t}{6} + \frac{\pi}{2}\right).$$

Sketch the graph of this function over one year. Find its period and amplitude and explain their practical significance.

49. The pressure, P (in lbs/ft^2), in a pipe varies over time. Five times an hour, the pressure oscillates from a low of 90 to a high of 230 and then back to a low 90. The pressure at $t = 0$ is 90.

(a) Sketch the graph of $P = f(t)$, where t is time in minutes. Label your axes.
(b) Find a possible formula for $P = f(t)$.
(c) By graphing $P = f(t)$ for $0 \le t \le 2$, estimate when the pressure first equals 115 lbs/ft^2.

50. As reported by the US census, the population of Somerville, MA, from 1920 to 1990 is given in Table 6.8.

(a) Graph the data in Table 6.8, with time in years on the horizontal axis.
(b) Based on the data, a researcher decides the population varies in an approximately periodic way with time. Do you agree?
(c) On your graph, sketch in a sine curve that fits your data as closely as possible. Your sketch should capture the overall trend of the data but need not pass through all the data points. [Hint: Start by choosing a midline.]
(d) Find a formula for the curve you drew in part (c).
(e) According to the US census, the population of Somerville in 1910 was 77,236. How well does this agree with the value given by your formula?

TABLE 6.8

Years since 1920	0	10	20	30	40	50	60	70
Population (thousands)	93	104	102	102	95	89	77	76

51. Table 6.9 shows the average daily maximum temperature in degrees Fahrenheit in Boston each month.[5]
    (a) Plot the average daily maximum temperature as a function of the number of months past January.
    (b) What is the amplitude of the function? What is the period?
    (c) Find a trigonometric approximation of this function.
    (d) Use your formula to estimate the daily maximum temperature for October. How well does this estimate agree with the data?

TABLE 6.9

Month	Jan	Feb	Mar	Apr	May	Jun	Jul	Aug	Sep	Oct	Nov	Dec
Temperature	36.4	37.7	45.0	56.6	67.0	76.6	81.8	79.8	72.3	62.5	47.6	35.4

52. Table 6.10 gives the average monthly temperature, y, in degrees Fahrenheit for the city of Fairbanks, Alaska, as a function of t, the month, where $t = 0$ indicates January.
    (a) Plot the data points. On the same graph, draw a curve which fits the data.
    (b) What kind of function best fits the data? Be specific.
    (c) Find a formula for a function, $f(t)$, that models the temperature data. Explain how you found your formula. [Note: There are many correct answers.]
    (d) Use your answer to part (c) to solve $f(t) = 32$. Interpret your results.
    (e) Check your results from part (d) graphically.
    (f) In the southern hemisphere, the times at which summer and winter occur are reversed, relative to the northern hemisphere. Modify your formula from part (c) so that it represents the average monthly temperature for a southern-hemisphere city whose summer and winter temperatures are similar to Fairbanks.

TABLE 6.10

t	0	1	2	3	4	5	6	7	8	9	10	11
y	−11.5	−9.5	0.5	18.0	36.9	53.1	61.3	59.9	48.8	31.7	12.2	−3.3

53. US petroleum imports for the years 1973–1992 are in Table 6.11. Suppose t is the number of years since 1900. Plot the data on the interval $73 \leq t \leq 92$ and find a trigonometric function $f(t)$ that approximates US petroleum imports as a function of t.

TABLE 6.11

Year	1973	1974	1975	1976	1977	1978	1979	1980	1981	1982
BTUs (quadrillion)	14.2	14.5	14	16.5	18	17	17	16	15	14

Year	1983	1984	1985	1986	1987	1988	1989	1990	1991	1992
BUTs (quadrillion)	13	15	12	14	16	17	18	17	16	16

54. A flight from La Guardia Airport in New York City to Logan Airport in Boston has to circle Boston several times before landing. Figure 6.74 shows the graph of the distance, d, of the plane from La Guardia as a function of time, t. Construct a function $f(t)$ whose graph approximates this one. Do this by using different formulas on the intervals $0 \leq t \leq 1$ and $1 \leq t \leq 2$.

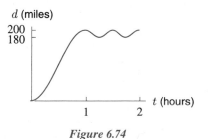

Figure 6.74

---
[5]Statistical Abstract of the United States

6.6 OTHER TRIGONOMETRIC FUNCTIONS

The Tangent Function

Another useful trigonometric function is called the tangent. Suppose $P = (x, y)$ is the point on the unit circle designated by the angle θ. We define the *tangent* of θ, or $\tan \theta$, by

$$\tan \theta = \frac{y}{x}.$$

The graphical interpretation of $\tan \theta$ is as a slope. In Figure 6.75, the slope of the line passing from the origin through P is given by

$$m = \frac{\Delta y}{\Delta x} = \frac{y - 0}{x - 0} = \frac{y}{x},$$

so

$$m = \tan \theta.$$

In words, $\tan \theta$ is the slope of the line passing through the origin and point P.

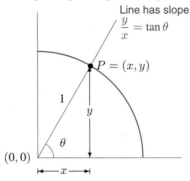

Figure 6.75: The slope of the line passing through the origin and point P is $\tan \theta$

Example 1   Find the slope of the line passing through the origin at an angle of (a) 30°   (b) $\pi/4$

Solution   (a)   A calculator set in degree mode gives

$$\tan 30° \approx 0.6.$$

Thus, the slope of the line is about 0.6.

(b)   A calculator set in radian mode gives $\tan(\pi/4) = 1$. This makes sense because an angle of $\pi/4$ describes a ray halfway between the x- and y-axes, or the line $y = x$, which has slope of 1.

Graph of the Tangent Function

Table 6.12 contains values of the tangent function. Figure 6.76 illustrates the connection between an angle θ on the unit circle and the graph of the tangent function. As the angle θ opens from 0 to $\pi/2$, the slope of the line, and therefore the tangent of θ, increases from 0 to $+\infty$. At $\pi/2$, this line is vertical, and its slope is undefined. Thus, $\tan(\pi/2)$ is undefined and the graph of $y = \tan \theta$ has a vertical asymptote at $\theta = \pi/2$.

TABLE 6.12   *Values of tangent function (rounded), with θ in degrees and radians*

θ	0°	30°	60°	90°	120°	150°	180°
θ	0	$\pi/6$	$\pi/3$	$\pi/2$	$2\pi/3$	$5\pi/6$	π
$\tan \theta$	0	0.6	1.7	Undefined	-1.7	-0.6	0

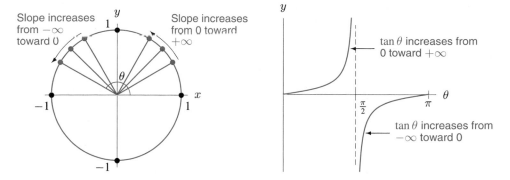

Figure 6.76: The connection between the unit circle and the graph of the tangent function

For θ between $\pi/2$ and π, the slopes are negative. The line is very steep near $\pi/2$, but becomes less steep as θ approaches π, where it is horizontal. Thus, $\tan\theta$ becomes less negative as θ increases in the second quadrant and reaches 0 at $\theta = \pi$.

For values of θ between π and 2π, observe that

$$\tan(\theta + \pi) = \tan\theta,$$

because the angles θ and $\theta + \pi$ determine the same line through the origin, and hence the same slope. Thus, $y = \tan\theta$ has period π. The completed graph of $y = \tan\theta$ is shown in Figure 6.77.

Notice the differences between the tangent function and the sinusoidal functions. The tangent function has a period of π, whereas the sine and cosine both have periods of 2π. The tangent function has vertical asymptotes at odd multiples of $\pi/2$; the sine and cosine have no asymptotes. Even though the tangent function is periodic, it does not have an amplitude or midline because it does not have a maximum or minimum value.

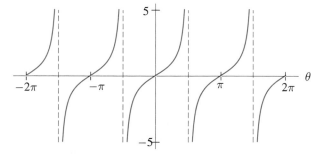

Figure 6.77: The graph of $y = \tan\theta$ in radians

Relationships Between the Trigonometric Functions

There is a relationship among the three trigonometric functions we have defined. If (x, y) are the coordinates of the point P determined by the angle θ on the unit circle, we defined

$$\cos\theta = x \qquad \text{and} \qquad \sin\theta = y.$$

Since $\tan\theta = y/x$, we have

$$\boxed{\tan\theta = \frac{\sin\theta}{\cos\theta}.}$$

Example 2   Calculate the exact value of $\tan(\pi/4)$.

Solution   Since we know $\cos(\pi/4) = \sin(\pi/4) = 1/\sqrt{2}$, we have

$$\tan\left(\frac{\pi}{4}\right) = \frac{\sin(\pi/4)}{\cos(\pi/4)} = \frac{1/\sqrt{2}}{1/\sqrt{2}} = 1.$$

Note that this is the same result as in Example 1.

Because of their definitions in terms of the unit circle, the sine and cosine functions are also related. We have seen that any point on the unit circle has coordinates $(\cos\theta, \sin\theta)$. But we also know that points on the unit circle must satisfy its equation

$$x^2 + y^2 = 1.$$

Substituting $x = \cos\theta$ and $y = \sin\theta$ gives

$$(\cos\theta)^2 + (\sin\theta)^2 = 1,$$

or, writing $\cos^2\theta$ instead of $(\cos\theta)^2$ and $\sin^2\theta$ instead of $(\sin\theta)^2$, we have

$$\boxed{\cos^2\theta + \sin^2\theta = 1.}$$

If we know which quadrant a given angle occupies and we know any one of its three trigonometric values, we can calculate the other two by using these relationships.

Example 3   Check that the relationships between $\sin\theta$, $\cos\theta$, and $\tan\theta$ are satisfied for $\theta = 40°$.

Solution   To three decimal places, a calculator (in degree mode) gives $\sin 40° = 0.643$, $\cos 40° = 0.766$, and $\tan 40° = 0.839$. The fact that

$$\frac{0.643}{0.766} = 0.839,$$

confirms that, to three decimal places,

$$\frac{\sin 40°}{\cos 40°} = \tan 40°.$$

Similarly, to three decimal places, we find that

$$(0.643)^2 + (0.766)^2 = 1,$$

confirming, to the accuracy with which we are working, that

$$\sin^2 40° + \cos^2 40° = 1.$$

Example 4   Suppose that $\cos\theta = 2/3$ and $3\pi/2 \le \theta \le 2\pi$. Find $\sin\theta$ and $\tan\theta$.

Solution   Use the relationship $\cos^2\theta + \sin^2\theta = 1$ to find $\sin\theta$. Substitute $\cos\theta = 2/3$:

$$\left(\frac{2}{3}\right)^2 + \sin^2\theta = 1$$

$$\frac{1}{9} + \sin^2\theta = 1$$

$$\sin^2\theta = 1 - \frac{4}{9} = \frac{5}{9}$$

$$\sin\theta = \pm\sqrt{\frac{5}{9}} = \pm\frac{\sqrt{5}}{3}.$$

Because θ is in the fourth quadrant, $\sin\theta$ is negative, so $\sin\theta = -\sqrt{5}/3$. To find $\tan\theta$, use the relationship

$$\tan\theta = \frac{\sin\theta}{\cos\theta} = \frac{-\sqrt{5}/3}{2/3} = -\frac{\sqrt{5}}{2}.$$

The Reciprocals of the Trigonometric Functions

The reciprocals of the trigonometric functions are given special names. Where the denominators are not equal to zero, we have

$$\text{secant }\theta = \sec\theta = \frac{1}{\cos\theta}.$$

$$\text{cosecant }\theta = \csc\theta = \frac{1}{\sin\theta}.$$

$$\text{cotangent }\theta = \cot\theta = \frac{1}{\tan\theta} = \frac{\cos\theta}{\sin\theta}.$$

The Pythagorean identity, $\cos^2\theta + \sin^2\theta = 1$, can be rewritten in terms of other trigonometric functions. Dividing through by $\cos^2\theta$ gives, provided $\cos\theta \neq 0$,

$$\frac{\cos^2\theta}{\cos^2\theta} + \frac{\sin^2\theta}{\cos^2\theta} = \frac{1}{\cos^2\theta}$$

$$1 + \left(\frac{\sin\theta}{\cos\theta}\right)^2 = \left(\frac{1}{\cos\theta}\right)^2$$

so

$$1 + \tan^2\theta = \sec^2\theta.$$

A similar identity relates $\cot\theta$ and $\csc\theta$. See Problem 30.

Example 5   Use a graph of $g(\theta) = \cos\theta$ to explain the shape of the graph of $f(\theta) = \sec\theta$.

Solution   Figure 6.78 shows the graphs of $\cos\theta$ and $\sec\theta$. In the first quadrant $\cos\theta$ decreases from 1 to 0, so the reciprocal of $\cos\theta$ increases from 1 towards $+\infty$. The values of $\cos\theta$ are negative in the second quadrant and decrease from 0 to -1, so the values of $\sec\theta$ increase from $-\infty$ to -1. The graph of $y = \cos\theta$ is symmetric about the vertical line $\theta = \pi$, so the graph of $f(\theta) = \sec\theta$ is symmetric about the same line. Thus, the graph of $f(\theta) = \sec\theta$ on the interval $\pi \leq \theta \leq 2\pi$ is the mirror image of the graph on $0 \leq \theta \leq \pi$. Note that $\sec\theta$ is undefined wherever $\cos\theta = 0$, namely, at $\theta = \pi/2$ and $\theta = 3\pi/2$. The graph of $f(\theta) = \sec\theta$ has vertical asymptotes at those values.

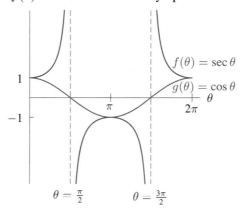

Figure 6.78: The graphs of $g(\theta) = \cos\theta$ and $f(\theta) = \sec\theta$

Problems for Section 6.6

1. Find exact values for $\sin 0°$, $\cos 0°$ and $\tan 0°$.

Find exact values for each of the quantities in Problems 2–13.

2. $\sin 90°$ 3. $\cos 90°$ 4. $\tan 90°$ 5. $\tan 135°$

6. $\tan 225°$ 7. $\cos 540°$ 8. $\tan 540°$ 9. $\tan \dfrac{\pi}{3}$

10. $\tan \dfrac{2\pi}{3}$ 11. $\tan \dfrac{5\pi}{4}$ 12. $\tan \dfrac{11\pi}{6}$ 13. $\tan \dfrac{13\pi}{6}$

14. Find $\tan \theta$ if $\sin \theta = -3/5$, and θ is in the fourth quadrant.

15. (a) $\cos \alpha = -\sqrt{3}/5$ and α is in the third quadrant. Find exact values for $\sin \alpha$ and $\tan \alpha$.
    (b) $\tan \beta = 4/3$ and β is in the third quadrant. Find exact values for $\sin \beta$ and $\cos \beta$.

16. (a) $\cos \phi = 0.4626$ and $3\pi/2 < \phi < 2\pi$. Find decimal approximations for $\sin \phi$ and $\tan \phi$.
    (b) $\sin \theta = -0.5917$ and $\pi < \theta < 3\pi/2$. Find decimal approximations for $\cos \theta$ and $\tan \theta$.

17. Suppose that $y = \sin \theta$ for $0° < \theta < 90°$. Evaluate $\cos \theta$ in terms of y.

Problems 18–21 give an expression for one of the three functions $\sin \theta$, $\cos \theta$, or $\tan \theta$ with θ is in the first quadrant. Find expressions for the other two functions. Your answers will be algebraic expressions in terms of x.

18. $\sin \theta = x/3$ 19. $\cos \theta = \dfrac{4}{x}$ 20. $x = 2\cos \theta$ 21. $x = 9\tan \theta$

22. (a) Find an equation for the line l in Figure 6.79.
    (b) Find the x-intercept of the line.

Figure 6.79   **Figure 6.80**

23. Use Figure 6.80 to find an equation for the line l in terms of x_0, y_0, and θ.

24. Graph $y = \tan \theta$ for $0 \leq \theta \leq 2\pi$. Describe the graph in words. Relate your description to the interpretation of $\tan \theta$ as the slope of a line passing through the origin at an angle of θ to the x-axis. Is this interpretation consistent with your graph?

25. (a) At what values of t does the graph of $y = \tan t$ have vertical asymptotes? What can you say about the graph of $y = \cos t$ at the same values of t?
    (b) At what values of t does the graph of $y = \tan t$ have t-intercepts? What can you say about the graph of $y = \sin t$ at the same values of t?

26. Graph $y = \cos x \cdot \tan x$. Is this function exactly the same as $y = \sin x$? Why or why not?

27. Graph $y = \sin \theta$ and $y = \tan \theta$, for $-20° \leq \theta \leq 20°$. Explain any similarities or differences between the two graphs.

Find exact values for the lengths of the labeled segments in Problems 28–29.

28.

29.

30. Show how to obtain the identity $\cot^2 \theta + 1 = \csc^2 \theta$ from the Pythagorean identity.
31. Graph the functions $y = \sec x$ and $y = \csc x$, and $y = \cot x$. Describe the behavior of each one.
32. For each of the six trigonometric functions, sine, cosine, tangent, cotangent, secant, and cosecant, indicate the intervals between 0 and 2π on which the function is increasing, decreasing, concave up, and concave down.

6.7 INVERSE TRIGONOMETRIC FUNCTIONS

Solving Trigonometric Equations Graphically

A trigonometric equation is an equation that involves trigonometric functions. Consider, for example, the rabbit population of Example 7 on page 263:

$$R = -5000 \cos\left(\frac{\pi}{6}t\right) + 10000.$$

To find when the population reaches 12,000, we need to solve the trigonometric equation

$$-5000 \cos\left(\frac{\pi}{6}t\right) + 10000 = 12{,}000.$$

We can find approximate solutions to trigonometric equations by using a graph. We start with a simpler example.

Example 1   Use a graph to approximate solutions to the equation

$$\cos t = 0.4$$

Solution   We draw a graph of $y = \cos t$ and trace along it on a calculator to find points at which $y = 0.4$ and read off the t-values at these points. In Figure 6.81, the points t_0, t_1, t_2, t_3 represent values of t satisfying $\cos t = 0.4$. If t is in radians, we find $t_0 \approx -1.16$, $t_1 \approx 1.16$, $t_2 \approx 5.12$, $t_3 \approx 7.44$. We can check these values by evaluating:

$$\cos(-1.16) \approx 0.40, \quad \cos(1.16) \approx 0.40, \quad \cos(5.12) \approx 0.40, \quad \cos(7.44) \approx 0.40.$$

Notice that because the cosine function is periodic, the equation $\cos t = 0.4$ has infinitely many solutions. However the symmetry of the graph suggests that the solutions are related in some way.

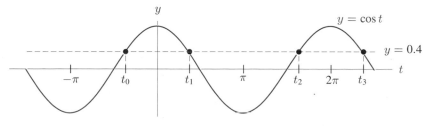

Figure 6.81: The points t_0, t_1, t_2, t_3 are solutions to the equation $\cos t = 0.4$

Solving Trigonometric Equations Using the Inverse Cosine

It is often difficult to obtain accurate answers by reading values from a graph; however there is an another method of solving a trigonometric equation. Finding a solution to

$$\cos t = 0.4$$

means finding an angle whose cosine is 0.4. In other words, given an output for the cosine function, we want to find the corresponding input. This is what the *inverse cosine*, labeled $\boxed{\cos^{-1}}$ on a calculator, gives us. Evaluating an inverse cosine produces an angle whose cosine is the given value.

A calculator in radian mode gives one of the t-values we found in Example 1:

$$\cos^{-1}(0.4) \approx 1.16, \qquad \text{where} \qquad \cos(1.16) \approx 0.4.$$

Notice that the $\boxed{\cos^{-1}}$ key gives only one solution to a trigonometric equation. We can find other solutions using the symmetry of the cosine graph. Since $t_1 = 1.16$, we see in Figure 6.82 that $t_0 \approx -1.16$ because the cosine function is symmetric about the y-axis. In addition, the arch of the cosine graph from $-\pi/2$ to $\pi/2$ is exactly the same shape as the arch from $3\pi/2$ to $5\pi/2$, so

$$t_2 \approx 2\pi - 1.16 = 5.12 \qquad \text{and} \qquad t_3 \approx 2\pi + 1.16 = 7.44.$$

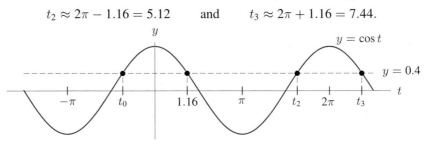

Figure 6.82: Symmetry of cosine graph shows relationship between solutions of $\cos t = 0.4$

Example 2   Estimate when the rabbit population, R, in Example 7 on page 263 reaches 12,000.

Solution   We use Figure 6.83 to estimate where the line $R = 12,000$ cuts the graph of

$$R = -5000\cos\left(\frac{\pi}{6}t\right) + 10000.$$

The first solution, t_1, occurs between the 3rd and 6th months, or between April and July (remember that January is month $t = 0$). The second solution, t_2, occurs somewhere between the 6th and 9th months, or between July and October. Tracing, we find

$$t_1 \approx 3.79 \qquad \text{and} \qquad t_2 \approx 8.21.$$

Thus the rabbit population is 12,000 in late April (month $t = 3$) and early September (month $t = 8$).

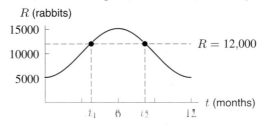

Figure 6.83: Solutions to the equation $-5000\cos(\frac{\pi}{6}t) + 10000 = 12{,}000$

Example 3   As an alternative solution to Example 2, use the inverse cosine to solve the equation:

$$-5000\cos\left(\frac{\pi}{6}t\right) + 10000 = 12000.$$

Solution   We first isolate the trigonometric expression $\cos\left(\frac{\pi}{6}t\right)$:

$$-5000\cos\left(\frac{\pi}{6}t\right) = 2000$$

$$\cos\left(\frac{\pi}{6}t\right) = -0.4$$

To solve for t, we need an angle whose cosine is -0.4. This is what $\cos^{-1}(-0.4)$ gives us. So

$$\frac{\pi}{6}t = \cos^{-1}(-0.4),$$

giving

$$t = \frac{6}{\pi}\cos^{-1}(-0.4).$$

Using a calculator to evaluate $\cos^{-1}(-0.4)$, we find

$$t \approx \frac{6}{\pi}(1.98) = 3.79.$$

This agrees with the value of t_1 we found graphically in Example 2. But how do we find the second solution, t_2? Notice that the graph of R in Figure 6.84 is symmetric about the line $t = 6$, so, as before,

$$t_2 = 12 - t_1 \approx 8.21.$$

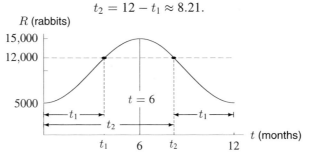

Figure 6.84: By symmetry about the line $t = 6$, we see that $t_2 = 12 - t_1$

The Inverse Cosine Function

As we saw in Figure 6.82 on page 274, the equation

$$\cos t = 0.4$$

has infinitely many solutions. However, the inverse cosine key on a calculator gives only one of them, namely

$$\cos^{-1}(0.4) \approx 1.16,$$

Why does the calculator select this particular solution?

From the graph in Figure 6.85, we see that the angles in the interval $0 \leq t \leq \pi$ produce all values in the range of $\cos t$ once and once only. By choosing output values for the inverse cosine in this interval, we define a new *function* that assigns just one angle to each possible value of the cosine.[6] (Recall that a function can have only one output for each input value.) This is the function that your calculator evaluates when you press the $\boxed{\cos^{-1}}$ key.

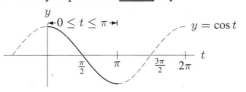

Figure 6.85: The solid portion of this graph represents a function that has only one input value for each output value

---

[6]Other intervals, such as $-\pi \leq t \leq 0$, could also be used. The interval $0 \leq t \leq \pi$ has become the agreed upon choice.

The inverse cosine function inputs values of y between -1 and 1 (all possible values of $\cos t$) and outputs angles between 0 and π. We interpret the value of $\cos^{-1}(y)$ as "the angle between 0 and π whose cosine is y." Because an angle in radians determines an arc of the same measure on a unit circle, the inverse cosine of y is sometimes called the *arccosine* of y. We summarize:

> The **inverse cosine** function, also called the **arccosine** function, is denoted by $\cos^{-1} y$ or $\arccos y$. We define
>
> $$t = \cos^{-1} y \qquad \text{provided that} \qquad y = \cos t \quad \text{and} \quad 0 \leq t \leq \pi.$$
>
> In other words, if $t = \arccos y$, then t is the angle between 0 and π whose cosine is y. The domain of the inverse cosine is $-1 \leq y \leq 1$ and its range is $0 \leq t \leq \pi$.

Example 4   Evaluate   (a) $\cos^{-1}(0)$ (b) $\arccos(1)$ (c) $\cos^{-1}(-1)$

Solution   (a) $\cos^{-1}(0)$ means the angle between 0 and π whose cosine is 0. Since $\cos(\pi/2) = 0$, we have $\cos^{-1}(0) = \pi/2$.

(b) $\arccos(1)$ means the angle between 0 and π whose cosine is 1. Since $\cos(0) = 1$, we have $\arccos(1) = 0$.

(c) $\cos^{-1}(-1)$ means the angle between 0 and π whose cosine is -1. Since $\cos(\pi) = -1$, we have $\cos^{-1}(-1) = \pi$.

Warning! It is important to realize that the notation $\cos^{-1} y$ does *not* indicate the reciprocal of $\cos y$. In other words, $\cos^{-1} y$ is not the same as $(\cos y)^{-1}$. For example,

$$\cos^{-1}(0) = \frac{\pi}{2} \quad \text{because} \quad \cos\left(\frac{\pi}{2}\right) = 0,$$

but

$$(\cos 0)^{-1} = \frac{1}{\cos 0} = \frac{1}{1} = 1.$$

The Inverse Sine and Inverse Tangent Functions

To solve equations involving sine or tangent, we need an *inverse sine* and an *inverse tangent* function. For example, suppose we want to solve

$$\sin t = 0.8, \qquad 0 \leq t \leq 2\pi.$$

The inverse sine of 0.8, or $\sin^{-1}(0.8)$, gives us an angle whose sine is 0.8. But there are many angles with a given sine and the inverse sine function specifies only one of them. Just as we did with the inverse cosine, we choose an interval on the t-axis that produces all values in the range of the sine function once and once only.

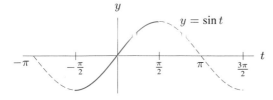

Figure 6.86: The graph of $y = \sin t$, $-\frac{\pi}{2} \leq t \leq \frac{\pi}{2}$

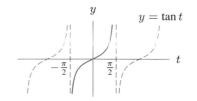

Figure 6.87: The graph of $y = \tan t$, $-\frac{\pi}{2} < t < \frac{\pi}{2}$

Figure 6.86 shows that we cannot choose the same interval that we chose for $\cos t$, which was $0 \le t \le \pi$. However, the interval $-\pi/2 \le t \le \pi/2$ includes a unique angle for each value of $\sin t$. This interval is chosen because it is the smallest interval around $t = 0$ that includes all values of $\sin t$. This same interval, except for the endpoints, is also used to define the inverse tangent function, as illustrated in Figure 6.87.

The **inverse sine** function, also called the **arcsine** function, is denoted by $\sin^{-1} y$ or $\arcsin y$. We define

$$t = \sin^{-1} y \quad \text{provided that} \quad y = \sin t \text{ and } -\frac{\pi}{2} \le t \le \frac{\pi}{2}.$$

The domain of the inverse sine is $-1 \le y \le 1$ and the range is $-\pi/2 \le t \le \pi/2$.

The **inverse tangent** function, also called the **arctangent** function, is denoted by $\tan^{-1} y$ or $\arctan y$. We define

$$t = \tan^{-1} y \quad \text{provided that} \quad y = \tan t \text{ and } \frac{\pi}{2} < t < \frac{\pi}{2}.$$

The domain of the inverse tangent is $-\infty < y < \infty$ and the range is $-\pi/2 < t < \pi/2$.

Example 5 Evaluate (a) $\sin^{-1}(1)$ (b) $\arcsin(-1)$ (c) $\tan^{-1}(0)$ (d) $\arctan(1)$

Solution (a) $\sin^{-1}(1)$ means the angle between $-\pi/2$ and $\pi/2$ whose sine is 1. Since $\sin(\pi/2) = 1$, we have $\sin^{-1}(1) = \pi/2$.
  (b) $\arcsin(-1) = -\pi/2$ since $\sin(-\pi/2) = -1$.
  (c) $\tan^{-1}(0)$ since $\tan 0 = 0$.
  (d) $\arctan(1) = \pi/4$ since $\tan(\pi/4) = 1$.

Example 6 Evaluate (a) $\sin^{-1}(-0.5)$ (b) $\arctan(-1)$

Solution (a) $\sin^{-1}(-0.5)$ is the angle between $-\pi/2 \le t \le \pi/2$ whose sine is -0.5. From page 253 we have $\sin(\pi/6) = 0.5$. The sine is an odd function, so $\sin(-\pi/6) = -0.5$, and $\sin^{-1}(-0.5) = -\pi/6$.
  (b) The tangent is also an odd function, and $\tan \pi/4 = 1$, so $\tan(-\pi/4) = -1$, and therefore $\arctan(-1) = -\pi/4$.

Example 7 While riding the ferris wheel, how much time during the first turn do you spend above 400 feet?

Solution Since your first turn takes 20 minutes and your height is given by $f(t) = 250 + 250 \sin\left(\frac{\pi}{10}t - \frac{\pi}{2}\right)$, we must solve the inequality

$$250 + 250 \sin\left(\frac{\pi}{10}t - \frac{\pi}{2}\right) \ge 400 \qquad \text{for } 0 \le t \le 20.$$

This can be simplified as follows:

$$250 \sin\left(\frac{\pi}{10}t - \frac{\pi}{2}\right) \ge 150$$

$$\sin\left(\frac{\pi}{10}t - \frac{\pi}{2}\right) \ge \frac{3}{5}.$$

From Figure 6.88, we see that the equation $\sin\left(\dfrac{\pi}{10}t - \dfrac{\pi}{2}\right) = \dfrac{3}{5}$ has two solutions on the interval $0 \leq t \leq 20$. One of them can be found by using the arcsine function:

$$\frac{\pi}{10}t - \frac{\pi}{2} = \arcsin\frac{3}{5}$$

$$\frac{\pi}{10}t = \frac{\pi}{2} + \arcsin\frac{3}{5}$$

$$t = \underbrace{\frac{10}{\pi}\left(\frac{\pi}{2} + \arcsin\frac{3}{5}\right)}_{\text{Exact solution}} \approx \underbrace{7.05.}_{\text{Approximate solution}}$$

Since the period of this function is 20, the other solution can be found by symmetry to be $20 - 7.05 = 12.95$ minutes. Thus, you spend $12.95 - 7.05 = 5.9$ minutes at a height of 400 feet or more.

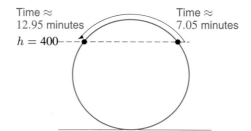

Figure 6.88: On the ferris wheel between $t = 7.05$ minutes and $t = 12.95$ minutes, you are above 400 feet

Solving Equations Using Reference Angles

Because of the symmetry of the unit circle, the values of sine, cosine, and tangent of angles in the first quadrant can be used to find values of these functions for angles in the other three quadrants.

Example 8   In Figure 6.89, point P is determined by the angle $\theta = 65°$. What are the coordinates of point P?

Solution   Rounding to two decimal places, a calculator gives $\cos 65° = 0.42$ and $\sin 65° = 0.91$. Thus, $P = (0.42, 0.91)$.

Example 9   Use Example 8 to find the sine and cosine of $-65°, 245°$, and $785°$.

Solution   Let $P = (0.42, 0.91)$ be the point on the unit circle given by the angle $65°$. In Figure 6.89, we see that $-65°$ gives a point labeled Q that is the reflection of P across the x-axis. Thus, the y-coordinate of Q is the negative of the y-coordinate of P, and so $Q = (0.42, -0.91)$. This means that

$$\sin(-65°) = -0.91 \quad \text{and} \quad \cos(-65°) = 0.42.$$

In Figure 6.90, we see that $245° = 180° + 65°$ gives point R that is diametrically opposite the point P. The coordinates of R are the negatives of the coordinates of P, so $R = (-0.42, -0.91)$. Thus,

$$\sin 245° = -0.91 \quad \text{and} \quad \cos 245° = -0.42.$$

Finally in Figure 6.91, we see that $785° = 720° + 65°$, so this angle specifies the same point as $65°$. This means that

$$\sin 785° = 0.91 \quad \text{and} \quad \cos 785° = 0.42.$$

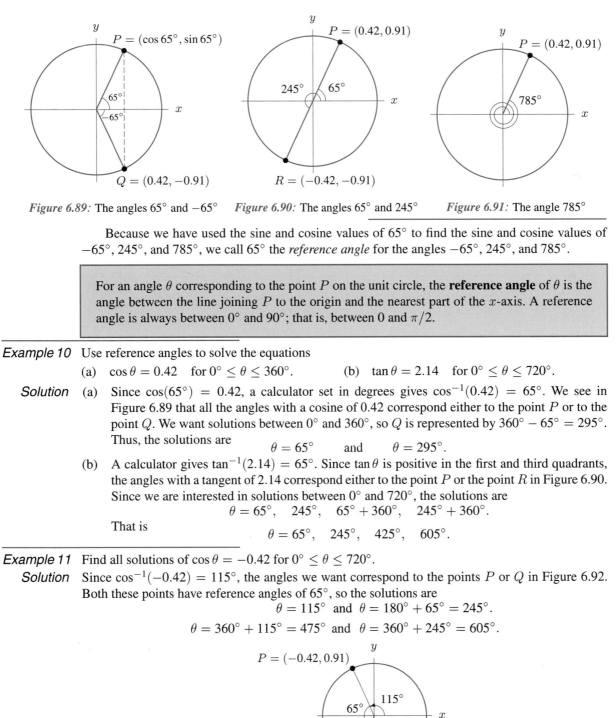

Figure 6.89: The angles $65°$ and $-65°$ *Figure 6.90:* The angles $65°$ and $245°$ *Figure 6.91:* The angle $785°$

Because we have used the sine and cosine values of $65°$ to find the sine and cosine values of $-65°$, $245°$, and $785°$, we call $65°$ the *reference angle* for the angles $-65°$, $245°$, and $785°$.

> For an angle θ corresponding to the point P on the unit circle, the **reference angle** of θ is the angle between the line joining P to the origin and the nearest part of the x-axis. A reference angle is always between $0°$ and $90°$; that is, between 0 and $\pi/2$.

Example 10 Use reference angles to solve the equations
   (a) $\cos \theta = 0.42$ for $0° \leq \theta \leq 360°$.   (b) $\tan \theta = 2.14$ for $0° \leq \theta \leq 720°$.

Solution   (a) Since $\cos(65°) = 0.42$, a calculator set in degrees gives $\cos^{-1}(0.42) = 65°$. We see in Figure 6.89 that all the angles with a cosine of 0.42 correspond either to the point P or to the point Q. We want solutions between $0°$ and $360°$, so Q is represented by $360° - 65° = 295°$. Thus, the solutions are
$$\theta = 65° \quad \text{and} \quad \theta = 295°.$$
   (b) A calculator gives $\tan^{-1}(2.14) = 65°$. Since $\tan \theta$ is positive in the first and third quadrants, the angles with a tangent of 2.14 correspond either to the point P or the point R in Figure 6.90. Since we are interested in solutions between $0°$ and $720°$, the solutions are
$$\theta = 65°, \quad 245°, \quad 65° + 360°, \quad 245° + 360°.$$
   That is
$$\theta = 65°, \quad 245°, \quad 425°, \quad 605°.$$

Example 11 Find all solutions of $\cos \theta = -0.42$ for $0° \leq \theta \leq 720°$.
Solution   Since $\cos^{-1}(-0.42) = 115°$, the angles we want correspond to the points P or Q in Figure 6.92. Both these points have reference angles of $65°$, so the solutions are
$$\theta = 115° \quad \text{and} \quad \theta = 180° + 65° = 245°.$$
$$\theta = 360° + 115° = 475° \quad \text{and} \quad \theta = 360° + 245° = 605°.$$

Figure 6.92: Points corresponding to angles with $\cos \theta = -0.42$

Example 12 Use reference angles to solve the equation from Example 7:

$$\sin\left(\frac{\pi}{10}t - \frac{\pi}{2}\right) = \frac{3}{5}.$$

Solution As in Example 7, the first solution is given by the arcsine. Solving for t in the equation

$$\frac{\pi}{10}t - \frac{\pi}{2} = \arcsin\frac{3}{5} = 0.644$$

gives $t = 7.05$ minutes. The second solution corresponds to another on the circle with the same reference angle, 0.644, and a positive value of the sine. This is in the second quadrant, so we have

$$\frac{\pi}{10}t - \frac{\pi}{2} = \pi - 0.644.$$

Solving for t in this equation gives $t = 12.95$ minutes, as in Example 7.

Problems for Section 6.7

1. (a) Use a graph of $y = \cos t$ to estimate two solutions to the equation $\cos t = -0.3$ for $0 \le t \le 2\pi$.
    (b) Solve the same equation using the inverse cosine.

2. (a) Find exact values of the solutions to the equation $\cos t = 1/2$ with $-2\pi \le t \le 2\pi$. Plot them on a graph of $y = \cos t$.
    (b) Using a unit circle, explain how many solutions to the equation $\cos t = 1/2$ you would expect in the interval $0 \le t \le 2\pi$.

3. For each angle in the set $\{0°, 15°, 30°, \cdots, 330°, 345°, 360°\}$, list the reference angle.

4. Make a table of exact values of $\sin\theta$ and $\cos\theta$ for $0°$, $30°$, $45°$, $60°$, and $90°$. Extend the table to include all angles $0° \le \theta < 360°$ which have $0°$, $30°$, $45°$, $60°$, or $90°$ as their reference angle. Include the radian measure of all these angles.

5. Use reference angles to find exact values for

    (a) $\cos 120°$ (b) $\sin 135°$ (c) $\cos 225°$ (d) $\sin 300°$

6. Evaluate the following expressions exactly, making use of reference angles if applicable:

    (a) $\sin 30°$ (b) $\sin 150°$ (c) $\cos 150°$ (d) $\cos 300°$

7. Use reference angles to find exact values for:

    (a) $\sin\left(\dfrac{2\pi}{3}\right)$ (b) $\cos\left(\dfrac{3\pi}{4}\right)$ (c) $\tan\left(-\dfrac{3\pi}{4}\right)$ (d) $\cos\left(\dfrac{11\pi}{6}\right)$

In Problem 8–10, use a graph to estimate all the solutions of the equations between 0 and 2π.

8. $\sin\theta = 0.65$ 9. $\cos t = -0.24$ 10. $\tan x = 2.8$

11. Use inverse functions to solve the equations in Problems 8–10.

Solve the equations in Problems 12–15 for t, assuming that t is between 0 and 2π. First obtain approximate answers from a graph, and then find exact answers.

12. $\cos(2t) = \dfrac{1}{2}$ 13. $\tan t = \dfrac{1}{\tan t}$ 14. $2\sin t \cos t - \cos t = 0$ 15. $3\cos^2 t = \sin^2 t$

16. (a) Solve $\tan\theta = 1$ exactly for $-\pi/2 < \theta < \pi/2$.
    (b) Give a solution correct to 3 decimal places for $\sin\theta = 0.95$ with $-360° \le \theta \le 0°$.

Find exact values for the solutions of each of the equations in Problems 17–19.

17. $\sin\theta = -\sqrt{2}/2$ 18. $\cos\theta = \sqrt{3}/2$ 19. $\tan\theta = -\sqrt{3}/3$

20. A company's sales are seasonal with the peak in mid-December and the lowest point in mid-June. The company makes $100,000 in sales in December, and only $20,000 in June.

    (a) Find a trigonometric function, $s = f(t)$, representing sales at time t months after mid-January.
    (b) What would you expect the sales to be for mid-April?
    (c) Find the t-value for which $s = 60,000$. Interpret your answer.

21. In a tidal river, the time between high tide and low tide is 6.2 hours. At high tide the depth of the water is 17.2 feet, while at low tide the depth is 5.6 feet. Assume the water depth is a trigonometric function of time.

    (a) Sketch a graph of the depth of the water over time if there is a high tide at 12:00 noon. Label your graph, indicating the high and low tide.
    (b) Write an equation for the curve you drew in part (a).
    (c) A boat requires a depth of 8 feet to sail, and is docked at 12:00 noon. What is the latest time in the afternoon it can set sail? Your answer should be accurate to the nearest minute.

22. Approximate the x-coordinates of points P and Q shown in Figure 6.93, assuming that the curve is a sine curve. [Hint: Find a formula for the curve.]

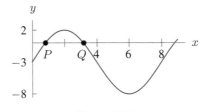

Figure 6.93

23. For what value(s) of θ does $\sin \theta = 3/4$, for $0 \le \theta \le \pi$?
24. Find the angle θ, in radians, in the second quadrant whose tangent is -3.
25. Solve for α exactly: $\sec^2 \alpha + 3 \tan \alpha = \tan \alpha$ with $0 < \alpha < 2\pi$.
26. In your own words, explain what each of the following expressions means. Evaluate each expression for $x = 0.5$. Give an exact answer if possible.

    (a) $\sin^{-1} x$ (b) $\sin(x^{-1})$ (c) $(\sin x)^{-1}$.

27. Evaluate the following expressions in radians. Give an exact answer if possible.

    (a) $\arccos(0.5)$ (b) $\arccos(-1)$ (c) $\arcsin(0.1)$

28. Use a graph to find all the solutions to the equation $12 - 4\cos 3t = 14$ between 0 and $2\pi/3$ (one cycle of the graph). How many solutions are there between 0 and 2π?

29. Approximate the zero(s) of $f(t) = 3 - 5\sin(4t)$ for $0 \le t < \pi/2$.

    (a) Graphically.   (b) Using the arcsine function.

30. The statement $e^2 = 7.3891$ is equivalent to the statement $\ln(7.3891) = 2$.

    (a) Given $\sin(\pi/8) = 0.3827$, write the equivalent statement using arcsine.
    (b) Given $\cos 0 = 1$, write the equivalent statement using arccosine.

31. Evaluate each expression that is defined. Give an exact answer if possible. If the expression is undefined, say so.

    (a) $\sin^{-1} 1$ (b) $\tan^{-1} 1$ (c) $\sin 1$ (d) $\tan 1$
    (e) $\sin^{-1} 2$ (f) $\tan^{-1} 2$ (g) $\sin 2$ (h) $\tan 2$

32. Use a calculator to find $\cos^{-1}(\cos x)$ for $x = 1, 2, 3, 4, 5, 6, 7$ radians. Explain your results.

33. Without a calculator, evaluate the following exactly.

    (a) $\cos^{-1}(1/2)$ (b) $\cos^{-1}(-1/2)$ (c) $\cos(\cos^{-1}(1/2))$ (d) $\cos^{-1}(\cos(5\pi/3))$

34. State the domain and range of the following functions and explain what your answers mean in terms of evaluating the functions.

    (a) $f(x) = \sin^{-1} x$ (b) $g(x) = \cos^{-1} x$ (c) $h(x) = \tan^{-1} x$

35. One of the following statements is always true; the other is true for some values of x and not for others. Which is which? Justify your answer with an example.

    I. $\arcsin(\sin x) = x$ II. $\sin(\arcsin x) = x$

36. Suppose that a is a number with $0 \le a \le \pi/2$. Suppose $b = \pi + a$.

    (a) What is $\arccos(\cos a)$?   (b) What is $\arccos(\cos b)$?

37. You are perched in the crow's nest, C, on top of the mast of a ship, S. See Figure 6.94. You want to figure out how far you can see when you are x meters above the surface of the ocean.

    (a) Find formulas for d, the distance you can see to the horizon, H, and l, the distance to the horizon along the earth's surface, in terms of x, the height of the ship's mast, and r, the radius of the earth.
    (b) How far is the horizon from the top of a 50-meter mast? How far, measured along the earth's surface, is the horizon from the ship's position on the ocean? Use $r = 6,370,000$ meters.

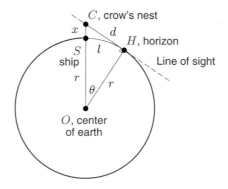

Figure 6.94

38. Let k be a positive constant and t be an angle measured in radians. Consider the equation
$$k \sin t = t^2.$$

    (a) Explain why any solution to the equation must be between $-\sqrt{k}$ and $\sqrt{k}$, inclusive.
    (b) Approximate every solution to the equation when $k = 2$.
    (c) Explain why the equation has more solutions for larger values of k than it does for small values.
    (d) Approximate the least value of k, if any, for which the equation has a negative solution.

REVIEW PROBLEMS FOR CHAPTER SIX

State whether each of the statements in Problems 1–12 is true or false.

1. $\sin(-x) = -\sin x$ **2.** $\cos(-x) = -\cos x$ **3.** $\sin(-x) = \sin x$

4. $\cos(-x) = \cos x$ **5.** $\sin(x + \pi) = -\sin x$ **6.** $\cos(x + \pi) = \sin x$

7 $\cos(x + 4\pi) = \cos x$ **8.** $2\cos(x) = \cos(2x)$ **9.** $\cos(x + 1) = \cos x + \cos 1$

10. $\sin x$ has period 2π **11.** $\sin(\pi x)$ has period π **12.** $\cos\left(\dfrac{1}{x}\right) = \dfrac{\cos 1}{\cos x}$

13. If you start at the point $(1, 0)$ on the unit circle and travel counterclockwise through the given angle (in radians), in which quadrant will you be?

    (a) 2   (b) 4   (c) 6   (d) 1.5   (e) 3.2

14.   Graph $y = \sin\theta$ for $0° \leq \theta \leq 360°$ where θ is in degrees and put ticks marks on the x-axis at $0°, 90°, 180°, 270°, 360°$. Use this graph to explain the sign of the sine function as θ goes through each of the four quadrants.

15.   Graph $y = \sin t$ for   (a) $-6° \leq t \leq 6°$ (degrees)   (b) $-6 \leq t \leq 6$ (radians)

16.   (a)   Evaluate and compare the following four pairs of values: $\cos(\pi/6)$ and $\cos(-\pi/6)$, $\cos(3\pi/4)$ and $\cos(-3\pi/4)$, $\cos(7\pi/6)$ and $\cos(-7\pi/6)$, $\cos(11\pi/6)$ and $\cos(-11\pi/6)$. What do you observe?
      (b)   Compare the following pairs of values: $\sin(\pi/6)$ and $\sin(-\pi/6)$, $\sin(3\pi/4)$ and $\sin(-3\pi/4)$, $\sin(7\pi/6)$ and $\sin(-7\pi/6)$, $\sin(11\pi/6)$ and $\sin(-11\pi/6)$. What do you observe?
      (c)   Based on parts (a) and (b) what can you say about $\tan(x)$ and $\tan(-x)$?

17.   Computers and calculators often show algebraically simple values as a decimal approximation, which you should be able to recognize. Without using a calculator, match the exact form to the decimal approximation.

   (a) $\pi/3$ (b) $\sqrt{2}/2$ (c) $\pi/6$
   (d) $\pi/4$ (e) $\sqrt{3}/2$ (f) $\pi/2$

   0.5235987756   1.570796327   1.047197551
   0.7853981634   0.7071067812   0.8660254038

Use a graph to find all of the solutions between 0 and 2π of each of the equations in Problems 18–21.

18. $\cos 2\theta = 0.63$ 19. $\sin 3\theta = -0.9$ 20. $\sin\theta/2 = 0.4$ 21. $\cos 4\theta = -1$

22.   Find all solutions to the following equations. Give your answers in degrees.

   (a) $\cos x = \dfrac{1}{\sqrt{2}}$ (exactly)   (b) $\tan x = -10.5$ (to 3 decimal places)

In Problems 23–30, solve the equations for α for $0 \leq \alpha < 2\pi$. Give exact answers if possible.

23. $2\cos\alpha = 1$ 24. $\tan\alpha = \sqrt{3} - 2\tan\alpha$ 25. $\sin(2\alpha) + 3 = 4$

26. $4\tan\alpha + 3 = 2$ 27. $3\sin^2\alpha + 4 = 5$ 28. $\tan^2\alpha = 2\tan\alpha$

29. $3\cos^2\alpha + 2 = 3 - 2\cos\alpha$ 30. $3\sin^2\alpha + 3\sin\alpha + 4 = 3 - 2\sin\alpha$

31.   How far does the tip of the minute hand of a clock move in 1 hour and 27 minutes if the hand is 2 inches long?

32.   A revolving door has three panels of length one which make equal angles with each other, as in Figure 6.95. What is the width, w, of the opening between A and B?

Figure 6.95   *Figure 6.96*

33.   A stationary sensor, S, is mounted on the wall containing a revolving door with 1 meter panels as shown in Figure 6.96. Suppose that a second, moving sensor, M, is mounted at the tip of a panel in such a way that it continually sends a beam perpendicular to the wall line.

   (a)   How far is the intersection point of the two beams from the sensor S when the panels are as shown in Figure 6.96?
   (b)   How far is the intersection point of the two beams from the wall sensor after the door has been rotated counterclockwise $75°$ from its position in Figure 6.96?

34. How many miles on the surface of the earth correspond to one degree of latitude? (The radius of the earth is 3960 miles.)

35. A person on earth is observing the moon, which is 238,860 miles away. The moon has a diameter of 2160 miles. What is the angle in degrees spanned by the moon in the eye of the beholder?

36. In order to measure angles, the ancient Babylonians divided the circumference of a circle into 360 equal parts, so that an angle measuring one degree spans an arc equal to one 360th of the circle's circumference. Since the more modern metric system is based on powers of ten, write a proposal to create a new metric system for measuring angles. (Use a different subdivision of the circle.)

37. A compact disk is 12 cm in diameter and rotates at 100 rpm (revolutions per minute) when being played. The hole in the center is 1.5 cm in diameter. Find the speed in cm/min of a point on the outer edge of the disk and the speed of a point on the inner edge.

38. Explain in words the difference between the expressions $\sin^2 \theta$ and $\sin \theta^2$. Give a specific numerical example illustrating the difference between these expressions.

39. In which quadrants do the following statements hold?
(a) $\sin \theta > 0$ and $\cos \theta > 0$ (b) $\tan \theta > 0$ (c) $\tan \theta < 0$
(d) $\sin \theta < 0$ and $\cos \theta > 0$ (e) $\cos \theta < 0$ and $\tan \theta > 0$

40. Find a formula for the function $f(t)$ that approximately fits the data in the following table.

t	0	1	2	3	4	5	6	7
$f(t)$	4.0	5.25	6.165	6.5	6.167	5.25	4.0	2.75
t	8	9	10	11	12	13	14	15
$f(t)$	1.835	1.5	1.8348	2.75	4.0	5.22	6.16	6.5

41. Without using a calculator, match the graphs in Figure 6.97 to the following functions:
(a) $y = \sin(2t)$ (b) $y = (\sin t) + 2$ (c) $y = 2\sin t$ (d) $y = \sin(t+2)$

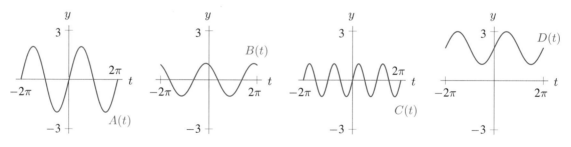

Figure 6.97

42. Find a possible formula for the trigonometric function in Figure 6.98.

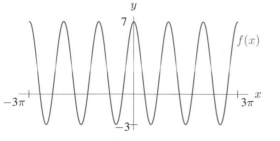

Figure 6.98

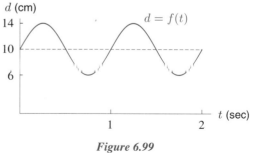

Figure 6.99

43. A weight is suspended from the ceiling by a spring. Figure 6.99 shows a graph of the distance from the ceiling to the weight, $d = f(t)$, as a function of time.

    (a) Find a possible formula for $f(t)$.
    (b) Solve $f(t) = 12$ exactly. Interpret your results.

In Problems 44–46, find a formula (using the sine) for your height above ground after t minutes on the London ferris wheel, which has diameter 500 feet and a period of 20 minutes. Passengers board the wheel at the bottom. Graph the function to check that it is correct.

44. The London ferris wheel has increased its rotation speed. The wheel completes one full revolution every ten minutes.

45. The London ferris wheel is built as planned, but with the wheel only 300 feet high.

46. The developers of the London ferris wheel build a loading tower of height 250 feet. Loading is done at the 3 o'clock position and you start by ascending.

47. An animal population increases from a low of 1200 in year $t = 0$, to a high of 3000 four years later and then decreases back to 1200 over the next four years. Model this behavior by a sinusoidal function.

48. Suppose the angle θ is in the first quadrant, and that $\tan \theta = \dfrac{3}{4}$. Since $\tan \theta = \dfrac{\sin \theta}{\cos \theta}$, does this mean that $\sin \theta = 3$ and $\cos \theta = 4$?

49. Consider the equation $\sin t = k$ for $-\pi \le t \le \pi$. Using graphs, explain why this equation:

    (a) Has one solution when $k = 1$.   (b) Has two solutions when $k = -1/2$.
    (c) Has three solutions when $k = 0$.   (d) Has no solution when $k = 2$.

50. If possible, find one exact zero of each of the following functions.

    (a) $f(t) = \cos(t + 2)$ (b) $f(t) = 2\cos(t)$ (c) $f(t) = \cos(2t)$ (d) $f(t) = \cos(t) + 2$

51. State exact values for all of the zeroes of the function $f(x) = 3\sin(\pi x - 1) + 1$.

52. Graph the following functions, labeling your axes. State the domain and range of each.

    (a) $f(x) = 2\cos^{-1}(x)$ (b) $g(x) = \sin^{-1}(\pi x)$

53. An electronic timer beeps every 3 minutes; another timer beeps every 4 minutes. At 9:03 am they beep at the same time.

    (a) What is the period of each of the timers (separately)?
    (b) How long is it before the two timers beep at the same time again?
    (c) What is the period at which they beep together?

54. For $f(t) = \tan t$ are there any limitations on:

    (a) The input values of $f(t)$?   (b) The output values of $f(t)$?

55. State whether each of the following statements is true or false.

    (a) For all angles θ in radians, $\arccos(\cos \theta) = \theta$.
    (b) For all values of x between -1 and 1, $\cos(\arccos x) = x$.
    (c) If $\tan A = \tan B$, then $\dfrac{A - B}{\pi}$ is an integer.
    (d) $\cos A = \cos B$ implies that $\sin A = \sin B$.

56. In a physics lab you collected the data in Table 6.13 on the height above the floor of a weight bobbing on a spring attached to the ceiling. Find a sine function to model this data.

TABLE 6.13

t, sec	0.0	0.1	0.2	0.3	0.4	0.5	0.6	0.7	0.8	0.9	1.0	1.1
y, cm	120	136	165	180	166	133	120	135	164	179	165	133

57. The average high temperature each month in Hong Kong is given in Table 6.14. With t representing time in months since the start of the year, give an approximate formula for average high temperature as a function of t.

TABLE 6.14

Month	Jan	Feb	Mar	Apr	May	Jun	Jul	Aug	Sep	Oct	Nov	Dec
Temp (°C)	19	19	21	25	28	30	31	31	30	28	24	21

58. In climates that require central heating, the setting of a household thermostat gives the temperature at which the furnace turns on. Water is boiled, and steam is forced into radiators that warm the house. Once the household temperature reaches the thermostat setting, the furnace shuts off. However, the radiators remain hot for some time, and so the household temperature continues rising for a while before it starts to drop. Once it drops back to the thermostat setting, the furnace relights, and the cycle begins anew. As it takes some time for the radiators to reheat, the house continues to cool even after the furnace has turned back on. The household temperature is represented by the graph in Figure 6.100.

(a) The household temperature, T, can be modeled by a trigonometric function. Assume the furnace spends as much time on as it does off. What is the setting on the thermostat?
(b) According to the graph, describe what is happening at $t = 0, 0.25, 0.5, 0.75, 1$.
(c) Give a formula for $T = f(t)$, the temperature in terms of time, t, in hours.
(d) Describe the physical significance of the period, the amplitude, and the midline.
(e) The house described in this problem takes as much time to heat up as it does to cool down. Suppose another house takes 15 minutes to heat up but 45 minutes to cool. Sketch (roughly) its temperature over one hour. Label any significant points (such as times the furnace turns on or off). Is this the graph of a trigonometric function?

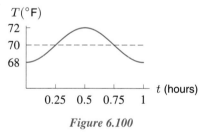

Figure 6.100

CHAPTER SEVEN

TRIGONOMETRY

Chapter 6 concentrated on the use of trigonometric functions to represent periodic phenomena. This chapter uses trigonometry to analyze triangles.

From the definitions of the trigonometric functions, we derive identities which relate the trigonometric functions to one another. These identities are used to simplify expressions, solve equations, and understand trigonometric graphs. The chapter concludes with applications of trigonometry to mathematical models and to polar coordinates.

7.1 RIGHT TRIANGLES

For most of recorded history, people have used the properties of triangles to make indirect measurements of the world around them. In the fourth century BC, the Greek mathematician Eudoxus used trigonometry (which means, literally, *triangle measurement*) to calculate the radius of the earth. Today scientists and engineers, surveyors, and contractors still use trigonometry.

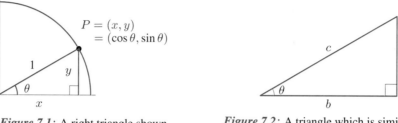

Figure 7.1: A right triangle shown with the unit circle

Figure 7.2: A triangle which is similar to the triangle in Figure 7.1

There is a relationship between the trigonometric functions and right triangles. An angle θ determines a point P on the unit circle in Figure 7.1. The angle θ also determines the right triangle with hypotenuse of length 1; the other sides of this triangle have lengths $x = \cos\theta$ and $y = \sin\theta$. The right triangle in Figure 7.2 also has an angle θ, so these two triangles are similar. Therefore, the ratios of the lengths of their corresponding sides are equal:

$$\frac{a}{c} = \frac{\sin\theta}{1} = \sin\theta \qquad \text{and} \qquad \frac{b}{c} = \frac{\cos\theta}{1} = \cos\theta.$$

In addition,

$$\frac{a}{b} = \frac{\sin\theta}{\cos\theta} = \tan\theta.$$

The side directly across from the angle θ is referred to as the *opposite* side, and the other side, which forms one side of the angle θ, is called the *adjacent* side. Using this terminology we have:

If θ is an angle in a right triangle (other than the right angle),

$$\sin\theta = \frac{\text{Opposite}}{\text{Hypotenuse}}, \qquad \cos\theta = \frac{\text{Adjacent}}{\text{Hypotenuse}}, \qquad \tan\theta = \frac{\text{Opposite}}{\text{Adjacent}}.$$

Example 1 A surveyor must measure the distance between the two banks of a straight river. (See Figure 7.3.) She sights a tree at point T on the opposite bank of the river and drives a stake into the ground (at point P) directly across from the tree. Then she walks 50 meters upstream and places a stake at point Q. She measures angle PQT and finds that it is $58°$. Find the width of the river.

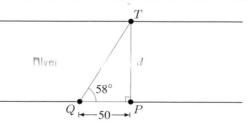

Figure 7.3: Measuring the width of a river

Solution   See Figure 7.3. Since the tangent is the length of the opposite side divided by the length of the adjacent side,

$$\tan 58° = \frac{d}{50}$$
$$d = 50 \tan 58° \approx 80.$$

The width of the river is about 80 meters.

Example 2   The ground crew for a hot air balloon is positioned 200 meters from the point of lift-off and monitors the ascent of the balloon. Express the height of the balloon as a function of the ground crew's angle of observation.

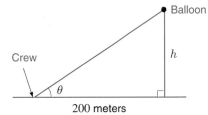

Figure 7.4: A model for the ascent of a balloon

Solution   The angle of observation is labeled θ in Figure 7.4. The distance 200 meters forms the adjacent side for this angle and the height of the balloon, h, is the opposite side. Thus,

$$\tan \theta = \frac{h}{200},$$

so

$$h = 200 \tan \theta.$$

Special Angles: *30°, 45°, 60°*

In Chapter 6, we calculated some values of the sine and cosine exactly and used a calculator for others. We now use right triangles to calculate exact values of the sine, cosine, and tangent of 30°, 45°, and 60°.

Example 3   Figure 7.5 shows the point $P = (x, y)$ corresponding to the angle 45° on the unit circle. A right triangle has been drawn in. The triangle is isosceles; it has two equal angles (both 45°) and two equal sides, so $x = y$. Therefore, the Pythagorean theorem $x^2 + y^2 = 1$ gives

$$x^2 + x^2 = 1$$
$$2x^2 = 1$$
$$x = \sqrt{\frac{1}{2}}.$$

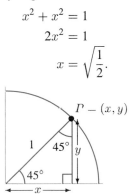

Figure 7.5: This triangle has two equal angles and two equal sides, so $x = y$

We know that x is positive because P is in the first quadrant. This quantity is often rewritten as

$$\sqrt{\frac{1}{2}} = \sqrt{\frac{1}{2} \cdot \frac{2}{2}} = \sqrt{\frac{2}{4}} = \frac{\sqrt{2}}{2}.$$

Since $x = y$, we see that $y = \sqrt{2}/2$ as well. Thus, since x and y are the coordinates of P, we have $\cos 45° = \sqrt{2}/2$, $\sin 45° = \sqrt{2}/2$, and $\tan 45° = 1$.

Example 4   Figure 7.6 shows the point $Q = (x, y)$ corresponding to the angle 30° on the unit circle. A right triangle has been drawn in, and a mirror image of this triangle is shown below the x-axis. Together these two triangles form the triangle $\triangle OQA$. This triangle has three equal 60° angles and so has three equal sides, each side of length 1. The length of side $\overline{QA}$ can also be written as $2y$, and so we have $2y = 1$, or $y = 1/2$. By the Pythagorean theorem,

$$x^2 + y^2 = 1$$

$$x^2 + \left(\frac{1}{2}\right)^2 = 1$$

$$x^2 = \frac{3}{4}$$

$$x = \sqrt{\frac{3}{4}} = \frac{\sqrt{3}}{2}.$$

Note that x is positive because Q is in the first quadrant. Since x and y are the coordinates of Q, this means that $\cos 30° = \sqrt{3}/2$, $\sin 30° = 1/2$, and $\tan 30° = 1/\sqrt{3}$.

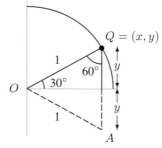

Figure 7.6: The triangle $\triangle OQA$ has three equal angles and three equal sides, so $2y = 1$

A similar argument shows that $\cos 60° = 1/2$, $\sin 60° = \sqrt{3}/2$, and $\tan 60° = \sqrt{3}$.

It is worth memorizing the values of sine and cosine for these special angles. As the following example shows, special angles and reference angles[1] can be used to calculate exact values of the trigonometric functions of some angles greater than 90° or less than 0°. However, the values of sine and cosine for most other angles have to be found using a calculator or computer.

Example 5   Find the exact coordinates of a point B designated by 315° on a circle of radius 6.

Solution   Point B is in the fourth quadrant, where $\cos 315°$ is positive and $\sin 315°$ is negative. See Figure 7.7. The reference angle for 315° is 45°, so $\cos 315° = \cos 45°$ and $\sin 315° = -\sin 45°$.

The coordinates of point B are given by

$$x = r \cos \theta \qquad \text{and} \qquad y = r \sin \theta$$

$$= 6 \cos 315° \qquad\qquad\qquad\qquad = 6 \sin 315°$$

$$= 6\left(\frac{\sqrt{2}}{2}\right) = 3\sqrt{2} \qquad\qquad\qquad = 6\left(\frac{-\sqrt{2}}{2}\right) = -3\sqrt{2}$$

[1] See page 280.

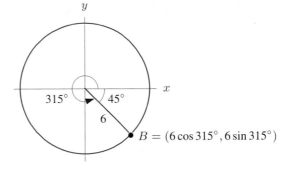

Figure 7.7: Exact coordinates of B found using reference angle of $45°$

Thus, the coordinates of B are $(3\sqrt{2}, -3\sqrt{2})$.

The conversion factor $\pi/180°$ gives the following radian measures for the special angles in the first quadrant:

$$30° = \frac{\pi}{6}, \qquad 45° = \frac{\pi}{4}, \qquad 60° = \frac{\pi}{3}.$$

Of course, whether the angles are expressed in degrees or in radians, their sine, cosine, and tangent values are the same. Thus,

$$\sin\frac{\pi}{6} = \frac{1}{2}, \qquad \cos\frac{\pi}{6} = \frac{\sqrt{3}}{2}, \qquad \tan\frac{\pi}{6} = \frac{1}{\sqrt{3}},$$

$$\sin\frac{\pi}{4} = \frac{\sqrt{2}}{2}, \qquad \cos\frac{\pi}{4} = \frac{\sqrt{2}}{2}, \qquad \tan\frac{\pi}{4} = 1,$$

$$\sin\frac{\pi}{3} = \frac{\sqrt{3}}{2}, \qquad \cos\frac{\pi}{3} = \frac{1}{2}, \qquad \tan\frac{\pi}{3} = \sqrt{3}.$$

Example 6 Evaluate exactly: $\cos^{-1}\left(-\dfrac{\sqrt{2}}{2}\right)$.

Solution $\cos^{-1}(-\sqrt{2}/2)$ is the angle between 0 and π whose cosine is $-\sqrt{2}/2$. Because the cosine is negative, the angle must be in the second quadrant, with reference angle $\pi/4$. This angle is $3\pi/4$. Thus,

$$\cos^{-1}\left(-\frac{\sqrt{2}}{2}\right) = \frac{3\pi}{4}.$$

Problems for Section 7.1

1. Consider the angle θ in Figure 7.8. Evaluate the following expressions:

    (a) $\tan\theta$ (b) $\sin\theta$ (c) $\cos\theta$

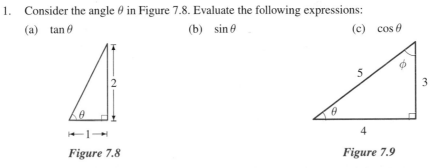

             Figure 7.8   **Figure 7.9**

2. Find the values of the sine, cosine and tangent for the angles θ and ϕ of the right triangle in Figure 7.9.

3. Given the triangle shown in Figure 7.10, find the following:

   (a) $\sin\theta$ (b) $\sin\phi$ (c) $\cos\theta$

   (d) $\cos\phi$ (e) $\tan\theta$ (f) $\tan\phi$

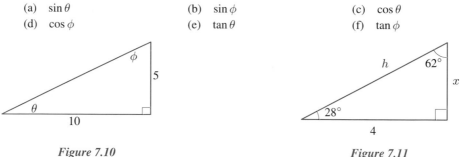

Figure 7.10   *Figure 7.11*

4. Find the lengths of the sides of the triangle in Figure 7.11.

5. Find the radian value of the angle labeled x in Figure 7.12.

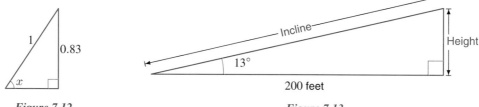

Figure 7.12   *Figure 7.13*

6. You have been asked to build a ramp for Dan's Daredevil Motorcycle Jump. The dimensions you are given are indicated in Figure 7.13. Find all the other dimensions.

7. A kite flyer wondered how high her kite was flying. She used a protractor to measure an angle of $38°$ from level ground to the kite string. If she used a full 100 yard spool of string, how high, in feet, was the kite? (Disregard the string sag and the height of the string reel above the ground.)

8. The top of a 200-foot vertical tower is to be anchored by cables that make an angle of $30°$ with the ground. How long must the cables be? How far from the base of the tower should anchors be placed?

9. A tree 50 feet tall casts a shadow 60 feet long. Find the angle of elevation θ of the sun.

10. A staircase is to rise 17.3 feet over a horizontal distance of 10 feet. At approximately what angle with respect to the floor should it be built?

11. Find approximately the acute angle formed by the line $y = -2x + 5$ and the x-axis.

12. The front door to the student union is 20 feet above the ground, and it is reached by a flight of steps. The school wants to build a wheel-chair ramp, with an incline of 15 degrees, from the ground to the door. How much horizontal distance is needed for the ramp?

13. A ladder 3 meters long leans against a house, making an angle α with the ground. How far is the base of the ladder from the base of the wall, in terms of α? Include a sketch.

14. A plane is flying at an elevation of 35,000 feet when the Gateway Arch in St. Louis, Missouri comes into view. The pilot wants to estimate her horizontal distance from the arch, so she notes the angle of depression, θ, between the horizontal and a line joining her eye to a point on the ground directly below the arch. Make a sketch. Express her horizontal distance to that point as a function of θ.

15. Hampton is a small town on a straight stretch of coast line running north and south. A lighthouse is located 3 miles off-shore directly east of Hampton. The light house has a revolving search light that makes two revolutions per minute. The angle that the beam makes with the east-west line through Hampton is called ϕ. Find the distance from Hampton to the point where the beam strikes the shore, as a function of ϕ. Include a sketch.

16. You see a friend, whose height you know is 5 feet 10 inches, some distance away. Using a surveying device called a transit, you determine the angle between the top of your friend's head and ground level

to be $8°$. You want to find the distance d between you and your friend.

(a) First assume you cannot find your calculator to evaluate trigonometric functions. Find d by approximating the arc length of an $8°$ angle with your friend's height. (See Figure 7.14.)

(b) Now assume that you have a calculator. Use it to find the distance d. (See Figure 7.15.)

(c) Would the difference between these two values for d increase or decrease if the angle were smaller?

Figure 7.14: Arc of 70 in with $8°$ angle   **Figure 7.15:** Triangle with side 70 in and angle $8°$

17. You are parasailing on a rope that is 125 feet long behind a boat. See Figure 7.16.

(a) At first, you stabilize at a height that forms a $45°$ angle with the water. What is that height?

(b) After enjoying the scenery, you encounter a strong wind that blows you down to a height that forms a $30°$ angle with the water. At what height are you now?

(c) Find c and d, i.e. the horizontal distances between you and the boat, in parts (a) and (b).

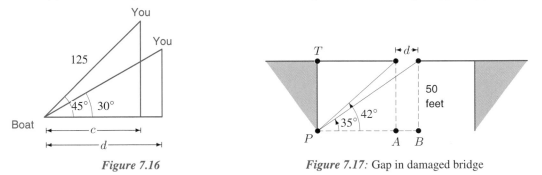

Figure 7.16   **Figure 7.17:** Gap in damaged bridge

18. A bridge over a river was damaged in an earthquake and you are called in to determine the length, d, of the steel beam needed to fill the gap. (See Figure 7.17.) You cannot be on the bridge, but you are able to drop a line from T, the beginning of the bridge, and measure a distance of 50 ft to the point P. From P you find the angles of elevation to the two ends of the gap to be $42°$ and $35°$. How wide is the gap?

7.2 NON-RIGHT TRIANGLES: LAWS OF SINES AND COSINES

Sines and cosines are useful because they relate the angles of a right triangle to its sides. Although more complicated for non-right triangles, such relationships exist for all triangles, not just right triangles.

The Law of Cosines

The Pythagorean theorem relates the three sides of a right triangle. The *Law of Cosines* relates the three sides of any triangle, not just right triangles.

Law of Cosines: For a triangle with sides a, b, c, and angle C opposite side c, we have

$$c^2 = a^2 + b^2 - 2ab \cos C$$

We use Figure 7.18 to derive the Law of Cosines. The dashed line of length h is at right angles to side a and it divides this side into two pieces, one of length x and one of length $a - x$.

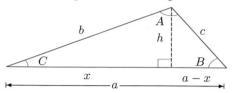

Figure 7.18: Triangle used to derive the Law of Cosines

Applying the Pythagorean theorem to the right-hand right triangle, we get

$$(a - x)^2 + h^2 = c^2$$
$$a^2 - 2ax + x^2 + h^2 = c^2.$$

Applying the Pythagorean theorem to the left-hand triangle, we get $x^2 + h^2 = b^2$. Substituting into the previous equation gives

$$a^2 - 2ax + \underbrace{x^2 + h^2}_{b^2} = c^2$$
$$a^2 + b^2 - 2ax = c^2.$$

But $\cos C = x/b$, so $x = b \cos C$. This gives the Law of Cosines:

$$a^2 + b^2 - 2ab \cos C = c^2.$$

Notice that if C happens to be a right angle, that is, if $C = 90°$, then $\cos C = 0$. In this case, the Law of Cosines reduces to the Pythagorean theorem:

$$a^2 + b^2 - 2ab \cdot 0 = c^2$$
$$a^2 + b^2 = c^2.$$

Therefore, the Law of Cosines is a generalization of the Pythagorean theorem that works for any triangle. Notice also that in Figure 7.18 we assumed that angle C is acute, that is, less than $90°$. Problem 16 concerns the derivation for the case where $90° < C < 180°$.

Example 1   A person leaves her home and walks 5 miles due east and then 3 miles northeast. How far has she walked? How far away from home is she?

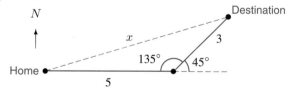

Figure 7.19: A person walks 5 miles east and then 3 miles northeast

Solution   One side of the triangle in Figure 7.19 is 5 miles long, while the second side is 3 miles long and forms an angle of $135°$ with the first. This is because when the person turns northeast, she turns through an angle of $45°$. Thus, we know two sides of this triangle, 5 and 3, and the angle between them, which is $135°$. The length of the third side, x, can be found by applying the Law of Cosines

$$x^2 = 5^2 + 3^2 - 2 \cdot 5 \cdot 3 \cos 135°$$
$$= 34 - 30 \left(-\frac{\sqrt{2}}{2} \right)$$
$$= 55.2.$$

This gives $x = \sqrt{55.2} = 7.43$ miles. Notice that this is less than her total distance walked, which is $5 + 3 = 8$ miles.

In the previous example, we used the Law of Cosines when two sides of a triangle and the angle between them were known. The Law of Cosines is also useful if all three sides of a triangle are known.

Example 2   At what angle must the person from Example 1 walk to go directly home?

Solution   According to Figure 7.20, if the person faces due west and then turns south through an angle of θ, she heads directly home. This angle θ is the same as the angle opposite the side of length 3 in the triangle. The Law of Cosines tells us that

$$5^2 + 7.43^2 - 2(5)(7.43)\cos\theta = 3^2$$
$$-74.3\cos\theta = -71.2$$
$$\cos\theta = 0.9583$$
$$\theta = \arccos(0.9583) = 16.6°.$$

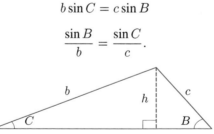

Figure 7.20: The person must face at an angle θ to the south of due west in order to head directly home

In Examples 1 and 2, notice that we used the Law of Cosines in two different ways for the same triangle.

The Law of Sines

Figure 7.21 shows the same triangle as in Figure 7.18. Since $\sin C = h/b$ and $\sin B = h/c$, we have $h = b\sin C$ and $h = c\sin B$. This means that

$$b\sin C = c\sin B$$

so

$$\frac{\sin B}{b} = \frac{\sin C}{c}.$$

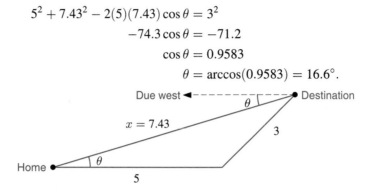

Figure 7.21: Triangle used to derive the Law of Sines

A similar type of argument (see Problem 17 on page 298 and Problem 10 on page 312) shows that

$$a\sin B = b\sin A,$$

which leads to the Law of Sines:

Law of Sines: For a triangle with sides a, b, c opposite angles A, B, C respectively:

$$\frac{\sin A}{a} = \frac{\sin B}{b} = \frac{\sin C}{c}.$$

The Law of Sines is useful when we know a side and the angle opposite it.

Example 3   An aerial tram starts at a point one half mile from the base of a mountain whose face has a 60° angle of elevation. (See Figure 7.22.) The tram ascends at an angle of 20°. What is the length of the cable from T to A?

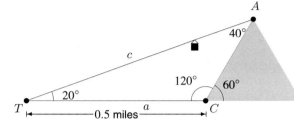

Figure 7.22

Solution   The Law of Cosines will not help us here because we only know the length of one side of the triangle. We do however know two angles in this diagram. Thus, we can use the Law of Sines. We have:

$$\frac{\sin A}{a} = \frac{\sin C}{c}$$

$$\frac{\sin 40°}{0.5} = \frac{\sin 120°}{c}$$

so $c = 0.5\left(\dfrac{\sin 120°}{\sin 40°}\right) \approx 0.6736$. Therefore, the cable from T to A is 0.6736 miles.

The Ambiguous Case

There is a drawback to using the Law of Sines for finding angles. The problem is that the Law of Sines does not tell us the angle, but only its sine, and there are two angles between 0° and 180° with a given sine. For example, if the sine of an angle is $1/2$, the angle may be either 30° or 150°.

Example 4   Solve the following triangles for θ and ϕ.

(a)

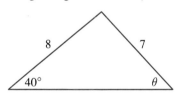

Figure 7.23

(b)

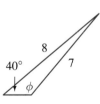

Figure 7.24

Solution   (a)   Using the Law of Sines in Figure 7.23, we have

$$\frac{\sin \theta}{8} = \frac{\sin 40°}{7}$$

$$\sin \theta = \frac{8}{7}\sin 40° = 0.7346$$

$$\theta = \sin^{-1}(0.7346) \approx 47.3°.$$

(b)   From Figure 7.24, we get

$$\frac{\sin \phi}{8} = \frac{\sin 40°}{7}$$

$$\sin \phi = \frac{8}{7}\sin 40° = 0.7346$$

This is the same equation we had for θ in part (a). However, judging from the figures, ϕ is not equal to θ. Knowing the sine of an angle is not enough to tell us the angle. In fact, there are *two* angles between $0°$ and $180°$ whose sine is 0.7346. One of them is $\theta = \sin^{-1}(0.7346) \approx 47.3°$, and the other is $\phi = 180° - \theta \approx 132.7°$.

Problems for Section 7.2

In Problems 1–2, solve for x.

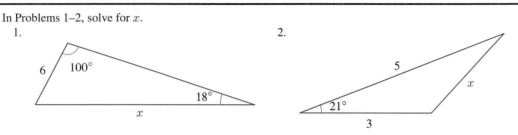

1.

2.

Find all sides and angles of the triangles in Problems 3–8. (Sides and angles are not necessarily to scale.)

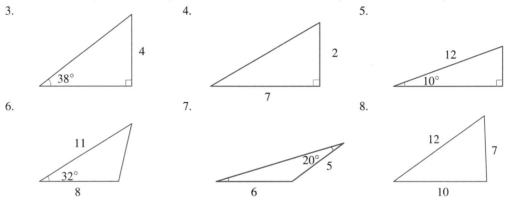

3.

4.

5.

6.

7.

8.

Find all the sides and angles in each of the triangles in Problem 9–12. Sketch each triangle. If there is more than one possible triangle, solve and sketch both. Note that α is the angle opposite side a, and β is the angle opposite side b, and γ is the angle opposite side c.

9. $b = 510.0$ ft, $c = 259.0$ ft, $\gamma = 30.0°$

10. $a = 16.0$ m, $b = 24.0$ m, $c = 20.0$ m

11. $a = 18.7$ cm, $c = 21.0$ cm, $\beta = 22°$

12. $a = 2.00$ m, $\alpha = 25.80°$, $\beta = 10.50°$

13.   A triangle has angles: $27°$, $32°$, and $121°$. The length of the side opposite the $121°$ angle is 8.

  (a)   Find the lengths of the two remaining sides.
  (b)   Calculate the area of the triangle.

14.   (a)   Find an expression for $\sin \theta$ (from Figure 7.25) and $\sin \phi$ (in Figure 7.26.)
  (b)   Explain how you can find θ and ϕ by using the inverse sine function. In what way does the method used for θ differ from the method used for ϕ?

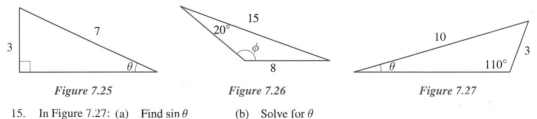

  Figure 7.25   *Figure 7.26*   *Figure 7.27*

15.   In Figure 7.27: (a)   Find $\sin \theta$ (b)   Solve for θ

  (c)   Find the area of the triangle if the lengths are given in centimenters

16.   Give a derivation of the Law of Cosines for the case when the angle C is obtuse, as in Figure 7.28.

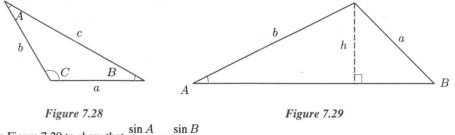

Figure 7.28   *Figure 7.29*

17.   Use Figure 7.29 to show that $\dfrac{\sin A}{a} = \dfrac{\sin B}{b}$.

18.   Consider a central angle of $2°$ in a circle of radius 5 feet. To six decimal places, find the lengths of the arc and the chord determined by this angle.

19.   Repeat Problem 18 for an angle of $30°$.

20.   A parcel of land is in the shape of an isosceles triangle. The base has a length of 425 feet; the other sides, which are of equal length, meet at an angle of $39°$. How long are they?

21.   Two fire stations are located 567 feet apart, at points A and B. There is a forest fire at point C. If $\angle CAB = 54°$ and $\angle CBA = 58°$, which fire station is closer? How much closer?

22.   Two airplanes leave Kennedy airport in New York at 11 am. The air traffic controller reports that they are traveling away from each other at an angle of $103°$. The DC-10 travels 504 mph and the L-1011 travels at 517 mph.[2] How far apart are they at 11:30 am?

23.   A tall monument cannot be measured by simply taking a long measuring tape and dropping it down from the top of the monument to the ground; instead, trigonometry is used. For example, to measure the height of the Eiffel Tower in Paris, a person stands away from the base and measures the angle of elevation to the top of the tower to be $60°$. Moving 210 feet closer, the angle of elevation to the top of the tower is $70°$. How tall is the Eiffel Tower?[3]

24.   In baseball, the four bases form a square with sides 90 feet long.[4] The distance between the pitcher's rubber and home plate is 60.5 feet. See Figure 7.30.

   (a)   The pitcher is standing on the pitching rubber and runners are coming to both first and second bases. Which is the shorter throw? How much shorter is this throw?

   (b)   A ball is hit from home plate to a point 30 feet past second base. An outfielder comes in to catch the ball. How far is the throw to home plate? To third base?

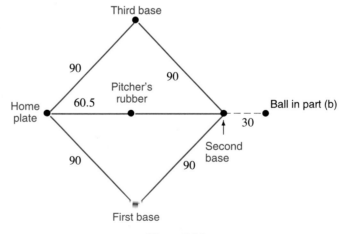

Figure 7.30

[2]The World Almanac Book of Facts, 1999 p. 217
[3]The World Almanac Book of Facts, 1999 p. 629
[4]www.majorleaguebaseball.com

25. A park director wants to build a bridge across a river to a bird sanctuary. He hires a surveyor to determine the length of the bridge, represented by AB in Figure 7.31. The surveyor places a transit (an instrument for measuring vertical and horizontal angles) at point A and measures angle BAC to be $93°$. The surveyor then measures 102 feet to point C, places the transit at point C and measures angle BCA to be $49°$. How long should the bridge be?

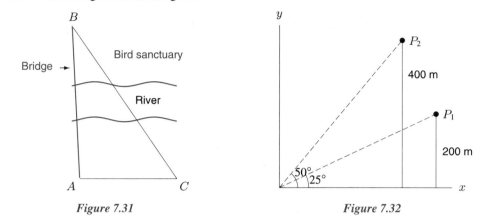

Figure 7.31 *Figure 7.32*

26. A UFO is first sighted at a point P_1 due east from an observer at an angle of $25°$ from the ground and at an altitude (vertical distance above ground) of 200 m. (See Figure 7.32.) The UFO is next sighted at a point P_2 due east at an angle of $50°$ and an altitude of 400 m. What is the distance from P_1 to P_2?

27. To estimate the width of an archeological mound, archeologists place two stakes on opposite ends of the widest point. See Figure 7.33. They set a third stake 82 feet from one stake and 97 feet from the other stake. The angle formed is $125°$. Find the width of the mound.

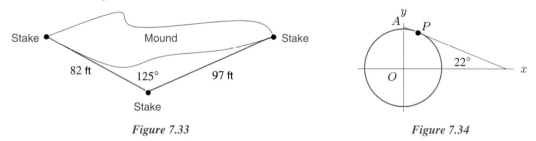

Figure 7.33 *Figure 7.34*

28. The line in Figure 7.34 has a y-intercept of 4, makes a $22°$ angle with the x-axis, and is tangent to the given circle at point P. Find the coordinates of P. [Hint: The tangent line is perpendicular to the radius of the circle at point P.]

29. Every triangle has three sides and three angles. Make a chart showing the set of all possible triangle configurations where three of these six measures are known, and the other three measures can be deduced (or partially deduced) from the three known measures.

7.3 TRIGONOMETRIC IDENTITIES

Equations Versus Identities

An equation that is true for all values of x is called an *identity*. When we solve an equation, we find a value of the variable that makes the equation true. For example, we can solve the equation

$$2(x - 1) = x,$$

to get

$$x = 2.$$

The equation $2(x-1) = x$ is true for $x = 2$, but is not true for any other value of x. Now consider the equation

$$2(x-1) = 2x - 2.$$

This equation is an identity because it is true for all values of x: the left and right sides always have the same value. If two expressions have equal values for all values of the variables for which they are defined, we say the expressions are *identically equal*.

We can get an idea whether an equation is an identity by graphing the function defined by each side of the equation. If we graph $f(x) = 2(x-1)$ and $g(x) = 2x - 2$, the two graphs look identical. This suggests that the expressions produce the same output for any input x. However, the graphs may look identical without being identical. Thus, to be sure that the equation is an identity, we use algebra. On the other hand, if we graph $f(x) = 2(x-1)$ and $g(x) = x$, the graphs intersect only at one point, namely where $x = 2$. This means that the expressions $2(x-1)$ and x are equal only for $x = 2$, and that $2(x-1) = x$ is not an identity.

The Tangent and Pythagorean Identities

There are many identities involving trigonometric functions. In Section 6.6 we saw that

$$\tan\theta = \frac{\sin\theta}{\cos\theta}.$$

This identity holds whenever $\cos\theta$ is not zero. We also derived the *Pythagorean identity*,

$$\sin^2\theta + \cos^2\theta = 1.$$

(Try graphing the function $f(\theta) = \sin^2\theta + \cos^2\theta$ to visualize this identity.) The Pythagorean identity is often used in one of two alternate forms,

$$\sin^2\theta = 1 - \cos^2\theta \qquad \text{or} \qquad \cos^2\theta = 1 - \sin^2\theta.$$

Why We Need Identities

Identities are useful because they allow us to replace one expression by an equivalent one that may be more convenient. The next two examples show equations that are easier to solve after being transformed by identities.

Example 1   Solve $2\sin\theta = \sqrt{2}\cos\theta$, for $0 \le \theta \le 2\pi$.

Solution   It is usually easier to solve a trigonometric equation if we can write it in terms of a single trigonometric function. If we divide both sides by $\cos\theta$, we find

$$2\frac{\sin\theta}{\cos\theta} = \sqrt{2}.$$

Replacing $\dfrac{\sin\theta}{\cos\theta}$ by $\tan\theta$, we get

$$2\tan\theta = \sqrt{2}$$
$$\tan\theta = \frac{\sqrt{2}}{2}.$$

One solution is $\theta = \tan^{-1}(\sqrt{2}/2) = 0.6155$. The tangent function has period π, so a second solution is $\theta = \pi + 0.6155 = 3.7571$.

We cannot divide both sides of an equation by 0, so dividing by $\cos \theta$ is not valid if $\cos \theta = 0$. We must check separately whether values of θ that make $\cos \theta = 0$, namely $\theta = \pi/2$ and $\theta = 3\pi/2$, are solutions. You can check that neither of these values satisfies the original equation.

Example 2   Solve $3 \sin^2 t = 5 - 5 \cos t$, for $0 \le t \le \pi$.

Solution   We can get approximate solutions to the equation by graphing $y = 3(\sin t)^2$ and $y = 5 - 5 \cos t$ and finding the points of intersection for $0 \le t \le \pi$. To solve this equation algebraically, we want to write the equation entirely in terms of $\sin t$ or entirely in terms of $\cos t$. Since $\sin^2 t = 1 - \cos^2 t$, we convert to $\cos t$, giving

$$3(1 - \cos^2 t) = 5 - 5 \cos t$$
$$3 - 3 \cos^2 t = 5 - 5 \cos t$$
$$3 \cos^2 t - 5 \cos t + 2 = 0.$$

The left-hand side factors:

$$(3 \cos t - 2)(\cos t - 1) = 0,$$

so

$$3 \cos t - 2 = 0 \quad \text{or} \quad \cos t - 1 = 0.$$

This gives $\cos t = 1$, with solution $t = 0$, and $\cos t = 2/3$, with solution $t = \cos^{-1}(2/3) = 0.84$. Note that both solutions fall in the interval $0 \le t \le \pi$.

Double-Angle Formula for Sine

We now find a formula for $\sin 2\theta$ in terms of $\sin \theta$ and $\cos \theta$. First, note that $\sin 2\theta$ is not the same as $2 \sin \theta$! For one thing, the graph of $y = \sin 2\theta$ has a different period and amplitude from the graph of $y = 2 \sin \theta$. A graph suggests that $\sin 2\theta$ and $2 \sin \theta \cos \theta$ could be the same function. In fact, they are, as we now show.

We derive our formula by using Figure 7.35. The right triangles OAB and OCB each have an angle θ at O. The lengths of OA and OC are 1; the length of AC is $2 \sin \theta$. Writing α for the angle at A and applying the Law of Sines to triangle OAC gives

$$\frac{\sin 2\theta}{2 \sin \theta} = \frac{\sin \alpha}{1}.$$

In triangle OAB, the length of side OB is $\cos \theta$, and the hypotenuse is 1, so

$$\sin \alpha = \frac{\text{Opposite}}{\text{Hypotenuse}} = \frac{\cos \theta}{1} = \cos \theta.$$

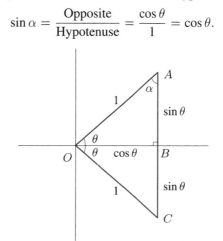

Figure 7.35: Triangle used to derive the double angle formula for sine

Thus, substituting $\cos\theta$ for $\sin\alpha$, we have

$$\frac{\sin 2\theta}{2\sin\theta} = \cos\theta$$

so

$$\boxed{\sin 2\theta = 2\sin\theta\cos\theta.}$$

This identity is known as the *double-angle formula for sine*.

Example 3   Find all solutions to $\sin 2t = \sin t$, for $0 \le t \le 2\pi$.

Solution   Using the double-angle formula $\sin 2t = 2\sin t \cos t$, we have

$$2\sin t \cos t = \sin t$$
$$2\sin t \cos t - \sin t = 0$$
$$\sin t(2\cos t - 1) = 0 \qquad \text{Factoring out } \sin t.$$

Thus,

$$\sin t = 0 \quad \text{or} \quad 2\cos t - 1 = 0.$$

Now, $\sin t = 0$ for $t = 0$, π, and 2π. We solve the second equation by writing

$$2\cos t - 1 = 0$$
$$\cos t = \frac{1}{2}.$$

We know that $\cos t = 1/2$ for $t = \pi/3$ and $t = 5\pi/3$. Thus there are five solutions to the original equation on the interval $0 \le t \le 2\pi$: $t = 0, \pi/3, \pi, 5\pi/3$, and 2π. Figure 7.36 illustrates these solutions graphically as the points where the graphs of $\sin 2t$ and $\sin t$ intersect.

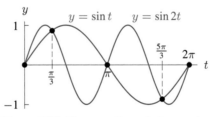

Figure 7.36: There are five solutions to the equation $\sin 2t = \sin t$, for $0 \le t \le 2\pi$

Double-Angle Formulas for Cosine and Tangent

There is also a double-angle formula for cosine, which we derive from the formula for sine. By the Pythagorean identity, we know that

$$(\cos 2t)^2 + (\sin 2t)^2 = 1,$$

so

$$(\cos 2t)^2 = 1 - (\sin 2t)^2.$$

Now use the double-angle formula for sine to replace $\sin 2t$ by $2\sin t\cos t$, giving

$$(\cos 2t)^2 = 1 - (2\sin t\cos t)^2$$
$$= 1 - 4\sin^2 t\cos^2 t.$$

Next use the Pythagorean identity again to replace $\cos^2 t$ by $1 - \sin^2 t$, to get

$$(\cos 2t)^2 = 1 - 4\sin^2 t(1 - \sin^2 t)$$
$$= 1 - 4\sin^2 t + 4\sin^4 t$$
$$= (1 - 2\sin^2 t)^2.$$

Finally, take the square root of both sides and check the signs, as in Problem 1, to get

$$\cos 2t = 1 - 2\sin^2 t.$$

This is the double-angle formula for cosine. In Problem 7, we use the Pythagorean identity to write the formula for $\cos 2t$ in two other ways:

$$\cos 2t = 2\cos^2 t - 1 \qquad \text{and} \qquad \cos 2t = \cos^2 t - \sin^2 t.$$

Finally, we can find a double-angle formula for tangent from the formulas for sine and cosine. Problem 8 asks you to derive the result

$$\tan 2t = \frac{2\tan t}{1 - \tan^2 t}.$$

Other Identities

Because the cosine is an even function and the sine and tangent are odd functions, we have

$$\cos(-t) = \cos t.$$
$$\sin(-t) = -\sin t \qquad \text{and} \qquad \tan(-t) = -\tan t.$$

Several other useful identities are suggested by graphing. In Figure 7.37 we see that shifting the graph of $y = \cos t$ to the right by $\pi/2$ units gives the graph of $y = \sin t$. This suggests, as is shown in Section 7.4, that

$$\sin t = \cos\left(t - \frac{\pi}{2}\right) \qquad \text{for all } t.$$

Figure 7.37: The graph of $y = \sin t$ can be obtained by shifting the graph of $y = \cos t$ to the right by $\pi/2$

Similarly, shifting the graph of $y = \sin t$ to the left by $\pi/2$ gives the graph of $y = \cos t$, so

$$\cos t = \sin\left(t + \frac{\pi}{2}\right) \qquad \text{for all } t.$$

Here is a summary of the identities in this section.

- **Tangent identity**:

$$\tan t = \frac{\sin t}{\cos t}$$

- **Pythagorean identity**:

$$\sin^2 t + \cos^2 t = 1$$

- **Double-angle formula for sine**:

$$\sin 2t = 2 \sin t \cos t$$

- **Double-angle formula for cosine** (expressed in three different ways):

$$\cos 2t = 1 - 2 \sin^2 t$$
$$\cos 2t = 2 \cos^2 t - 1$$
$$\cos 2t = \cos^2 t - \sin^2 t$$

- **Negative angle identities**:

$$\sin(-t) = -\sin t \quad \cos(-t) = \cos t \quad \tan(-t) = -\tan t$$

- **Identities relating sine and cosine**:

$$\sin t = \cos\left(t - \frac{\pi}{2}\right) \qquad \cos t = \sin\left(t + \frac{\pi}{2}\right)$$

Problems for Section 7.3

1. Use graphs to check that $\cos 2t$ and $1 - 2 \sin^2 t$ have the same sign for all values of t.

2. Complete the following table, using exact values where possible. State the trigonometric identities that relate the quantities in the table, and check these identities numerically.

θ (radians)	$\sin^2 \theta$	$\cos^2 \theta$	$\sin 2\theta$	$\cos 2\theta$
1				
$\pi/2$				
2				
$5\pi/6$				

3. Simplify the expression: $(\cos(2\theta))^2 + (\sin(2\theta))^2$.

4. Using the Pythagorean identity, give an expression for $\sin \theta$ in terms of $\cos \theta$.

5. Describe the similarities and differences between the functions $y = \cos(2x)$ and $y = \cos(x^2)$.

6. Write a memo to a classmate describing what your graphing calculator draws for each of the following:

   (a) $y = \sin\left(x^2\right)$ (b) $y = (\sin x)^2$ (c) $y = \sin^2 x$

7. Use the Pythagorean identity to write the double angle formula for cosine in these two alternate forms:

   (a) $\cos 2t = 2(\cos t)^2 - 1$ (b) $\cos 2t = (\cos t)^2 - (\sin t)^2$

8. Use the fact that $\tan 2t = \dfrac{\sin 2t}{\cos 2t}$ to derive a double angle formula for tangent. [Hint: Use the form for $\cos 2t$ from Problem 7 (b).]

9. Use graphs to find five pairs of expressions which appear to be identically equal.

(a) $2\cos^2 t + \sin t + 1$ (b) $\cos^2 t$ (c) $\cos(2t)$

(d) $1 - 2\sin^2 t$ (e) $2\sin t \cos t$ (f) $\cos^2 t - \sin^2 t$

(g) $\sin(2t)$ (h) $\sin(3t)$ (i) $-2\sin^2 t + \sin t + 3$

(j) $\sin(2t)\cos t + \cos(2t)\sin t$ (k) $\dfrac{1 - \sin t}{\sin t}$ (l) $\dfrac{1 + \cos(2t)}{2}$

(m) $\dfrac{\left(\dfrac{1}{\cos t} - 1\right)}{1 - \cos t}$

For Problems 10–19, use graphs to decide which of the following equations are identities. If the equation is an identity, prove it algebraically. If it is not an identity, find a value of x for which the equation is false.

10. $\dfrac{1}{2 + x} = \dfrac{1}{2} + \dfrac{1}{x}$ 11. $\sin\left(\dfrac{1}{x}\right) = \dfrac{1}{\sin x}$ 12. $\sin(2x) = \dfrac{2\tan x}{1 + (\tan x)^2}$

13. $\sqrt{64 - x^2} = 8 - x$ 14. $\sin(2x) = 2\sin x$ 15. $\cos(x^2) = (\cos x)^2$

16. $\sin x \tan x = \dfrac{1 - (\cos x)^2}{\cos x}$ 17. $\tan x = \dfrac{\sin(2x)}{1 + \cos(2x)}$ 18. $\cos(2x) = \dfrac{1 - (\tan x)^2}{1 + (\tan x)^2}$

19. $\cos\left(x + \dfrac{\pi}{3}\right) = \cos x + \cos\dfrac{\pi}{3}$

For Problems 20–22, show algebraically that the identity is true.

20. $\dfrac{\sin t}{1 - \cos t} = \dfrac{1 + \cos t}{\sin t}$ 21. $\dfrac{\cos x}{1 - \sin x} - \tan x = \dfrac{1}{\cos x}$

22. $\dfrac{\sin x \cos y + \cos x \sin y}{\cos x \cos y - \sin x \sin y} = \dfrac{\tan x + \tan y}{1 - \tan x \tan y}$

23. Find all the exact solutions to the following equations.

(a) $\cos 2\theta + \cos \theta = 0$, $\quad 0 \le \theta < 360°$ (b) $2\cos^2 \theta = 3\sin \theta + 3$, $\quad 0 \le \theta \le 2\pi$

24. Use Figure 7.38 to evaluate the following expressions in terms of θ.

(a) y (b) $\cos \varphi$ (c) $1 + y^2$ (d) The triangle's area

25. Let x be the side adjacent to the angle θ, $0 < \theta < \pi/4$, in a right triangle whose hypotenuse is 1. (See Figure 7.39.) Express the following functions in terms of x.

(a) $\cos \theta$ (b) $\cos(\frac{\pi}{2} - \theta)$ (c) $\tan^2 \theta$

(d) $\sin(2\theta)$ (e) $\cos(4\theta)$ (f) $\sin(\cos^{-1} x)$

Figure 7.38   Figure 7.39   Figure 7.40

26. Let y be the side opposite to the angle θ in a right triangle whose hypotenuse is 1. (See Figure 7.40.) Find a formula in terms of y for each of the following expressions.

(a) $\cos \theta$ (b) $\tan \theta$ (c) $\cos(2\theta)$

(d) $\sin(\pi - \theta)$ (e) $\sin^2(\cos^{-1}(y))$

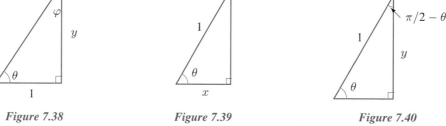

27.   Suppose that $\sin \theta = 3/5$ and θ is in the second quadrant. Find $\sin(2\theta)$, $\cos(2\theta)$, and $\tan(2\theta)$ exactly.

28.   If $x = 3 \cos \theta$, $0 < \theta < \pi/2$, express $\sin(2\theta)$ in terms of x.

29.   If $x + 1 = 5 \sin \theta$, $0 < \theta < \pi/2$, express $\cos(2\theta)$ in terms of x.

30.   Express in terms of x without trigonometric functions. [Hint: Let $\theta = \cos^{-1} x$.]

   (a) $\tan(2 \cos^{-1} x)$ (b) $\sin(2 \tan^{-1} x)$

31.   Find an identity for $\cos(4\theta)$ in terms of $\cos \theta$. (You need not simplify your answer.)

32.   Use trigonometric identities to find an identity for $\sin(4\theta)$ in terms of $\sin \theta$ and $\cos \theta$.

33.   The identity $\sin^2 x + \cos^2 x = 1$ can be used to eliminate radicals from algebraic expressions. For example, suppose $f(x) = \sqrt{4 - x^2}$. We can transform the formula for $f(x)$ by making the substitution $x = 2 \sin u$ and $u = \sin^{-1}(x/2)$. Observe that

$$4 - x^2 = 4 - (2 \sin u)^2 = 4 - 4 \sin^2 u = 4(1 - \sin^2 u) = 4 \cos^2 u,$$

so

$$f(x) = \sqrt{4 - x^2} = \sqrt{4 \cos^2 u} = 2|\cos u| = 2 \left| \cos \left(\sin^{-1} \left(\frac{x}{2} \right) \right) \right|.$$

Notice that the choice of 2 in $x = 2 \sin u$ comes from $2 = \sqrt{4}$. Rewrite the following in a form which does not use a radical.

   (a) $f(x) = \sqrt{9 - x^2}$ (b) $g(x) = \sqrt{5 - x^2}$

7.4   SUM AND DIFFERENCE FORMULAS FOR SINE AND COSINE

The double-angle formulas introduced in Section 7.3 are special cases of a group of more general trigonometric identities, the sum and difference formulas. If we know the sine and cosine values for two angles, say, θ and ϕ, can we use those values to find the sine and cosine of the angle $\theta + \phi$? For example, can we use the exact values for the sine and cosine of $30°$ and $45°$ to find an exact value for the sine of $75°$? You can use your calculator to check that the sine of $75°$ is *not* equal to $\sin 30° + \sin 45°$, and, in general, $\sin(\theta + \phi) \neq \sin \theta + \sin \phi$. At the end of this section, we show that the following formula holds for any angles θ and ϕ:

$$\boxed{\sin(\theta + \phi) = \sin \theta \cos \phi + \sin \phi \sin \theta.}$$

Before proving this result on page 309–311, we see how it is used.

Example 1   Find an exact value for $\sin 75°$.

Solution   Use the sum formula with $\theta = 30°$ and $\phi = 45°$. Then

$$\sin 75° = \sin(30° + 45°)$$
$$= \sin 30° \cos 45° + \sin 45° \cos 30°$$
$$= \frac{1}{2} \cdot \frac{\sqrt{2}}{2} + \frac{\sqrt{2}}{2} \cdot \frac{\sqrt{3}}{2} = \frac{\sqrt{2} + \sqrt{6}}{4}.$$

Rewriting $\sin t + \cos t$

The sum formula can be used to simplify sums of sines and cosines. Consider the graph of $f(t) = \sin t + \cos t$ in Figure 7.41. This graph looks remarkably like the graph of the sine function itself, having the same period but a larger amplitude (somewhere between 1 and 2). It appears to have a phase shift of about $\pi/4$.

We would like to determine whether f is actually sinusoidal, that is, whether f can be written in the form $f(t) = A \sin(Bt + \phi) + k$. If so, Figure 7.41 suggests that $B = 1$, that the phase shift $\phi = \pi/4$, and that $k = 0$. What about the value of A?

We use the identity $\sin(\theta + \phi) = \sin \theta \cos \phi + \sin \phi \cos \theta$ to study $\sin(t + \pi/4)$. Substituting $\theta = t$ and $\phi = \pi/4$, the identity becomes

$$\sin\left(t + \frac{\pi}{4}\right) = \sin t \cos \frac{\pi}{4} + \sin \frac{\pi}{4} \cos t$$

$$= \frac{1}{\sqrt{2}} \sin t + \frac{1}{\sqrt{2}} \cos t.$$

Multiplying both sides of this equation by $\sqrt{2}$ gives

$$\sqrt{2} \sin\left(t + \frac{\pi}{4}\right) = \sin t + \cos t.$$

Thus,

$$f(t) = \sqrt{2} \sin\left(t + \frac{\pi}{4}\right).$$

We see that f is indeed a sine function with period 2π, amplitude $A = \sqrt{2} \approx 1.4$, and phase shift $\phi = \pi/4$. These values can be seen in the graph in Figure 7.41.

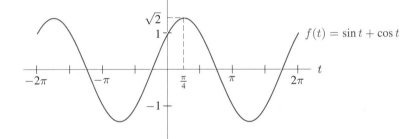

Figure 7.41: A graph of $f(t)$ reflecting sinusoidal behavior, with amplitude $\sqrt{2}$ and phase shift $\pi/4$

Rewriting $a_1 \sin Bt + a_2 \cos Bt$

The function $f(t) = \sin t + \cos t$ has just been rewritten as a single sine function. We now show that this result can be extended: The sum of any two cosine and sine functions having the same periods can be written as a single sine function.[5] We start with

$$a_1 \sin(Bt) + a_2 \cos(Bt),$$

where a_1, a_2, and B are constants. We want to write this in the form $A \sin(Bt + \phi)$. To do this, we show that we can find the constants A and ϕ for any a_1 and a_2. We use the identity $\sin(\theta + \phi) = \sin \theta \cos \phi + \sin \phi \cos \theta$ with $\theta = Bt$. Multiplying by A, we have

$$A \sin(Bt + \phi) = A \sin(Bt) \cos \phi + A \sin \phi \cos(Bt)$$

$$= A \cos \phi \sin(Bt) + A \sin \phi \cos(Bt).$$

This looks very much like $a_1 \sin(Bt) + a_2 \cos(Bt)$ provided $a_1 = A \cos \phi$ and $a_2 = A \sin \phi$. If so,

---

[5]The sum of a sine and a cosine with the same periods can also be written as a single cosine function.

can we solve for A and ϕ? Observe that

$$a_1^2 + a_2^2 = A^2 \cos^2 \phi + A^2 \sin^2 \phi = A^2,$$

so we have

$$A = \sqrt{a_1^2 + a_2^2}.$$

From $a_1 = A \cos \phi$ and $a_2 = A \sin \phi$, we get $\cos \phi = a_1/A$ and $\sin \phi = a_2/A$. If $a_1 \neq 0$, then

$$\frac{a_2}{a_1} = \frac{A \sin \phi}{A \cos \phi} = \frac{\sin \phi}{\cos \phi} = \tan \phi.$$

These formulas allow A and ϕ to be determined from a_1 and a_2. Applying them to the example $\sin t + \cos t$, where $a_1 = 1$ and $a_2 = 1$, gives, as before, $A = \sqrt{2}$ and $\cos \phi = \sin \phi = 1/\sqrt{2}$, so $\tan \phi = 1$, and $\phi = \pi/4$. In summary:

Provided their periods are equal, the sum of a sine function and a cosine function can be written as a single sine function. We have

$$a_1 \sin(Bt) + a_2 \cos(Bt) = A \sin(Bt + \phi)$$

where

$$A = \sqrt{a_1^2 + a_2^2} \qquad \text{and} \qquad \tan \phi = \frac{a_2}{a_1}.$$

The angle ϕ is determined by the equations $\cos \phi = a_1/A$ and $\sin \phi = a_2/A$.

To find the angle ϕ, find its quadrant from the signs of $\cos \phi = a_1/A$ and $\sin \phi = a_2/A$. Then find the angle ϕ in that quadrant with $\tan \phi = a_2/a_1$.

Example 2   If $g(t) = 2 \sin 3t + 5 \cos 3t$, write $g(t)$ as a single sine function.

Solution   We have $a_1 = 2$, $a_2 = 5$, and $B = 3$. Thus,

$$A = \sqrt{2^2 + 5^2} = \sqrt{29} \qquad \text{and} \qquad \tan \phi = \frac{5}{2}.$$

Since $\cos \phi = 2/\sqrt{29}$ and $\sin \phi = 5/\sqrt{29}$ are both positive, ϕ is in the first quadrant. Since $\arctan(5/2) = 1.19$ is in the first quadrant, we take $\phi = 1.19$. Therefore,

$$g(t) = \sqrt{29} \sin(3t + 1.19).$$

We now list the four identities concerning the sums and differences of angles. They are derived on page 309–311.

Sum-of-angle and **difference-of-angle** formulas for sine and cosine:

$$\sin(\theta + \phi) = \sin \theta \cos \phi + \sin \phi \cos \theta$$

$$\sin(\theta - \phi) = \sin \theta \cos \phi - \sin \phi \cos \theta$$

and

$$\cos(\theta + \phi) = \cos \theta \cos \phi - \sin \theta \sin \phi$$

$$\cos(\theta - \phi) = \cos \theta \cos \phi + \sin \theta \sin \phi.$$

Sums and Differences of Sines and Cosines

The four formulas for the sines and cosines of the sums and differences of angles lead to four other formulas for the sums and differences of the trigonometric functions themselves.

We begin with the fact that

$$\cos(\theta + \phi) = \cos \theta \cos \phi - \sin \theta \sin \phi$$

and $$\cos(\theta - \phi) = \cos \theta \cos \phi + \sin \theta \sin \phi.$$

If we add these two equations, the terms involving the sine function cancel out, leaving only cosines:

$$\cos(\theta + \phi) + \cos(\theta - \phi) = 2 \cos \theta \cos \phi.$$

The left-hand side of this equation can be rewritten by making the substitutions

$$u = \theta + \phi \qquad \text{and} \qquad v = \theta - \phi.$$

This gives

$$\cos \underbrace{(\theta + \phi)}_{u} + \cos \underbrace{(\theta - \phi)}_{v} = 2 \cos \theta \cos \phi.$$

The right-hand side of the equation can also be rewritten in terms of u and v. To do this, solve for θ and ϕ by adding and subtracting the equations $u = \theta + \phi$ and $v = \theta - \phi$, giving $u + v = 2\theta$ and $u - v = 2\phi$, so that

$$\theta = \frac{u + v}{2} \qquad \text{and} \qquad \phi = \frac{u - v}{2}.$$

With these substitutions, we have

$$\cos u + \cos v = 2 \cos \frac{u + v}{2} \cos \frac{u - v}{2}.$$

This identity relates the sum of two cosine functions to the product of two new cosine functions.

There are similar formulas for the sum of two sine functions, and the difference between sine functions and cosine functions. See Problems 11, 12, 13. In summary:

Sum and difference of sine and cosine:

$$\cos u + \cos v = 2 \cos \frac{u + v}{2} \cos \frac{u - v}{2} \qquad \sin u + \sin v = 2 \sin \frac{u + v}{2} \cos \frac{u - v}{2}$$

$$\cos u - \cos v = -2 \sin \frac{u + v}{2} \sin \frac{u - v}{2} \qquad \sin u - \sin v = 2 \cos \frac{u + v}{2} \sin \frac{u - v}{2}$$

Derivation of Identities

Justification of cos $(\theta - \phi) =$ cos θcos $\phi +$ sin θsin ϕ. To show why this identity for $\cos(\theta - \phi)$ is true, we find the distance between points A and B in Figure 7.42 in two ways. Points A and B correspond to the angles θ and ϕ on the unit circle, so their coordinates are $A = (\cos \theta, \sin \theta)$ and $B = (\cos \phi, \sin \phi)$. The angle AOB is $(\theta - \phi)$, so by the Law of Cosines, the distance AB is given by

$$AB^2 = 1^2 + 1^2 - 2 \cdot 1 \cdot 1 \cos(\theta - \phi) = 2 - 2 \cos(\theta - \phi).$$

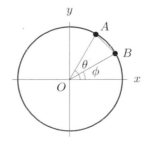

Figure 7.42: Figure to justify the identity for $\cos(\theta - \phi)$, find the distance between A and B two ways

Finding the distance AB by the distance formula and multiplying out, we get

$$
\begin{aligned}
AB^2 &= (\cos\theta - \cos\phi)^2 + (\sin\theta - \sin\phi)^2 \\
&= \cos^2\theta - 2\cos\theta\cos\phi + \cos^2\phi + \sin^2\theta - 2\sin\theta\sin\phi + \sin^2\phi \\
&= \cos^2\theta + \sin^2\theta + \cos^2\phi + \sin^2\phi - 2\cos\theta\cos\phi - 2\sin\theta\sin\phi \\
&= 2 - 2(\cos\theta\cos\phi + \sin\theta\sin\phi).
\end{aligned}
$$

Setting the two distances equal gives

$$
2 - 2\cos(\theta - \phi) = 2 - 2(\cos\theta\cos\phi + \sin\theta\sin\phi).
$$

So

$$
\boxed{\cos(\theta - \phi) = \cos\theta\cos\phi + \sin\theta\sin\phi.}
$$

Obtaining other identities from the identity for cos $(\theta - \phi)$. Letting $\phi = \pi/2$, we obtain

$$
\cos\left(\theta - \frac{\pi}{2}\right) = \cos\theta\cos\frac{\pi}{2} + \sin\theta\sin\frac{\pi}{2} = \cos\theta \cdot 0 + \sin\theta \cdot 1 = \sin\theta.
$$

So

$$
\boxed{\sin\theta = \cos\left(\theta - \frac{\pi}{2}\right).}
$$

Replacing θ by $(\theta - \pi/2)$ in this identity gives

$$
\sin\left(\theta - \frac{\pi}{2}\right) = \cos\left(\theta - \frac{\pi}{2} - \frac{\pi}{2}\right) = \cos(\theta - \pi).
$$

Substituting $\phi = \pi$ in the identity for $\cos(\theta - \phi)$ gives

$$
\cos(\theta - \pi) = \cos\theta\cos\pi + \sin\theta\sin\pi = \cos\theta(-1) + \sin\theta \cdot 0 = -\cos\theta.
$$

So

$$
\boxed{\sin\left(\theta - \frac{\pi}{2}\right) = -\cos\theta.}
$$

Justification of Other Sum and Difference Formulas

Using $\sin\theta = \cos(\theta - \pi/2)$ with θ replaced by $(\theta + \phi)$ gives

$$
\sin(\theta + \phi) = \cos\left(\theta + \phi - \frac{\pi}{2}\right).
$$

Rewriting $(\theta + \phi - \pi/2)$ as $(\theta - \pi/2 - (-\phi))$, we use the identity for $\cos(\theta - \phi)$ with θ replaced by $(\theta - \pi/2)$ and ϕ replaced by $-\phi$:

$$
\begin{aligned}
\cos\left(\theta + \phi - \frac{\pi}{2}\right) &= \cos\left(\theta - \frac{\pi}{2} - (-\phi)\right) \\
&= \cos\left(\theta - \frac{\pi}{2}\right)\cos(-\phi) + \sin\left(\theta - \frac{\pi}{2}\right)\sin(-\phi).
\end{aligned}
$$

Since $\cos(-\phi) = \cos\phi$ and $\sin(-\phi) = -\sin\phi$, and since $\cos(\theta - \pi/2) = \sin\theta$ and $\sin(\theta - \pi/2) = -\cos\theta$, we have

$$\cos\left(\theta + \phi - \frac{\pi}{2}\right) = \sin\theta\cos\phi + (-\cos\theta)(-\sin\theta).$$

So

$$\boxed{\sin(\theta + \phi) = \sin\theta\cos\phi + \cos\theta\sin\phi.}$$

Replacing ϕ by $-\phi$ in the formula for $\sin(\theta + \phi)$ gives

$$\sin(\theta - \phi) = \sin(\theta + (-\phi)) = \sin\theta\cos(-\phi) + \cos\theta\sin(-\phi),$$

and since $\cos(-\phi) = \cos\phi$ and $\sin(-\phi) = -\sin\phi$,

$$\boxed{\sin(\theta - \phi) = \sin\theta\cos\phi - \cos\theta\sin\phi.}$$

Replacing ϕ by $-\phi$ in the formula for $\cos(\theta - \phi)$ gives

$$\cos(\theta + \phi) = \cos(\theta - (-\phi)) = \cos\theta\cos(-\phi) + \sin\theta\sin(-\phi).$$

$$\boxed{\cos(\theta + \phi) = \cos\theta\cos\phi - \sin\theta\sin\phi.}$$

Problems for Section 7.4

1. Test each of the following identities first by evaluating both sides to see that they are numerically equal when $u = 15°$ and $v = 42°$. Then let $u = x$ and $v = 20°$ and compare the graphs of each side of the identity as functions of x. For example, in part (a) graph $y = \sin(x + 20)$ and $y = \sin x\,\cos 20 + \sin 20\,\cos x$. Be sure to use degree mode on your calculator.

   (a) $\sin(u + v) = \sin u \cos v + \sin v \cos u$ (b) $\sin(u - v) = \sin u \cos v - \sin v \cos u$

   (c) $\cos(u + v) = \cos u \cos v - \sin v \sin u$ (d) $\cos(u - v) = \cos u \cos v + \sin v \sin u$

2. Test each of the following identities by first evaluating both sides to see that they are numerically equal when $u = 35°$ and $v = 40°$. Then let $u = x$ and $v = 25°$ and compare the graphs of each side of the identity as functions of x. For example, in part (a) graph $y = \cos x + \cos 25$ and $y = 2\cos((x + 25)/2)\cos((x - 25)/2)$. Be sure to use degree mode on your calculator.

   (a) $\cos u + \cos v = 2\cos\left(\dfrac{u + v}{2}\right)\cos\left(\dfrac{u - v}{2}\right)$

   (b) $\cos u - \cos v = -2\sin\left(\dfrac{u + v}{2}\right)\sin\left(\dfrac{u - v}{2}\right)$

   (c) $\sin u + \sin v = 2\sin\left(\dfrac{u + v}{2}\right)\cos\left(\dfrac{u - v}{2}\right)$

   (d) $\sin u - \sin v = 2\cos\left(\dfrac{u + v}{2}\right)\sin\left(\dfrac{u - v}{2}\right)$

3. Use identities and the exact values of $\sin\theta$ and $\cos\theta$ for $\theta = 30°, 45°, 60°$ to find exact values of $\sin\theta$ and $\cos\theta$ for $\theta = 15°$ and $\theta = 75°$.

Write the functions in Problems 4–7 in the form $A\sin(Bt + \phi)$.

4. $8\sin t + 6\cos t$ 5. $8\sin t - 6\cos t$ 6. $-\sin t + \cos t$ 7. $-2\sin 3t + 5\cos 3t$

8. Starting from the addition formulas for sine and cosine, derive the following identities:

   (a) $\sin t = \cos(t - \pi/2)$ (b) $\cos t = \sin(t + \pi/2)$

Example 1   A utility company serves two different cities. Let P_1 be the power requirement in megawatts (mw) for City 1 and P_2 be the requirement for City 2. Both P_1 and P_2 are functions of t, the number of hours elapsed since midnight. Suppose P_1 and P_2 are given by the following formulas:

$$P_1 = 40 - 15 \cos\left(\frac{\pi}{12}t\right) \quad \text{and} \quad P_2 = 50 + 10 \sin\left(\frac{\pi}{12}t\right).$$

(a)   Describe the power requirements of each city in words.

(b)   What is the maximum total power the utility company must be prepared to provide?

Solution   (a)   The power requirement of City 1 is at a minimum of $40 - 15 = 25$ mw at $t = 0$, or midnight. It rises to a maximum of $40 + 15 = 55$ mw at noon and falls back to 25 mw by the following midnight. The power requirement of City 2 is at a maximum of 60 mw at 6 am. It falls to a minimum of 40 mw by 6 pm but by the following morning has climbed back to 60 mw, again at 6 am. Figure 7.46 shows P_1 and P_2 over a two-day period.

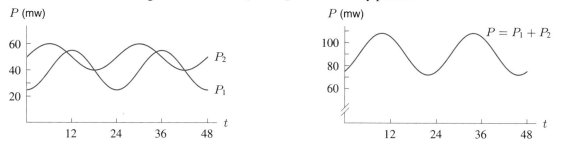

Figure 7.46: Power requirements for cities 1 and 2   *Figure 7.47:* Total power demand for both cities combined

(b)   The utility company must provide enough power to satisfy the needs of both cities. The total power required is given by

$$P = P_1 + P_2 = 90 + 10 \sin\left(\frac{\pi}{12}t\right) - 15 \cos\left(\frac{\pi}{12}t\right).$$

The graph of total power in Figure 7.47 looks like a sinusoidal function. It varies between about 108 mw and 72 mw, giving it an amplitude of roughly 18 mw. Since the maximum value of P is about 108 mw, the utility company must be prepared to provide at least this much power at all times. Since the maximum value of P_1 is 55 and the maximum value of P_2 is 60, you might have expected that the maximum value of P would be $55 + 60 = 115$. The reason that this isn't true is that the maximum values of P_1 and P_2 occur at different times. However, the midline of P is a horizontal line at 90, which does equal the midline of P_1 plus the midline of P_2. The period of P is 24 hours because the values of both P_1 and P_2 begin repeating after 24 hours, so their sum repeats that frequently as well.

The result on page 308 allows us to find an expression for the total power in the form:

$$P = P_1 + P_2 = 90 + 10 \sin\left(\frac{\pi}{12}t\right) - 15 \cos\left(\frac{\pi}{12}t\right) = 90 + A \sin(Bt + \phi),$$

where

$$A = \sqrt{10^2 + (-15)^2} \approx 18.0 \text{ mw}.$$

Since $\cos \phi = 10/18 = 0.56$ and $\sin \phi = -15/18 = -0.83$, we know ϕ must be in the fourth quadrant. Also,

$$\tan \phi = \frac{15}{10} = -1.5 \quad \text{and} \quad \tan^{-1}(-1.5) = -0.98,$$

and since -0.98 is in the fourth quadrant, we take $\phi = -0.98$. Thus, the sinusoidal formula for P is

$$P = P_1 + P_2 = 90 + 18 \sin\left(\frac{\pi}{12}t - 0.98\right).$$

The next example shows that the sum of sine or cosine functions is not always a sinusoidal function.

Example 2   Sketch and describe the graph of $y = \sin 2x + \sin 3x$.

Solution   Figure 7.48 shows that the function $y = \sin 2x + \sin 3x$ is not sinusoidal. It is, however, periodic. Its period seems to be 2π, since it repeats twice on the interval of length 4π shown in the figure.

We can see that the function $y = \sin 2x + \sin 3x$ has period 2π by looking at the periods of $\sin 2x$ and $\cos 3x$. Since the period of $\sin 2x$ is π and the period of $\sin 3x$ is $2\pi/3$, on any interval of length 2π, the function $y = \sin 2x$ completes two cycles and the function $y = \sin 3x$ completes three cycles. Both functions are at the beginning of a new cycle after an interval of 2π, so their sum begins to repeat at this point. (See Figure 7.49.) Notice that even though the maximum value of each of the functions $\sin 2x$ and $\sin 3x$ is 1, the maximum value of their sum is not 2; it is a little less than 2. This is because $\sin 2x$ and $\sin 3x$ achieve their maximum values for different x values.

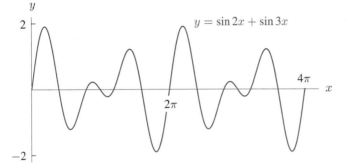

Figure 7.48: A graph of $y = \sin 2x + \sin 3x$, $0 \le x \le 4\pi$

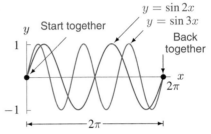

Figure 7.49: The graphs of $y = \sin 2x$ and $y = \sin 3x$ on an interval of 2π

Damped Oscillation

In Problems 13–16 in Section 6.1, we considered a weight attached to the ceiling by a spring. If the weight is disturbed, it begins bobbing up and down; we modeled the weight's motion using a trigonometric function. Figure 7.50 shows such a weight at rest. Suppose d is the displacement in centimeters from the weight's position at rest. For instance, if $d = 5$ then the weight is 5 cm above its rest position; if $d = -5$, then the weight is 5 cm below its rest position.

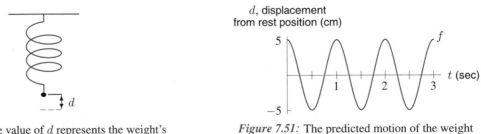

Figure 7.50: The value of d represents the weight's displacement from its at-rest position, $d = 0$

Figure 7.51: The predicted motion of the weight for the first 3 seconds

Imagine that we raise the weight 5 cm above its rest position and release it at time $t = 0$, where t is in seconds. Suppose the weight bobbed up and down once every second for the first few seconds. We could model this behavior by the function

$$d = f(t) = 5\cos(2\pi t).$$

One full cycle is completed each second, so the period is 1, and the amplitude is 5. Figure 7.51 gives a graph of d for the first three seconds of the weight's motion.

This trigonometric model of the spring's motion is flawed, however, because it predicts that the weight will bob up and down forever. In fact, we know that as time passes, the amplitude of the bobbing diminishes and eventually the weight comes to rest. How can we alter our formula to model this kind of behavior? We need an amplitude which decreases over time.

For example, the amplitude of the spring's motion might decrease at a constant rate, so that after 5 seconds, the weight stops moving. In other words, we could imagine that the amplitude is a decreasing linear function of time. Using A to represent the amplitude, this means that $A = 5$ at $t = 0$, and that $A = 0$ at $t = 5$. Thus, $A(t) = 5 - t$. Then, instead of writing

$$d = f(t) = \underbrace{\text{Constant amplitude}}_{5} \cdot \cos 2\pi t,$$

we write

$$d = f(t) = \underbrace{\text{Decreasing amplitude}}_{(5-t)} \cdot \cos 2\pi t$$

so that the formula becomes

$$d = f(t) = (5 - t) \cdot \cos 2\pi t.$$

Figure 7.52 shows a graph of this function.

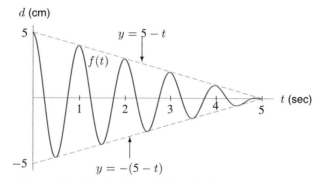

Figure 7.52: Graph of the weight's displacement assuming amplitude is a decreasing linear function of time

While the function $d = f(t) = (5 - t) \cos 2\pi t$ is a better model for the behavior of the spring for $t < 5$, Figure 7.53 shows that this model does not work for $t > 5$. The breakdown in the model occurs because the magnitude of $A(t) = 5 - t$ starts to increase when t grows larger than 5.

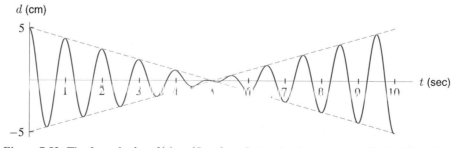

Figure 7.53: The formula $d = f(t) = (5 - t) \cos 2\pi t$ makes inaccurate predictions for values of t larger than 5

To improve the model, we keep the idea of representing the amplitude as a decreasing function of time, but we pick a different decreasing function. The problem with the linear function $A(t) - 5 \quad t$ is two-fold. It approaches zero too abruptly, and, after attaining zero, it becomes negative. We want a function that approaches zero gradually and does not become negative.

Let's try a decreasing exponential function. Suppose that the amplitude of the spring's motion is halved each second. Then at $t = 0$ the amplitude is 5, and the amplitude at time t is given by

$$A(t) = 5 \left(\frac{1}{2} \right)^t .$$

Thus, a formula for the motion is

$$d = f(t) = \underbrace{5 \left(\frac{1}{2} \right)^t}_{\text{Decreasing amplitude}} \cdot \cos 2\pi t.$$

Figure 7.54 shows a graph of this function. The dashed curves in Figure 7.54 show the decreasing exponential function. (There are two curves, $y = 5(\frac{1}{2})^t$ and $y = -5(\frac{1}{2})^t$, because the amplitude measures distance on both sides of the midline.)

Figure 7.54 predicts that the weight's oscillations diminish gradually, so that at time $t = 5$ the weight is still oscillating slightly. Figure 7.55 shows that the weight continues to make small oscillations long after 5 seconds have elapsed.

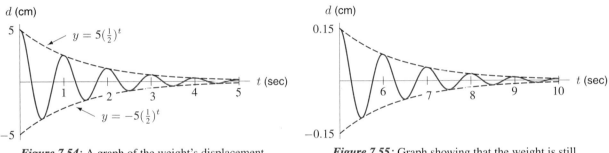

Figure 7.54: A graph of the weight's displacement assuming that the amplitude is a decreasing exponential function of time

Figure 7.55: Graph showing that the weight is still oscillating for $t \geq 5$, but with extremely small oscillations. Note the reduced scale on the d-axis

In general, provided $k > 0$, a function of the form

$$y = A_0 e^{-kt} \cos(Bt) + C \qquad \text{or} \qquad y = A_0 e^{-kt} \sin(Bt) + C$$

can be used to model an oscillating quantity whose amplitude decreases exponentially. Note that A_0, B, and C, and k are constants. Here, A_0 is the initial amplitude, and the amplitude decreases exponentially with time according to $A(t) = A_0 e^{-kt}$. We can recognize that our model for the displacement of a weight is in this form by using $k = \sin 2$.

Oscillation With a Rising Midline

In the next example, we consider an oscillating quantity which does not have a horizontal midline, but whose amplitude of oscillation is in some sense constant.

Example 3   In Section 6.5, Example 7, we considered a rabbit population which was undergoing seasonal size fluctuations. We modeled this population by the function

$$R = f(t) = 10000 - 5000 \cos\left(\frac{\pi}{6}t\right),$$

where R is the size of the rabbit population t months after January. Figure 7.56 gives a graph of the rabbit population over a five–year period. We see that the rabbit population varies periodically about the midline, $y = 10000$. The average number of rabbits is 10,000, but, depending on the time of year, the actual number may be above or below the average.

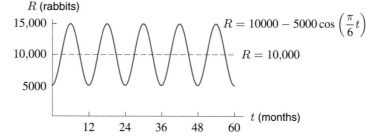

Figure 7.56: A graph of the rabbit population over a 5-year (60 month) period

This suggests the following way of thinking about the formula for R:

$$R = \underbrace{10{,}000}_{\text{Average value}} - \underbrace{5000 \cos\left(\frac{\pi}{6}t\right)}_{\text{Seasonal variation}}.$$

Notice that we can't say the average value of the rabbit population is 10,000 unless we look at the population over year-long units. For example, if we looked at the population over the first two months, an interval on which it is always below 10,000, then the average would be less than 10,000.

But what if the average, even over long periods of time, does not remain constant? For example, suppose that, due to conservation efforts, there is a steady increase of 50 rabbits per month in the average rabbit population. Thus, instead of writing

$$P = f(t) = \underbrace{10{,}000}_{\text{Constant midline}} - \underbrace{5000 \cos\left(\frac{\pi}{6}t\right)}_{\text{Seasonal variation}},$$

we could write

$$P = f(t) = \underbrace{10{,}000 + 50t}_{\substack{\text{Midline population increasing} \\ \text{by 50 every month}}} - \underbrace{5000 \cos\left(\frac{\pi}{6}t\right)}_{\text{Seasonal variation}}.$$

Figure 7.57 gives a graph of this new function over a five-year period.

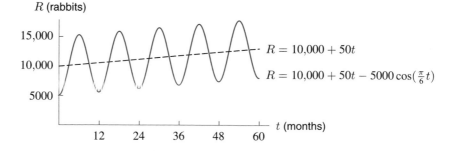

Figure 7.57: A graph of the gradually increasing rabbit population

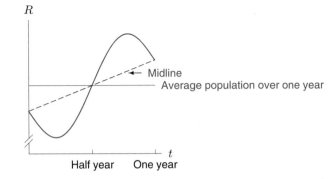

Figure 7.58: Although the rabbit population varies in a regular manner about its midline, the average population over a one-year period is not given by the midline

The population now oscillates above and below the line $R = 10000 + 50t$. How can we think about the "average"? Consider Figure 7.58, which gives a graph of the gradually increasing rabbit population over a one-year period. On this interval, although the population varies in a regular manner about a diagonal midline, its average value can still be indicated by a horizontal line.

Acoustic Beats

By international agreement, on a perfectly tuned piano, the A above middle C has a frequency of 440 cycles per second, also written 440 hertz (hz). The lowest-pitched note on the piano (the key at the left-most end) has frequency 55 hertz. Suppose a frequency of 55 hertz is struck on a tuning fork together with a note on an out-of-tune piano, whose frequency is 61 hertz. The intensities, I_1 and I_2, of these two tones can be modeled by the functions

$$I_1 = \cos(2\pi f_1 t) \qquad \text{and} \qquad I_2 = \cos(2\pi f_2 t),$$

where $f_1 = 55$, and $f_2 = 61$, and t is in seconds. If both tones are sounded at the same time, then their combined intensity is the sum of their separate intensities:

$$I = I_1 + I_2 = \cos(2\pi f_1 t) + \cos(2\pi f_2 t).$$

From Figure 7.59, we see that the graph of this function resembles a rapidly varying sinusoidal function except that its amplitude goes up and down. The ear perceives this variation in amplitude as a variation in loudness, and so the tone appears to waver (or *beat*) in a regular way. This is an example of the phenomenon known as *acoustic beats*. By adjusting a piano until the beats fade, the piano can be tuned.

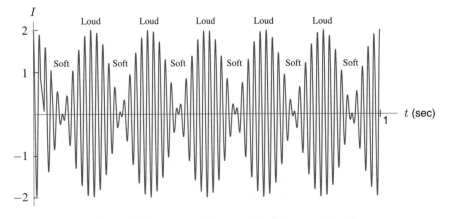

Figure 7.59: A graph of $I = \cos(2\pi f_1 t) + \cos(2\pi f_2 t)$

How can we explain the graph in Figure 7.59 in terms of what we know about the cosine function? We use an identity and rewrite the intensity as

$$I = \cos(2\pi f_2 t) + \cos(2\pi f_1 t) = 2\cos\frac{2\pi f_2 t + 2\pi f_1 t}{2} \cdot \cos\frac{2\pi f_2 t - 2\pi f_1 t}{2}$$

$$= 2\cos\frac{2\pi \cdot (61+55)t}{2} \cdot \cos\frac{2\pi \cdot (61-55)t}{2}$$

$$= 2\cos(2\pi \cdot 58t)\cos(2\pi \cdot 3t).$$

We can rewrite this formula in the following way:

$$I = 2\cos(2\pi \cdot 3t) \cdot \cos(2\pi \cdot 58t)$$

$$= A(t)p(t),$$

where $A(t) = 2\cos(2\pi \cdot 3t)$ gives a (slowly) changing amplitude and $p(t) = \cos(2\pi \cdot 58t)$ gives a pure tone of 58 hz. Thus, we can think of the tone described by I as having a pitch of 58 hz, which is midway between the tones sounded by the tuning fork and the out-of-tune piano. As the amplitude rises and falls, the tone grows louder and softer, but its pitch remains a constant 58 hz. (See Figure 7.60.)

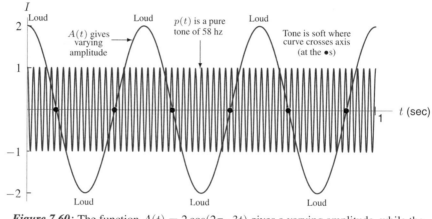

Figure 7.60: The function $A(t) = 2\cos(2\pi \cdot 3t)$ gives a varying amplitude, while the function $p(t) = \cos(2\pi \cdot 58t)$ gives a pure tone of 58 hz

Notice from Figure 7.60 that the function $A(t)$ completes three full cycles on the interval $0 \le t \le 1$. The tone is loudest when $A(t) = 1$ or $A(t) = -1$. Since both of these values occur once per cycle, the tone grows loud six times every second.

Problems for Section 7.5

1. Graph the function $f(x) = e^{-x}(2\cos x + \sin x)$ on the interval $0 \le x \le 12$. Use a calculator or computer to find the maximum and minimum values of the function.

2. Graph $y = t + 5\sin t$ and $y = t$ on $0 \le t \le 2\pi$. Where do the two graphs intersect if t is not restricted to $0 \le t \le 2\pi$?

3. A power company serves two different cities, City A and City B. The power requirements of both cities vary in a predictable fashion over the course of a typical day.

   (a) At midnight, the power requirement of City A is at a minimum of 40 megawatts. (A megawatt is a unit of power.) By noon the city has reached its maximum power consumption of 90 megawatts and by midnight it once again requires only 40 megawatts. This pattern repeats every day. Find a possible formula for $f(t)$, the power, in megawatts, required by City A as a function of t, the number of hours since midnight.

(b) The power requirements of City B differ from those of City A. Let $g(t)$ be the power in megawatts required by City B as a function of t, the number of hours since midnight. Suppose

$$g(t) = 80 - 30 \sin\left(\frac{\pi}{12}t\right).$$

State the amplitude and the period of $g(t)$, and give physical interpretations of these quantities.

(c) Graph and find all t such that

$$f(t) = g(t), \qquad 0 \le t < 24,$$

and interpret your solution(s) in terms of power usage.

(d) The power company is interested in the maximum value of the function

$$h(t) = f(t) + g(t), \qquad 0 \le t < 24,$$

Why should the power company be interested in the function h? What is the approximate maximum of this function, and approximately when does it attain this maximum?

(e) Find a formula for $h(t)$ as a single sine function. What is the exact maximum of this function?

4. An amusement park has a giant double ferris wheel as in Figure 7.61. The double ferris wheel has a 30-meter rotating arm attached at its center to a 25-meter main support. At each end of the rotating arm is attached a ferris wheel measuring 20 meters in diameter. Each ferris wheel rotates in the direction shown in Figure 7.61. It takes the rotating arm 6 minutes to complete one full revolution, and it takes 4 minutes for each wheel to complete a revolution about that wheel's hub. At time $t = 0$ the rotating arm is parallel to the ground and your seat is at the 3 o'clock position of the rightmost wheel.

(a) Find a formula for $h = f(t)$, your height above the ground in meters, as a function of time elapsed in minutes. [Hint: Your height above ground equals the height of your wheel hub above ground plus your height above that hub.]

(b) Sketch the graph of $f(t)$. Is $f(t)$ a periodic function? If so, what is its period?

(c) Approximate the least value of t such that h is at a maximum value. What is this maximum value?

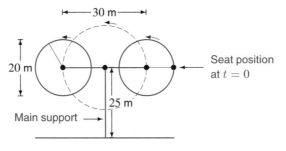

Figure 7.61

5. John wants to develop a mathematical model for predicting the value of a certain stock traded on the New York Stock Exchange. He has made two observations from the past behavior of the stock: 1) its value seems to have a cyclical component which increases for the first three months of each year, falls for the next six and then rises again for the last three; 2) inflation adds a linear component to the stock's price. For these reasons John is seeking a model of the form

$$f(t) = mt + b + A \sin\frac{\pi t}{6},$$

where t represents the time in months after Jan 1, 1990. He has the following data:

Date	1/1/90	4/1/90	7/1/90	10/1/90	1/1/91
Stock price	$20.00	$37.50	$35.00	$32.50	$50.00

(a) Find values of m, b, and A so that f fits the data.

(b) During which month(s) does this stock appreciate the most?

(c) During what period each year is this stock actually losing value?

6. Let $f(x) = \sin\left(\dfrac{1}{x}\right)$ for $x > 0$, x in radians.

    (a) $f(x)$ has a horizontal asymptote as $x \to \infty$. Find the equation for the asymptote and explain carefully why $f(x)$ has this asymptote.

    (b) Describe the behavior of $f(x)$ as $x \to 0$. Explain why $f(x)$ behaves in this way.

    (c) Is $f(x)$ a periodic function?

    (d) Let z_1 be the greatest zero of $f(x)$. Find the exact value of z_1.

    (e) How many zeros do you think the function $f(x)$ has?

    (f) Suppose a is a zero of f. Find a formula for b, the largest zero of f less than a.

7. Let $f(t) = \cos(e^t)$, where t is measured in radians.

    (a) Note that $f(t)$ has a horizontal asymptote as $t \to -\infty$. Find its equation, and explain why f has this asymptote.

    (b) Describe the behavior of f as $t \to \infty$. Explain why f behaves this way.

    (c) Find the vertical intercept of f.

    (d) Let t_1 be the least zero of f. Find t_1 exactly. [Hint: What is the smallest positive zero of the cosine function?]

    (e) Find an expression for t_2, the least zero of f greater than t_1.

8. A rigid metal bar of length l_0 is tapped sharply on one end at time $t = 0$. The bar begins to ring, which means that it is oscillating very rapidly along its length. Let Δl be the initial change in the bar's length, so that $l_0 - \Delta l$ is the length of the contracted bar at time $t = 0$. (See Figure 7.62.) Suppose that the bar rings at a frequency of 250 hertz, or 250 cycles per second, and that after 1 second the amplitude of the ringing has diminished by a factor of 10,000. It continues to decrease by this factor each second. Find a formula for $l(t)$, the length of the bar as a function of time, t, in seconds.

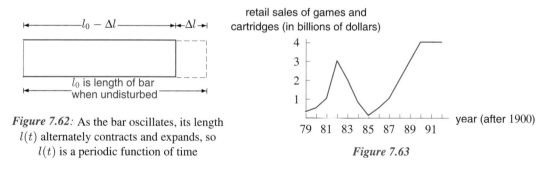

Figure 7.62: As the bar oscillates, its length $l(t)$ alternately contracts and expands, so $l(t)$ is a periodic function of time

Figure 7.63

9. In the July 1993 issue of *Standard and Poors Industry Surveys* the editors stated:[6]

    The strength (of sales) of video games, seven years after the current fad began, is amazing .... What will happen next year is anything but clear. While video sales ended on a strong note last year, the toy industry is nothing if not cyclical.

    The graph for this industry is shown in Figure 7.63.

    (a) Why might sales of video games be cyclical?

    (b) Would $s(t) = a \sin(bt)$, where t is time, serve as a reasonable model for this sales graph? What about $s(t) = a \cos(bt)$?

    (c) Think of a way to modify your choice in part (b) to provide for the higher amplitude in the years 1985–1992 as compared to 1979–1982.

    (d) Graph the function created in part (c) and compare your results to Figure 7.63. Modify your function to improve your approximation.

    (e) Use your function to predict sales for 1993.

    ---
    [6]Source: Nintendo of America

10. Derive the following identity used in an electrical engineering text[7] to represent the received AM signal function. Note that A, ω_c, ω_d are constants and $M(t)$ is a function of time, t.

$$r(t) = [A + M(t)]\cos\omega_c t + I\cos(\omega_c + \omega_d)t$$
$$= \{[A + M(t)] + I\cos\omega_d t\}\cos\omega_c t - I\sin\omega_d t\sin\omega_c t$$

11. Imagine a rope with one free end. If we give the free end a small upward shake, a wiggle travels down the length of the rope. Suppose that we repeatedly shake the free end so that a periodic series of wiggles travels down the rope. This situation can be described by a wave function:

$$y(x, t) = A\sin(kx - \omega t).$$

Here, x is the distance along the rope in meters; y is the displacement distance perpendicular to the rope; t is time in seconds; A is the amplitude; $2\pi/k$ is the distance from peak to peak, called the wavelength, λ, measured in meters; and $2\pi/\omega$ is the time in seconds for one wavelength to pass by. Suppose $A = 0.06$, $k = 2\pi$, and $\omega = 4\pi$.

(a) What is the wavelength of the motion?
(b) How many peaks of the wave pass by a given point each second?
(c) Construct the graph of this wave from $x = 0$ to $x = 1.5$ m when t is fixed at 0.
(d) What other values of t would give the same graph as the one found in part (d)?

7.6 POLAR COORDINATES

A point, P, in the plane is often labeled with *Cartesian coordinates* (x, y), where x is the horizontal distance of the point from the origin and y is the vertical distance. Alternatively, we can identify the point, P, by specifying its disance, r, from the origin and the angle, θ, shown in Figure 7.64. The angle, θ, is measured counterclockwise from the positive x-axis to the line joining P to the origin. The labels r and θ are called the *polar coordinates* of point P.

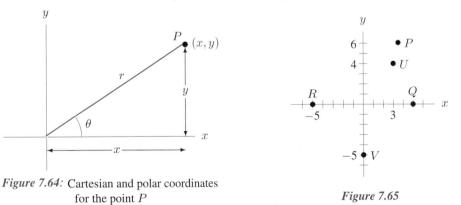

Figure 7.64: Cartesian and polar coordinates for the point P

Figure 7.65

Relation Between Cartesian and Polar Coordinates

By applying trigonometry to the right triangle in Figure 7.64, we see that
- $x = r\cos\theta$ and $y = r\sin\theta$
- $r = \sqrt{x^2 + y^2}$ and $\tan\theta = \dfrac{y}{x}$, $x \neq 0$

The angle θ is determined by the equations $\cos\theta = x/\sqrt{x^2 + y^2}$ and $\sin\theta = y/\sqrt{x^2 + y^2}$.

Warning: In general $\theta \neq \tan^{-1}(y/x)$. It is not possible to determine which quadrant θ is in from the value of $\tan\theta$ alone.

[7]*Modern Digital and Analog Communication Systems* by B. P. Lathi, 2nd ed., page 269.

Example 1 (a) Give Cartesian coordinates for the points with polar coordinates (r, θ) given by $P = (7, \pi/3)$, $Q = (5, 0)$, $R = (5, \pi)$.

(b) Give polar coordinates for the points with Cartesian coordinates (x, y) given by $U = (3, 4)$ and $V = (0, -5)$.

Solution (a) See Figure 7.65. Point P is a distance of 7 from the origin. The angle $\theta = \pi/3$ radians (60°). The Cartesian coordinates of P are

$$x = r \cos \theta = 7 \cos \frac{\pi}{3} = \frac{7}{2} \quad \text{and} \quad y = r \sin \theta = 7 \sin \frac{\pi}{3} = \frac{7\sqrt{3}}{2}.$$

Point Q is located a distance of 5 units along the positive x-axis with Cartesian coordinates

$$x = r \cos \theta = 5 \cos 0 = 5 \quad \text{and} \quad y = r \sin \theta = 5 \sin 0 = 0.$$

For point R, which is on the negative x-axis,

$$x = r \cos \theta = 5 \cos \pi = -5 \quad \text{and} \quad y = r \sin \theta = 5 \sin \pi = 0.$$

(b) For $U = (3, 4)$, we have $r = \sqrt{3^2 + 4^2} = 5$ and $\tan \theta = 4/3$. A possible value for θ is $\theta = \arctan 4/3 = 0.93$ radians, or about 53.1°. There are other possibilities, because the angle θ can be allowed to wrap around the origin more than once. Similarly, since the point $V = (0, -5)$ falls on the y-axis, we can choose $r = 5$, $\theta = 3\pi/2$, or instead $r = 5$, $\theta = -\pi/2$. However, we usually pick θ between 0 and 2π. Notice that for point V we cannot use our conversion formulas to find θ, because the tangent of any angle with one side along the y-axis is undefined:

$$\tan \theta = \frac{y}{x} = \frac{-5}{0}, \quad \text{so } \tan \theta \text{ is undefined.}$$

Graphing Equations in Polar Coordinates

The equations for certain graphs are much simpler when expressed in polar coordinates than in Cartesian coordinates. On the other hand, some graphs that have simple equations in Cartesian coordinates have complicated equations in polar coordinates.

Example 2 (a) Describe in words the graphs of the equation $y = 1$ (in Cartesian coordinates) and the equation $r = 1$ (in polar coordinates).

(b) Write the equation $r = 1$ using Cartesian coordinates. Write the equation $y = 1$ using polar coordinates.

Solution (a) The equation $y = 1$ describes a horizontal line. Since the equation $y = 1$ places no restrictions on the value of x, it describes every point having a y-value of 1, no matter what the value of its x-coordinate. Similarly, the equation $r = 1$ places no restrictions on the value of θ. Thus, it describes every point having an r-value of 1, that is, having a distance of 1 from the origin. This set of points is the unit circle. See Figure 7.66.

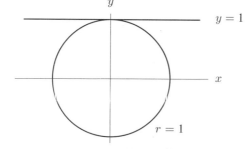

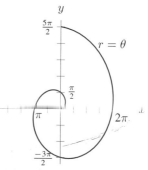

Figure 7.66: The graph of the equation $r = 1$ is the unit circle because $r = 1$ for every point regardless of the value of θ. The graph of $y = 1$ is a horizontal line since $y = 1$ for any x

Figure 7.67: A graph of the Archimedean spiral $r = \theta$

(b)   Since $r = \sqrt{x^2 + y^2}$, we rewrite the equation $r = 1$ using Cartesian coordinates as $\sqrt{x^2 + y^2} = 1$, or, squaring both sides, as $x^2 + y^2 = 1$. We see that the equation for the unit circle is simpler in polar coordinates than it is in Cartesian coordinates.

On the other hand, since $y = r \sin \theta$, we can rewrite the equation $y = 1$ in polar coordinates as $r \sin \theta = 1$, or, dividing both sides by $\sin \theta$, as $r = 1/\sin \theta$. We see that the equation for this horizontal line is simpler in Cartesian coordinates than it is in polar coordinates.

Example 3   Graph the equation $r = \theta$. The graph is called an *Archimedean* spiral after the Greek mathematician Archimedes who described its properties (although not by using polar coordinates).

Solution   To construct this graph, use the values in Table 7.1. To help us visualize the shape of the spiral, we convert the angles in Table 7.1 to degrees and the r-values to decimals. See Table 7.2.

TABLE 7.1   *Points on the Archimedean spiral $r = \theta$, with θ in radians*

θ	0	$\frac{\pi}{6}$	$\frac{\pi}{3}$	$\frac{\pi}{2}$	$\frac{2\pi}{3}$	$\frac{5\pi}{6}$	π	$\frac{7\pi}{6}$	$\frac{4\pi}{3}$	$\frac{3\pi}{2}$
r	0	$\frac{\pi}{6}$	$\frac{\pi}{3}$	$\frac{\pi}{2}$	$\frac{2\pi}{3}$	$\frac{5\pi}{6}$	π	$\frac{7\pi}{6}$	$\frac{4\pi}{3}$	$\frac{3\pi}{2}$

TABLE 7.2   *Points on the Archimedean spiral $r = \theta$, with θ in degrees*

θ	0	30°	60°	90°	120°	150°	180°	210°	240°	270°
r	0.00	0.52	1.05	1.57	2.09	2.62	3.14	3.67	4.19	4.71

Notice that as the angle θ increases, points on the curve move farther from the origin. At $0°$, the point is at the origin. At $30°$, it is 0.52 units away from the origin, at $60°$ it is 1.05 units away, and at $90°$ it is 1.57 units away. As the angle winds around, the point traces out a curve that moves away from the origin, giving a spiral. (See Figure 7.67.)

Polar coordinates can be used with inequalities to describe regions that are obtained from circles. Such regions are often much harder to represent in Cartesian coordinates.

Example 4   Using inequalities, describe a compact disk with an outer diameter of 120 mm and an inner diameter of 15 mm.

Solution   The compact disk lies between two circles of radius 7.5 mm and 60 mm. See Figure 7.68. Thus, if the origin is at the center, the disk is represented by

$$7.5 \leq r \leq 60 \quad \text{and} \quad 0 \leq \theta \leq 2\pi.$$

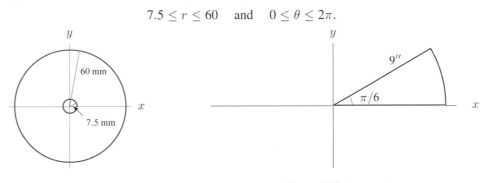

Figure 7.68: Compact disk   *Figure 7.69:* Pizza slize

Example 5   An 18 inch pizza is cut into 12 slices. Use inequalities to describe one of the slices.

Solution   The pizza has radius 9 inches; the angle at the center is $2\pi/12 = \pi/6$. See Figure 7.69. Thus, if the origin is at center of the original pizza, the slice shown in the figure is represented by

$$0 \leq r \leq 9 \quad \text{and} \quad 0 \leq \theta \leq \frac{\pi}{6}.$$

Problems for Section 7.6

For Problems 1–8, the origin is at the center of a clock, with the positive x-axis going through 3 and the positive y-axis going through 12. The hour hand is 3 cm long and the minute hand is 4 cm long. What are the Cartesian coordinates and polar coordinates of the tips of the hour hand and minute hand, H and M, respectively, at the following times?

1.   12 noon
2.   3 pm
3.   9 am
4.   11 am
5.   1:30 pm
6.   7 am
7.   3:30 pm
8.   9:15 am

In Problems 9–11, give inequalities for r and θ which describe the following regions in polar coordinates.

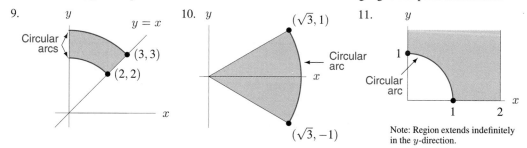

12.   (a)   Make a table of values for the equation $r = 1 - \sin\theta$. Include $\theta = 0, \pi/3, \pi/2, 2\pi/3, \pi, \cdots$.
      (b)   Use the table to graph the equation $r = 1 - \sin\theta$ in the xy-plane. This curve is called a *cardioid*.
      (c)   At what point(s) does the cardioid $r = 1 - \sin\theta$ intersect a circle of radius 1/2 centered at the origin?
      (d)   Graph the curve $r = 1 - \sin 2\theta$ in the xy-plane. Compare this graph to the cardioid $r = 1 - \sin\theta$.

13.   Graph the equation $r = 1 - \sin(n\theta)$, for $n = 1, 2, 3$ and 4. What is the relationship between the value of n and the shape of the graph?

14.   Graph the equation $r = 1 - \sin\theta$, with $0 \le \theta \le n\pi$, for $n = 2, 3$ and 4. What is the relationship between the value of n and the shape of the graph?

15.   Graph the equation $r = 1 - n\sin\theta$, for $n = 2, 3$ and 4. What is the relationship between the value of n and the shape of the graph?

16.   Graph the equation $r = 1 - \cos\theta$. Describe its relationship to $r = 1 - \sin\theta$.

17.   Give inequalities that describe the flat surface of a washer that is one inch in diameter and has an inner hole with a diameter of 3/8 inch.

18.   Graph the equation $r = 1 - \sin(2\theta)$ for $0 \le \theta \le 2\pi$. There are two loops. For each loop, give a restriction on θ that shows all of that loop and none of the other loop.

19.   A slice of pizza is one eighth of a circle of radius 1 foot. The slice is in the first quadrant, with one edge along the x-axis, and the center of the pizza at the origin. Give inequalities describing this region using:
      (a)   Polar coordinates
      (b)   Rectangular coordinates

REVIEW PROBLEMS FOR CHAPTER SEVEN

1.   Find the angles of the triangle in Figure 7.70.

2.   Figure 7.71 gives a right triangle ABC with sides a, b, and c.
      (a)   If angle $A = 30°$ and $b = 2\sqrt{3}$, find the lengths of the other sides and find the other angles of triangle ABC.
      (b)   Repeat part (a), this time assuming only that $a = 25$ and $c = 24$.

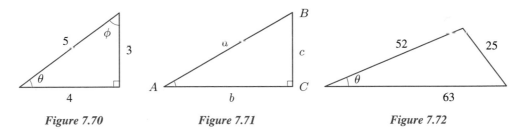

Figure 7.70 Figure 7.71 Figure 7.72

3. Find the value of the angle θ shown in Figure 7.72.

Find all sides and angles of the triangles in Problems 4–6. (Not necessarily drawn to scale.)

4. **5.** **6.**

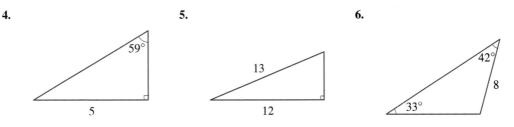

7. One student maintained that $\sin 2\theta \neq 2 \sin \theta$, but another student said let $\theta = \pi$ and then $\sin 2\theta = 2 \sin \theta$. Who is right?

8. Suppose $\sin \theta = 1/7$ for $\pi/2 < \theta < \pi$.
(a) Use the Pythagorean identity to find $\cos \theta$. (b) What is θ?

9. Suppose that $\cos(2\theta) = 2/7$ and θ is in the first quadrant. Find $\cos \theta$ exactly.

10. Use a trigonometric identity to find all exact solutions to: $\cos 2\theta = \sin \theta$, $0 \leq \theta < 2\pi$.

11. Solve for α exactly: $\cos(2\alpha) = -\sin \alpha$ with $0 \leq \alpha < 2\pi$.

12. Simplify the expression $\sin \left(2 \cos^{-1} \left(\dfrac{5}{13} \right) \right)$ to a rational number.

13. Graph the two functions $f(t) = \cos t$ and $g(t) = \sin(t + \pi/2)$. Explain what you see.

14. (a) Graph $g(\theta) = \sin \theta - \cos \theta$.
(b) Write $g(\theta)$ as a sine function without using cosine. Then write $g(\theta)$ as a cosine function without using sine.

15. To check the calibration of their transit, two student surveyors used the set-up shown (in meters) in Figure 7.73. What angles in degrees for α and β should they get if their transit is accurate?

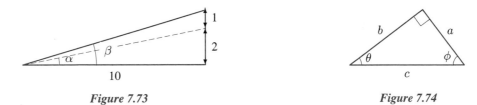

Figure 7.73 Figure 7.74

16. (a) Find expressions in terms of a, b, and c for the sine, the cosine, and the tangent of the angle ϕ in Figure 7.74.
(b) Based on your answers in part (a), show that $\sin \phi = \cos \theta$ and $\cos \phi = \sin \theta$.

17. Two surveyors want to check their transit using a setup similar to Problem 15, but with only one test angle. They wish to have the test angle (in degrees) come out to an integer value. What triangle could they set up?

18. Knowing the height of the Columbia Tower in Seattle, determine the height of the Seafirst Tower and the distance between the towers. See Figure 7.75.

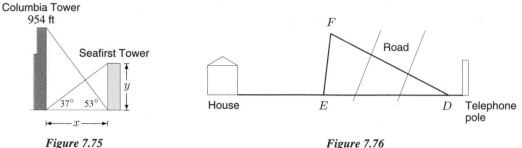

Figure 7.75   **Figure 7.76**

19. The telephone company needs to run a wire from the telephone pole across a street to a new house. See Figure 7.76. Since they cannot measure across the busy street, they hire a surveyor to determine the amount of wire needed. The surveyor measures a distance of 23.5 feet from the telephone pole to a stake, D, which she sets in the ground. She measures 145.3 feet from the house to a second stake, E. She sets a third stake, F, at a distance of 105.2 feet from the second stake. Using a transit, the surveyor measures angle DEF to be $83°$ and angle EFD to be $68°$. A total of 20 feet of wire is needed to make connections at the end, and wire is sold in 100 feet rolls. How many rolls of wire are needed?

20. Find the range for each of the following functions.

   (a) $y = \arccos x$

   (b) $y = \dfrac{3}{2 - \cos x}$

   (c) $y = \sin\left(\dfrac{1}{1 + x^2}\right)$

   (d) $y = 4 \sin x \cos x$

   (e) $y = 3^{3 + \sin x}$

21. Sketch graphs of each of the following functions for $0 \le x \le 2\pi$.

   (a) $y = x \sin x$

   (b) $y = (1/x) \sin x$

   (c) $y = \sin(x^2)$

   (d) $y = (\sin x)^2$

   (e) $y = |\sin x|$

   (f) $y = e^x \sin x$

22. (a) Explain why $r = 2 \cos \theta$ and $(x - 1)^2 + y^2 = 1$ are equations for the same circle.

   (b) Give Cartesian and polar coordinates for the 12, 3, 6, and 9 o'clock positions on the circle described in part (a).

23. For each of the following statements, write *true* if the statement must be true, *false* otherwise.

   (a) One solution of $3e^t \sin t - 2e^t \cos t = 0$ is $t = \arctan\left(\frac{2}{3}\right)$.

   (b) $0 \le \cos^{-1}(\sin(\cos^{-1} x)) \le \frac{\pi}{2}$ for all x in the domain of $\cos^{-1} x$.

   (c) If $a > b > 0$, then the domain of $f(t) = \ln(a + b \sin t)$ is all real numbers t.

   (d) For all $t \le 0$, $\cos(e^t) + \sin(e^t) \le \sin 1$.

24. The planet Betagorp travels around its sun, Trigosol, in an elliptical orbit. The distance of Betagorp from Trigosol at time θ is given by

$$r = \frac{5.35 \times 10^7}{1 - 0.234 \cos \theta}.$$

   Find the least positive θ (in decimal degrees to 3 significant digits) so that Betagorp is 5.12×10^7 miles from Trigosol.

25. For each of the following expressions, find a line segment in Figure 7.77 with length equal to the value of the expression.

   (a) $\sin \theta$ (b) $\cos \theta$ (c) $\tan \theta$ (d) $\dfrac{1}{\sin \theta}$ (e) $\dfrac{1}{\cos \theta}$ (f) $\dfrac{1}{\tan \theta}$

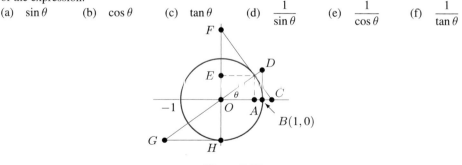

Figure 7.77

CHAPTER EIGHT

COMPOSITIONS, INVERSES, AND COMBINATIONS OF FUNCTIONS

In Chapter 5, we studied transformations of functions. The composite functions in this chapter are generalizations of these transformations. In Chapter 3, we introduced the inverse function and its notation, f^{-1}. In Chapter 4, we defined the inverse of the exponential function (that is, the logarithm) and in Chapter 6, we defined the inverses of the trigonometric functions. In this chapter, we consider inverse functions in more detail.

8.1 COMPOSITION OF FUNCTIONS

The Effect of a Drug on Heart Rates

A therapeutic drug has the side effect of raising a patient's heart rate. Table 8.1 gives the relationship between Q, the amount of drug in the patient's body (in milligrams), and r, the patient's heart rate (in beats per minute). We see that the higher the drug level, the faster the heart rate.

TABLE 8.1 *Heart rate, $r = f(Q)$, as a function of drug level, Q*

Q, drug level (mg)	0	50	100	150	200	250
r, heart rate (beats per minute)	60	70	80	90	100	110

A patient is given a 250 mg injection of the drug. Over time, the level of drug in the patient's bloodstream falls. Table 8.2 gives the drug level, Q, as a function of time, t.

TABLE 8.2 *Drug level, $Q = g(t)$, as a function of time, t, since the medication was given*

t, time (hours)	0	1	2	3	4	5	6	7	8
Q, drug level (mg)	250	200	160	128	102	82	66	52	42

Since heart rate depends on the drug level and drug level depends on time, the heart rate also depends on time. Tables 8.1 and 8.2 can be combined to give the patient's heart rate, r, as a function of t. For example, according to Table 8.2, at time $t = 0$ the drug level is 250 mg. According to Table 8.1, at this drug level, the patient's heart rate is 110 beats per minute. So $r = 110$ when $t = 0$. The results of similar calculations have been compiled in Table 8.3. Note that many of the entries, such as $r = 92$ when $t = 2$, are estimates.

TABLE 8.3 *Heart rate, $r = h(t)$, as a function of time, t*

t, time (hours)	0	1	2	3	4	5	6	7	8
r, heart rate (beats per minute)	110	100	92	86	80	76	73	70	68

Now, since

$$r = f(Q) \quad \text{or} \quad \underbrace{\text{Heart rate}}_{r} = f(\underbrace{\text{drug level}}_{Q}),$$

and

$$Q = g(t) \quad \text{or} \quad \underbrace{\text{Drug level}}_{Q} = g(\underbrace{\text{time}}_{t}),$$

we can substitute $Q = g(t)$ into $r = f(Q)$, giving

$$r = f(\underset{g(t)}{Q}) = f(g(t)).$$

The function h in Table 8.3 is said to the *composition* of the function f and g, written

$$h(t) = f(g(t)).$$

This formula represents the process that we used to find the values of $r = h(t)$ in Table 8.3.

Example 1   Use Tables 8.1 and 8.2 to estimate the values of: (a) $h(0)$ (b) $h(4)$

Solution   (a)   If $t = 0$, then

$$r = h(0) = f(g(0)).$$

Table 8.2 shows that $g(0) = 250$, so

$$r = h(0) = f(\underbrace{250}_{g(0)}).$$

We see from Table 8.1 that $f(250) = 110$. Thus,

$$r = h(0) = \underbrace{110}_{f(250)}.$$

As before, this tells us that the patient's heart rate at time $t = 0$ is 110 beats per minute.

(b)   If $t = 4$, then

$$h(4) = f(g(4)).$$

Working from the inner set of parentheses outward, we start by evaluating $g(4)$. Table 8.2 shows that $g(4) = 102$. Thus,

$$h(4) = f(\underbrace{102}_{g(4)}).$$

Table 8.1 does not have a value for $f(102)$. But since $f(100) = 80$ and $f(150) = 90$, we estimate that $f(102)$ is close to 80. Thus, we let

$$h(4) \approx \underbrace{80}_{f(102)}.$$

This indicates that four hours after the injection, the patient's heart rate is approximately 80 beats per minute.

> The function $h(t) = f(g(t))$ is said to be a **composition** of f with g. The function h is defined by using the output of the function g as the input to f.

Formulas for Composite Functions

A possible formula for $r = f(Q)$, the heart rate as a function of drug level is

$$r = f(Q) = 60 + 0.2Q.$$

A possible formula for $Q = g(t)$, the drug level as a function of time is

$$Q = g(t) = 250(0.8)^t.$$

To find a formula for $r = h(t) = f(g(t))$, the heart rate as a function of time, the function $g(t)$ is the input to f. Thus,

$$r = f(\underbrace{\text{input}}_{g(t)}) = 60 + 0.2(\underbrace{\text{input}}_{g(t)}),$$

so

$$r = f(g(t)) = 60 + 0.2g(t).$$

Now, substitute the formula for $g(t)$. This gives

$$r = h(t) = f(g(t)) = 60 + 0.2 \cdot \underbrace{250(0.8)^t}_{g(t)}$$

so
$$r = h(t) = 60 + 50(0.8)^t.$$

We can check the formula against Table 8.3. For example, if $t = 4$

$$h(4) = 60 + 50(0.8)^4 = 80.48.$$

This result is in agreement with the value $h(4) \approx 80$ that we estimated in Table 8.3.

Example 2 Let $p(x) = 2x + 1$ and $q(x) = x^2 - 3$. Suppose $u(x) = p(q(x))$ and $v(x) = q(p(x))$.
(a) Calculate $u(3)$ and $v(3)$.
(b) Find formulas for $u(x)$ and $v(x)$.

Solution (a) We want

$$u(3) = p(q(3)).$$

We start by evaluating $q(3)$. The formula for q gives $q(3) = 3^2 - 3 = 6$, so

$$u(3) = p(6).$$

The formula for p gives $p(6) = 2 \cdot 6 + 1 = 13$, so

$$u(3) = 13.$$

To calculate $v(3)$, we have

$$v(3) = q(p(3))$$
$$= q(7) \qquad \text{Because } p(3) = 2 \cdot 3 + 1 = 7$$
$$= 46 \qquad \text{Because } q(7) = 7^2 - 3$$

Notice that, $v(3) \neq u(3)$. The functions $v(x) = q(p(x))$ and $u(x) = p(q(x))$ are different.

(b) In the formula for u,

$$u(x) = p(\underbrace{q(x)}_{\text{Input for } p})$$
$$= 2q(x) + 1 \qquad \text{Because } p(\text{Input}) = 2 \cdot \text{Input} + 1$$
$$= 2(x^2 - 3) + 1 \qquad \text{Substituting } q(x) = x^2 - 3$$
$$= 2x^2 - 5.$$

Check this formula by evaluating $u(3)$ which we know to be 13:

$$u(3) = 2 \cdot 3^2 - 5 = 13.$$

In the formula for v,

$$v(x) = q(\underbrace{p(x)}_{\text{Input for } q})$$
$$= q(2x + 1) \qquad \text{Because } p(x) = 2x + 1$$
$$= (2x + 1)^2 - 3 \qquad \text{Because } q(\text{Input}) = \text{Input}^2 - 3$$
$$= 4x^2 + 4x - 2.$$

Check this formula by evaluating $v(3)$, which we know to be 46:

$$v(3) = 4 \cdot 3^2 + 4 \cdot 3 - 2 = 46.$$

So far we have considered examples of two functions composed together, but there is no limit on the number of functions that can be composed. Functions can even be composed with themselves.

Example 3   Let $p(x) = 2x + 1$ and $q(x) = x^2 - 3$. Find a formula in terms of x for $w(x) = p(p(q(x)))$.

Solution   We work from inside the parentheses outward. First we find $p(q(x))$, and then input the result to p.

$$w(x) = p(p(q(x)))$$
$$= p(\underbrace{2x^2 - 5}_{\text{Input for } p}) \qquad \text{Because we found in Example 2 that } p(q(x)) = 2x^2 - 5$$
$$= 2(2x^2 - 5) + 1 \qquad \text{Because } p(\text{Input}) = 2(\text{Input}) + 1$$
$$= 4x^2 - 9.$$

Composition of Functions Defined by Graphs

So far we have composed functions defined by tables and formulas. In the next example, we compose functions defined by graphs.

Example 4   Let u and v be two functions defined by the graphs in Figure 8.1. Evaluate:

(a) $v(u(-1))$ (b) $u(v(5))$ (c) $v(u(0)) + u(v(4))$

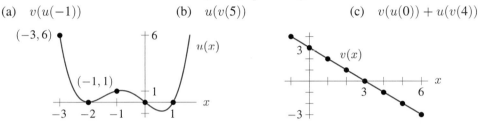

Figure 8.1: Evaluate the composition of functions u and v defined by their graphs

Solution   (a)   To evaluate $v(u(-1))$, start with $u(-1)$. From Figure 8.1, we see that $u(-1) = 1$. Thus,
$$v(u(-1)) = v(1).$$
From the graph we see that $v(1) = 2$, so
$$v(u(-1)) = 2.$$

(b)   Since $v(5) = -2$, we have $u(v(5)) = u(-2) = 0$.

(c)   Since $u(0) = 0$, we have $v(u(0)) = v(0) = 3$.
Since $v(4) = -1$, we have $u(v(4)) = u(-1) = 1$.
Thus $v(u(0)) + u(v(4)) = 3 + 1 = 4$.

Decomposition of Functions

Sometimes we reason backward to find the functions which went into a composition. This process is called *decomposition*.

Example 5   Let $h(x) = f(g(x)) = \sqrt{x+1}$. Find possible formulas for $f(x)$ and $g(x)$.

Solution   In the formula $h(x) = \sqrt{x+1}$, the expression $x + 1$ is under the square root sign. Thus, we can take the inside function to be $g(x) = x + 1$. This means that we can write
$$h(x) = \sqrt{\underbrace{x + 1}_{g(x)}} = \sqrt{g(x)}.$$

Then the outside function is $f(x) = \sqrt{x}$. We check that composing f and g gives h:

$$f(g(x)) = f(x + 1) = \sqrt{x + 1} = h(x).$$

There are many possible solutions to Example 5. For example, we might choose $f(x) = \sqrt{x - 1}$ and $g(x) = x + 2$. Then

$$f(g(x)) = \sqrt{(x + 2) - 1} = \sqrt{x + 1} = h(x).$$

Alternatively, we might choose $f(x) = \sqrt{x + 1}$ and $g(x) = x$. Although this satisfies the condition that $h(x) = f(g(x))$, it is not very useful, because f is the same as h. This kind of decomposition is referred to as *trivial*. Another example of a trivial decomposition of $h(x)$ is $f(x) = x$ and $g(x) = \sqrt{x + 1}$.

Example 6   The vertex formula for the family of quadratic functions is

$$p(x) = a(x - h)^2 + k.$$

Decompose the formula into three simple functions. That is, find formulas for u, v, and w where

$$p(x) = u(v(w(x))),$$

Solution   We work from inside the parentheses outward. In the formula $p(x) = a(x - h)^2 + k$, we have the expression $x - h$ inside the parentheses. In the formula $p(x) = u(v(w(x)))$, the innermost function is $w(x)$. Thus, we let

$$w(x) = x - h.$$

In the formula $p(x) = a(x - h)^2 + k$, the first operation done to $x - h$ is squaring. Thus, we let

$$v(\text{Input}) = \text{Input}^2$$
$$v(x) = x^2.$$

So we have

$$v(w(x)) = v(x - h) = (x - h)^2.$$

Finally, to obtain $p(x) = a(x - h)^2 + k$, we multiply $(x - h)^2$ by a and add k. Thus, we let

$$u(\text{Input}) = a \cdot \text{Input} + k$$
$$u(x) = ax + k.$$

To check, we compute

$$
\begin{aligned}
u(v(w(x))) &= u(v(x - h)) && \text{Since } w(x) = x - h \\
&= u(\underbrace{(x - h)^2}_{\text{Input for } u}) && \text{Since } v(x - h) = (x - h)^2 \\
&= a(x - h)^2 + k && \text{Since } u((x - h)^2) = a(x - h)^2 + k.
\end{aligned}
$$

Functions of Related Quantities

Composition can be used to relate two quantities by means of an intermediate quantity.

Example 7   The formula for the volume of a cube with side s is $V = s^3$. The formula for the surface area of a cube is $A = 6s^2$. Express the volume of a cube, V, as a function of its surface area, A.

Solution   We express V as a function of A by exploiting the fact that V is a function of s, which in turn is expressed as a function of A. To write s as a function of A, we solve $A = 6s^2$ for s

$$s^2 = \frac{A}{6} \quad \text{so} \quad s = +\sqrt{\frac{A}{6}} \qquad \text{Because the length of a side of a cube is positive.}$$

Substituting $s = \sqrt{A/6}$ in the formula $V = s^3$ gives V as a function of A:

$$V = s^3 = \left(\sqrt{\frac{A}{6}}\right)^3.$$

Problems for Section 8.1

1. Table 8.4 contains values of the functions p and q. Construct a table of values for $r(x) = p(q(x))$.

TABLE 8.4

x	0	1	2	3	4	5
$p(x)$	1	0	5	2	3	4
$q(x)$	5	2	3	1	4	8

2. Let p and q be the functions in Problem 1. Construct a table of values for $s(x) = q(p(x))$.

3. Complete Table 8.5 with values of the functions f, g, and h, given the following conditions:

   (a) f is symmetric about the y-axis (even).
   (b) g is symmetric about the origin (odd).
   (c) h is the composition of g with f, that is, $h(x) = g(f(x))$.

TABLE 8.5

x	-3	-2	-1	0	1	2	3
$f(x)$	0	2	2	0			
$g(x)$	0	2	2	0			
$h(x)$							

4. Values for $f(x)$ and $g(y)$ are given in Tables 8.6 and 8.7. Complete Table 8.8:

TABLE 8.6

x	$f(x)$
0	0
$\pi/6$	$1/2$
$\pi/4$	$\sqrt{2}/2$
$\pi/3$	$\sqrt{3}/2$
$\pi/2$	1

TABLE 8.7

y	$g(y)$
0	$\pi/2$
$1/4$	π
$\sqrt{2}/4$	0
$1/2$	$\pi/3$
$\sqrt{2}/2$	$\pi/4$
$3/4$	0
$\sqrt{3}/2$	$\pi/6$
1	0

TABLE 8.8

x	$g(f(x))$
0	
$\pi/6$	
$\pi/4$	
$\pi/3$	
$\pi/2$	

5. Suppose that $h(x) = f(g(x))$. Complete the following table.

done

x	$f(x)$	$g(x)$	$h(x)$
0	1	2	5
1	9	0	
2		1	

6. Complete the following tables given that $h(x) = g(f(x))$.

x	$f(x)$
-2	4
-1	
0	
1	5
2	1

x	$g(x)$
1	
2	1
3	2
4	0
5	-1

x	$h(x)$
-2	
-1	1
0	2
1	
2	-2

7. Let $f(x) = 3x - 2$ and $h(x) = 3x^2 - 5x + 2$. Find a formula for $f(h(x))$.

8. Let $f(x) = 1 - x$. Evaluate the following and simplify your answer.

   (a) $2f(x)$ (b) $f(x) + 1$ (c) $f(1 - x)$ (d) $(f(x))^2$ (e) $f(1)/x$ (f) $\sqrt{f(x)}$

9. Let $g(x) = x^2 + x$. Evaluate the following and simplify your answer.

   (a) $-3g(x)$ (b) $g(1) - x$ (c) $g(x) + \pi$

   (d) $\sqrt{g(x)}$ (e) $g(1)/(x + 1)$ (f) $(g(x))^2$

Find and simplify formulas for the functions in Problems 10–15. Let $f(x) = x^2 + 1$, $g(x) = \dfrac{1}{x - 3}$, and $h(x) = \sqrt{x}$.

10. $f(g(x))$ 11. $g(f(x))$ 12. $f(h(x))$

13. $h(f(x))$ 14. $g(g(x))$ 15. $g(f(h(x)))$

Find formulas for the functions in Problems 16–25. Let $m(x) = \dfrac{1}{x - 1}$, $k(x) = x^2$, and $n(x) = \dfrac{2x^2}{x + 1}$.

done   done   done

16. $k(m(x))$ 17. $m(k(x))$ 18. $k(n(x))$ 19. $n(k(x))$ 20. $m(n(x))$

21. $n(m(x))$ 22. $m(m(x))$ 23. $(m(x))^2$ 24. $m(x^2)$ 25. $n(n(x))$

In Problems 26–30, find a simplified formula for the difference quotient

$$\frac{f(x + h) - f(x)}{h}.$$

26. $f(x) = x^2 - 1$ 27. $f(x) = x^2 + x$ 28. $f(x) = \sqrt{x}$ 29. $f(x) = \dfrac{1}{x}$ 30. $f(x) = 2^x$

31. Using the following definition of $f(x)$, find $f(f(1))$. Show your reasoning.

$$f(x) = \begin{cases} 2 & \text{if } x \le 0 \\ 3x + 1 & \text{if } 0 < x < 2 \\ x^2 - 3 & \text{if } x \ge 2 \end{cases}$$

32. Let $f(x) = 1/x$. For n a positive integer, define $f_n(x)$ as the composition of f with itself n times. For example, $f_2(x) = f(f(x))$ and $f_3(x) = f(f(f(x)))$.

    (a) Evaluate $f_7(2)$. (b) Evaluate $f_{23}(f_{22}(5))$.

Decompose each of the functions in Problems 33–46 into two new functions, u and v, where v is the inside function.

33. $f(x) = \sqrt{3 - 5x}$ 34. $F(x) = (2x + 5)^3$ 35. $f(x) = \sqrt{x + 8}$ 36. $h(x) = x^4 + x^2$

37. $j(x) = 1 - \sqrt{x}$ 38. $k(x) = \sqrt{1 - x}$ 39. $g(x) = \dfrac{1}{x^2}$ 40. $l(x) = 2 + \dfrac{1}{x}$

41. $g(x) = \dfrac{1}{1 - x}$ 42. $G(x) = \dfrac{2}{1 + \sqrt{x}}$ 43. $H(x) = 3^{2x - 1}$ 44. $J(x) = 8 - 2|x|$

45. $K(x) = \sqrt{1 - 4x^2}$ 46. $L(x) = x^6 - 2x^3 + 1$

Decompose each of the functions in Problems 47–52 into three new functions, u, v, and w, where u is the outside function and w is the inside function.

47. $m(x) = \sqrt{1 - x^2}$ 48. $n(x) = 1/(1 - 2x)$ 49. $o(x) = 1 - \sqrt{x - 1}$

50. $p(x) = \sqrt[3]{5 - \sqrt{x}}$ 51. $q(x) = \left(1 + \dfrac{1}{x}\right)^2$ 52. $r(x) = \dfrac{1}{1 + \dfrac{1}{x + 1}}$

For Problems 53–59, let $k(x) = x^2 + 2$ and $g(x) = x^2 + 3$. Find a possible formula for the function named.

53. $h(x)$ if $h(k(x)) = (x^2 + 2)^3$.

54. $j(x)$ if $k(j(t)) = \left(\dfrac{1}{t}\right)^2 + 2$.

55. $f(x)$ if $f(k(v)) = \dfrac{1}{v^2 + 2}$.

56. $m(x)$ if $m(k(\pi)) = -\dfrac{1}{\sqrt{\pi^2 + 2}}$.

57. $f(x)$ if $g(f(x)) = (x + 1)^2 + 3$.

58. $h(x)$ if $h(g(x)) = \dfrac{1}{x^2 + 3} + 5x^2 + 15$.

59. $j(x)$ if $j(x) = g(g(x))$.

60. Suppose $p(x) = (1/x) + 1$ and $q(x) = x - 2$.

    (a) Let $r(x) = p(q(x))$. Find a formula for $r(x)$ and simplify it.
    (b) Write formulas for $s(x)$ and $t(x)$ such that $p(x) = s(t(x))$, where $s(x) \neq x$ and $t(x) \neq x$.
    (c) Let a be different from 0 and -1. Find a simplified expression for $p(p(a))$.

61. If $s(x) = 5 + \dfrac{1}{x + 5} + x$, $k(x) = x + 5$, and $s(x) = v(k(x))$, what is $v(x)$?

62. Suppose $u(v(x)) = \dfrac{1}{x^2 - 1}$ and $v(u(x)) = \dfrac{1}{(x - 1)^2}$. Find formulas for $u(x)$ and $v(x)$.

63. Assume that $f(x) = 3 \cdot 9^x$ and that $g(x) = 3^x$.

    (a) If $f(x) = h\left(g(x)\right)$, find a formula for $h(x)$.
    (b) If $f(x) = g\left(j(x)\right)$, find a formula for $j(x)$.

64. You have two money machines, both of which increase any money inserted into them. The first machine doubles your money. The second adds five dollars. The money that comes out is described by $d(x) = 2x$, in the first case, and $a(x) = x + 5$, in the second, where x is the number of dollars inserted. The machines can be hooked up so that the money coming out of one machine goes into the other. Find formulas for each of the two possible composition machines. Is one machine more profitable than the other?

65. Currency traders often move investments from one country to another in order to make a profit. Table 8.9 gives exchange rates for US dollars, Japanese yen, and the European Union's euro.[1] In March, 1999, for example, 1 US dollar purchases 84.62 Japanese yen or 1.0908 European euros. Similarly, 1 European euro purchases 77.58 Japanese yen or 0.625 US dollar. Suppose

$$f(x) = \text{Number of yen one can buy with } x \text{ dollars}$$

$$g(x) = \text{Number of euros one can buy with } x \text{ dollars}$$

$$h(x) = \text{Number of euros one can buy with } x \text{ yen}$$

(a) Find formulas for f, g, and h.

(b) Evaluate $h(f(1000))$ and interpret this in terms of currency.

TABLE 8.9 *Exchange rate for US dollars, Japanese yen and European euro, March 22, 1999*

Amount invested	Dollars purchased	Yen purchased	European euro purchased
1 dollar	1.0000	84.62	1.0908
1 yen	0.0118	1.0000	0.029
1 euro	0.9168	77.58	1.0000

8.2 INVERSE FUNCTIONS

Inverse functions were introduced in Section 3.2. In Section 4.4, we defined the logarithm as the inverse function of the exponential function. In Section 6.7, we defined the arccosine as the inverse of cosine. We now study inverse functions in general.

Definition of Inverse Function

Recall that the statement $f^{-1}(50) = 20$ means that $f(20) = 50$. In fact, the values of f^{-1} are determined in just this way. In general,

> Suppose $Q = f(t)$ is a function with the property that each value of Q determines exactly one value of t. Then f has an **inverse function**, f^{-1} and
>
> $$f^{-1}(Q) = t \quad \text{if and only if} \quad Q = f(t).$$
>
> If a function has an inverse, it is said to be **invertible**.

The definitions of the logarithm and of the inverse cosine have the same form as the definition of f^{-1}. Since $y = \log x$ is the inverse function of $y = 10^x$, we have

$$x = \log y \quad \text{if and only if} \quad y = 10^x,$$

and since $y = \cos^{-1} t$ is the inverse function of $y = \cos t$,

$$t = \cos^{-1} y \quad \text{if and only if} \quad y = \cos t.$$

Example 1 Solve the equation $\sin x = 0.8$ using an inverse function.

Solution The solution is $x = \sin^{-1}(0.8)$. A calculator (set in radians) gives $x = \sin^{-1}(0.8) \approx 0.93$.

---

[1] Bloomberg's Cross Currency Rates, 3:48 pm EST, March 22, 1999. Currency exchange rates fluctuate constantly.

Example 2   Suppose that g is an invertible function, with $g(10) = -26$ and $g^{-1}(0) = 7$. What other values of g and g^{-1} do you know?

Solution   Because $g(10) = -26$, we know that $g^{-1}(-26) = 10$; because $g^{-1}(0) = 7$, we know that $g(7) = 0$.

Example 3   A population is given by the formula $P = f(t) = 20 + 0.4t$ where P is the number of people (in thousands) and t is the number of years since 1970. Evaluate the following quantities. Explain in words what each tells you about the population.

(a) $f(25)$ (b) $f^{-1}(25)$

(c)   Show how to estimate $f^{-1}(25)$ from a graph of f.

Solution   (a)   Substituting $t = 25$, we have

$$f(25) = 20 + 0.4 \cdot 25 = 30.$$

Thus, in 1995 (year $t = 25$), we have $P = 30$, so the population was 30,000 people.

(b)   We have $t = f^{-1}(P)$. Thus, in $f^{-1}(25)$, the 25 is a population. So $f^{-1}(25)$ is the year in which the population reaches 25 thousand. We find t by solving the equation

$$20 + 0.4t = 25$$
$$0.4t = 5$$
$$t = 12.5.$$

Therefore, $f^{-1}(25) = 12.5$, which means that the population reached 25,000 people 12.5 years after 1970, or midway into 1982.

(c)   We can estimate $f^{-1}(25)$ by reading the graph of $P = f(t)$ backwards as shown in Figure 8.2.

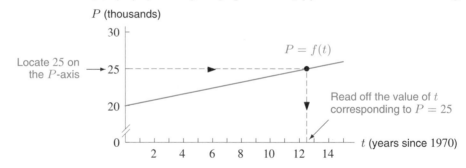

Figure 8.2: Using a graph of the function $P = f(t)$ to read off values of the inverse function $f^{-1}(P)$

Finding a Formula for an Inverse Function

It is sometimes possible to find a formula for an inverse function, f^{-1} from a formula for f. If the function $P = f(t)$ gives the population (in thousands) of a town in year t, then $f^{-1}(P)$ is the year in which the population reaches the value P. In Example 3 we found $f^{-1}(P)$ for $P = 25$. We now perform the same calculations for a general P. Since

$$P = 20 + 0.4t,$$

solving for t gives

$$0.4t = P - 20$$
$$t = \frac{P - 20}{0.4},$$

TABLE 8.10   **TABLE 8.11**

t	$P = f(t)$
0	20
5	22
10	24
15	26
20	28

P	$t = f^{-1}(P)$
20	0
22	5
24	10
26	15
28	20

so

$$f^{-1}(P) = 2.5P - 50.$$

The values in Table 8.10 were calculated using the formula $P = f(t) = 20 + 0.4t$; the values in Table 8.11 were calculated using the formula for $t = f^{-1}(P) = 2.5P - 50$. The table for f^{-1} can be obtained from the table for f by interchanging its columns, reflecting the fact that the inverse function reverses the roles of inputs and outputs.

Example 4   Suppose you deposit \$500 into a savings account that pays 4% interest compounded annually. The balance, in dollars, in the account after t years is given by $B = f(t) = 500(1.04)^t$.

(a)   Find a formula for $t = f^{-1}(B)$.

(b)   What does the inverse function represent in terms of the account?

Solution   (a)   To find a formula for f^{-1}, we solve for t in terms of B:

$$B = 500(1.04)^t$$

$$\frac{B}{500} = (1.04)^t$$

$$\log\left(\frac{B}{500}\right) = t \log 1.04 \qquad \text{Taking logs of both sides}$$

$$t = \frac{\log(B/500)}{\log 1.04}.$$

Thus, a formula for the inverse function is

$$t = f^{-1}(B) = \frac{\log(B/500)}{\log 1.04}.$$

(b)   The function $t = f^{-1}(B)$ gives the number of years for the balance to grow to \$B.

Noninvertible Functions: Horizontal Line Test

Not every function has an inverse function. A function $Q = f(t)$ has no inverse if it returns the same Q-value for two different t-values. When that happens, the value of t cannot be uniquely determined from the value of Q.

For example, if $q(x) = x^2$ then $q(-3) = 9$ and $q(+3) = 9$. This means that we can not say what the value $q^{-1}(9)$ would be. (Is it $+3$ or -3?). Thus, q is not invertible. In Figure 8.3, notice that the horizontal line $y = 9$ intersects the graph of $q(x) = x^2$ at two different points: $(-3, 9)$ and $(3, 9)$. This corresponds to the fact that the function q returns $y = 9$ for two different x-values, $x = +3$ and $x = -3$.

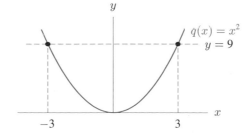

Figure 8.3: The graph of $q(x) = x^2$ fails the horizontal line test

We have the following general result:

The Horizontal Line Test   If there is a horizontal line which intersects a function's graph in more than one point, then the function does not have an inverse. If every horizontal line intersects a function's graph at most once, then the function has an inverse.

Evaluating an Inverse Function Graphically

Finding a formula for an inverse function can be difficult. However, this does not mean that the inverse function does not exist. Even without a formula, it may be possible to find values of the inverse function.

Example 5   Let $u(x) = x^3 + x + 1$. Explain why a graph suggests the function u is invertible. Assuming u has an inverse, estimate $u^{-1}(4)$.

Solution   To show that u is invertible, we could try to find a formula for u^{-1}. To do this, we would solve the equation $y = x^3 + x + 1$ for x. Unfortunately, this is difficult. However, the graph in Figure 8.4 suggests that u passes the horizontal line test and therefore that u is invertible. To estimate $u^{-1}(4)$, we find an x-value such that

$$x^3 + x + 1 = 4.$$

In Figure 8.4, the graph of $y = u(x)$ and the horizontal line $y = 4$ intersect at the point $x \approx 1.213$. Thus, tracing along the graph, we estimate $u^{-1}(4) \approx 1.213$.

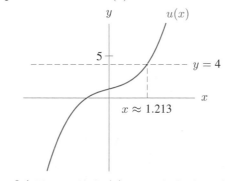

Figure 8.4: The graph of $u(x)$ passes the horizontal line test.
Since $u(1.213) \approx 4$, we have $u^{-1}(4) \approx 1.213$

In Example 5, even without a formula for u^{-1}, we can approximate $u^{-1}(a)$ for any value of a.

Example 6   Let $P(x) = 2^x$.

(a) Show that P is invertible.
(b) Find a formula for $P^{-1}(x)$.
(c) Sketch the graphs of P and P^{-1} on the same axes.
(d) What are the domain and range of P and P^{-1}?

Solution (a) Since P is an exponential function with base 2, it is always increasing, and therefore passes the horizontal line test. (See the graph of P in Figure 8.5.) Thus, P has an inverse function.

(b) To find a formula for $P^{-1}(x)$, we solve for x in the equation

$$2^x = y.$$

We can take the log of both sides to get

$$\log 2^x = \log y$$
$$x \log 2 = \log y$$
$$x = P^{-1}(y) = \frac{\log y}{\log 2}.$$

Thus, we have a formula for P^{-1} with y as the input. To graph P and P^{-1} on the same axes, we write P^{-1} as a function of x:

$$P^{-1}(x) = \frac{\log x}{\log 2} = \frac{1}{\log 2} \cdot \log x = 3.32 \log x.$$

(c) Table 8.12 gives values of $P(x)$ for $x = -3, -2, \ldots, 3$. Interchanging the columns of Table 8.12 gives Table 8.13 for $P^{-1}(x)$. We use these tables to sketch Figure 8.5.

TABLE 8.12 *Values of $P(x) = 2^x$*

x	$P(x) = 2^x$
-3	0.125
-2	0.25
-1	0.5
0	1
1	2
2	4
3	8

TABLE 8.13 *Values of $P^{-1}(x)$*

x	$P^{-1}(x)$
0.125	-3
0.25	-2
0.5	-1
1	0
2	1
4	2
8	3

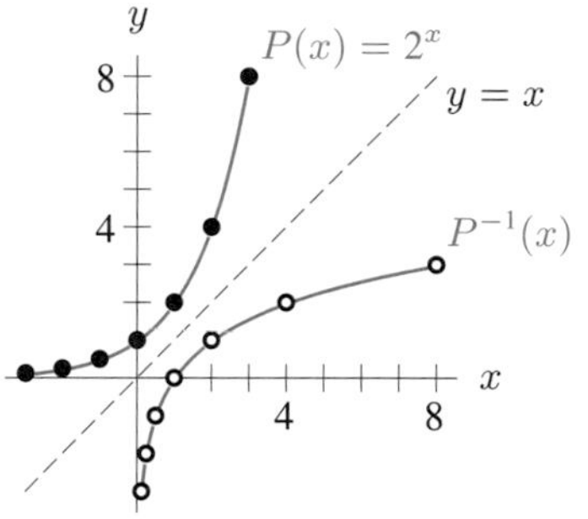

Figure 8.5: The graphs of $P(x) = 2^x$ and its inverse are symmetrical across the line $y = x$

(d) The domain of P, an exponential function, is all real numbers, and its range is all positive numbers. The domain of P^{-1}, a logarithmic function, is all positive numbers and its range is all real numbers.

The Graph, Domain, and Range of an Inverse Function

In Figure 8.5, we see that the graph of P^{-1} is the mirror image of the graph of P across the line $y = x$. To understand why this occurs, consider how a function is related to its inverse.

If f is an invertible function with, for example, $f(2) = 5$, then $f^{-1}(5) = 2$. Thus, the point $(2, 5)$ is on the graph of f and the point $(5, 2)$ is on the graph of f^{-1}. Generalizing, if (a, b) is any point on the graph of f, then (b, a) is a point on the graph of f^{-1}. Figure 8.6 shows how reflecting the point (a, b) across the line $y = x$ gives the point (b, a). Consequently, the graph of f^{-1} is the

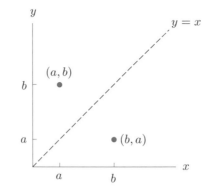

Figure 8.6: The reflection of the point (a, b) across the line $y = x$ is the point (b, a)

reflection of the graph of f across the line $y = x$.

Notice that outputs from a function are inputs to its inverse function. Similarly, outputs from the inverse function are inputs to the original function. This is expressed in the statement $f^{-1}(b) = a$ if and only if $f(a) = b$, and also in the fact that we can obtain a table for f^{-1} by interchanging the columns of a table for f. Consequently, the domain and range for f^{-1} are obtained by interchanging the domain and range of f. In other words,

$$\text{Domain of } f^{-1} = \text{Range of } f \qquad \text{and} \qquad \text{Range of } f^{-1} = \text{Domain of } f.$$

In Example 6, the function P has all real numbers as its domain and all positive numbers as its range; the function P^{-1} has all positive numbers as its domain and all real numbers as its range.

A Property of Inverse Functions

The fact that Tables 8.12 and 8.13 contain the same values, but with the columns switched, reflects the special relationship between the values of $P(x)$ and $P^{-1}(x)$. For the population function in Example 6

$$P^{-1}(2) = 1 \quad \text{and} \quad P(1) = 2 \quad \text{so} \quad P^{-1}(P(1)) = 1,$$

and

$$P^{-1}(0.25) = -2 \quad \text{and} \quad P(-2) = 0.25 \quad \text{so} \quad P^{-1}(P(-2)) = -2.$$

This result holds for any input x, so in general,

$$P^{-1}(P(x)) = x.$$

In addition, $P(P^{-1}(2)) = 2$ and $P(P^{-1}(0.25)) = 0.25$, and for any x

$$P(P^{-1}(x)) = x.$$

Similar reasoning holds for any other invertible function, suggesting the general result:

If $y = f(x)$ is an invertible function and $y = f^{-1}(x)$ is its inverse, then
- $f^{-1}(f(x)) = x$ for all values of x for which $f(x)$ is defined,
- $f(f^{-1}(x)) = x$ for all values of x for which $f^{-1}(x)$ is defined.

This property tell us that composing a function and its inverse function returns the original value as the end result. We can use this property to decide whether two functions are inverses.

Example 7   (a)   Check that $f(x) = \dfrac{x}{2x + 1}$ and $f^{-1}(x) = \dfrac{x}{1 - 2x}$ are inverse functions of one another.

(b)   Graph f and f^{-1} on the same axes. What are the domains and ranges of f and f^{-1}?

Solution   (a)   To check that these functions are inverses, we compose

$$f^{-1}(f(x)) = \frac{f(x)}{1 - 2f(x)} = \frac{\dfrac{x}{2x + 1}}{1 - 2\dfrac{x}{2x + 1}}$$

$$= \frac{\dfrac{x}{2x + 1}}{\dfrac{2x + 1}{2x + 1} - \dfrac{2x}{2x + 1}}$$

$$= \frac{\dfrac{x}{2x + 1}}{\dfrac{1}{2x + 1}}$$

$$= x.$$

Similarly, you can check that $f(f^{-1}(x)) = x$.

(b)   The graphs of f and f^{-1} in Figure 8.7 are symmetric about the line $y = x$.

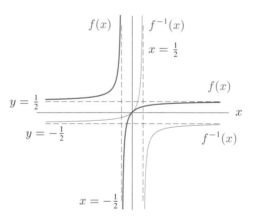

Figure 8.7: The graph of $f(x) = x/(2x + 1)$ and the inverse $f^{-1}(x) = x/(1 - 2x)$

The function $f(x) = x/(2x + 1)$ is undefined at $x = -1/2$, so its domain consists of all real numbers except $-1/2$. Figure 8.7 suggests that f has a horizontal asymptote at $y = 1/2$ which it does not cross and that its range is all real numbers except $1/2$.

Because the inverse function $f^{-1}(x) = x/(1 - 2x)$ is undefined at $x = 1/2$ its domain is all real numbers except $1/2$. Note that this is the same as the range of f. The graph of f^{-1} appears to have a horizontal asymptote which it does not cross at $y = -1/2$ suggesting that its range is all real numbers except $-1/2$. Note that this is the same as the domain of f.

The ranges of the functions f and f^{-1} can be confirmed algebraically.

Restricting the Domain

A function that fails the horizontal line test is not invertible. For this reason, the function $f(x) = x^2$ does not have an inverse function. However, by considering only part of the graph of f, we can eliminate the duplication of y-values. Suppose we consider the half of the parabola with $x \geq 0$. See Figure 8.8. This part of the graph does pass the horizontal line test because there is only one (positive) x-value for each y-value in the range of f.

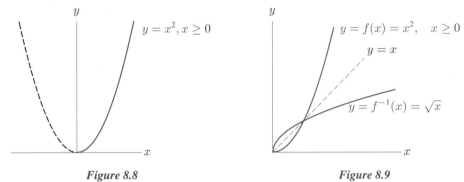

Figure 8.8   Figure 8.9

We can find an inverse for $f(x) = x^2$ on its restricted domain,[2] $x \geq 0$. Using the fact that $x \geq 0$ and solving $y = x^2$ for x gives

$$x = \sqrt{y}.$$

Thus a formula for the inverse function is

$$x = f^{-1}(y) = \sqrt{y}.$$

Rewriting the formula for f^{-1} with x as the input, we have

$$f^{-1}(x) = \sqrt{x}.$$

The graphs of f and f^{-1} are shown in Figure 8.9. Note that the domain of f is the the range of f^{-1}, and the domain of f^{-1} $(x \geq 0)$ is the range of f.

In Section 6.7 we restricted the domains of the sine, cosine, and tangent functions in order to define their inverse functions:

$$y = \sin^{-1} x \quad \text{if and only if} \quad x = \sin y \quad \text{and} \quad -\frac{\pi}{2} \leq y \leq \frac{\pi}{2}$$
$$y = \cos^{-1} x \quad \text{if and only if} \quad x = \cos y \quad \text{and} \quad 0 \leq y \leq \pi$$
$$y = \tan^{-1} x \quad \text{if and only if} \quad x = \tan y \quad \text{and} \quad -\frac{\pi}{2} < y < \frac{\pi}{2}.$$

The graphs of each of the inverse trigonometric functions are shown in Figures 8.10-8.12. Note the symmetry about the line $y = x$ for each trigonometric function and its inverse.

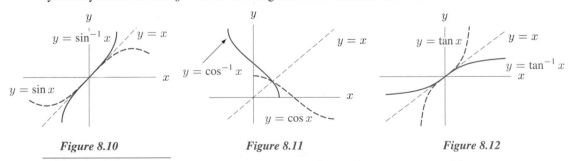

Figure 8.10   Figure 8.11   Figure 8.12

[2]Technically, changing the domain results in a new function, but we will continue to call it $f(x)$.

In each case, the restricted domain of the function is the range of the inverse function. In addition, the domain of the inverse is the range of the original function. For example

$$y = \sin x \quad \text{has restricted domain} \quad -\frac{\pi}{2} \leq x \leq \frac{\pi}{2} \quad \text{and} \quad \text{range} \quad -1 \leq y \leq 1$$

$$y = \sin^{-1} x \quad \text{has domain} \quad -1 \leq x \leq 1 \quad \text{and} \quad \text{range} \quad -\frac{\pi}{2} \leq y \leq \frac{\pi}{2}.$$

Problems for Section 8.2

1. Let $R = f(T) = 150 + 5T$ give the resistance of a circuit element as a function of temperature.
    (a) Find a formula for $T = f^{-1}(R)$.
    (b) Make a table of values showing values of f. Make another table showing values of f^{-1}. Explain the relationship between the tables.

In Problems 2–7, use a graph to decide whether or not the function is invertible.

2. $y = x^4 - 6x + 12$ 3. $y = x^3 + 7$ 4. $y = x^6 + 2x^2 - 10$
5. $y = x^5$ 6. $y = \sqrt{x} + x$ 7. $y = |x|$

8. Graph the population function $P = f(t) = 20 + 0.4t$ and its inverse, $t = f^{-1}(P) = 2.5P - 50$ on the same axes. Geometrically, what is the relationship between the graphs? Explain why this occurs.

9. Find the inverse of $f(x) = x^3$ and sketch a graph of f and f^{-1} on the same axes.

10. Let f and g be defined by the values in Table 8.14. Based on this table, answer the following questions:
    (a) Is $f(x)$ invertible? If not, explain why; if so, construct a table of values of $f^{-1}(x)$ for all values of x for which $f^{-1}(x)$ is defined.
    (b) Answer the same question as in part (a) for $g(x)$.
    (c) Make a table of values for $h(x) = f(g(x))$, with $x = -3, -2, -1, 0, 1, 2, 3$.
    (d) Explain why you cannot define a function $j(x)$ by the formula $j(x) = g(f(x))$.

TABLE 8.14

x	-3	-2	-1	0	1	2	3
$f(x)$	9	7	6	-4	-5	-8	-9
$g(x)$	3	1	3	2	-3	-1	3

11. Suppose that $f(x)$ is an invertible function and that both f and f^{-1} are defined for all values of x. Suppose also that $f(2) = 3$ and $f^{-1}(5) = 4$. Evaluate each of the following expressions, or, if the given information is insufficient, write unknown.   (a) $f^{-1}(3)$ (b) $f^{-1}(4)$ (c) $f(4)$

12. Suppose that $j(x) = h^{-1}(x)$ and that both j and h are defined for all values of x. Suppose also that $h(4) = 2$ and $j(5) = -3$. Evaluate the following expressions if possible.
    (a) $j(h(4))$ (b) $j(4)$ (c) $h(j(4))$ (d) $j(2)$ (e) $h^{-1}(-3)$
    (f) $j^{-1}(-3)$ (g) $h(5)$ (h) $\left(h(-3)\right)^{-1}$ (i) $\left(h(2)\right)^{-1}$

13. Figure 8.13 defines the function f. Rank the following quantities in order from least to greatest: $0, f(0), f^{-1}(0), 3, f(3), f^{-1}(3)$.

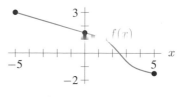

Figure 8.13

14. Check that the functions $R = f(T) = 150 + 5T$ and $T = f^{-1}(R) = \frac{1}{5}R - 30$, satisfy the identities

$$f^{-1}(f(T)) = T \quad \text{and} \quad f(f^{-1}(R)) = R.$$

In Problems 15–18, check that each pair of functions are inverses.

15. $h(x) = \sqrt{2x}$ and $k(t) = \dfrac{t^2}{2}$, for $x, t \geq 0$

16. $f(x) = \dfrac{x}{4} - \dfrac{3}{2}$ and $g(t) = 4\left(t + \dfrac{3}{2}\right)$

17. $f(x) = 1 + 7x^3$ and $f^{-1}(x) = \sqrt[3]{\dfrac{x - 1}{7}}$

18. $g(x) = 1 - \dfrac{1}{x - 1}$ and $g^{-1}(x) = 1 + \dfrac{1}{1 - x}$

19. Find a formula for $g^{-1}(x)$, the inverse function for $g(x) = 2x + 5$.

Find the inverses of the functions in Problems 20–40.

20. $f(x) = -2x - 7$

21. $h(x) = 12x^3$

22. $g(x) = \dfrac{1}{x - 3}$

23. $k(x) = \dfrac{x + 2}{x - 2}$

24. $h(x) = \dfrac{x}{2x + 1}$

25. $f(x) = 10^x$

26. $g(x) = e^x$

27. $k(x) = 3 \cdot e^{2x}$

28. $g(x) = e^{3x+1}$

29. $n(x) = \log(x - 3)$

30. $h(x) = \ln(1 - 2x)$

31. $j(x) = \log(x - 1) + 2$

32. $h(x) = \dfrac{\sqrt{x}}{\sqrt{x} + 1}$

33. $f(x) = \dfrac{3 + 2x}{2 - 5x}$

34. $f(x) = \sqrt{\dfrac{4 - 7x}{4 - x}}$

35. $f(x) = \dfrac{1}{9 - \sqrt{x - 4}}$

36. $f(x) = \dfrac{\sqrt{x} + 3}{11 - \sqrt{x}}$

37. $m(x) = \sqrt{\dfrac{x + 1}{x}}, x > 0$

38. $p(x) = 2\ln\left(\dfrac{1}{x}\right)$

39. $q(x) = \ln(x + 3) - \ln(x - 5)$

40. $s(x) = \dfrac{3}{2 + \log x}$

41. (a) Explain how to find the formula for an inverse function from the formula for the original function. Illustrate your explanation with an example.
     (b) Explain how to obtain the graph of the inverse function from the graph of the original function. Illustrate your explanation with an example.
     (c) Describe what kind of symmetry is found between the graph of a function and the graph of its inverse. Illustrate your explanation with at least one example.

42. The function $y = \sin t$ is not invertible because it fails the horizontal line test.
     (a) Explain in your own words the way that the domain of $\sin t$ is restricted in order to define the inverse function, $\sin^{-1} y$.
     (b) Find some other restriction of the domain and explain why it would be a valid one to use.

Solve the equations in Problems 43–48 exactly. Use an inverse function when appropriate.

43. $7\sin(3x) = 2$

44. $2^{x+5} = 3$

45. $x^{1.05} = 1.09$

46. $\ln(x + 3) = 1.8$

47. $\dfrac{2x + 3}{x + 3} = 8$

48. $\sqrt{x + \sqrt{x}} = 3$

49. (a) What is the formula for the area of a circle in terms of its radius?
     (b) Graph this function for the domain all real numbers.
     (c) What is the domain that actually applies in this situation? On separate axes, graph the function for this domain.
     (d) Find a formula for the inverse of the function in part (c).
     (e) Graph the inverse function on the domain you gave in part (c) on the same axes used in part (c).
     (f) If area is a function of the radius, is radius a function of area? Explain carefully.

50. Let $p(x) = 3\pi x^3$ and $q(x) = \dfrac{x^2}{2\pi}$. Suppose $A = p(x)$ and $B = q(x)$. Express B as a function of A.

51. Let $y = P(t)$ give the population of a town (in thousands) t years after 1980. Suppose that P is a linear function and that the population was 18,000 in 1985 and 21,000 in 1989.

    (a) Find a formula for $P(t)$.
    (b) Evaluate and interpret $P(20)$ and $P(-10)$ using the formula from part (a).
    (c) Interpret the slope and the intercepts of P in terms of population.
    (d) Find a formula for $P^{-1}(y)$.
    (e) Evaluate and interpret the quantities $P^{-1}(20)$ and $P^{-1}(5)$.
    (f) Interpret the slope of P^{-1} in terms of population.

52. A company believes there is a linear relationship between the consumer demand for its products and the price charged. When the price was \$3 per unit, the quantity demanded was 500 units per week. When the unit price was raised to \$4, the quantity demanded dropped to 300 units per week. Define $D(p)$ as the quantity per week demanded by consumers at a unit price of \$p. Assume that $D(p)$ is linear.

    (a) Estimate and interpret $D(5)$.
    (b) Find a formula for $D(p)$ in terms of p.
    (c) Calculate and interpret $D^{-1}(5)$.
    (d) Give an interpretation of the slope of $D(p)$ in terms of demand.
    (e) Currently, the company can produce 400 units every week. What should the price of the product be if the company wants to sell all 400 units?
    (f) If the company produced 500 units per week instead of 400 units per week, would its weekly revenues increase, and if so, by how much?

53. Let $P = f(t) = 37.8(1.044)^t$ be the population of a town (in thousands) in year t.

    (a) Describe the town's population in words.
    (b) Evaluate $f(50)$. What does this quantity tell you about the population?
    (c) Find a formula for $f^{-1}(P)$ in terms of P.
    (d) Evaluate $f^{-1}(50)$. What does this quantity tell you about the population?

54. Table 8.15 gives the number of cows in a herd.

    (a) Find an exponential function that approximates the data.
    (b) Find the inverse function of the function in part (a).
    (c) When do you predict that the herd will contain 400 cows?

    TABLE 8.15

t (years)	0	1	2
$P(t)$ (cows)	150	165	182

55. Suppose $P = f(t)$ is the population (in thousands) in year t, and that $f(7) = 13$ and $f(12) = 20$,

    (a) Find a formula for $f(t)$ assuming f is exponential.
    (b) Find a formula for $f^{-1}(P)$.
    (c) Evaluate $f(25)$ and $f^{-1}(25)$. Explain what these expressions mean in terms of population.

56. A gymnast at Ringling Brothers, Barnum, & Bailey Circus is fired straight up in the air from a cannon. While she is in the air, a trampoline is moved into the spot where the cannon was. Figure 8.14 is a graph of the gymnast's height h as a function of time t.

    (a) Approximately what is her maximum height?
    (b) Approximately when does she land on the trampoline?
    (c) Take the function $h(t)$ and restrict its domain so that it has an inverse which is a function. That is, pick a piece of the graph which does have an inverse. Graph this new restricted function.
    (d) Change the story to go with your new graph in part (c).
    (e) Graph the inverse of the function you graphed in part (c). Explain in your story why it makes sense

that the inverse is a function.

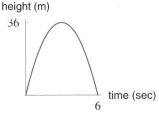

Figure 8.14

57. A 100 ml solution contains 99% alcohol and 1% water. Define $y = C(x)$ to be the concentration of alcohol in the solution after x ml of alcohol are removed, so

$$C(x) = \frac{\text{Amount of alcohol}}{\text{Amount of solution}}.$$

(a) What is $C(0)$?
(b) Find a formula in terms of x for $C(x)$.
(c) Find a formula in terms of y for $C^{-1}(y)$.
(d) Explain the physical significance of $C^{-1}(y)$.

58. Suppose that in Problem 57 you need an alcohol solution that is 98% alcohol, instead of 99%.

(a) How much alcohol do you think should be removed from the 99% solution described in Problem 57 in order to obtain a 98% solution? (Make a guess.)
(b) Express the exact answer to part (a) using the function C^{-1}.
(c) Determine the exact answer to part (a). Are you surprised by your result?

59. (a) Using the data in Table 8.16, find a formula $f(x)$ that expresses the amount of taxes owed as a function of taxable income. [Hint: Use a piecewise defined function.]
(b) What is the domain and range of $f(x)$?
(c) Graph $f(x)$.
(d) Is $f(x)$ invertible? How do you know?

TABLE 8.16   *Federal taxes for a single person, 1999*

Taxable income	Taxes owed
$0–$25,750	15% of the taxable income
$25,750–$62,450	$3,862.50 + 28% of the excess over $25,750
$62,450–$130,250	$14,138.50 + 31% of the excess over $62,450
$130,250–$283,150	$35,156.50 + 36% of the excess over $130,250
Over $283,150	$90,200.50 + 39.6% of the excess over $283,150

60. Use the data given in Problem 59 and the following definition of average tax rate:

$$\text{Average tax rate} = \frac{\text{Taxes owed}}{\text{Annual taxable income}}.$$

(a) Find the average tax rate for a person with a taxable income of $30,000.
(b) The average tax rate can be interpreted as the slope of the line through $(0, 0)$ and $(x, f(x))$. Sketch this line for the person with a taxable income of $30,000.
(c) From the shape of your graph of $f(x)$, do you expect the average tax rate for $60,000 to be larger or smaller than the average tax rate for $30,000?
(d) Using your graph, if the average tax rate is 20%, estimate the annual taxable income and the amount of taxes owed.
(e) Is the average tax rate function invertible?

61. Suppose that f, g, and h are all invertible functions, and that

$$f(x) = g(h(x)).$$

Find a formula for $f^{-1}(x)$ in terms of $g^{-1}(x)$ and $h^{-1}(x)$.

62. Find the inverse of each of the following functions. You need not state the domains of the inverses, and you may assume that the given functions are defined on domains on which they are invertible.

(a) $f(x) = \arcsin\left(\dfrac{3x}{2-x}\right)$ (b) $g(x) = \ln(\sin x) - \ln(\cos x)$ (c) $h(x) = \cos^2 x + 2\cos x + 1$

63. Simplify the expression $\cos^2(\arcsin t)$, using the property that inverses "undo" each other.

8.3   COMBINATIONS OF FUNCTIONS

Like numbers, functions can be combined using addition, subtraction, multiplication, and division.

The Difference of Two Functions Defined by Formulas: A Measure of Prosperity

We can define new functions as the sum or difference of two functions. In Chapter 4, we discussed Thomas Malthus, who predicted widespread food shortages because he believed that human populations increase exponentially, whereas the food supply increases linearly. We considered a country with population $P(t)$ million in year t. The population is initially 2 million and grows at the rate of 4% per year, so

$$P(t) = 2(1.04)^t.$$

Let $N(t)$ be the number of people (in millions) that the country can feed in year t. The annual food supply is initially adequate for 4 million people and it increases by enough for an additional 0.5 million people every year. Thus

$$N(t) = 4 + 0.5t.$$

This country first experiences shortages in about 78 years. (See Figure 8.15.) When is it most prosperous?

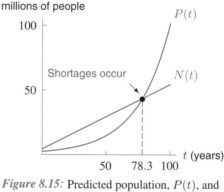

Figure 8.15: Predicted population, $P(t)$, and number of people who can be fed, $N(t)$, over a 100-year period

The answer depends on how we decide to measure prosperity. We could measure prosperity in one of the following ways:

- By the food surplus—that is, the amount of food the country has over and above its needs. This surplus food could be warehoused or exported in trade.

- By the per capita food supply—that is, how much food there is per person. (The term *per capita* means per person, or literally, "per head.") This indicates the portion of the country's wealth each person might enjoy.

First, we choose to measure prosperity in terms of food surplus, $S(t)$, in year t, where

$$S(t) = \underbrace{\text{Number of people that can be fed}}_{N(t)} - \underbrace{\text{Number of people living in the country}}_{P(t)}$$

so

$$S(t) = N(t) - P(t).$$

For example, to determine the surplus in year $t = 25$, we evaluate

$$S(25) = N(25) - P(25).$$

Since $N(25) = 4 + 0.5(25) = 16.5$ and $P(25) = 2(1.04)^{25} \approx 5.3$, we have

$$S(25) \approx 16.5 - 5.3 = 11.2.$$

Thus, in year 25 the food surplus could feed 11.2 million additional people.

We use the formulas for N and P to find a formula for S:

$$S(t) = \underbrace{N(t)}_{4+0.5t} - \underbrace{P(t)}_{2(1.04)^t} ,$$

so

$$S(t) = 4 + 0.5t - 2(1.04)^t.$$

A graph of S is shown in Figure 8.16. The maximum surplus occurs sometime during the 48th year. In that year, there is surplus food sufficient for an additional 14.9 million people.

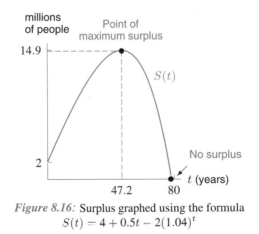

Figure 8.16: Surplus graphed using the formula
$S(t) = 4 + 0.5t - 2(1.04)^t$

The Sum and Difference of Two Functions Defined by Graphs

How does the graph of the surplus function S, shown in Figure 8.16, relate to the graphs of N and P in Figure 8.15? Since

$$S(t) = N(t) - P(t),$$

the value of $S(t)$ is represented graphically as the vertical distance between the graphs of $N(t)$ and $P(t)$. See Figure 8.17. Figure 8.18 shows the surplus plotted against time.

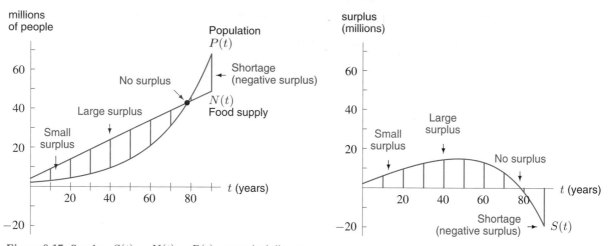

Figure 8.17: Surplus, $S(t) = N(t) - P(t)$, as vertical distance between $N(t)$ and $P(t)$ graphs

Figure 8.18: Surplus as a function of time

From year $t = 0$ to $t \approx 78.3$, the food supply is more than the population needs. Therefore the surplus, $S(t)$, is positive on this time interval. At time $t = 78.3$, the food supply is exactly sufficient for the population, so $S(t) = 0$, resulting in the horizontal intercept $t = 78.3$ on the graph of $S(t)$ in Figure 8.18. For times $t > 78.3$, the food supply is less than the population needs. Therefore the surplus is negative, representing a food shortage.

In the next example we consider a sum of two functions.

Example 1   Let $f(x) = x$ and $g(x) = \dfrac{1}{x}$. By adding vertical distances on the graphs of f and g, sketch

$$h(x) = f(x) + g(x) \quad \text{for } x > 0.$$

Solution   The graphs of f and g are shown in Figure 8.19. For each value of x, we add the vertical distances that represent $f(x)$ and $g(x)$ to get a point on the graph of $h(x)$. Compare the graph of $h(x)$ to the values shown Table 8.17.

TABLE 8.17   *Adding function values*

x	$\frac{1}{4}$	$\frac{1}{2}$	1	2	4
$f(x) = x$	$\frac{1}{4}$	$\frac{1}{2}$	1	2	4
$g(x) = 1/x$	4	2	1	$\frac{1}{2}$	$\frac{1}{4}$
$h(x) = f(x) + g(x)$	$4\frac{1}{4}$	$2\frac{1}{2}$	2	$2\frac{1}{2}$	$4\frac{1}{4}$

Note that as x increases, $g(x)$ decreases towards zero, so the values of $h(x)$ get closer to the values of $f(x)$. On the other hand, as x approaches zero, $h(x)$ gets closer to $g(x)$.

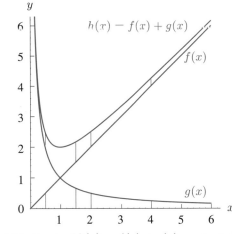

Figure 8.19: Graph of $h(x) = f(x) + g(x)$ constructed by adding vertical distances under f and g

Factoring a Function's Formula into a Product

It is often useful to be able to express a given function as a product of functions.

Example 2   Find exactly all the zeros of the function

$$p(x) = 2^x \cdot 6x^2 - 2^x \cdot x - 2^{x+1}.$$

Solution   We could approximate the zeros by finding the points where the graph of the function p crosses the x-axis. Unfortunately, these solutions are not exact. Alternatively, we can express p as a product. Using the fact that

$$2^{x+1} = 2^x \cdot 2^1 = 2 \cdot 2^x,$$

we rewrite the formula for p as

$$p(x) = 2^x \cdot 6x^2 - 2^x x - 2 \cdot 2^x$$
$$= 2^x(6x^2 - x - 2) \qquad \text{Factoring out } 2^x$$
$$= 2^x(2x + 1)(3x - 2) \qquad \text{Factoring the quadratic.}$$

Thus p is the product of the exponential function 2^x and two linear functions. Since p is a product, it equals zero if one or more of its factors equals zero. But 2^x is never equal to 0, so $p(x)$ equals zero if and only if one of the linear factors is zero:

$$(2x + 1) = 0 \qquad \text{or} \qquad (3x - 2) = 0$$
$$x = -\frac{1}{2} \qquad\qquad\qquad x = \frac{2}{3}.$$

The Quotient of Functions Defined by Formulas and Graphs: Prosperity

Now let's think about our second proposed measure of prosperity, the per capita food supply, $R(t)$. With this definition of prosperity

$$R(t) = \frac{\text{Number of people that can be fed}}{\text{Number of people living in the country}} = \frac{N(t)}{P(t)}.$$

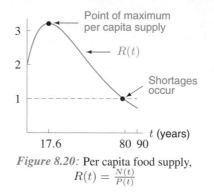

Figure 8.20: Per capita food supply,
$$R(t) = \frac{N(t)}{P(t)}$$

For example,

$$R(25) = \frac{N(25)}{P(25)} = \frac{16.5}{5.3} \approx 3.1.$$

This means that in year 25, everybody in the country could, on average, have more than three times as much food as he or she needs. The formula for $R(t)$ is

$$R(t) = \frac{N(t)}{P(t)} = \frac{4 + 0.5t}{2(1.04)^t}.$$

From the graph of R in Figure 8.20, we see that the maximum per capita food supply occurs during the 18th year. Notice this maximum prosperity prediction is different from the one made using the surplus function $S(t)$.

However, both prosperity models predict that shortages begin after time $t = 78.3$. This is not a coincidence. The food surplus model predicts shortages when $S(t) = N(t) - P(t) < 0$, or $N(t) < P(t)$. The per capita food supply model predicts shortages when $R(t) < 1$, meaning that the amount of food available per person is less than the amount necessary to feed 1 person. Since $R(t) = \frac{N(t)}{P(t)} < 1$ is true only when $N(t) < P(t)$, the same condition leads to shortages.

The Quotient of Functions Defined by Tables: Per Capita Crime Rate

Table 8.18 gives the number of violent crimes committed in two cities between 1994 and 1999. It appears that crime in both cities is on the rise and that there is less crime in City B than in City A.

TABLE 8.18   *Number of violent crimes committed each year in two cities*

Year	1994	1995	1996	1997	1998	1999
t, years since 1994	0	1	2	3	4	5
Crimes in City A	793	795	807	818	825	831
Crimes in City B	448	500	525	566	593	652

Table 8.19 gives the population for these two cities from 1994 to 1999. The population of City A is larger than that of City B and both cities are growing.

TABLE 8.19   *Population of the two cities*

Year	1994	1995	1996	1997	1998	1999
t, years since 1994	0	1	2	3	4	5
Population of City A	61,000	62,100	63,220	64,350	65,510	66,690
Population of City B	28,000	28,588	29,188	29,801	30,427	31,066

Can we attribute the growth in crime in both cities to the population growth? Can we attribute the larger number of crimes in City A to its larger population? To answer these questions, we consider the per capita crime rate in each city.

Let's define $N_A(t)$ to be the number of crimes in City A during year t (where $t = 0$ means 1994). Similarly, let's define $P_A(t)$ to be the population of City A in year t. Then the per capita crime rate in City A, $r_A(t)$, is given by

$$r_A(t) = \frac{\text{Number of crimes in year } t}{\text{Number of people in year } t} = \frac{N_A(t)}{P_A(t)}.$$

We have defined a new function, $r_A(t)$, as the quotient of $N_A(t)$ and $P_A(t)$. For example, the data in Tables 8.18 and 8.19 shows that the per capita crime rate for City A in year $t = 0$ is

$$r_A(0) = \frac{N_A(0)}{P_A(0)} = \frac{793}{61,000} = 1.30\%.$$

Similarly, the per capita crime rate for the year $t = 1$ is

$$r_A(1) = \frac{N_A(1)}{P_A(1)} = \frac{795}{62,100} = 1.28\%.$$

Thus, the per capita crime rate in City A actually decreased from 1.30% in 1994 to 1.28% in 1995.

Example 3   (a)   Make a table of values for $r_A(t)$ and $r_B(t)$, the per capita crime rates of Cities A and B.
(b)   Use the table to decide which city is more dangerous.

Solution   (a)   Table 8.20 gives values of $r_A(t)$ for $t = 0, 1, \ldots, 5$. The per capita crime rate in City A declined between 1994 and 1999 despite the fact that the total number of crimes rose during this period. Table 8.20 also gives values of $r_B(t)$, the per capita crime rate of City B, defined by

$$r_B(t) = \frac{N_B(t)}{P_B(t)},$$

where $N_B(t)$ is the number of crimes in City B in year t and $P_B(t)$ is the population of City B in year t. For example, the per capita crime rate in City B in year $t = 0$ is

$$r_B(0) = \frac{N_B(0)}{P_B(0)} = \frac{448}{28,000} = 1.6\%.$$

TABLE 8.20   *Values of $r_A(t)$ and $r_B(t)$, the per capita violent crime rates of Cities A and B*

Year	1994	1995	1996	1997	1998	1999
t, years since 1994	0	1	2	3	4	5
$r_A(t) = N_A(t)/P_A(t)$	1.300%	1.280%	1.276%	1.271%	1.259%	1.246%
$r_B(t) = N_B(t)/P_B(t)$	1.600%	1.749%	1.799%	1.899%	1.949%	2.099%

(b)   From Table 8.20, we see that between 1994 and 1999, City A has a lower per capita crime rate than City B. The crime rate of City A is decreasing, whereas the crime rate of City B is increasing. Thus, even though Table 8.18 indicates that there are more crimes committed in City A, Table 8.20 tells us that City B is, in some sense, more dangerous. Table 8.20 also tells us that, even though the number of crimes is rising in both cities, City A is getting safer, while City B is getting more dangerous.

Problems for Section 8.3

1. Let $f(x)$ and $g(x)$ be defined by Table 8.21. Make tables of values for $x = -1, 0, 1, 2, 3, 4$ for the following functions.

   (a) $h(x) = f(x) + g(x)$ (b) $j(x) = 2f(x)$

   (c) $k(x) = (g(x))^2$ (d) $m(x) = g(x)/f(x)$

   TABLE 8.21

x	-1	0	1	2	3	4
$f(x)$	-4	-1	2	5	8	11
$g(x)$	4	1	0	1	4	9

2. Let $r(x)$, $s(x)$, and $t(x)$ be defined by Table 8.22. Make tables of values for the following functions.

   (a) $f(x) = r(x) + t(x)$ (b) $g(x) = 4 - 2s(x)$ (c) $h(x) = r(x)t(x)$

   (d) $j(x) = \dfrac{r(x) - t(x)}{s(x)}$ (e) $k(x) = r(x)^2$ (f) $l(x) = r(x) + s(x)t(x)$

   TABLE 8.22

x	-2	-1	0	1	2	3
$r(x)$	4	5	6	7	8	9
$s(x)$	-2	2	-2	2	-2	2
$t(x)$	8	5	7	-3	2	13

3. Use the graph of g in Figure 8.21 to graph the following functions.

   (a) $y = g(x) - 3$ (b) $y = g(x) + x$

   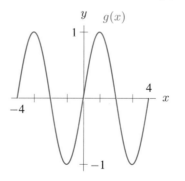

   Figure 8.21

4. Let the functions f and g be defined as in Table 8.23. Make tables of values for the following functions.

   (a) $n(x) = f(x) + g(x)$ (b) $p(x) = 2f(x)g(x) - f(x)$ (c) $q(x) = g(x)/f(x)$

   TABLE 8.23

x	1	2	3	4
$f(x)$	3	4	1	2
$g(x)$	2	1	4	3

5. Let $f(x) = x + 1$ and $g(x) = x^2 - 1$. In parts (a)–(e), write a formula in terms of $f(x)$ and $g(x)$ for the function. Then evaluate the formula for $x = 3$. Write a formula in terms of x for each function. Check your formulas for $x - 3$.

    (a) $h(x)$ is the sum of $f(x)$ and $g(x)$.
    (b) $j(x)$ is the difference between $g(x)$ and two times $f(x)$.
    (c) $k(x)$ is the product of $f(x)$ and $g(x)$.
    (d) $m(x)$ is the ratio of $g(x)$ to $f(x)$.
    (e) $n(x)$ is defined by the equation $n(x) = (f(x))^2 - g(x)$.

In Problems 6–11, find a simplified formula for each function, using
$$u(x) = 2x - 1, \quad v(x) = 1 - x, \quad \text{and} \quad w(x) = \frac{1}{x}.$$

6. $f(x) = u(x) + v(x)$

7. $g(x) = v(x)w(x)$

8. $h(x) = 2u(x) - 3v(x)$

9. $j(x) = \dfrac{u(x)}{w(x)}$

10. $k(x) = v(x)^2$

11. $l(x) = u(x) - v(x) - w(x)$

12. Let $f(t)$ be the number of males and $g(t)$ be the number of females in Canada in year t. Let $h(t)$ be the average income, in Canadian dollars, of females in Canada in year t.

     (a) Find the function $p(t)$ which gives the number of people in Canada in year t.
     (b) Find the total amount of money $m(t)$ earned by Canadian females in year t.

13. Table 8.24 gives $N(t)$, the number of existing warheads, and $P(t)$, the world's population, both as functions of t, the time elapsed in years since 1980.

TABLE 8.24

t, in years	$N(t)$, warheads	$P(t)$, population
0	30,000	
5		4,500,000,000
10	21,000	
15		5,695,300,000

     (a) Assume that $P(t)$ is an exponential function. Based on Table 8.24, find a formula for $P(t)$.
     (b) According to your formula for $P(t)$, by what percent does the population change each year?
     (c) Assume that $N(t)$ is a linear function of time. Based on Table 8.24, find a formula for $N(t)$.
     (d) Construct a table of values for $t = 0, 5, 10, 15$ for the function f defined by the formula:
     $$f(t) = \frac{N(t)}{P(t)}.$$

     (e) Is f an exponential function, a linear function, both, or neither? Justify your answer.
     (f) In practical terms, what does the function $f(t)$ represent?

14. (a) Graph the functions $f(x) = (x - 4)^2 - 2$ and $g(x) = -(x - 2)^2 + 8$ on the same set of axes.
     (b) Make a table of values for each function for $x = 0, 1, 2, ..., 6$.
     (c) Make a table of values for the function $y = f(x) - g(x)$ for $x = 0, 1, 2, ..., 6$.
     (d) On your graph, sketch the vertical line segment of length $f(x) - g(x)$ for each integer value of x from 0 to 6. Check that the segment lengths agree with the table values from part (c).
     (e) Plot the values from your table for the function $y = f(x) - g(x)$ on your graph.
     (f) Simplify the algebraic formulas for $f(x)$ and $g(x)$ given in part (a). Find a formula for the function $y - f(x) - g(x)$.
     (g) Use your answer to part (f) to graph the function $y = f(x) - g(x)$ on the same axes as graphs. Does the graph pass through the points you plotted in part (e)?

15. Graphs of $f(x)$ and $g(x)$ are given in Figure 8.22. Sketch a graph of $h(x) = g(x) - f(x)$. On the graph of $h(x)$, label the points whose x-coordinates are $x = a$, $x = b$, and $x = c$. Label the y-intercept.

Figure 8.22   Figure 8.23

16. The graphs of two functions, $f(x)$ and $g(x)$, are shown in Figure 8.23.

    (a) Find possible formulas for the functions.
    (b) Let $h(x) = f(x) \cdot g(x)$. Graph $f(x)$, $g(x)$ and $h(x)$ on the same set of axes.

17. Sketch two linear functions whose product is the function f graphed in Figure 8.24(a). Explain why this is not possible for the function q graphed in Figure 8.24(b).

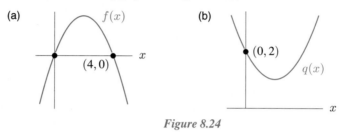

Figure 8.24

18. An average of 50,000 people visit Riverside Park each day in the summer. The park charges $15.00 for admission. Consultants predict that for each $1.00 increase in the entrance price, the park would lose an average of 2500 customers per day. Express the daily revenue from ticket sales as a function of the number of $1.00 price increases. What ticket price maximizes the revenue from ticket sales?

19. Figure 8.25 shows the graphs of two functions, $a(x)$ and $b(x)$. Let $c(x) = a(x) \cdot b(x)$. Sketch a rough graph of $c(x)$ on the same axes as $a(x)$ and $b(x)$. [Hint: There is not enough information to determine formulas for these functions but you can use the method of Problem 16.]

Figure 8.25   Figure 8.26

20. In Figure 8.26, the line l_2 is fixed and the point P moves along l_2. Define $f(\theta)$ as the y-coordinate of P.

    (a) Find a formula for $f(\theta)$ if $0 < \theta < \pi/2$. [Hint: Use the equation for l_2.]
    (b) Graph $y = f(\theta)$ on the interval $-\pi \leq \theta \leq \pi$.
    (c) How does the y-coordinate of P change as θ changes? Is $y = f(\theta)$ periodic?

21. Is the following statement true or false? If $f(x) \cdot g(x)$ is an odd function, then both $f(x)$ and $g(x)$ are odd functions. Explain your answer.

22. (a) Is the sum of two even functions even, odd, or neither? Justify your answer.
    (b) Is the sum of two odd functions even, odd, or neither? Justify your answer.
    (c) Is the sum of an even and an odd functions even, odd, or neither? Justify your answer.

23. Graph $y = \sec\theta$, $y = \csc\theta$ and $y = \cot\theta$ for $-\pi \leq \theta \leq \pi$. Describe and compare the graphs.

24. Describe the similarities and differences between the graphs of $y = \sin(1/x)$ and $y = 1/\sin x$.

25. Let $f(x) = kx^2 + B$ and $g(x) = C^{2x}$ and
$$h(x) = kx^2 C^{2x} + BC^{2x} + C^{2x}.$$
Suppose $f(3) = 7$ and $g(3) = 5$. Evaluate $h(3)$.

REVIEW PROBLEMS FOR CHAPTER EIGHT

1. Find formulas, in terms of x, for the following functions, given that
$$f(x) = x^2 + x, \qquad g(x) = 2x - 3 \qquad \text{and} \qquad h(x) = \frac{x}{1 - x}.$$
   (a) $f(2x)$ (b) $g(x^2)$ (c) $h(1 - x)$ (d) $(f(x))^2$
   (e) $g^{-1}(x)$ (f) $(h(x))^{-1}$ (g) $f(x) \cdot g(x)$ (h) $h(f(x))$

2. Using Tables 8.25 and 8.26, evaluate and interpret the following expressions:
   (a) $g(f(23))$ (b) $f(g(5))$

TABLE 8.25   *The temperature Celsius, $y = f(x)$, as a function of the temperature Fahrenheit, x*

Temperature in °F, x	-4	5	14	23	32	41	50	59	68	77	86
Temperature in °C, y	-20	-15	-10	-5	0	5	10	15	20	25	30

TABLE 8.26   *The temperature Fahrenheit, $y = g(x)$, as a function of the temperature Celsius, x*

Temperature in °C, x	-20	-15	-10	-5	0	5	10	15	20	25	30
Temperature in °F, y	-4	5	14	23	32	41	50	59	68	77	86

3. The graph of $y = f(x)$ is given in Figure 8.27. Graph $y = f(x) + f(-x)$. Label the y-intercept.

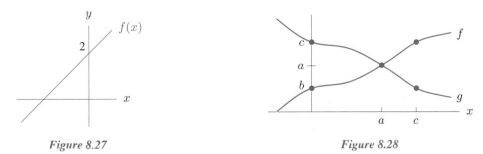

Figure 8.27   *Figure 8.28*

4. The graphs of f and g are given in Figure 8.28.
   (a) Evaluate $f(g(a))$.
   (b) Evaluate $g(f(c))$.
   (c) Evaluate $f^{-1}(b) - g^{-1}(b)$.
   (d) For what positive value(s) of x is $f(x) \leq g(x)$?

5. A research facility on the Isle of Shoals has a limited quantity (800 gallons) of fresh water which it must conserve over a two-month period.

  (a) There are 7 members of the research team and each is allotted 2 gallons of water per day for cooking and drinking. Find a formula for $f(t)$, the amount of fresh water left on the island after t days has elapsed.

  (b) Evaluate and interpret the following expressions

    (i) $f(0)$ (ii) $f^{-1}(0)$ (iii) t if $f(t) = \dfrac{1}{2}f(0)$ (iv) $800 - f(t)$

6. A company finds that there is a linear relationship between the number of units of its product it sells and the amount of money that it spends on advertising. If the company spends \$25,000 on advertising, it sells 400 units, and for each \$5,000 more or less spent, it sells 20 units more or less, respectively. Let $N(x)$ be the number of units sold as a function of the amount spent on advertising, x.

  (a) Calculate and interpret $N(20,000)$.
  (b) Find a formula for $N(x)$ in terms of x.
  (c) Give interpretations of the slope and the x- and y-intercepts of $N(x)$ if possible.
  (d) Calculate and interpret $N^{-1}(500)$.
  (e) An internal audit reveals that the profit made by the company on the sale of 10 units of its product, before advertising costs have been accounted for, is \$2,000. What are the implications regarding the company's advertising campaign? Discuss.

7. The population of a rapidly growing town triples every seven years.

  (a) If the initial population is P_0, find a formula for $P(t)$, the population after t years.
  (b) By approximately what percent does the town's population increase each year?
  (c) Find a formula for $P^{-1}(x)$. How is this function useful?
  (d) Calculate the doubling time of $P(t)$.

8. A hot brick is removed from a kiln. Initially, the temperature of the brick is 200°C above room temperature. Over time, the brick cools off. After 2 hours have elapsed, the brick is 20°C above room temperature. Let t be the time in hours since the brick was removed from the kiln. Let $y = H(t)$ be the difference between the brick's and the room's temperature at time t. Assume that $H(t)$ is an exponential function.

  (a) Find a formula for $H(t)$.
  (b) How many degrees does the brick's temperature drop during the first quarter hour? During the next quarter hour?
  (c) Find a formula for and give an interpretation of $H^{-1}(y)$.
  (d) How much time elapses before the brick's temperature is 5°C above room temperature?
  (e) Interpret the physical meaning of the horizontal asymptote of $H(t)$.

The graph of $y = g(x)$ is given in Figure 8.29. Graph the functions in Problems 9–17. Label intercepts and other relevant points.

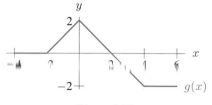

Figure 8.29

9. $y = g(2x)$ **10.** $y = g(-2x)$ **11.** $y = 2g(x)$

12. $y = -\frac{1}{2}g(x+1)$ **13.** $y = 3g(x) - 1$ **14.** $y = \dfrac{1}{g(x)}$

15. $y = (g(x))^2$ **16.** $y = g(g(x))$ **17.** $x = g(y)$

18. Figure 8.30 shows the graph of $f(x)$. For each function in (a)–(g), indicate the graph in (I)–(IV) to which it corresponds. There may be some functions whose graphs are not shown.

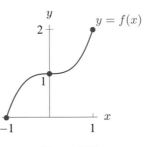

Figure 8.30

(a) $y = -f(x)$ (b) $y = f(-x)$ (c) $y = f(-x) - 2$ (d) $y = f^{-1}(x)$

(e) $y = -f^{-1}(x)$ (f) $y = f(x+1)$ (g) $y = -(f(x) - 2)$

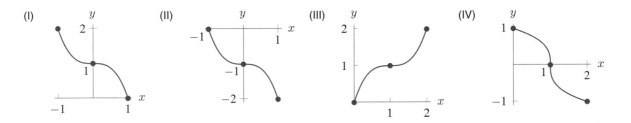

19. Using Figure 8.31, sketch the following transformations of $h(x)$. For each graph, label the points corresponding to the points P and Q. (Note that the graph of h has a horizontal asymptote at $y = 1.25$.)

(a) $y = -2h(x)$ (b) $y = h(-x)$ (c) $y = h(-\frac{1}{2}x)$ (d) $y = h^{-1}(x)$

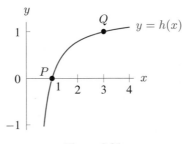

Figure 8.31

Using the graphs of f and g in Figures 8.32 and 8.33, graph the functions in Problems 20–23.

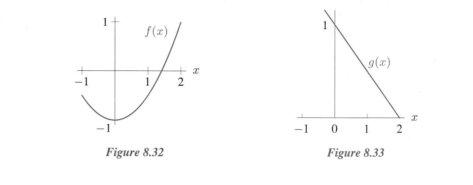

Figure 8.32   Figure 8.33

20. $f(x) - g(x)$ **21.** $f(g(x))$ **22.** $g(f(x))$ **23.** $g(f(x - 2))$

Using the graphs of f and g in Figure 8.34, graph the functions in Problems 24–29.

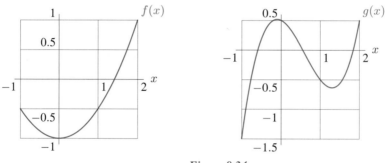

Figure 8.34

24. $f(x) + g(x)$ **25.** $2g(x)$ **26.** $g(f(x))$

27. $f(x) - g(x)$ **28.** $g(f(x - 2))$ **29.** $f(x + 1) + g(x + 1) + 0.5$

30. Graph $f(x) = \ln(|x - 3|)$ and $g(x) = \ln(|x|)$. Find the vertical asymptotes of both functions.

31. Let $f(x) = \dfrac{1}{x + 1}$. Find and simplify $f\left(\dfrac{1}{x}\right) + \dfrac{1}{f(x)}$.

32. Let $f(x) = 12 - 4x$, $g(x) = 1/x$, and $h(x) = \sqrt{x - 4}$. Find the domain of the functions:

(a) $g(f(x))$ (b) $h(f(x))$

33. Many college students work to pay tuition. The number of hours worked usually affects the number of credits taken. In turn, study time in preparation for class is dependent on credits taken. Tables 8.27 and 8.28 show these relationships.

(a) Construct a table showing the relationship between number of hours worked and number of hours of study time.

(b) Make a graph of this relationship. Is it linear?

(c) Is there a situation in which a student reduces the hours worked, with the result that the number of credits increases and there is also more time available for leisure activities?

TABLE 8.27

Number of hours worked per week	Number of credits taken
$0 \leq h < 4$	18
$4 \leq h < 8$	17
$8 \leq h < 12$	16
$12 \leq h < 16$	15
$16 \leq h < 20$	14
$20 \leq h < 24$	13
$24 \leq h < 28$	12

TABLE 8.28

Number of credits taken	Number of hours of study time per week
12	12
13	14
14	16
15	20
16	25
17	31
18	39

In Problems 34–40, you decide to hire either Ace Construction or Space Contractors to build some office space. Let $f(x)$ be the average total cost in dollars of building x square feet of office space, as estimated by Ace. Let $h(x)$ be the total number of square feet of office space you can build with x dollars, as estimated by Space.

34. Describe in words what the following statement tells you: $f(2000) = 200{,}000$

35. Let $g(x) = f(x)/x$. Using the information from Problem 34, evaluate $g(2000)$, and describe in words what $g(2000)$ represents. [Hint: Think about the units.]

36. One of Ace's contractors tells you that, due to the economies of scale, "Building twice as much office space always costs less than twice as much." Express the contactor's statement symbolically, in terms of f and x. [Hint: If you are building x square feet, how would you represent the cost? How would you represent twice the cost? How would you represent the cost of building twice as many square feet?]

37. Suppose that $q > p$ and $p > 1$. Assuming that the contractor's statement in Problem 36 is correct, rank the following in increasing order, using inequality signs: $f(p), g(p), f(q), g(q)$.

38. Describe what the following statement tells you: $h(200{,}000) = 1500$.

39. Let $j(x)$ be the average cost in dollars per square foot of office space. Give a formula for $j(x)$. (Your formula will have $h(x)$ in it.)

40. Research which reveals that $h(f(x)) < x$ for every value of x you check. Explain the implications of this statement. [Hint: Which construction company seems more economical?]

In Problems 41–46, suppose that $f(x) = g(h(x))$. Find possible formulas for $g(x)$ and $h(x)$ (There may be more than one possible answer. Assume $g(x) \neq x$ and $h(x) \neq x$.)

41. $f(x) = 2x + 1$
42. $f(x) = (x + 3)^2$
43. $f(x) = \sqrt{1 + \sqrt{x}}$
44. $f(x) = 9x^2 + 3x$
45. $f(x) = \dfrac{1}{x^2 + 8x + 16}$
46. $f(x) = \dfrac{1}{x^2 + 8x + 17}$

In Problems 47–50 let $f(x) = x - 3$ and $g(x) = 2x + 5$. Find a formula for $h(x)$.

47. $f(h(x)) = \sqrt{x}$
48. $g(h(x)) = 6x - 7$
49. $h(g(x)) = \dfrac{2x + 5}{1 + \sqrt{2x + 5}}$
50. $h(g(x)) = \dfrac{2x + 6}{1 + \sqrt{2x + 4}}$

In Problems 51–63, find a formula for the inverse function. Assume these functions are defined on domains on which they are invertible.

51. $f(x) = 3x - 7$

52. $g(x) = \dfrac{1}{x} - 2$

53. $j(x) = \sqrt{1 + \sqrt{x}}$

54. $h(x) = \dfrac{2x + 1}{3x - 2}$

55. $k(x) = \dfrac{3 - \sqrt{x}}{\sqrt{x} + 2}$

56. $l(x) = \dfrac{2 - (1/x)}{3 - (2/x)}$

57. $f(x) = \dfrac{3 \cdot 2^x + 1}{3 \cdot 2^x + 3}$

58. $g(x) = \dfrac{\ln x - 5}{2 \ln x + 7}$

59. $h(x) = \log\left(\dfrac{x + 5}{x - 4}\right)$

60. $f(x) = \cos\sqrt{x}$

61. $g(x) = 2^{\sin x}$

62. $h(x) = \sin(2x)\cos(2x)$

63. $j(x) = \dfrac{\sin x}{2 - \sin x}$

64. For any positive integer x, let $f(x) = 1$ if x is odd and $f(x) = 0$ if x is even. Let $g(x) = 1$ if x is even and $g(x) = 0$ if x is odd. Which of the following statements are always true? [Hint: Recall that ab is even unless both a and b are odd.]

(i) $f(ab) = f(a)f(b)$
(ii) $g(ab) = g(a)g(b)$
(iii) $f(g(x)) = f(x)$
(iv) $g(f(x)) = g(x)$
(v) $f^{-1}(x)$ does not exist.
(vi) $f(x) \neq 1 - g(x)$

In Problems 65–70, suppose f and g are invertible functions defined for all values of x. For each of the following statements, write *true* if the statement must be true, and justify your response; otherwise, write *false*, and give an example for which the statement does not hold.

Example: The statement $f(x) + g(x) = g(x) + f(x)$ is true because the expressions $f(x)$ and $g(x)$ represent numbers, and the order in which you add numbers does not matter.
Example: The statement $f(a + b) = f(a) + f(b)$ is not necessarily true. For example, if $f(x) = x^2$, then $f(a + b) = (a + b)^2 = a^2 + 2ab + b^2$, whereas $f(a) + f(b) = a^2 + b^2$. Since $a^2 + b^2$ need not equal $a^2 + 2ab + b^2$, we see that $f(a + b)$ need not equal $f(a) + f(b)$.

65. $f(a) = g(b)$ implies $f(b) = g(a)$

66. $f(g(x)) = g(f(x))$

67. $f(ab) = f(a) \cdot f(b)$

68. $f(a) = f(b)$ implies $a = b$

69. $f(x^2) = (f(x))^2$

70. $f(x) \neq f^{-1}(x)$

In Problems 71–76, let $f(x)$ be an increasing function and let $g(x)$ be a decreasing function. Are the following functions increasing, decreasing, or is it impossible to tell? Explain your reasoning.

71. $f(f(x))$

72. $f(g(x))$

73. $g(g(x))$

74. $g(f(x))$

75. $f(x) + g(x)$

76. $f(x) - g(x)$

77. For a positive integer x, let $f(x)$ be the remainder obtained by dividing x by 3. For example, $f(6) = 0$, because 6 divided by 3 equals 2 with a remainder of 0. Likewise, $f(7) = 1$, because 7 divided by 3

(a) Evaluate $f(8)$, $f(17)$, $f(29)$, $f(99)$.
(b) Find a formula for $f(3x)$.
(c) Is $f(x)$ invertible?
(d) Find a formula in terms of $f(x)$ for $f(f(x))$.
(e) Does $f(x + y)$ necessarily equal $f(x) + f(y)$?

CHAPTER NINE

POLYNOMIAL AND RATIONAL FUNCTIONS

This chapter begins with power functions and compares them to exponential and logarithmic functions. Sums and differences of power functions lead to the family of polynomial functions. Ratios of polynomials lead to the family of rational functions.

9.1 POWER FUNCTIONS

What is a Power Function?

Section 1.2 introduced functions in which one quantity is proportional to a constant power of another quantity. For instance, we considered the following examples.

Example 1   The area, A, of a circle is proportional to the square of its radius, r:

$$A = \pi r^2.$$

Example 2   The weight of an object is inversely proportional to the square of the object's distance from the earth's center. If w is the object's weight and d is its distance from the center of the earth, then

$$w = \frac{k}{d^2} = kd^{-2}.$$

These functions are examples of power functions. Generalizing, we make the following definition:

A **power function** is a function of the form

$$f(x) = kx^p$$

where k and p are constants.

Example 3   Which of the following functions are power functions? For each power function, state the value of the constants k and p in the formula $y = kx^p$.

(a) $f(x) = 13\sqrt[3]{x}$ (b) $g(x) = 2(x+5)^3$ (c) $u(x) = \sqrt{\dfrac{25}{x^3}}$ (d) $v(x) = 6 \cdot 3^x$

Solution   The functions f and u are power functions; the functions g and v are not.

(a)   The function $f(x) = 13\sqrt[3]{x}$ is a power function because we can write its formula as

$$f(x) = 13x^{1/3}.$$

Here, $k = 13$ and $p = 1/3$.

(b)   Although the value of $g(x) = 2(x+5)^3$ is proportional to the cube of $x+5$, it is *not* proportional to a power of x. We cannot write $g(x)$ in the form $g(x) = kx^p$; thus, g is not a power function.

(c)   We can rewrite the formula for $u(x) = \sqrt{25/x^3}$ as

$$u(x) = \frac{\sqrt{25}}{\sqrt{x^3}} = \frac{5}{(x^3)^{1/2}} = \frac{5}{x^{3/2}} = 5x^{-3/2}.$$

Thus, u is a power function. Here, $k = 5$ and $p = -3/2$.

(d)   Although the value of $v(x) = 6 \cdot 3^x$ is proportional to a power of 3, the power is not a constant it is the variable. To find $v(1)$ is is an exponential function, not a power function. Notice that the similar-looking function $y = 6 \cdot x^3$ is a power function. However, $6 \cdot x^3$ and $6 \cdot 3^x$ are quite different.

The Effect of the Power p

In this section, we compare power functions to one another. We begin by studying functions whose constant of proportionality is $k = 1$ so that we can focus on the effect of the power p.

Graphs of the Special Cases $y = x^0$ and $y = x^1$

The power functions corresponding to $p = 0$ and $p = 1$ are both linear. The graph of $y = x^0 = 1$ is a horizontal line through the point $(1, 1)$. The graph of $y = x^1 = x$ is a line through the origin with slope $+1$.

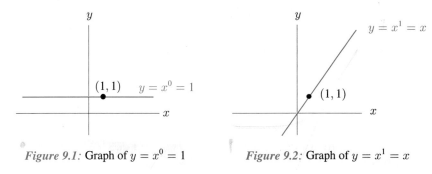

Figure 9.1: Graph of $y = x^0 = 1$ **Figure 9.2:** Graph of $y = x^1 = x$

Graphs of Positive Even Integer Powers: $y = x^2, y = x^4, y = x^6, \ldots$

The graphs of all power functions with p a positive even integer have the same characteristic $\bigsqcup$-shape. For instance, Figure 9.3 shows the graphs of $y = x^2$ and $y = x^4$. Both graphs are similar in shape, although the graph of $y = x^4$ is flatter near the origin than the graph of $y = x^2$, and away from the origin $y = x^4$ is steeper.

The graphs of all functions $y = x^p$, where p is even and positive:
- Pass through $(0, 0)$ and $(1, 1)$ and $(-1, 1)$.
- Decrease for negative values of x and increase for positive values of x.
- Are symmetric about the y-axis because the functions are even.
- Are concave up on every interval.

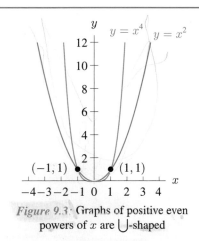

Figure 9.3: Graphs of positive even powers of x are $\bigsqcup$-shaped

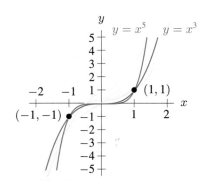

Figure 9.4: Graphs of positive odd powers of x are "chair"-shaped

Graphs of Positive Odd Integer Powers: $y = x^3, y = x^5, y = x^7, \ldots$

The graphs of power functions with p a positive odd integer resemble the side view of a chair. Figure 9.4 shows the graphs of $y = x^3$ and $y = x^5$. The graph of $y = x^5$ is flatter near the origin and steeper far from the origin than the graph of $y = x^3$.

The graphs of all functions $y = x^p$, where p is odd and positive:

- Pass through $(0, 0)$ and $(1, 1)$ and $(-1, -1)$.
- Increase on every interval.
- Are symmetric about the origin because the functions are odd.
- Are concave down for negative values of x and concave up for positive values of x.

Negative Integer Powers: $y = x^{-1}, x^{-3}, x^{-5}, \ldots$ and $y = x^{-2}, x^{-4}, x^{-6}, \ldots$

For negative powers, if we rewrite

$$y = x^{-1} = \frac{1}{x}$$

and

$$y = x^{-2} = \frac{1}{x^2},$$

then it is clear that as $x > 0$ increases, the denominators increase and the functions decrease. The graphs of power functions with odd negative powers, $y = x^{-3}, x^{-5}, \ldots$ resemble the graph of $y = x^{-1} = 1/x$. The graphs of even integer powers, $y = x^{-4}, x^{-6}, \ldots$ are similar in shape to the graph of $y = x^{-2} = 1/x^2$. See Figures 9.5 and 9.6.

The graph of $y = x^{-1}$:

- Passes through $(1, 1)$ and $(-1, -1)$ and does not have a y-intercept.
- Is decreasing everywhere that it is defined.
- Is symmetric about the origin because the function is odd.
- Is concave down for negative values of x and concave up for positive values of x.
- Has the x-axis as a horizontal asymptote and the y-axis as a vertical asymptote.

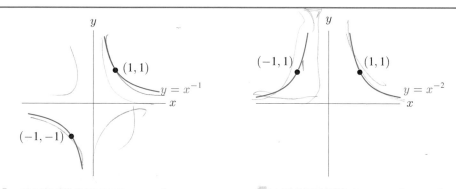

Figure 9.5: Graph of $y = x^{-1} = 1/x$ **Figure 9.6:** Graph of $y = x^{-2} = 1/x^2$

The graph of $y = x^{-2}$:

- Passes through $(1, 1)$ and $(-1, 1)$ and does not have a y-intercept.
- Is increasing for negative values of x and decreasing for positive values of x.
- Is symmetric about the y-axis because the function is even.
- Is concave up everywhere that it is defined.
- Has the x-axis as a horizontal asymptote and the y-axis as a vertical asymptote.

Let's consider the asymptotes more carefully. The values of $1/x$ and $1/x^2$ can be made as close to zero as we like by choosing a sufficiently large x. See Table 9.1. Graphically, this means that the curves $y = 1/x$ and $y = 1/x^2$ get closer and closer to the x-axis for large values of x.

TABLE 9.1 *Values of x^{-1} and x^{-2} approach zero as x grows large*

x	0	10	20	30	40	50
$y = 1/x$	Undefined	0.1	0.05	0.033	0.025	0.02
$y = 1/x^2$	Undefined	0.01	0.0025	0.0011	0.0006	0.0004

On the other hand, as x gets close to zero, the values of $1/x$ and $1/x^2$ get very large. See Table 9.2. Graphically, this means that the curves $y = 1/x$ and $y = 1/x^2$ get very close to the y-axis, as x get close to zero.

TABLE 9.2 *Values of x^{-1} and x^{-2} grow large as x approaches zero from the positive side*

x	0.1	0.05	0.01	0.001	0.0001	0
$y = 1/x$	10	20	100	1000	10,000	Undefined
$y = 1/x^2$	100	400	10,000	1,000,000	100,000,000	Undefined

Graphs of Positive Fractional Powers: $y = x^{1/2}, x^{1/3}, x^{1/4}, \ldots$

Figure 9.7 shows the graphs of $y = x^{1/2}$ and $y = x^{1/4}$. These graphs have the same shape, although $y = x^{1/4}$ is steeper near the origin and flatter away from the origin than $y = x^{1/2}$. The same can be said about the graphs of $y = x^{1/3}$ and $y = x^{1/5}$ in Figure 9.8. In general, if n is a positive integer, then the graph of $y = x^{1/n}$ resembles the graph of $y = x^{1/2}$ if n is even; if n is odd, the graph resembles the graph of $y = x^{1/3}$.

Notice that the graphs of $y = x^{1/2}$ and $y = x^{1/3}$ bend in a direction opposite to that of the graphs of $y = x^2$ and x^3. For example, the graph of $y = x^2$ is concave up, but the graph of $y = x^{1/2}$ is concave down. However, all these functions become infinitely large as x increases.

The Effect of the Size of p for Positive Powers in the First Quadrant

In Figure 9.9 for large x, the higher power of x, the faster the function climbs. Not only are the higher powers larger, but they are *much* larger. This is because if $x = 100$, for example, 100^5 is one

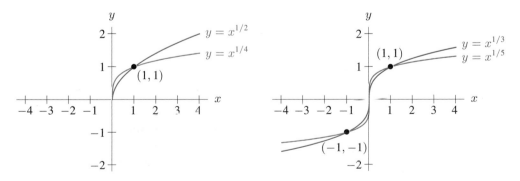

Figure 9.7: The graphs of $y = x^{1/2}$ and $y = x^{1/4}$ *Figure 9.8:* The graphs of $y = x^{1/3}$ and $y = x^{1/5}$

hundred times as big as 100^4 which is one hundred times as big as 100^3. As x gets larger (written as $x \to \infty$), any positive power of x grows much faster than all lower powers of x. We say that, as $x \to \infty$, higher powers of x *dominate* lower powers.

As x approaches zero (written $x \to 0$), the story is entirely different. See Figure 9.10, which is a close-up view near the origin. For x between 0 and 1, x^3 is bigger than x^4, which is bigger than x^5. (Try $x = 0.1$ to confirm this.) For values of x near zero, smaller powers dominate.

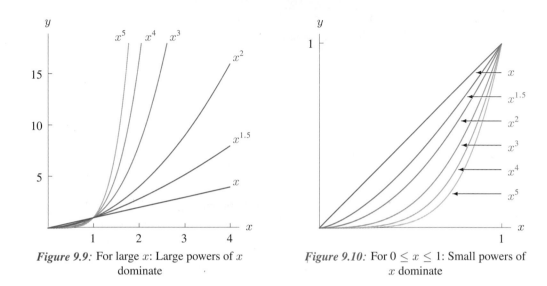

Figure 9.9: For large x: Large powers of x dominate

Figure 9.10: For $0 \le x \le 1$: Small powers of x dominate

The Effect of the Size of p for Negative Powers in the First Quadrant

For $y = x^p$, larger positive values of p indicate more rapid growth in the first quadrant. On the other hand, larger *negative* values of p indicate that the graph of $y = x^p$ approaches its horizontal asymptote more rapidly. Furthermore, larger negative values of p indicate the function $y = x^p$ climbs faster near its vertical asymptote. Figures 9.11 and 9.12 show that the asymptotic behavior of $y = x^{-3}$ is more extreme than that of $y = x^{-2}$ and $y = x^{-1}$. The graph of $y = x^{-3}$ dies off more rapidly towards $y = 0$ and "blows up" more explosively near $x = 0$.

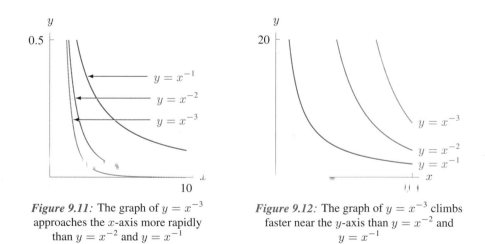

Figure 9.11: The graph of $y = x^{-3}$ approaches the x-axis more rapidly than $y = x^{-2}$ and $y = x^{-1}$

Figure 9.12: The graph of $y = x^{-3}$ climbs faster near the y-axis than $y = x^{-2}$ and $y = x^{-1}$

Problems for Section 9.1

1. Without a calculator, match the following functions with the graphs in Figure 9.13.

   (i) $y = x^5$ D   (ii) $y = x^2$ B   (iii) $y = x$ A   (iv) $y = x^3$ C

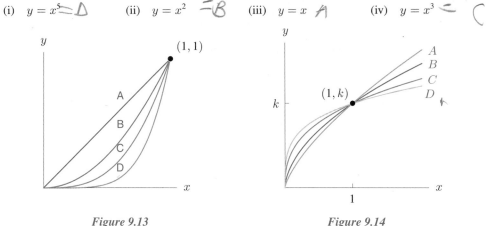

Figure 9.13   Figure 9.14

2. Match the graphs in Figure 9.14 with the functions $y = kx^{9/16}$, $y = kx^{3/8}$, $y = kx^{5/7}$, $y = kx^{3/11}$.

3. Compare the graphs of $y = x^2$, $y = x^4$, and $y = x^6$. Describe the similarities and differences.

4. Compare the graphs of $y = x^{-2}$, $y = x^{-4}$, and $y = x^{-6}$. Describe the similarities and differences.

5. Describe the behavior of the functions $y = x^{-10}$ and $y = -x^{10}$ as

   (a) $x \to 0$ (b) $x \to \infty$ (c) $x \to -\infty$

6. Describe the behavior of the functions $y = x^{-3}$ and $y = x^{1/3}$ as

   (a) $x \to 0$ from the right   (b) $x \to \infty$

7. If $f(x) = kx^p$, p an integer, show that f is an even function if p is even, and an odd function if p is odd.

8. (a) Match the functions x, x^2, x^3, $x^{1/2}$, $x^{1/3}$, $x^{3/2}$ with the graphs in Figure 9.15. Justify your choice.

   (b) What is the relationship between the concavity of $y = x^2$ and $y = x^{1/2}$? Between the concavity of $y = x^3$ and $y = x^{1/3}$ for $x > 0$? Explain why this happens.

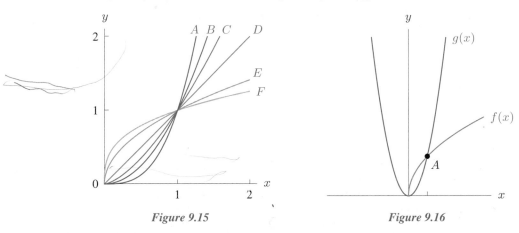

Figure 9.15   Figure 9.16

9. (a) Figure 9.16 gives the graphs of functions f and g. One of these functions is $y = x^n$ and the other is $y = x^{1/n}$, where n is a positive integer. Which is which? How do you know?

   (b) What are the coordinates of point A?

10. Two oil tankers crash in the Pacific ocean. The spreading oil slick has a circular shape, and the radius of the circle is increasing at 200 meters per hour.

    (a) Express the radius of the spill, r, as a power function of time, t, in hours since the crash.
    (b) Express the area of the spill, A, as a power function of time, t.
    (c) Clean-up efforts begin 7 hours after the spill. How large an area is covered by oil at that time?

11. The radius of a sphere is directly proportional to the cube root of its volume. If a sphere of radius 18.2 cm has a volume of 25,252.4 cm^3, what is the radius of a sphere whose volume is 30,000 cm^3?

12. The following questions involve the behavior of the power function $y = x^{-p}$, for p a positive integer. If a distinction between even and odd values of p is significant, the significance should be indicated.

    (a) What is the domain of $y = x^{-p}$? What is the range?
    (b) What symmetries are exhibited by the graph of $y = x^{-p}$?
    (c) What is the behavior of $y = x^{-p}$ as $x \to 0$?
    (d) What is the behavior of $y = x^{-p}$ for large positive values of x? for large negative values of x?

13. Assume that $f(x) = 16x^4$ and $g(x) = 4x^2$.
    (a) If $f(x) = g\left(h(x)\right)$ find a possible formula for $h(x)$, assuming $h(x) \le 0$ for all x.
    (b) If $f(x) = j\left(2g(x)\right)$ find a possible formula for $j(x)$, assuming $j(x)$ is a power function.

14. When an aircraft flies horizontally, its *stall velocity* (the minimum speed required to keep the aircraft aloft) is directly proportional to the square root of the quotient of its weight by its wing area. If a breakthrough in materials science allowed the construction of an aircraft with the same weight but twice the wing area, would the stall velocity increase or decrease? By what percent?

15. Consider the power function $y = t(x) = k \cdot x^{p/3}$ where p is any integer, $p \ne 0$.
    (a) For what values of p does $t(x)$ have domain restrictions? What are those restrictions?
    (b) What is the range of $t(x)$ if p is even?
    (c) What is the range of $t(x)$ if p is odd?
    (d) What symmetry does the graph of $t(x)$ exhibit if p is even? If p is odd?

16. A person's weight, w, on a planet of radius d is given by the formula
$$w = kd^{-2}, \quad k > 0,$$
where the constant k depends on the masses of the person and the planet; it does not depend on d. (Note that there is a distinction being made here between mass and weight. For example, an astronaut in orbit may be weightless, but he still has mass.)

    (a) A man weighs 180 lb on the surface of the earth. How much does he weigh on the surface of a planet as massive as the earth, but whose radius is three times as large? One-third as large?
    (b) What fraction of the earth's radius must an equally massive planet have if the surface-weight of the man in part (a) is one ton?

17. One of Johannes Kepler's three laws of planetary motion states that the square of the period, P, of a body orbiting the sun is proportional to the cube of its average distance, d, from the sun. Thus, we have
$$P^2 = kd^3.$$
    Solving for P gives
$$P = \sqrt{kd^3} = \sqrt{k}d^{3/2} = k_1 d^{3/2}.$$

    The earth has a period of 365 days and its distance from the sun is approximately 93,000,000 miles. With this information
$$365 = k_1(93{,}000{,}000)^{3/2},$$
so
$$k_1 = \frac{365}{(93{,}000{,}000)^{3/2}}.$$
This gives
$$P = \frac{365}{(93{,}000{,}000)^{3/2}} \cdot d^{3/2} = 365\frac{d^{3/2}}{(93{,}000{,}000)^{3/2}} = 365\left(\frac{d}{93{,}000{,}000}\right)^{3/2}.$$

Given that the planet Jupiter has an average distance from the sun of 483,000,000 miles, how long in earth days is a Jupiter year?

9.2 COMPARING POWER, EXPONENTIAL, AND LOG FUNCTIONS

The Effect of the Parameter k

In Chapter 5 we saw the effect of k on the graph of $f(x) = kx^p$. The coefficient k stretches or compresses the graph vertically; if k is negative, the graph is reflected across the x-axis. How does the value of k affect the long-term growth rate of $f(x) = kx^p$? Is the growth of a power function affected more by the size of the coefficient or by the size of the power?

Example 1   Let $f(x) = 100x^3$ and $g(x) = x^4$ for $x > 0$. Sketch graphs comparing these two functions. Discuss the long-term behavior of the functions and explain your reasoning.

Solution   For $x < 10$, Figure 9.17 suggests that f is growing faster than g and that f dominates g. Eventually, however, the fact that g has a higher power than f asserts itself. Figure 9.18 shows that $g(x)$ has caught up to $f(x)$ at $x = 100$. Figure 9.19 shows that for x larger than 100, values of g are larger than values of f.

Could the graphs of f and g intersect again for some value of $x > 100$? To show that this cannot be the case, solve the equation $g(x) = f(x)$:

$$x^4 = 100x^3$$
$$x^4 - 100x^3 = 0$$
$$x^3(x - 100) = 0.$$

Since the only solutions to this equation are $x = 0$ and $x = 100$, the graphs of f and g do not cross for $x > 100$.

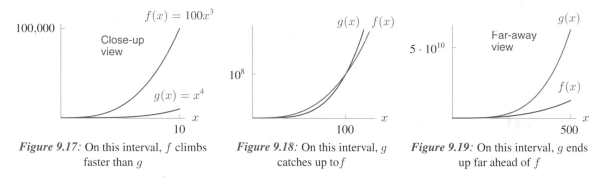

Figure 9.17: On this interval, f climbs faster than g

Figure 9.18: On this interval, g catches up to f

Figure 9.19: On this interval, g ends up far ahead of f

Comparing Exponential Functions and Power Functions

Both power functions and exponential functions can increase at phenomenal rates. For example, Table 9.3 shows values of $f(x) = x^4$ and $g(x) = 2^x$.

TABLE 9.3   *The exponential function $g(x) = 2^x$ eventually grows faster than the power function $f(x) = x^4$*

x	0	5	10	15	20
$f(x) = x^4$	0	625	10,000	50,625	160,000
$g(x) = 2^x$	1	32	1024	32,768	1,048,576

Despite the impressive growth in the value of the power function $f(x) = x^4$, in the long run $g(x) = 2^x$ grows faster. By the time $x = 20$, the value of $g(20)$ is over six times as large as $f(20)$. Figure 9.20 shows the exponential function $g(x) = 2^x$ catching up to $f(x) = x^4$.

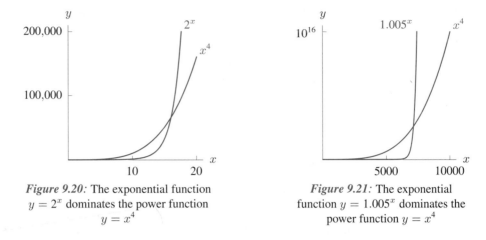

Figure 9.20: The exponential function $y = 2^x$ dominates the power function $y = x^4$

Figure 9.21: The exponential function $y = 1.005^x$ dominates the power function $y = x^4$

But what about a more slowly growing exponential function? After all, $y = 2^x$ increases at a 100% growth rate. Figure 9.21 compares $y = x^4$ to the exponential function $y = 1.005^x$. Despite the fact that this exponential function creeps along at a 0.5% growth rate, at around $x = 7000$, it overtakes the power function. In summary,

> *Any* positive increasing exponential function eventually grows faster than *any* power function.

Decreasing Exponential Functions and Decreasing Power Functions

Just as an increasing exponential function eventually outpaces any increasing power function, an exponential decay function wins the race towards the x-axis. In general:

> *Any* positive decreasing exponential function eventually approaches the horizontal axis faster than any positive decreasing power function.

For example, let's compare the long term behavior of the decreasing exponential function $y = 0.5^x$ with the decreasing power function $y = x^{-2}$. By rewriting

$$y = 0.5^x = \left(\frac{1}{2}\right)^x = \frac{1}{2^x} \quad \text{and} \quad y = x^{-2} = \frac{1}{x^2}$$

we can see the comparison more easily. In the long run, the smallest of these two fractions is the one with the largest denominator. The fact that 2^x is eventually larger than x^2 means that $1/2^x$ is eventually smaller than $1/x^2$.

Figure 9.22 shows the graphs of $y = 0.5^x$ and $y = x^{-2}$. Both graphs have the x-axis as a horizontal asymptote. As x increases, the exponential function $y = 0.5^x$ approaches the x-axis faster than the power function $y = x^{-2}$. Figure 9.23 shows what happens for large values of x. The exponential function approaches the x-axis so rapidly that it becomes invisible compared to $y = x^{-2}$.

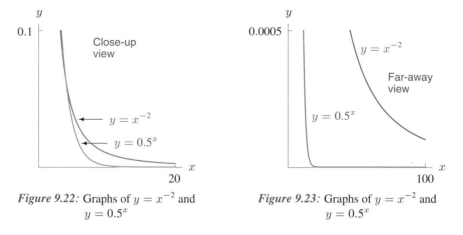

Figure 9.22: Graphs of $y = x^{-2}$ and $y = 0.5^x$

Figure 9.23: Graphs of $y = x^{-2}$ and $y = 0.5^x$

Comparing Log and Power Functions

Power functions like $y = x^{1/2}$ and $y = x^{1/3}$ grow quite slowly. However, they grow rapidly in comparison to log functions. In fact:

> *Any* positive increasing power function eventually grows more rapidly than $y = \log x$ and $y = \ln x$.

For example, Figure 9.24 shows the graphs of $y = x^{1/2}$ and $y = \log x$. The fact that exponential functions grow so fast should alert you to the fact that their inverses, the logarithms, grow very slowly. This is illustrated in Figure 9.25.

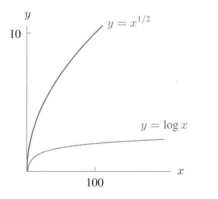

Figure 9.24: Graphs of $y = x^{1/2}$ and $y = \log x$

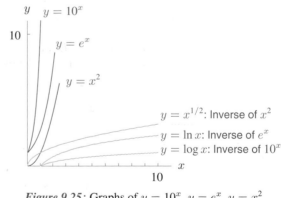
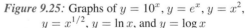

Figure 9.25: Graphs of $y = 10^x$, $y = e^x$, $y = x^2$, $y = x^{1/2}$, $y = \ln x$, and $y = \log x$

Finding the Formula for a Power Function

As is the case for linear and exponential functions, the formula of a power function can be found from two points on its graph.

Example 2   Water is leaking out of a container which has a hole in the bottom. Torricelli's Law states that at any instant, the velocity v with which water escapes from the container is a power function of d, the depth of the water at that moment. Experiments show that when $d = 1$ foot, then $v = 8$ ft/sec; that when $d = 1/4$ foot, then $v = 4$ ft/sec. Express v as a function of d.

Solution   Torricelli's Law tells us that $v = kd^p$, where k and p are constants. The fact that $v = 8$ when $d = 1$, gives $8 = k(1)^p$, so $k = 8$, and therefore $v = 8d^p$. Now use the fact that $v = 4$ when $d = 1/4$ to get

$$4 = 8 \left(\frac{1}{4} \right)^p.$$

Rewriting $(1/4)^p = 1/4^p$, we have

$$4 = 8 \cdot \frac{1}{4^p}.$$

Solving for 4^p gives

$$4^p = \frac{8}{4} = 2.$$

Since $4^{1/2} = 2$, we must have $p = 1/2$. Therefore we have $v = 8d^{1/2}$. *Note*: Toricelli's Law is often written in the form $v = \sqrt{2gd}$, where $g = 32$ ft/sec^2 is the acceleration due to gravity.

Problems for Section 9.2

1. Determine if the formulas of the functions in parts (a) through (h) can be written either in the form of an exponential function or a power function. If not, explain why the function does not fit either form.

   (a) $h(x) = 3(-2)^{3x}$ (b) $j(x) = 3(-3)^{2x}$ (c) $m(x) = 3(3x+1)^2$ (d) $n(x) = 3 \cdot 2^{(3x+1)}$

   (e) $p(x) = (5^x)^2$ (f) $q(x) = 5^{(x^2)}$ (g) $r(x) = 2 \cdot 3^{-2x}$ (h) $s(x) = \dfrac{4}{5x^{-3}}$

2. Let $f(x) = 3^x$ and $g(x) = x^3$.

   (a) Complete the following table of values:

x	-3	-2	-1	0	1	2	3
$f(x)$							
$g(x)$							

   (b) Describe the long-run behaviors of f and g as $x \to -\infty$ and as $x \to +\infty$.

3. The functions $y = x^{-3}$ and $y = 3^{-x}$ both approach zero as $x \to \infty$. Which function approaches zero faster? Support your conclusion numerically.

4. The functions $y = x^{-3}$ and $y = e^{-x}$ both approach zero as $x \to \infty$. Which function approaches zero faster? Support your conclusion numerically.

5. Figure 9.26 gives the graphs, for $x \geq 0$, of $f(x) = x^2$, $g(x) = 2x^2$, and $h(x) = x^3$.

   (a) Match these functions to their graphs shown in Figure 9.26.

   (b) Does graph (B) intersect graph (A) for $x > 0$? If so, for what value(s) of x? If not, explain how you know.

   (c) Does graph (C) intersect graph (A) for $x > 0$? If so, for what value(s) of x? If not, explain how you know.

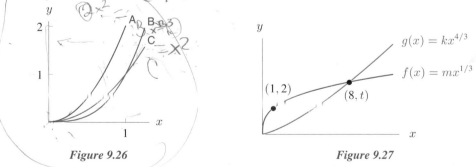

Figure 9.26   *Figure 9.27*

6. In Figure 9.27, find the values of m, t, and k.

7. (a) Given $t(x) = x^{-2}$ and $r(x) = 40x^{-3}$, find v such that $t(v) = r(v)$.
    (b) For $0 < x < v$, which is greater, $t(x)$ or $r(x)$?
    (c) For $x > v$, which is greater, $t(x)$ or $r(x)$?

8. Let $f(x) = x^x$. Is f a power function, an exponential function, both, or neither? Discuss.

In Problems 9–12, find possible formulas for the power functions whose values are given.

9.

x	0	1	2	3
$j(x)$	0	2	16	54

10.

x	2	3	4	5
$f(x)$	12	27	48	75

11.

x	-6	-2	3	4
$g(x)$	36	4/3	$-9/2$	$-32/3$

12.

x	-2	$-1/2$	1/4	4
$h(x)$	$-1/2$	-8	-32	$-1/8$

In Problems 13–15, find possible formulas for the power functions with the properties given.

13. $f(1) = \frac{3}{2}$ and $f(2) = \frac{3}{8}$ 14. $g\left(-\frac{1}{5}\right) = 25$ and $g(2) = -\frac{1}{40}$ 15. $g(3) = \frac{1}{3}$ and $g\left(\frac{1}{3}\right) = 27$

In Problems 16–18, find a possible formula for f assuming f is

(a) a linear function.   (b) an exponential function.   (c) a power function.

16. $f(1) = 18$ and $f(3) = 1458$ 17. $f(1) = 16$ and $f(2) = 128$ 18. $f(-1) = \frac{3}{4}$ and $f(2) = 48$

19. Table 9.4 gives approximate values for three functions, f, g, and h. One is exponential, one is trigonometric, and one is a power function. Determine which is which and find possible formulas for each.

TABLE 9.4

x	-2	-1	0	1	2
$f(x)$	4	2	4	6	4
$g(x)$	20.0	2.5	0.0	-2.5	-20.0
$h(x)$	1.3333	0.6667	0.3333	0.1667	0.0833

20. Figure 9.28 gives the graph of $g(x)$, a mystery power function.

(a) If you learn that the point $(-1, 3)$ lies on its graph, do you have enough information to write a formula for $g(x)$?
(b) If you are told that the point $(1, -3)$ also lies on the graph, what new deductions can you make?
(c) If the point $(2, -96)$ lies on the graph g, in addition to the points already given, state three other points which also lie on it.

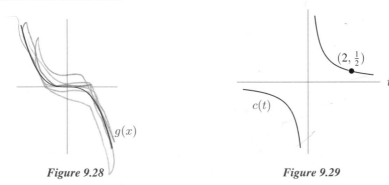

Figure 9.28   **Figure 9.29**

21. Figure 9.29 is a graph of the power function $y = c(t)$. Is $c(t) = 1/t$ the only possible formula for c? Could there be others?

22. The period, p, of the orbit of a planet whose average distance (in millions of miles) from the sun is d, is given by $p = kd^{3/2}$, where k is a constant. The average distance from the earth to the sun is 93 million miles.

     (a) If the period of the earths orbit were twice the current 365 days, what would be the average distance from the sun to earth?
     (b) Is there a planet in our solar system whose period is approximately twice the earth's?

23. Refer to Problem 22. If the distance between the sun and the earth were halved, how long (in earth days) would a "year" be?

24. (a) The functions in Table 9.5 are of the form $y = a \cdot r^{3/4}$ and $y = b \cdot r^{5/4}$. Explain how you can tell which is which from the values in the table.
     (b) Determine the constants a and b for the functions from part (a).

TABLE 9.5

r	2.5	3.2	3.9	4.6
$y = g(r)$	15.9	19.1	22.2	25.1
$y = h(r)$	9.4	12.8	16.4	20.2

25. Values of the functions f and g are in Table 9.6 and 9.7. One of the functions is of the form $y = a \cdot d^{p/q}$ with $p > q$; the other is of the form $y = b \cdot d^{p/q}$ with $p < q$. Which is which? How can you tell?

TABLE 9.6

d	2	2.2	2.4	2.6	2.8
$f(d)$	151.6	160.5	169.1	177.4	185.5

TABLE 9.7

d	10	10.2	10.4	10.6	10.8
$g(d)$	7.924	8.115	8.306	8.498	8.691

9.3 POLYNOMIAL FUNCTIONS

A *polynomial function* is a sum of power functions, whose exponents are nonnegative integers. We use what we learned about power functions to study polynomials.

Example 1   Suppose you make five separate deposits of $1000 each into a savings account, one deposit per year, beginning today. What annual interest rate must you earn if you want the account to have a balance of $6000 five years from today? (Assume the interest rate is constant over these five years.)

Solution   Let r be the annual interest rate. Our goal is to determine what value of r gives you $6000 in 5 years. You start in year $t = 0$ by making a $1000 deposit. In one year, you have $1000 plus the interest earned on that amount. At that point, you add another $1000.

To picture how this works, imagine the account pays 5% annual interest, compounded annually. Then, after one year, your balance will be

$$\text{Balance} = (100\% \text{ of Initial deposit}) + (5\% \text{ of Initial deposit}) + \text{Second deposit}$$

$$= 105\% \text{ of } \underbrace{\text{Initial deposit}}_{\$1000} + \underbrace{\text{Second deposit}}_{\$1000}$$

$$= 1.05(1000) + 1000.$$

Let x represent the annual growth factor, $1 + r$. For example, if the account paid 5% interest, then $x = 1 + 0.05 = 1.05$. We write the balance after one year in terms of x:

$$\text{Balance after one year} = 1000x + 1000.$$

After two years, you will have earned interest on the first-year balance. This gives

$$\text{Balance after earning interest} = \underbrace{(1000x + 1000)}_{\text{First-year balance}}x = 1000x^2 + 1000x.$$

The third \$1000 deposit brings your balance to

$$\text{Balance after two years} = 1000x^2 + 1000x + \underbrace{1000.}_{\text{Third deposit}}$$

A year's worth of interest on this amount, plus the fourth \$1000 deposit, brings your balance to

$$\text{Balance after three years} = \underbrace{(1000x^2 + 1000x + 1000)}_{\text{Second-year balance}}x + \underbrace{1000}_{\text{Fourth deposit}}$$

$$= 1000x^3 + 1000x^2 + 1000x + 1000.$$

The pattern is this: Each of the \$1000 deposits grows to \$$1000x^n$ by the end of its nth year in the bank. Thus,

$$\text{Balance after five years} = 1000x^5 + 1000x^4 + 1000x^3 + 1000x^2 + 1000x.$$

If the interest rate is chosen correctly, then the balance will be \$6000 in five years. This gives us

$$1000x^5 + 1000x^4 + 1000x^3 + 1000x^2 + 1000x = 6000.$$

Dividing by 1000 and moving the 6 to the left side, we have the equation

$$x^5 + x^4 + x^3 + x^2 + x - 6 = 0.$$

Solving this equation for x determines how much interest we must earn. Using a computer or calculator, we find where the graph of $Q(x) = x^5 + x^4 + x^3 + x^2 + x - 6$ crosses the x-axis. Figure 9.30 shows that this occurs at $x \approx 1.0614$. Since $x = 1 + r$, this means $r = 0.0614$. So the account must earn 6.14% annual interest[1] for the balance to be \$6000 at the end of five years.

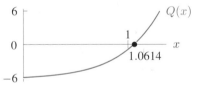

Figure 9.30: Finding where $Q(x)$ crosses the x-axis, for $x \geq 0$

You may wonder if Q crosses the x-axis more than once. For $x \geq 0$, graphing Q on a larger scale suggests that Q increases for all values of x and that it crosses the x-axis only once. For $x > 1$, we expect Q to be an increasing function, because larger values of x indicate higher interest rates and therefore larger values of $Q(x)$. Having crossed the axis once, the graph of Q does not "turn around" to cross it again.

The function $Q(x) = x^5 + x^4 + x^3 + x^2 + x - 6$ is the sum of power functions; Q is called a *polynomial*. (Note that the expression -6 can be written as $-6x^0$, so that it, too, is a power function.)

A General Formula for the Family of Polynomial Functions

The general formula for a polynomial function can be written as

$$p(x) = a_n x^n + a_{n-1}x^{n-1} + ... + a_1 x + a_0,$$

where n is called the *degree* of the polynomial and a_n is the *leading coefficient*.

---

[1]This is 6.14% interest per year, compounded annually.

For example, the function

$$g(x) = 3x^2 + 4x^5 + x - x^3 + 1,$$

is a polynomial of degree 5 because the term with the highest power is $4x^5$. It is customary to write a polynomial with the powers in decreasing order from left to right:

$$g(x) = 4x^5 - x^3 + 3x^2 + x + 1.$$

The function g has one other term, $0 \cdot x^4$, which we don't bother to write down. The values of g's coefficients are $a_5 = 4, a_4 = 0, a_3 = -1, a_2 = 3, a_1 = 1$, and $a_0 = 1$.

In summary:

The general formula for the family of polynomial functions can be written as

$$p(x) = a_n x^n + a_{n-1} x^{n-1} + \ldots + a_1 x + a_0,$$

where n is a positive integer called the **degree** of p and where $a_n \neq 0$.

- Each member of this sum, $a_n x^n$, $a_{n-1} x^{n-1}$, and so on, is called a **term**.
- The constants $a_n, a_{n-1}, \ldots, a_0$ are called **coefficients**.
- The term a_0 is called the **constant term**. The highest-powered term, $a_n x^n$, is called the **leading term**.
- To write a polynomial in **standard form**, we arrange its terms from highest power to lowest power, going from left to right.

Like the power functions from which they are built, polynomials are defined for all values of x. Except for polynomials of degree zero (which have graphs that are horizontal lines), the graphs of polynomials do not have horizontal or vertical asymptotes. The shape of the graph depends on its degree; typical graphs are shown in Figure 9.31.

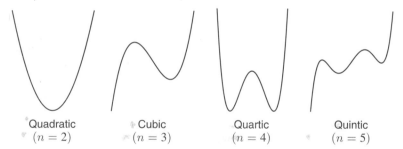

Quadratic   Cubic   Quartic   Quintic
$(n = 2)$ $(n = 3)$ $(n = 4)$ $(n = 5)$

Figure 9.31: Graphs of typical polynomials of degree n

The Long-Run Behavior of Polynomial Functions

We have seen that, as x grows large, $y = x^2$ increases fast, $y = x^3$ increases faster, and $y = x^4$ increases faster still. In general, power functions with larger positive powers eventually grow much faster than those with smaller powers. This tells us about the behavior of polynomials for large x. For instance, consider the polynomial $g(x) = 4x^5 - x^3 + 3x^2 + x + 1$. Provided x is large enough, the value of the term $4x^5$ is much larger (either positive or negative) than the value of the other terms combined. For example, if $x = 100$,

$$4x^5 = 4(100)^5 = 40{,}000{,}000{,}000,$$

and the other terms in $g(x)$ are

$$-x^3 + 3x^2 + x + 1 = -(100)^3 + 3(100)^2 + 100 + 1$$
$$= -1,000,000 + 30,000 + 100 + 1 = -969,899.$$

Therefore $p(100) = 39,999,030,101$, which is approximately equal to the value of the $4x^5$ term. In general, if x is large enough, the most important contribution to the value of a polynomial p is made by the leading term; we can ignore the lower-powered terms.

When viewed on a large enough scale, the graph of the polynomial $p(x) = a_n x^n + a_{n-1} x^{n-1} + \cdots + a_1 x + a_0$ looks like the graph of the power function $y = a_n x^n$. This behavior is called the **long-run behavior** of the polynomial.

Example 2 Show that the graph of $f(x) = x^3 + x^2$ resembles the power function $y = x^3$ on a large scale.

Solution Figure 9.32 gives the graphs of $f(x) = x^3 + x^2$ and $y = x^3$. On this scale, f does not look like a power function. On the larger scale in Figure 9.33, the graph of f resembles the graph of $y = x^3$. On this larger scale, the "bumps" in the graph of f are too small to be seen. On an even larger scale, as in Figure 9.34, the graph of f is indistinguishable from the graph of $y = x^3$.

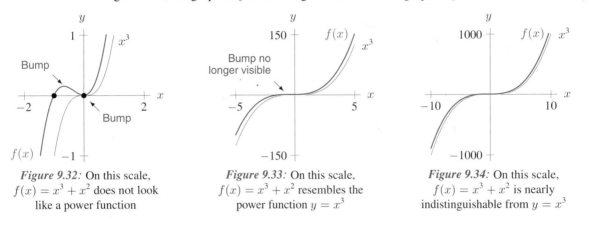

Figure 9.32: On this scale, $f(x) = x^3 + x^2$ does not look like a power function

Figure 9.33: On this scale, $f(x) = x^3 + x^2$ resembles the power function $y = x^3$

Figure 9.34: On this scale, $f(x) = x^3 + x^2$ is nearly indistinguishable from $y = x^3$

The data in Table 9.8 lead us to the same conclusion as the graphs in Figures 9.32-9.34. For large values of x, the values of the polynomial $f(x) = x^3 + x^2$ are close to the values of x^3.

TABLE 9.8 *Values of $f(x) = x^3 + x^2$. For large x, the x^3 term dominates the x^2 term*

x	x^2	x^3	$f(x) = x^3 + x^2$
-10	100	-1000	-900
-5	25	-125	-100
-1	1	-1	0
1	1	1	2
5	25	125	150
10	100	1000	1100

Zeros of Polynomials

The *zeros* of a polynomial p are values of x for which $p(x) = 0$. These values are also called the x-intercepts, because they tell us where the graph of p crosses the x-axis. Using algebra to find the zeros of a polynomial can be quite difficult (if not impossible). The numerical and graphical method used in Example 1 is often the only practical way of finding the zeros of a polynomial. However, the long-run behavior of the polynomial can give us clues as to how many zeros (if any) there may be.

Example 3 Given the polynomial

$$q(x) = 3x^6 - 2x^5 + 4x^2 - 1,$$

where $q(0) = -1$, is there a reason to expect a solution to the equation $q(x) = 0$? If not, explain why not. If so, how do you know?

Solution The equation $q(x) = 0$ must have at least two solutions. We know this because on a large scale, q looks like the power function $y = 3x^6$. (See Figure 9.35.) The function $y = 3x^6$ takes on large positive values as x grows large (either positive or negative). Since the graph of q is smooth and unbroken, it must cross the x-axis at least twice to get from $q(0) = -1$ to the positive values it attains as $x \to \infty$ and $x \to -\infty$.

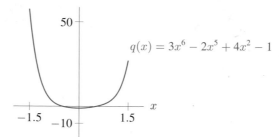

Figure 9.35: Graph must cross x-axis at least twice since $q(0) = -1$ and $q(x)$ looks like $3x^2$ for large x

A sixth degree polynomial such as q in Example 3 can have as many as six real zeros. We consider the zeros of a polynomial in more detail in Section 9.4.

Problems for Section 9.3

1. Show that the function $u(x) = x(x - 3)(x + 2)$ is a polynomial. What is its degree?
2. Show that the function $y = (x^2 - 4)(x^2 - 2x - 3)$ is a polynomial. What is its degree?

For the polynomials in Problems 3–5, state the degree, the number of terms, and describe the long-run behavior.

3. $y = 2x^3 - 3x + 7$ 4. $y = (x + 4)(2x - 3)(5 - x)$ 5. $y = 1 - 2x^4 + x^3$

6. Let $u(x) = -\frac{1}{5}(x - 3)(x + 1)(x + 5)$ and $v(x) = -\frac{1}{5}x^2(x - 5)$.

   (a) Draw graphs of u and v on the screen $-10 \le x \le 10$, $-10 \le y \le 10$. How are the graphs similar? How are they different?

   (b) Compare the graphs of u and v on the window $-20 \le x \le 20$, $-1600 \le y \le 1600$, the window $-50 \le x \le 50$, $-25{,}000 \le y \le 25{,}000$, and the window $-500 \le x \le 500$, $-25{,}000{,}000 \le y \le 25{,}000{,}000$. Discuss.

7. Compare the graphs of $f(x) = x^3 + 5x^2 - x - 5$ and $g(x) = -2x^3 - 10x^2 + 2x + 10$ on a window that shows all intercepts. How are the graphs similar? Different? Discuss.

8.   Estimate the zeros of the polynomial $f(x) = x^4 - 3x^2 - x + 2$.

9.   Estimate the minimum value of $g(x) = x^4 - 3x^3$ 8 to two decimal places.

10. Find the equation of the line through the y-intercept of $y = x^4 - 3x^5 - 1 + x^2$ and the x-intercept of $y = 2x - 4$.

11. Let $f(x) = \left(\dfrac{1}{50{,}000}\right)x^3 + \left(\dfrac{1}{2}\right)x$.

   (a)   For small values of x, which term of f is more important? Explain your answer.
   (b)   Sketch a graph of $y = f(x)$ for $-10 \le x \le 10$, $-10 \le y \le 10$. Is this graph linear? How does the appearance of this graph agree with your answer to part (a)?
   (c)   How large a value of x is required for the cubic term of f to be equal to the linear term?

12. The polynomial function $f(x) = x^3 + x + 1$ is invertible—that is, this function has an inverse.

   (a)   Sketch a graph of $y = f(x)$. Explain how you can tell from a graph whether f is invertible.
   (b)   Find $f(0.5)$ and an approximate value for $f^{-1}(0.5)$.

13. If $f(x) = x^2$ and $g(x) = (x+2)(x-1)(x-3)$, find all x for which $f(x) < g(x)$.

14. Let V represent the volume in liters of air in the lungs during a 5-second respiratory cycle. If t is time in seconds, V is given by
$$V = 0.1729t + 0.1522t^2 - 0.0374t^3.$$

   (a)   Graph this function for $0 \le t \le 5$.
   (b)   What is the maximum value of V on this interval? What is the practical significance of the maximum value?
   (c)   Explain the practical significance of the t- and V-intercepts on the interval $0 \le t \le 5$.

15. Let $C(x)$ be a firm's total cost, in millions of dollars, for producing a given quantity x, in thousands of units, of an item.

   (a)   Graph $C(x) = (x-1)^3 + 1$.
   (b)   Let $R(x)$ be the revenue to the firm (in millions of dollars) for selling a quantity, x, in thousands of units, of the good. Suppose $R(x) = x$. What does this tell you about the price of each unit?
   (c)   Profit equals revenue minus cost. For what values of x does the firm make a profit? Break even? Lose money?

16. The town of Smallsville was founded in 1900. Its population y (in hundreds) is given by the equation
$$y = -0.1x^4 + 1.7x^3 - 9x^2 + 14.4x + 5,$$
where x is the number of years since 1900. Use a the graph in the window $0 \le x \le 10$, $-2 \le y \le 13$.

   (a)   What was the population of Smallsville when it was founded?
   (b)   When did Smallsville become a ghost town (nobody lived there anymore)? Give the year and the month.
   (c)   What was the largest population of Smallsville after 1905? When did Smallsville reach that population? Again, include the month and year. Explain your method.

17. The volume, V, in milliliters, of 1 kg of water as a function of the temperature T is given by:
$$V - 999.87 - 0.06426T + 0.0085143T^2 - 0.0000679T^3, \quad \text{for } 0 \le T \le 30°\text{C}.$$

   (a)   Sketch a graph of V.
   (b)   Describe the shape of the graph. Does V increase or decrease? Does it curve upward or downward? What does the graph tell us about how the volume varies with temperature?
   (c)   At what temperature does water have the maximum density? How does that appear on your graph? (Density = Mass/Volume. In this problem, the mass of the water is 1 kg.)

18.   Let f and g be polynomial functions. Are the compositions

$$f(g(x)) \qquad \text{and} \qquad g(f(x))$$

also polynomial functions? Explain your answer.

19.   A function that is not a polynomial can often be approximated by a polynomial. For example, for certain x-values, the function $f(x) = e^x$ can be approximated by the fifth-degree polynomial

$$p(x) = 1 + x + \frac{x^2}{2} + \frac{x^3}{6} + \frac{x^4}{24} + \frac{x^5}{120}.$$

(a)   Show that $p(1) \approx f(1) = e$. How good is the estimate?
(b)   Calculate $p(5)$. How well does $p(5)$ approximate $f(5)$?
(c)   Graph $p(x)$ and $f(x)$ together on the same set of axes. Based on your graph, for what range of values of x do you think $p(x)$ gives a good estimate for $f(x)$?

20.   Let $f(x) = x - \dfrac{x^3}{6} + \dfrac{x^5}{120}$.

(a)   Sketch graphs of $y = f(x)$ and $y = \sin x$ for $-2\pi \le x \le 2\pi$, $-3 \le y \le 3$.
(b)   The graph of f resembles the graph of $\sin x$ on a small interval. Based on the graphs you made in part (a), give the approximate interval.
(c)   Your calculator uses a function similar to f in order to evaluate the sine function. How reasonable an approximation does f give for $\sin(\pi/8)$?
(d)   Explain how you could use the function f to approximate the value of $\sin\theta$, where $\theta = 18$ radians. [Hint: Use the fact that the sine function is periodic.]

21.   Suppose f is a polynomial function of degree n, where n is a positive even integer. For each of the following statements, write *true* if the statement is always true, *false* otherwise. If the statement is false, give an example that illustrates why it is false.

(a) f is an even function.
(b) f has an inverse.
(c) f cannot be an odd function.
(d)   If $f(x) \to +\infty$ as $x \to +\infty$, then $f(x) \to -\infty$ as $x \to -\infty$.

22.   A woman opens a bank account with an initial deposit of $1000. At the end of each year thereafter, she deposits an additional $1000.

(a)   The account earns 6% annual interest, compounded annually. Complete Table 9.9.
(b)   Does the balance of this account grow linearly, exponentially, or neither? Justify your answer.

TABLE 9.9

Number of years elapsed	Start-of-year balance	End-of-year deposit	End-of-year interest
0	$1000.00	$1000	$60.00
1	$2060.00	$1000	$123.60
2	$3183.60	$1000	
3		$1000	
4		$1000	
5		$1000	

23.   Suppose the annual percentage rate (APR) paid by the account in Problem 22 is r, where r does not necessarily equal 6%. Define $p_n(r)$ as the balance of the account after n years have elapsed. (For example, $p_2(0.06) = \$3183.60$, because, according to Table 9.9, the balance after 2 years is $3183.60 if the APR is 6%.)

(a)   Find formulas for $p_5(r)$ and $p_{10}(r)$.
(b)   What is APR if the woman in Problem 22 has $10,000 in 5 years? [Hint: Use a graphing calculator.]

9.4 THE SHORT-RUN BEHAVIOR OF POLYNOMIALS

The long-run behavior of a polynomial is determined by its leading term. However, polynomials with the same leading term may have very different short-run behaviors.

Example 1   Compare the graphs of the polynomials f, g, and h given by

$$f(x) = x^4 - 4x^3 + 16x - 16, \quad g(x) = x^4 - 4x^3 - 4x^2 + 16x, \quad h(x) = x^4 + x^3 - 8x^2 - 12x.$$

Solution   Each of these functions is a fourth-degree polynomial, and each has x^4 as its leading term. Thus, all their graphs resemble the graph of x^4 on a large scale. See Figure 9.36.

   However, on a smaller scale, the functions look different. See Figure 9.37. Two of the graphs go through the origin while the third does not. The graphs also differ from one another in the number of bumps each one has and in the number of times each one crosses the x-axis. Thus, polynomials with the same leading term look similar on a large scale, but may look dissimilar on a small scale.

Figure 9.36: On a large scale, the polynomials f, g, and h resemble the power function $y = x^4$

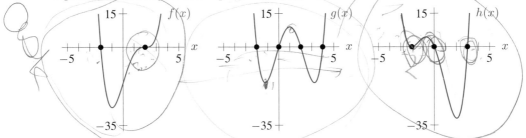

Figure 9.37: On a smaller scale, the polynomials f, g, and h look quite different from each other

Factored Form Shows the Zeros of a Polynomial

To predict the long-run behavior of a polynomial, we write it in standard form. However, to determine the zeros of a polynomial, we write it in factored form, as a product of other polynomials. Some, but not all, polynomials can be factored.

Example 2   Rewrite the third-degree polynomial $u(x) = x^3 - x^2 - 6x$ as a product of linear factors.

Solution   By factoring out an x and then factoring the quadratic, $x^2 - x - 6$, we rewrite $u(x)$ as

$$u(x) = x^3 - x^2 - 6x = x(x^2 - x - 6) = x(x - 3)(x + 2).$$

The expression $x(x - 3)(x + 2)$ is the product of three first-degree polynomials, x, $x - 3$, and $x + 2$.

The advantage of factored form is that we can easily see the zeros of the polynomial. To find these values, we use the following rule:

> If a and b are numbers and if $a \cdot b = 0$, then a or b (or both) must equal 0.

This rule also applies to products of three or more numbers: If such a product equals zero, then at least one of the members of the product must equal zero.

The number 0 is unique in this regard. For example, knowing that $a \cdot b = 12$ doesn't tell us that $a = 12$ or $b = 12$ (or both). For instance, $a = 3$ and $b = 4$ would work, as would $a = 2$ and $b = 6$, or $a = \sqrt{12}$ and $b = \sqrt{12}$, and so on.

Example 3   Find the zeros of $u(x) = x(x - 3)(x + 2)$.

Solution   The polynomial equals zero if and only if at least one of its factors is zero. We solve the equation:

$$x(x - 3)(x + 2) = 0.$$

Thus, either

$$x = 0, \quad \text{or} \quad x - 3 = 0, \quad \text{or} \quad x + 2 = 0,$$

so

$$x = 0, \quad \text{or} \quad x = 3, \quad \text{or} \quad x = -2.$$

These are the zeros, or x-intercepts, of u. To check, evaluate $u(x)$ for these x-values; you should get 0. There are no other zeros.

In Example 3, each linear factor produced a zero of the polynomial. Now suppose that we do not know the polynomial p, but we do know that it has zeros at $x = 0, -12, 31$. Then we know that the factored form of the polynomial must include the factors $(x - 0)$ or x, and $(x - (-12))$ or $(x + 12)$, and $(x - 31)$. It may include other factors too. In summary:

Suppose p is a polynomial. If the formula for p has a **linear factor**, that is, a factor of the form $(x - k)$, then p has a zero at $x = k$.
Conversely, if p has a **zero** at $x = k$, then p has a linear factor of the form $(x - k)$.

If a polynomial doesn't have a zero, this doesn't mean it can't be factored. For example, the polynomial $y = x^4 + 5x^2 + 6$ has no zeros. To see this, notice that x^4 and $5x^2$ are never negative. Thus, y is never less than 6 so it cannot equal 0. See Figure 9.38. However, this polynomial can be factored as $y = (x^2 + 2)(x^2 + 3)$. The point is that a polynomial with a zero has a linear factor, and a polynomial without a zero does not.

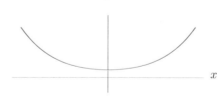

Figure 9.38: The polynomial $y = (x^2 + 2)(x^2 + 3)$ has no zeros

Factored Form and the Short-Run Behavior of Polynomials

On a large enough scale, the graph of the polynomial $u(x) = x^3 - x^2 - 6x$ from Example 2 looks like the graph of its leading term, $y = x^3$. Using factored form, $u(x) = x(x - 3)(x + 2)$, we can investigate the function's short-run behavior.

Example 4   Describe the graph of $y = u(x) = x^3 - x^2 - 6x$. Where does it cross the x-axis? the y-axis? Where is u positive? Negative?

Solution   To describe the graph of u, we give the x- and y-intercepts, and the long-run behavior.

The factored form, $u(x) = x(x-3)(x+2)$, shows that the graph crosses the x-axis at $x - 0$, 3, -2. The graph of u crosses the y-axis at $u(0) = 0^3 - 0^2 - 6 \cdot 0 = 0$; that is, at $y = 0$. For large values of x, the graph of $y = u(x)$ resembles the graph of $y = x^3$, its leading term. Figure 9.39 shows where u is positive and where u is negative.

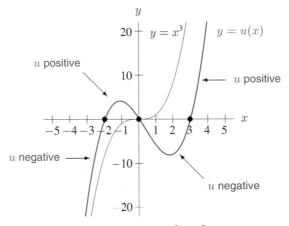

Figure 9.39: The graph of $u(x) = x^3 - x^2 - 6x$ has zeros at $x = -2$, 0, and 3. Its long-run behavior resembles $y = x^3$

Example 5   Analyze the behavior of the following polynomial, p. Explain the relationship between the graph of p and its formula:

$$p(x) = x^4 - 2x^3 - 7x^2 + 8x + 12 = (x+2)(x+1)(x-2)(x-3).$$

Solution   We find the zeros of p, its y-intercept, and its long-run behavior. Since

$$p(x) = (x+2)(x+1)(x-2)(x-3),$$

the zeros of p are given by $x = -2$, $x = -1$, $x = 2$, and $x = 3$.

To find the y-intercept of p, we use the standard form, $p(x) = x^4 - 2x^3 - 7x^2 + 8x + 12$. At $x = 0$ every term vanishes except for the constant term, so $p(0) = 12$. Figure 9.40 shows the short-run behavior. The long-run behavior of p resembles $y = x^4$ and is shown in Figure 9.41.

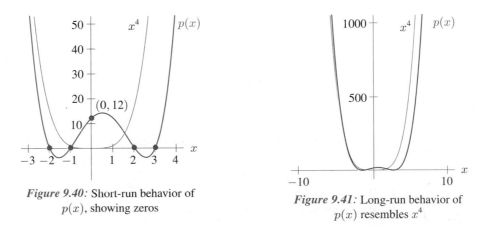

Figure 9.40: Short-run behavior of $p(x)$, showing zeros

Figure 9.41: Long-run behavior of $p(x)$ resembles x^4

The Number of Factors, Zeros, and Bumps

The number of linear factors is always less than or equal to the degree of a polynomial. For example, a fourth degree polynomial can have no more than four linear factors. This makes sense because if we had another factor in the product and multiplied out, the highest power of x would be greater than four. Since each zero corresponds to a linear factor, the number of zeros is less than or equal to the degree of the polynomial.

We can now say that there is a maximum number of bumps in the graph of a polynomial of degree n. Between any two consecutive zeros, there is a bump because the graph changes direction. In Figure 9.40, the graph, which rises at $x = -1$, must come back down in order to cross the x-axis at $x = 2$. In summary:

> The graph of an nth degree polynomial has at most n zeros and turns at most $(n-1)$ times.

Multiple Zeros

The functions $s(x) = (x-4)^2$ and $t(x) = (x+1)^3$ are both polynomials in factored form. Each is a horizontal shift of a power function. We refer to the zeros of s and t as *multiple zeros*, because in each case the factor contributing the value of $y = 0$ is repeated more than once. For instance, we say that $x = 4$ is a *double zero* of s, since

$$s(x) = (x-4)^2 = \underbrace{(x-4)(x-4)}_{\text{Repeated twice}},$$

Likewise, we say that $x = -1$ is a *triple zero* of t, since

$$t(x) = (x+1)^3 = \underbrace{(x+1)(x+1)(x+1)}_{\text{Repeated three times}}.$$

The graphs of s and t in Figures 9.42 and 9.43 show typical behavior near multiple zeros.

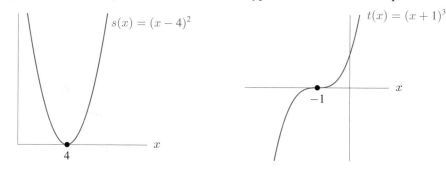

Figure 9.42: Double zero at $x = 4$ *Figure 9.43:* Triple zero at $x = -1$

In general:

> If p is a polynomial with a repeated linear factor, then p has a **multiple zero.**
> - If the factor $(x-k)$ is repeated an even number of times, the graph of $y = p(x)$ does not cross the x-axis at $x = k$, but "bounces" off the x-axis at $x = k$. (See Figure 9.42.)
> - If the factor $(x-k)$ is repeated an odd number of times, the graph of $y = p(x)$ crosses the x-axis at $x = k$, but it looks flattened there. (See Figure 9.43.)

Example 6 Describe in words the zeros of the 4th-degree polynomials $f(x)$, $g(x)$, and $h(x)$, in Figure 9.44.

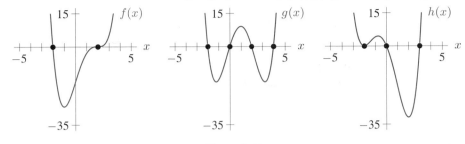

Figure 9.44

Solution The graph suggests that f has a single zero at $x = -2$. The flattened appearance near $x = 2$ suggests that f has a multiple zero there. Since the graph crosses the x-axis at $x = 2$ (instead of bouncing off it), this zero must be repeated an odd number of times. Since f is 4th degree, f has at most 4 factors, so there must be a triple zero at $x = 2$.

The graph of g has four single zeros. The graph of h has two single zeros and a double zero at $x = -2$. The multiplicity of the zero at $x = -2$ is not higher than two because h is of degree $n = 4$.

Finding the Formula for a Polynomial from its Graph

The graph of a polynomial often enables us to find a possible formula for the polynomial.

Example 7 Find a possible formula for the polynomial function f graphed in Figure 9.45.

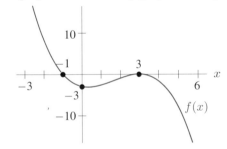

Figure 9.45: Features of the graph lead to a possible formula for this polynomial

Solution Based on its long-run behavior, f is of odd degree greater than or equal to 3. The polynomial has zeros at $x = -1$ and $x = 3$. We see that $x = 3$ is a multiple zero of even power, because the graph bounces off the x-axis here instead of crossing it. Therefore, we try the formula

$$f(x) = k(x + 1)(x - 3)^2$$

where k represents a stretch factor. The shape of the graph shows that k must be negative.

To find k, we use the fact that $f(0) = -3$, so

$$f(0) = k(0 + 1)(0 - 3)^2 = -3$$

which gives

$$9k = -3 \qquad \text{so} \qquad k = -\frac{1}{3}.$$

Thus, $f(x) = -\frac{1}{3}(x + 1)(x - 3)^2$ is a possible formula for this polynomial.

The formula for f we found in Example 7 is the polynomial of least degree we could have chosen. However, there are other polynomials, such as $y = -\frac{1}{27}(x + 1)(x - 3)^4$, with the same overall behavior as the function shown in Figure 9.45.

Problems for Section 9.4

1. Without using a calculator, decide which of the equations best describes the polynomial in Figure 9.46.

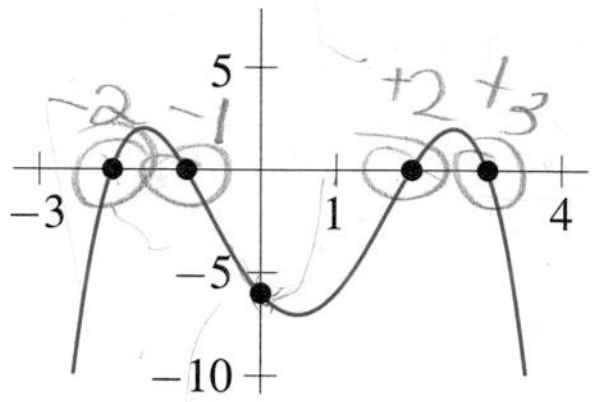

A	$y = (x+2)(x+1)(x-2)(x-3)$
B	$y = x(x+2)(x+1)(x-2)(x-3)$
C	$y = -\frac{1}{2}(x+2)(x+1)(x-2)(x-3)$
D	$y = \frac{1}{2}(x+2)(x+1)(x-2)(x-3)$
E	$y = -(x+2)(x+1)(x-2)(x-3)$

Figure 9.46

Without a calculator, graph the polynomials in Problems 2–3. Label all the x-intercepts and y-intercepts.

2. $f(x) = -5(x^2 - 4)(25 - x^2)$ 3. $g(x) = 5(x-4)(x^2 - 25)$

4. Use the graph of $f(x)$ in Figure 9.37 on page 385 to determine the factored form of
$$f(x) = x^4 - 4x^3 + 16x - 16.$$

5. Use the graph of $h(x)$ in Figure 9.37 on page 385 to determine the factored form of
$$h(x) = x^4 + x^3 - 8x^2 - 12x.$$

Give a possible formula for the polynomials graphed in Problems 6–17.

6.

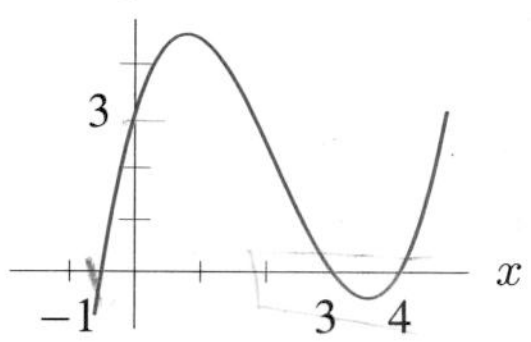

7.

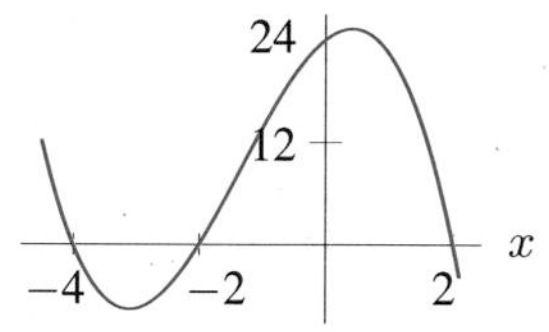

8.

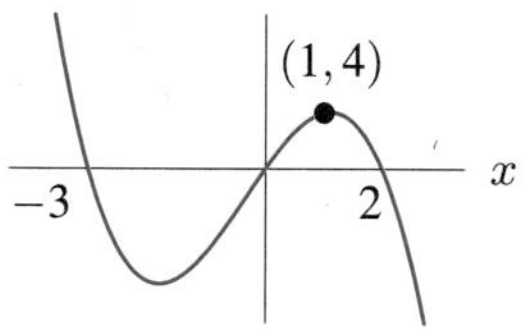

9.

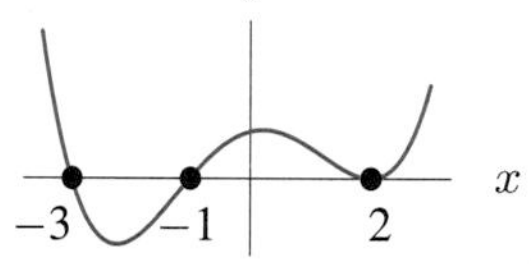

10.

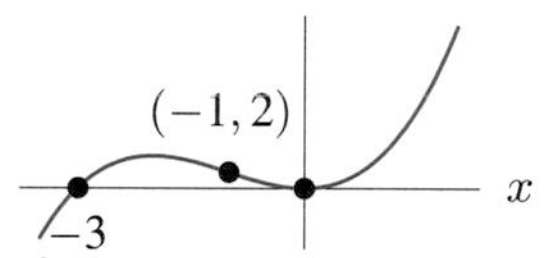

11.

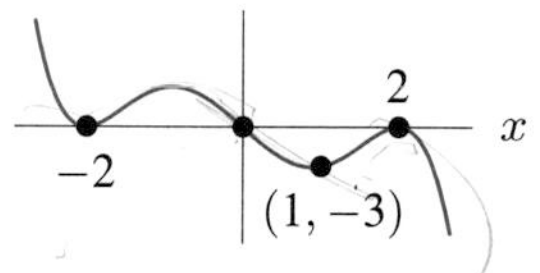

12.

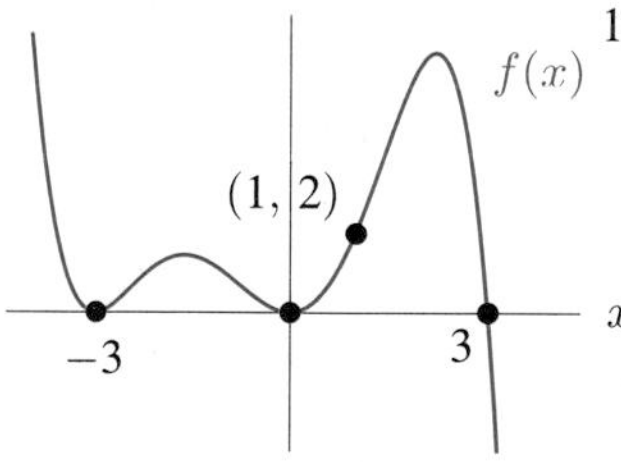

13.

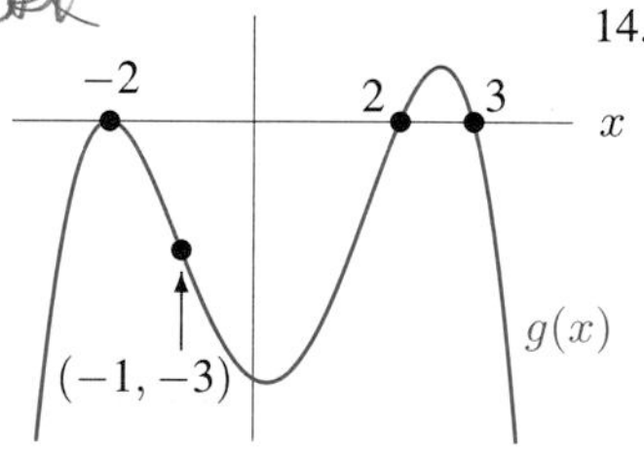

14.

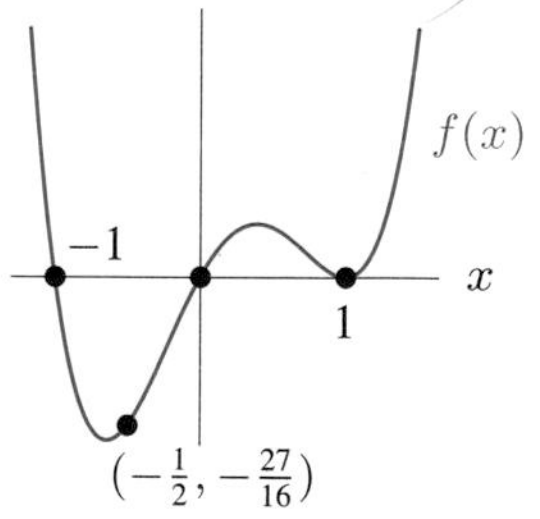

15.

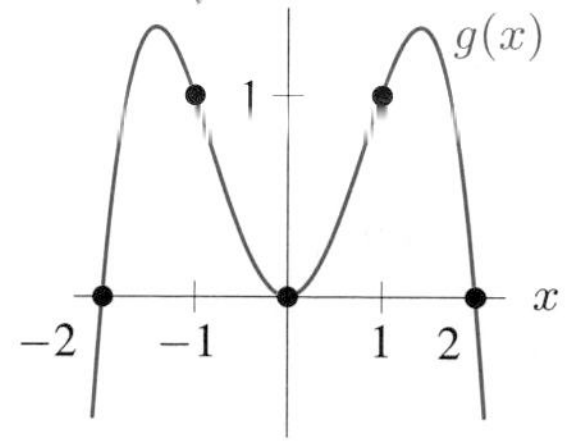

16.

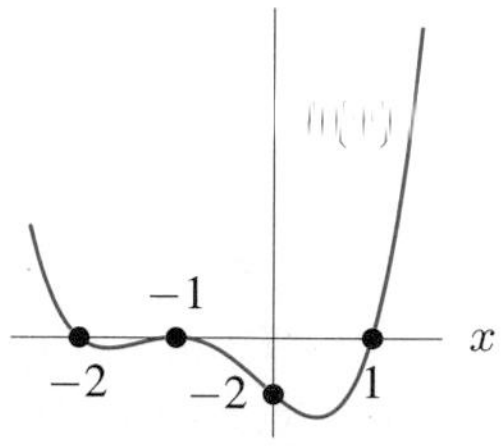

17.

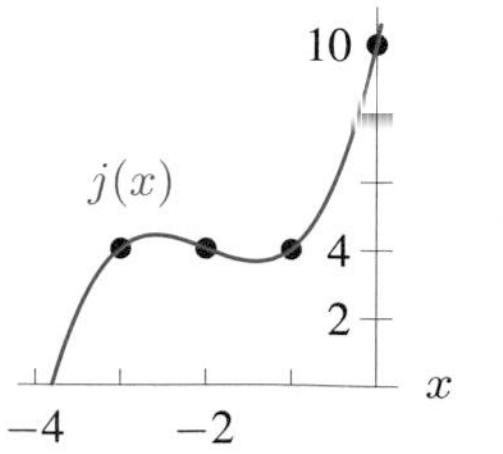

18. Let $u(x) = \frac{1}{8}x^3$ and $v(x) = \frac{1}{8}x(x-0.01)^2$. Do v and u have the exact same graph? Sketch u and v in the window $-10 \leq x \leq 10$, $-10 \leq y \leq 10$. Now do you think that v and u have the same graph? If so, explain why their formulas are different, if not, find a viewing window on which their graphs' differences are prominent.

19. Factor $f(x) = 8x^3 - 4x^2 - 60x$ completely, and determine the zeros of f.

In Problems 20–25, find possible formulas for the polynomials with the given properties.

20. f has degree ≤ 2, $f(0) = f(1) = f(2) = 1$.

21. f has degree ≤ 2, $f(0) = f(2) = 0$ and $f(3) = 3$.

22. f has degree ≤ 2, $f(0) = 0$ and $f(1) = 1$.

23. f is third degree with $f(-3) = 0$, $f(1) = 0$, $f(4) = 0$, and $f(2) = 5$.

24. g is fourth degree, g has a double zero at $x = 3$, $g(5) = 0$, $g(-1) = 0$, and $g(0) = 3$.

25. Least possible degree through the points $(-3, 0)$, $(1, 0)$, and $(0, -3)$.

26. Which of these functions have inverses that are functions? Discuss.
    (a) $f(x) = (x-2)^3 + 4$.   (b) $g(x) = x^3 - 4x^2 + 2$.

For Problems 27–32, find the real zeros (if any) of the polynomials.

27. $y = x^2 + 5x + 6$

28. $y = x^4 + 6x^2 + 9$

29. $y = 4x^2 - 1$

30. $y = 4x^2 + 1$

31. $y = 2x^2 - 3x - 3$

32. $y = 3x^5 + 7x + 1$

33. An open-top box is to be constructed from a 6 in by 8 in rectangular sheet of tin by cutting out squares of equal size at each corner, then folding up the resulting flaps. Let x denote the length of the side of each cut-out square. Assume negligible thickness.
    (a) Find a formula for the volume of the box as a function of x.
    (b) For what values of x does the formula from part (a) make sense in the context of the problem?
    (c) Sketch a graph of the volume function.
    (d) What, approximately, is the maximum volume of the box?

34. You wish to pack a cardboard box inside a wooden crate. In order to have room for the packing materials, you need to leave a 0.5-ft space around the front, back, and sides of the box, and a 1-ft space around the top and bottom of the box. If the cardboard box is x feet long, $(x+2)$ feet wide, and $(x-1)$ feet deep, find a formula in terms of x for the amount of packing material needed.

35. Take an 8.5 by 11-inch piece of paper and cut out four equal squares from the corners. Fold up the sides to create an open box. Find the dimensions of the box that has maximum volume.

36. Consider the function $a(x) = x^5 + 2x^3 - 4x$.
    (a) Without using a calculator or computer, what can you say about the graph of a?
    (b) Use a calculator or a computer to determine the zeros of this function to three decimal places.
    (c) Explain why you think that you have all the possible zeros.
    (d) What are the zeros of $b(x) = 2x^5 + 4x^3 - 8x$? Does your answer surprise you?

37. (a) Sketch a graph of $f(x) = x^4 - 17x^2 + 36x - 20$ for $-10 < x \leq 10$, $-10 \leq y \leq 10$.
    (b) Your graph should appear to have a vertical asymptote at $x = -5$. Does f actually have a vertical asymptote here? Explain.
    (c) How many zeros does f have? Can you find a window in which all of the zeros of f are clearly visible? Discuss.
    (d) Write the formula of f in factored form.
    (e) How many turning points does the graph of f have? Can you find a window in which all the turning points of f are clearly visible? Discuss.

What Causes Asymptotes?

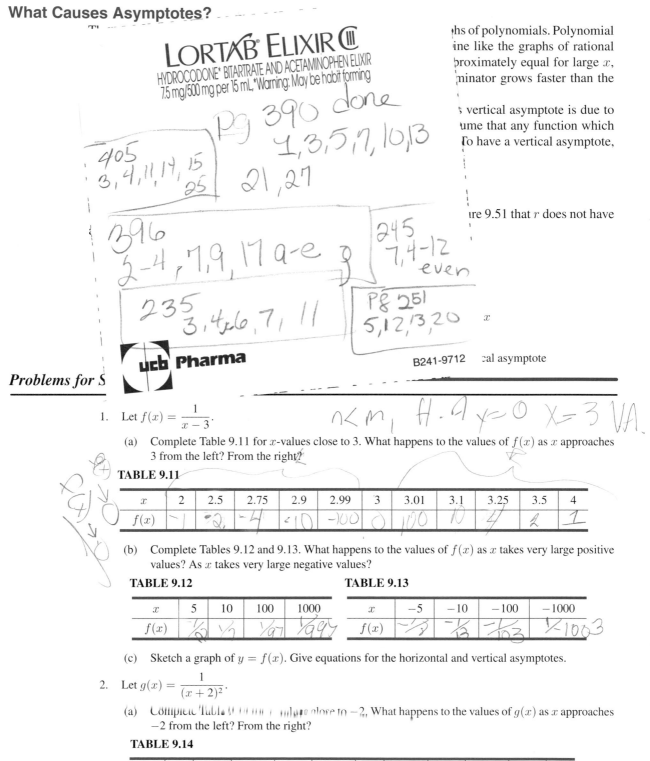

...hs of polynomials. Polynomial ...ine like the graphs of rational ...proximately equal for large x, ...minator grows faster than the

...s vertical asymptote is due to ...ume that any function which ...To have a vertical asymptote,

...re 9.51 that r does not have

...cal asymptote

Problems for S

1. Let $f(x) = \dfrac{1}{x-3}$.

    (a) Complete Table 9.11 for x-values close to 3. What happens to the values of $f(x)$ as x approaches 3 from the left? From the right?

    TABLE 9.11

x	2	2.5	2.75	2.9	2.99	3	3.01	3.1	3.25	3.5	4
$f(x)$											

    (b) Complete Tables 9.12 and 9.13. What happens to the values of $f(x)$ as x takes very large positive values? As x takes very large negative values?

    TABLE 9.12

x	5	10	100	1000
$f(x)$				

    TABLE 9.13

x	-5	-10	-100	-1000
$f(x)$				

    (c) Sketch a graph of $y = f(x)$. Give equations for the horizontal and vertical asymptotes.

2. Let $g(x) = \dfrac{1}{(x+2)^2}$.

    (a) Complete Table 9.14 for x-values close to -2. What happens to the values of $g(x)$ as x approaches -2 from the left? From the right?

    TABLE 9.14

x	-3	-2.5	-2.25	-2.1	-2.01	-2	-1.99	-1.9	-1.75	-1.5	-1
$g(x)$											

(b)   Complete Tables 9.15 and 9.16. What happens to the values of $g(x)$ as x takes very large positive values? As x takes very large negative values?

TABLE 9.15

x	5	10	100	1000
$g(x)$				

TABLE 9.16

x	-5	-10	-100	-1000
$g(x)$				

(c)   Sketch a graph of $y = g(x)$. Give equations for the horizontal and vertical asymptotes.

3.   Let $F(x) = \dfrac{x^2 - 1}{x^2}$.

(a)   Complete Table 9.17 for x-values close to 0. What happens to the values of $F(x)$ as x approaches 0 from the left? From the right?

TABLE 9.17

x	-1	-0.5	-0.25	-0.1	-0.01	0	0.01	0.1	0.25	0.5	1
$F(x)$											

(b)   Complete Tables 9.18 and 9.19. What happens to the values of $F(x)$ as x takes very large positive values? As x takes very large negative values?

TABLE 9.18

x	5	10	100	1000
$F(x)$				

TABLE 9.19

x	-5	-10	-100	-1000
$F(x)$				

(c)   Sketch a graph of $y = F(x)$. Give equation for the horizontal and vertical asymptotes.

4.   Let $G(x) = \dfrac{2x}{x + 4}$.

(a)   Complete Table 9.20 for x-values close to -4. What happens to the values of $G(x)$ as x approaches -4 from the left? From the right?

TABLE 9.20

x	-5	-4.5	-4.25	-4.1	-4.01	-4	-3.99	-3.9	-3.75	-3.5	-3
$G(x)$											

(b)   Complete Tables 9.21 and 9.22. What happens to the values of $G(x)$ as x takes very large positive values? As x takes very large negative values?

TABLE 9.21

x	5	10	100	1000
$G(x)$				

TABLE 9.22

x	-5	-10	-100	-1000
$G(x)$				

(c)   Sketch a graph of $y = G(x)$. Give equations for the horizontal and vertical asymptotes.

5.   Compare and discuss the long-run behaviors of the following functions:

$$f(x) = \frac{x^2 + 1}{x^2 + 5}, \qquad g(x) = \frac{x^3 + 1}{x^2 + 5}, \qquad h(x) = \frac{x + 1}{x^2 + 5}.$$

6.   Give examples of rational functions with even symmetry, odd symmetry, and neither. How does the symmetry of $f(x) = p(x)/q(x)$ depend on the symmetry of $p(x)$ and $q(x)$?

Finding a Formula for a Rational Function from its Graph

The graph of a rational function can give a good idea of its formula. Zeros of the function correspond to factors in the numerator and vertical asymptotes correspond to factors in the denominator.

Example 6   Find a possible formula for the rational function, $g(x)$, graphed in Figure 9.60.

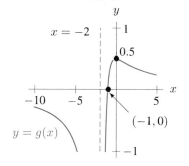

Figure 9.60: The graph of $y = g(x)$ a rational function

Solution   From the graph, we see that g has a zero at $x = -1$ and a vertical asymptote at $x = -2$. This means that the numerator of g has a zero at $x = -1$ and the denominator of g has a zero at $x = -2$. The zero of g does not seem to be a multiple zero because the graph crosses the x-axis instead of bouncing and does not have a flattened appearance. Thus, we conclude that the numerator of g has one factor of $(x + 1)$.

The behavior of g near its vertical asymptote is more like the behavior of $y = 1/(x + 2)^2$ than like $y = 1/(x + 2)$. We conclude that the denominator of g has a factor of $(x + 2)^2$. This tells us

$$g(x) = k \cdot \frac{x + 1}{(x + 2)^2},$$

where k is a stretch factor. To find the value of k, use the fact that $g(0) = 0.5$. So

$$0.5 = k \cdot \frac{0 + 1}{(0 + 2)^2}$$

$$0.5 = k \cdot \frac{1}{4}$$

$$k = 2.$$

Thus, a possible formula for g is $g(x) = \dfrac{2(x + 1)}{(x + 2)^2}$.

When Numerator and Denominator Have the Same Zeros: Holes

The rational function $h(x) = \dfrac{x^2 + x - 2}{x - 1}$ is undefined at $x = 1$ because the denominator equals zero at $x = 1$. However, the graph of h does not have a vertical asymptote at $x = 1$ because the numerator of h also equals zero at $x = 1$. At $x = 1$,

$$h(1) = \frac{x^2 + x - 2}{x - 1} = \frac{1^2 + 1 - 2}{1 - 1} = \frac{0}{0},$$

and this ratio is undefined. What does the graph of h look like? Factoring the numerator of h gives

$$h(x) = \frac{(x - 1)(x + 2)}{x - 1} = \frac{x - 1}{x - 1}(x + 2).$$

For any $x \neq 1$, we can cancel $(x - 1)$ top and bottom and rewrite the formula for h as

$$h(x) = x + 2, \qquad \text{provided } x \neq 1.$$

Thus, the graph of h is the line $y = x + 2$ except at $x = 1$, where h is undefined. The line $y = x + 2$ contains the point $(1, 3)$, but that the graph of h does not. Therefore, we say that the graph of h has a *hole* in it at the point $(1, 3)$. See Figure 9.61.

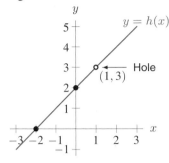

Figure 9.61: The graph of $y = h(x)$ is the line $y = x + 2$, except at the point $(1, 3)$, where it has a hole

Problems for Section 9.6

For each rational function in Problems 1–4, find all zeros and vertical asymptotes and describe its long-run behavior. Then sketch a graph of the function without using a calculator.

1. $y = \dfrac{x + 3}{x + 5}$
2. $y = \dfrac{x + 3}{(x + 5)^2}$
3. $y = \dfrac{x - 4}{x^2 - 9}$
4. $y = \dfrac{x^2 - 4}{x - 9}$

5. Sketch a graph of $y = \dfrac{2x^2 - 10x + 12}{x^2 - 16}$ without using a calculator.

6. Let f and g be polynomials given by $f(x) = x^2 + 5x + 6$ and $g(x) = x^2 + 1$.

    (a) What are the zeros of f and g?
    (b) Let r be a rational function given by $r(x) = f(x)/g(x)$. Sketch a graph of r. Does r have zeros? Vertical asymptotes? What is its long-run behavior as $x \to \pm\infty$?
    (c) Let s be a rational function given by $s(x) = g(x)/f(x)$. If you graph s in the window $-10 \le x \le 10$, $-10 \le y \le 10$, it appears to have a zero near the origin. Does it? Does s have a vertical asymptote? What is its long-run behavior?

7. Without using a calculator, match the functions in (a) – (f) with their graphs in (i) – (vi) by finding the zeros, asymptotes, and end behavior for each function.

    (a) $f(x) = \dfrac{-1}{x^2 - 10x + 25} - 1$
    (b) $f(x) = \dfrac{x - 2}{(x + 1)(x - 3)}$
    (c) $f(x) = \dfrac{2x + 4}{x - 1}$

    (d) $f(x) = \dfrac{1}{x + 1} + \dfrac{1}{x - 3}$
    (e) $f(x) = \dfrac{1 - x^2}{x - 2}$
    (f) $f(x) = \dfrac{1 - 4x}{2x + 2}$

(i)

(ii)

(iii)

(iv)

(v)

(vi)

8. Suppose that n is a constant and that $f(x)$ is a function defined when $x = n$. Complete the following sentences.

    (a) If $f(n)$ is large, then $\dfrac{1}{f(n)}$ is ...

    (b) If $f(n)$ is small, then $\dfrac{1}{f(n)}$ is ...

    (c) If $f(n) = 0$, then $\dfrac{1}{f(n)}$ is ...

    (d) If $f(n)$ is positive, then $\dfrac{1}{f(n)}$ is ...

    (e) If $f(n)$ is negative, then $\dfrac{1}{f(n)}$ is ...

9. (a) Use the results of Problem 8 to graph $y = 1/f(x)$ given the graph of $y = f(x)$ in Figure 9.62.

    (b) Find a possible formula for the function in Figure 9.62. Use this formula to check your graph for part (a).

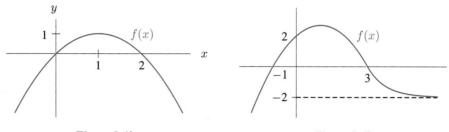

Figure 9.62

Figure 9.63

10. Use the graph of f in Figure 9.63 to graph   (a) $y = -f(-x) + 2$ (b) $y = \dfrac{1}{f(x)}$

Problems 11–13 each show a graph of a translation of the function $y = 1/x$. In each problem

(a) Find a possible formula that represents the graph.

(b) Write the formula from part (a) as the ratio of two linear polynomials.

(c) Find the coordinates of the intercepts of the graph.

11.   12.   13.

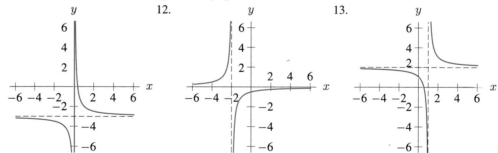

Problems 14–16 each show a graph of a translation of $y = 1/x^2$. In each case:

(a) Find a formula that represents the graph.

(b) Write the formula from part (a) as the ratio of two polynomials.

(c) Find the coordinates of any intercepts of the graph.

14.   15.   16.

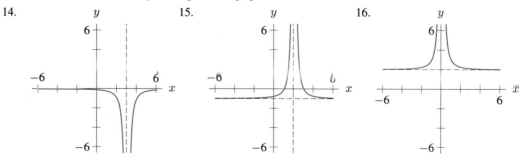

Each of the functions in Problems 17–19 is a transformation of $y = 1/x^p$. For each function, determine p, describe the transformation in words, and graph the function and label any intercepts and asymptotes.

17. $f(x) = \dfrac{1}{x-3} + 4$

18. $g(x) = -\dfrac{1}{(x-2)^2} - 3$

19. $h(x) = \dfrac{1}{x-1} + \dfrac{2}{1-x} + 2$

Problems 20–23 give values of translations of either $y = 1/x$ or $y = 1/x^2$. In each case

(a) Determine if the values are from a translation of $y = 1/x$ or $y = 1/x^2$. Explain your reasoning.

(b) Find a possible formula for the function given by the table.

20.

x	y
2.7	12.1
2.9	101
2.95	401
3	Undefined
3.05	401
3.1	101
3.3	12.1

21.

x	y
-1000	0.499
-100	0.490
-10	0.400
10	0.600
100	0.510
1000	0.501

22.

x	y
-1000	1.000001
-100	1.00001
-10	1.01
10	1.01
100	1.0001
1000	1.000001

23.

x	y
1.5	-1.5
1.9	-9.5
1.95	-19.5
2	Undefined
2.05	20.5
2.1	10.5
2.5	2.5

Find possible formulas for the functions graphed in Problems 24–31.

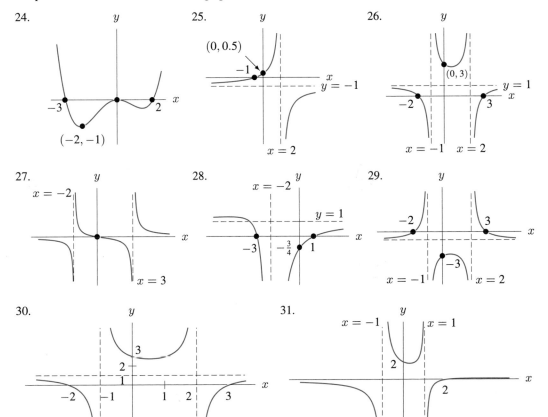

24.

25.

26.

27.

28.

29.

30.

31.

In Problems 32–34, find a possible formula for the rational functions.

32. The graph of $y = f(x)$ has one vertical asymptote, at $x = -1$, and a horizontal asymptote at $y = 1$. The graph of f crosses the y-axis at $y = 3$ and crosses the x-axis once, at $x = -3$.

33. The graph of $y = g(x)$ has two vertical asymptotes: one at $x = -2$ and one at $x = 3$. It has a horizontal asymptote of $y = 0$. The graph of g crosses the x-axis once, at $x = 5$.

34. The graph of $y = h(x)$ has two vertical asymptotes: one at $x = -2$ and one at $x = 3$. It has a horizontal asymptote of $y = 1$. The graph of h touches the x-axis once, at $x = 5$.

35. Cut four equal squares from the corners of a $8.5'' \times 11''$ piece of paper. Fold up the sides to create an open box. Find the dimensions of the box with the maximum volume per surface area.

36. The graph of the rational function $f(x) = \dfrac{18 - 11x + x^2}{x - 2}$ is a line with a hole in it. What is the equation of the line? What are the coordinates of the hole?

37. The graph of the rational function $g(x) = \dfrac{x^3 + 5x^2 + x + 5}{x + 5}$ is a parabola with a hole in it. What is the equation of the parabola? What are the coordinates of the hole?

38. Write a formula for a function, $h(x)$, whose graph is identical to the graph of $y = x^3$, except that the graph of h has a hole at $(2, 8)$. Express the formula as a ratio of two polynomials

REVIEW PROBLEMS FOR CHAPTER NINE

In Problems 1–6, give the type of functions—linear, exponential, logarithmic, trigonometric, power, or polynomial—that fits the data. Find a possible formula for the function whose values are in each table.

1.

x	-2	-1	0	1	2	3
y	3	-1	3	-1	3	-1

2.

x	1	2	3	4	5
y	8	3	-2	-7	-12

3.

x	-3	-2	-1	0	1	2	3
y	-40	0	12	8	0	0	20

4.

x	-2	-1	0	1	2
y	-24	-3	0	3	24

5.

x	-2	-1	0	1	2	3
y	0.02	0.10	0.5	2.50	12.5	62.5

6.

x	$1/100$	$1/10$	1	10	100
y	-2	-1	0	1	2

7. Without using a calculator, for each graph, find the function in Table 9.23 that it represents.

TABLE 9.23

(A) $y = 0.5\sin(2x)$	(G) $y = 0.5\sin(0.5x)$	(M) $y = 1/(x - 6)$
(B) $y = -\ln x$	(H) $y = \ln(x + 1)$	(N) $y = (x - 2)/(x^2 - 9)$
(C) $y = 10(0.6)^x$	(I) $y = 7(2.5)^x$	(O) $y = 1/(x^2 - 4)$
(D) $y = 2\sin(2x)$	(J) $y = 2\sin(0.5x)$	(P) $y = x/(x - 3)$
(E) $y = \ln(-x)$	(K) $y = \ln(x - 1)$	(Q) $y = (x - 1)/(x + 3)$
(F) $y = -15(3.1)^x$	(L) $y = 2e^{-0.2x}$	(R) $y = 1/(x^2 + 4)$

Find possible formulas for the polynomials in Problems 8–13.

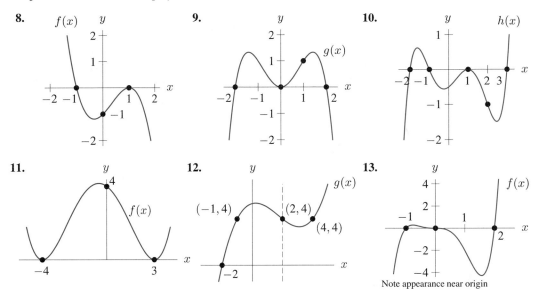

14. Without using a calculator, for each graph, find the function in Table 9.24 that it represents.

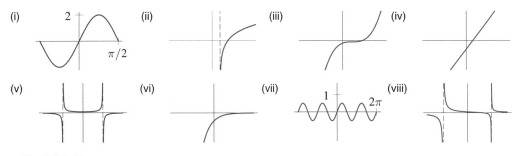

TABLE 9.24

(A) $y = 0.5\sin(2x)$	(I) $y = -\ln x$	(Q) $y = 3e^{-x}$
(B) $y = 2\sin(2x)$	(J) $y = \ln(-x)$	(R) $y = -3e^{x}$
(C) $y = 0.5\sin(0.5x)$	(K) $y = \ln(x+1)$	(S) $y = -3e^{-x}$
(D) $y = 2\sin(0.5x)$	(L) $y = \ln(x-1)$	(T) $y = 3e^{-x^2}$
(E) $y = (x-2)/(x^2-9)$	(M) $y = (x+3)/(x^2-4)$	(U) $y = 1/(4-x^2)$
(F) $y = (x-3)/(x^2-1)$	(N) $y = (x^2-4)/(x^2-1)$	(V) $y = 1/(x^2+4)$
(G) $y = (x-1)^3 - 1$	(O) $y = (x+1)^3 - 1$	(W) $y = (x+1)^3 + 1$
(H) $y = 2x - 4$	(P) $y = -2x - 4$	(X) $y = 2(x+2)$

15. Assume that $x = a$ and $x = b$ are zeros of the second degree polynomial, $y = q(x)$.

(a) Explain what you know and don't know about the graph of q. (Intercepts, vertex, end behavior.)

(b) Explain why a possible formula for $q(x)$ is $q(x) = k(x-a)(x-b)$, with k unknown.

16. (a) Suppose $f(x) = ax^2 + bx + c$. What must be true about the coefficients if f is an even function?

(b) Suppose $g(x) = ax^3 + bx^2 + cx + d$. What must be true about the coefficients if g is an odd function?

17. The gravitational force exerted by a planet is inversely proportional to the square of the distance to the center of the planet. Thus, the weight, w, of an object at a distance, r, from a planet's center is given by

$$w = \frac{k}{r^2},$$

where the constant k depends on the masses of the object and the planet.

A gravitational force of one ton (2000 lbs) will kill a 150-pound person. Suppose the earth's radius were to shrink with its mass remaining the same. What is the smallest radius at which the 150-pound person could survive? Give your answer as a percentage of the earth's radius.

18. Tradition holds that the ancient Greeks placed great importance on a number known as the *golden ratio*, ϕ (read "phi"). This number can be defined geometrically. Starting with a square, add a rectangle to one side of the square, so that the resulting rectangle has the same proportions as the rectangle that was added. The golden ratio is defined as the ratio of either rectangle's length to its width. (See Figure 9.64.) Given this information, show that $\phi = (1 + \sqrt{5})/2$.

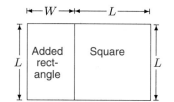

Figure 9.64: Large rectangle has the same proportions as small rectangle

19. The number ϕ, found in Problem 18, has an unusual property:

$$\phi^k + \phi^{k+1} = \phi^{k+2}.$$

(a) Check that the property holds true for $k = 3$ and $k = 10$.
(b) Show that this property follows from the definition of ϕ.

20. An alcohol solution consists of 5 gallons of pure water and x gallons of alcohol, $x > 0$. Let $f(x)$ be the ratio of the volume of alcohol to the total volume of liquid. [Note that $f(x)$ is the concentration of the alcohol in the solution.]

(a) Find a possible formula for $f(x)$.
(b) Evaluate and interpret $f(7)$ in the context of the mixture.
(c) What is the zero of f? Interpret your result in the context of the mixture.
(d) Find an equation for the horizontal asymptote of f. Explain its significance in the context of the mixture.

21. The function $f(x)$ defined in Problem 20 gives the concentration of a solution of x gallons of alcohol and 5 gallons of water.

(a) Find a formula for $f^{-1}(x)$.
(b) Evaluate and interpret $f^{-1}(0.2)$ in the context of the mixture.
(c) What is the zero of f^{-1}? Interpret your result in the context of the mixture.
(d) Find an equation for the horizontal asymptote of f^{-1}. Explain its significance in the context of the mixture.

22. For each of the following functions, state whether it is even, odd, or neither.

(a) $f(x) = x^2 + 3$ (b) $g(x) = x^3 + 3$ (c) $h(x) = 5/x$ (d) $j(x) = |x - 4|$
(e) $k(x) = \log x$ (f) $l(x) = \log(x^2)$ (g) $m(x) = 2^x + 2$ (h) $n(x) = \cos x + 2$

Problems 23–26 show graphs of rational functions of the form

$$y = \frac{(x - A)(x - B)}{(x - C)(x - D)},$$

for different values of the constants A, B, C, and D, with $A \le B$ and $C \le D$. In each case, rank the constants 0, A, B, C, D in order, from least to greatest.

23.   **24.**

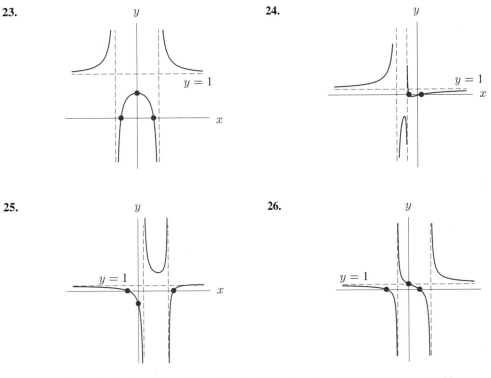

25.   **26.**

Find possible formulas for the polynomials and rational functions described in Problems 27–29.

27.   This function has zeros at $x = -3$, $x = 2$, $x = 5$, and a double-zero at $x = 6$. It has a y-intercept of 7.

28.   This function has zeros at $x = -3$ and $x = 2$, and vertical asymptotes at $x = -5$ and $x = 7$. It has a horizontal asymptote of $y = 1$.

29.   This function has zeros at $x = 2$ and $x = 3$. It has a vertical asymptote at $x = 5$. It has a horizontal asymptote of $y = -3$.

30.   Let $f(x) = (x - 3)^2$, $g(x) = x^2 - 4$, $h(x) = x + 1$, and $j(x) = x^2 + 1$. Without a calculator, match the functions described in (a) – (f) to the functions in (i) – (vi). Some of the descriptions may have no matching function or more than one matching function.

    (i) $p(x) = \dfrac{f(x)}{g(x)}$ (ii) $q(x) = \dfrac{h(x)}{g(x)}$ (iii) $r(x) = f(x)h(x)$

    (iv) $s(x) = \dfrac{g(x)}{j(x)}$ (v) $t(x) = \dfrac{1}{h(x)}$ (vi) $v(x) = \dfrac{j(x)}{f(x)}$

    (a)   Two zeros, no vertical asymptotes, and a horizontal asymptote.
    (b)   Two zeros, no vertical asymptote, and no horizontal asymptote.
    (c)   One zero, one vertical asymptote, and a horizontal asymptote.
    (d)   One zero, two vertical asymptotes, and a horizontal asymptote.
    (e)   No zeros, one vertical asymptote, and a horizontal asymptote at $y = 1$.
    (f)   No zeros, one vertical asymptote, and a horizontal asymptote at $y = 0$.

31. Suppose we want to launch an unpowered spacecraft, initially in free space, so that it hits a planet of radius R. When viewed from a distance, the planet looks like a disk of area πR^2. In the absence of gravity, we would need to aim the spacecraft directly towards this disk. However, because of gravity, the planet draws the spacecraft towards it, so that even if we are somewhat off the mark, the spacecraft might still hit its target. This means that the area we must aim for is actually larger than the apparent area of the planet. This area is called the planet's *capture cross-section*. The more massive the planet, the larger its capture cross-section, because it exerts a stronger pull on passing objects.

A spacecraft with a large initial velocity has a greater chance of slipping past the planet even if its aim is only slightly off, whereas a spacecraft with a low initial velocity has a good chance of drifting into the planet even if its aim is poor. Thus, the planet's capture cross-section is a function of the initial velocity of the spacecraft, v. If the planet's capture cross-section is denoted by A and M is the planet's mass, then it can be shown that, for a positive constant G,

$$A(v) = \pi R^2 \left(1 + \frac{2MG/R}{v^2} \right).$$

(a) Show from the formula for $A(v)$ that, for any initial velocity v,

$$A(v) > \pi R^2.$$

Explain why this makes sense physically.

(b) Consider two planets: The first is twice as massive as the second, and the radius of the second is twice the radius of the first. Which has the larger capture cross-section?

(c) The graph of $A(v)$ has both horizontal and vertical asymptotes. Find their equations, and explain their physical significance.

32. A group of x people is tested for the presence of a certain virus. Unfortunately, the test is imperfect and incorrectly identifies some healthy people as being infected and some sick people as being noninfected. There is no way to know when the test is right and when it is wrong. Since the disease is so rare, only 1% of those tested are actually infected.

(a) Write expressions in terms of x for the number of people tested that are actually infected and the number who are not.

(b) The test correctly identifies 98% of all infected people as being infected. (It incorrectly identifies the remaining 2% as being healthy.) Write, in terms of x, an expression representing the number of infected people who are correctly identified as being infected. (This group is known as the *true-positive* group.)

(c) The test incorrectly identifies 3% of all healthy people as being infected. (It correctly identifies the remaining 97% as being healthy.) Write, in terms of x, an expression representing the number of noninfected people who are incorrectly identified as being infected. (This group is known as the *false-positive* group.)

(d) Write, in terms of x, an expression representing the total number of people the test identifies as being infected, including both true- and false-positives.

(e) Write, in terms of x, an expression representing the fraction of those testing positive who are actually infected. Can this expression be evaluated without knowing the value of x?

(f) Suppose you are among the group of people who test positive for the presence of the virus. Based on your test result, do you think it is likely that you are actually infected?

33. Physicists use *energy diagrams* in the study of planetary motion. An energy diagram is a graph of the *effective potential energy*, U_{eff}, of a sun-planet system, as a function of the planet's distance from the sun, r. By convention, a negative value of U_{eff} signifies an attractive potential, and a positive value of U_{eff} signifies a repulsive potential. Typically, the effective potential of a sun-planet system has both an attractive term, due to gravity, and a repulsive term, due to the circular motion. For some sun-planet system, suppose that

$$U_{\text{eff}} = -\frac{3}{r} + \frac{1}{r^2}, \qquad r > 0.$$

(a) Which of the terms in this expression is attractive? Which is repulsive?

    (b)   Sketch an energy diagram for this sun-planet system.

    (c)   At what distance is the planet most strongly attracted to the sun?

    (d)   Give the equation of the horizontal asymptote of U_{eff}. Explain its physical significance; that is, describe how the planet's attraction to the sun varies as $r \to \infty$.

    (e)   Give the equation of the vertical asymptote of U_{eff}. Explain its physical significance.

34.   The orbit of a planet around the sun is determined by the effective potential energy of the sun-planet system, U_{eff}, defined in Problem 33, and by the system's *total energy*, E, which is a constant. The planet's motion about the sun is confined to those distances r for which $E > U_{\text{eff}}$. Suppose the sun-planet system described in Problem 33 has a total energy $E = -1$.

    (a)   Sketch a graph of the system's total energy, E, on the same set of axes as its effective potential, U_{eff}. [Hint: The total energy is a constant, so its graph is a horizontal line.]

    (b)   Using the fact that the planet is confined to those distances r for which $E > U_{\text{eff}}$, determine the planet's *perihelion* — the smallest distance it can come to the sun.

    (c)   Determine the planet's *aphelion* — the largest distance it can go from the sun.

35.   The *total energy* of a sun-planet system is defined in Problem 34.

    (a)   Using the fact that E must exceed U_{eff}, determine the minimum possible total energy, E_{min}, of the sun-planet system described in Problem 34.

    (b)   Suppose the total energy of this system is E_{min}. Describe the orbit of the planct.

36.   (a)   If the total energy E of a sun-planet system, defined in Problem 34, is negative, then the system is referred to as *bound*. The total energy of the system described in Problem 34 is negative. Explain why, based on your answer to Problem 34, it makes sense to refer to this system as bound.

    (b)   If the total energy E of a sun-planet system is positive, the system is referred to as *unbound*. Suppose the system described in Problem 33 has a positive total energy of $E = 0.5$. Sketch an energy diagram of this system that shows both E and U_{eff}. Explain why it makes sense to refer to this system as unbound.

37.   Ship designers usually construct scale models before building a real ship. The formula that relates the speed u to the hull length l of a ship is

$$u = k\sqrt{l},$$

where k is a positive constant. This constant k varies depending on the ship's design, but scale models of a real ship have the same k as the real ship after which they are modeled.[2]

    (a)   How fast should a scale model with hull length 4 meters travel to simulate a real ship with hull length 225 meters traveling 9 meters/sec?

    (b)   A new ship is to be built whose speed is to be 10% greater than the speed of an existing ship with the same design. What is the relationship between the hull lengths of the new ship and the existing ship?

38.   The cruising speed V of birds at sea-level (in meters/sec) is determined [3] by the mass M of the bird (in grams), and the surface area S of the wings exposed to the air (in square meters). It is given by

$$V = 0.164\sqrt{\frac{M}{S}}.$$

    (a)   The mass of a partridge is half the mass of a hawk. Their wing surface areas are typically 0.043 and 0.166 square meters, respectively. Which bird has the faster cruising speed? The cruising speed of the partridge is 15.6 meters/sec. What are the masses of the partridge and the hawk?

    (b)   The wing surface area of a Canadian goose is typically 12 times that of an American robin, whereas the mass of the goose is 70 times that of the robin. Which bird has the faster cruising speed? The mass and cruising speed of the American robin are typically 80 grams and 9.5

---

[2] R. McNeill Alexander, *Dynamics of Dinosaurs and Other Extinct Giants.* (New York: Columbia University Press, 1989).

[3] H. Tennekes, *The Simple Science of Flight.* (Cambridge: MIT Press, 1996).

meters/sec, respectively. What are the wing surface areas of the Canadian goose and the American robin?

(c)   Use a graphing calculator or computer to plot V against different masses M, for birds with the same surface area of 0.01 square meters—swallows, martins, swifts, and so on. Describe the graph in words. What happens to the cruising speed as the mass increases?

(d)   Use a graphing calculator or computer to plot V against different wing surface areas S, for birds with the same mass 784 grams—falcons, hawks, and so on. Describe the graph in words. What happens to the cruising speed as the wing surface area increases?

(e)   When a bird dives it draws in its wings. What happens to its cruising speed? Is this realistic?

CHAPTER TEN

VECTORS

Vectors are used to represent quantities, such as displacement and velocity, which have both magnitude and direction. This chapter shows how vectors are added, subtracted, and multiplied in two different ways.

10.1 VECTORS

Distance Versus Displacement

If you start at home and walk 3 miles east and then 4 miles north, you have walked a total of 7 miles. However, your distance from home is not 7 miles, but $5 = \sqrt{3^2 + 4^2}$ miles, by the Pythagorean Theorem. See Figure 10.1. As this example illustrates, there is a distinction between *distance* and *displacement*. Distance is a number that measures separation, while displacement consists of separation and direction. For example, walking 3 miles north and walking 3 miles east give different displacements, but both correspond to a distance of 3 miles.

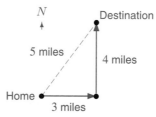

Figure 10.1: After walking 3 miles east then 4 miles north, your distance from home is 5 miles

Adding Displacements Using Triangles

Displacements are added by treating them as arrows called *vectors*. The length of each arrow indicates the magnitude of the displacement, while the orientation of the arrow (which way the arrow points) indicates the direction. To add two displacements, join the tail of the second arrow to the head of the first arrow. Together, the arrows form two sides of a triangle whose third side represents their sum. The sum is the *net displacement* resulting from the two displacements.

Example 1   A person leaves her home and walks 5 miles due east and then 3 miles northeast. How far has she walked? How far away from home is she? What is her net displacement?

Solution   In Figure 10.2, the first arrow is 5 units long because she walks 5 miles, while the second arrow is 3 units long and at an angle of $45°$ to the first. The length of the third side, x, of the triangle can be found by the Law of Cosines:

$$x^2 = 5^2 + 3^2 - 2 \cdot 5 \cdot 3 \cos 135°$$
$$= 34 - 30 \left(-\frac{\sqrt{2}}{2} \right)$$
$$= 55.2.$$

This gives $x = \sqrt{55.2} = 7.43$ miles for the distance from home to destination, which is less than the total distance walked ($5 + 3 = 8$ miles). To specify her net displacement we use the angle θ in Figure 10.2. We find θ using the Law of Sines:

$$\frac{\sin \theta}{3} = \frac{\sin 135°}{7.43}$$

giving

$$\sin \theta = 3 \cdot \frac{\sqrt{2}/2}{7.43} = 0.285.$$

Taking the arcsin of both sides, we obtain $\theta = \sin^{-1}(0.285) = 16.6°$. Thus, the net displacement is 7.43 miles in a direction $16.6°$ north of east.

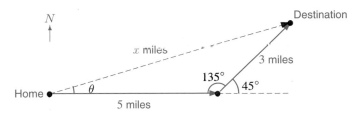

Figure 10.2: A person walks 5 miles east and then 3 miles northeast

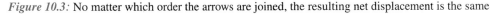

The Order of Addition Does Not Affect the Sum

When two displacements are added, the sum is independent of the order in which the two arrows are joined. For example, in Figure 10.3 the 5-mile displacement followed by the 3-mile displacement leads to the same destination as the 3-mile displacement followed by the 5-mile displacement.

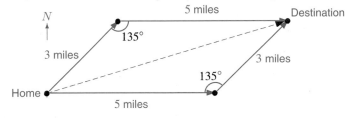

Figure 10.3: No matter which order the arrows are joined, the resulting net displacement is the same

Vectors

Many physical quantities add in the same way as displacements. Such quantities are known as *vectors*. Like displacements, vectors are often represented as arrows. A vector has *magnitude* (the arrow's length) and *direction* (which way the arrow points). Examples of vectors include:

- **Velocity** This is the speed and direction of travel.
- **Force** This is, loosely speaking, the strength and direction of a push or a pull.
- **Magnetic fields** A vector gives the direction and intensity of a magnetic field at a point.
- **Vectors in economics** Vectors are used to keep track of prices and quantities.
- **Vectors in computer animation** Computers generate animations by performing enormous numbers of vector-based calculations.
- **Population vectors** In biology, populations of different animals or age groups can be represented using vectors.

Vector Notation

In this book we write vectors as variables with arrows over them: $\vec{v}$. The notations is intended to ensure that a vector $\vec{v}$ is not mistaken for a *scalar*, which is another name for an ordinary number.

Notation for the Magnitude of a Vector

The length or magnitude of a vector $\vec{v}$ is written $||\vec{v}||$. This notation looks like the absolute value notation, $|x|$, used for scalars. The absolute value of a number is its size without regard to sign, so $|-10| = |+10| = 10$. Similarly, the vectors $\vec{u}$ and $\vec{v}$ in Figure 10.4 are both of length 5, even though they point in different directions. Therefore, $||\vec{u}|| = ||\vec{v}|| = 5$, (although $\vec{u} \neq \vec{v}$).

Figure 10.4: The vectors $\vec{u}$ and $\vec{v}$ are both of magnitude 5, but they point in different directions

Addition of Vectors

As with displacements, the sum of two vectors $\vec{u}$ and $\vec{v}$ represented by arrows is found by joining the tail of $\vec{v}$ to the head of $\vec{u}$. Then $\vec{w} = \vec{u} + \vec{v}$ is represented by the arrow drawn from the tail of of $\vec{u}$ to the head of $\vec{v}$. See Figure 10.5.[1]

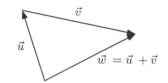

Figure 10.5: Constructing the vector sum $\vec{w} = \vec{u} + \vec{v}$

Example 2   Spacecraft such as *Voyager* and *Galileo* experience gravitational forces exerted by the sun, earth, and other planets. Vectors are used to represent the strength and direction of the gravitational forces.

Suppose a Jupiter-bound spacecraft is coasting between the orbits of Mars and Jupiter. Assume that Jupiter exerts a gravitational force of 8 units on the spacecraft, directed toward Jupiter, and that Mars exerts a force of 3 units on the spacecraft, directed toward Mars. The force $\vec{F}_M$ exerted by Mars is at an angle of $70°$ to the force $\vec{F}_J$ exerted by Jupiter, as shown in Figure 10.6. (For simplicity, we ignore the forces due to the sun and other planets.)

(a)   Find $||\vec{F}_{MJ}||$, the magnitude of the net force on the spacecraft due to Mars and Jupiter.
(b)   What is the direction of $\vec{F}_{MJ}$?
(c)   Without engine power, will the spacecraft stay on course?

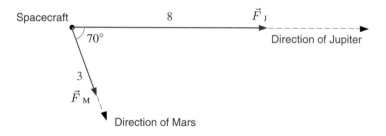

Figure 10.6: Forces $\vec{F}_M$ and $\vec{F}_J$ exerted by Mars and Jupiter, respectively

Solution   (a)   The net force, $\vec{F}_{MJ}$, is the sum of $\vec{F}_M$ and $\vec{F}_J$, the forces due to Mars and Jupiter, so

$$\vec{F}_{MJ} = \vec{F}_M + \vec{F}_J.$$

Figure 10.7 shows $\vec{F}_{MJ}$, obtained by joining the tail of $\vec{F}_J$ to the head of $\vec{F}_M$. The Law of Cosines tells us that $||\vec{F}_{MJ}||$, the length of the resulting vector, is given by

$$||\vec{F}_{MJ}||^2 = 8^2 + 3^2 - 2 \cdot 8 \cdot 3 \cos 110°$$
$$= 73 - 48 \cos 110°$$
$$= 89.4$$

so $||\vec{F}_{MJ}|| = \sqrt{89.4} = 9.46$ units. Notice that 9.46 is less than the sum of the magnitudes of the individual forces:

Force due to Mars + Force due to Jupiter $= 3 + 8 = 11$ units.

---

[1]If $\vec{u}$ and $\vec{v}$ are parallel, the vectors are added by the same method. However, in this case the figure is not a triangle.

The magnitude of the net force is less than 11 units because the forces exerted by Mars and Jupiter are not perfectly aligned.

(b) The direction of $\vec{F}_{MJ}$, denoted by θ in Figure 10.7, can be found using the Law of Sines:

$$\frac{\sin\theta}{3} = \frac{\sin 110°}{\|\vec{F}_{MJ}\|}$$

$$\sin\theta = 3\left(\frac{\sin 110°}{\|\vec{F}_{MJ}\|}\right) = 3\left(\frac{0.939}{9.46}\right) = 0.298,$$

so

$$\theta = \arcsin(0.298) = 17.3°.$$

(c) The gravity of Mars is pulling the spacecraft off its Jupiter-bound course by about 17.3°. The spacecraft will veer off course if it does not use its engines.

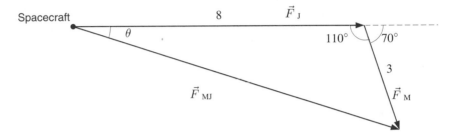

Figure 10.7: The net force, $\vec{F}_{MJ}$, on the spacecraft

Subtraction of Vectors

We have seen how vectors are added. Now we see how to subtract one vector from another. If

$$\vec{w} = \vec{u} - \vec{v},$$

then it is reasonable to assume that by adding $\vec{v}$ to both sides we obtain

$$\vec{w} + \vec{v} = \vec{u}.$$

In other words, $\vec{w}$ is the vector that when added to $\vec{v}$ gives $\vec{u}$. This is illustrated in Figure 10.8.

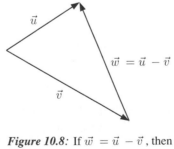

Figure 10.8: If $\vec{w} = \vec{u} - \vec{v}$, then $\vec{w}$ is the vector that when added to $\vec{v}$ gives $\vec{u}$

Figure 10.8 suggests the following rule: To find $\vec{w} = \vec{u} - \vec{v}$, join the tails of $\vec{v}$ and $\vec{u}$. Then $\vec{w}$ is the vector drawn from the head of $\vec{v}$ to the head of $\vec{u}$.

Example 3   Suppose the spacecraft in Example 2 can fire its engine thrusters in any direction, but that the strength of the thrust is always 4 units. In what direction should the spacecraft aim its engine thrusters in order to stay on course towards Jupiter?

Solution   The engine-thrust vector $\vec{F}_{\text{engines}}$ must be chosen so as to push the spacecraft back on course. The combined force on the craft due to Mars and Jupiter is given by $\vec{F}_{\text{MJ}}$. When the engines are on, the net force due to Mars, Jupiter, and the engines is given by

$$\vec{F}_{\text{net}} = \vec{F}_{\text{MJ}} + \vec{F}_{\text{engines}}.$$

We choose $\vec{F}_{\text{engines}}$ so that $\vec{F}_{\text{net}}$ points directly towards Jupiter. Solving for $\vec{F}_{\text{engines}}$ gives

$$\vec{F}_{\text{engines}} = \vec{F}_{\text{net}} - \vec{F}_{\text{MJ}}.$$

We do not know the length of $\vec{F}_{\text{net}}$, but we do know that it points towards Jupiter. From Example 2, the length of $\vec{F}_{\text{MJ}}$ is 9.46 units and the vector is directed at a 17.3° angle clockwise from $\vec{F}_{\text{net}}$. Figure 10.9 shows $\vec{F}_{\text{engines}} = \vec{F}_{\text{net}} - \vec{F}_{\text{MJ}}$. (The force $\vec{F}_{\text{engines}}$ is actually applied to the spacecraft, as shown in Figure 10.9. However, we can move a vector from one point to another when adding.)

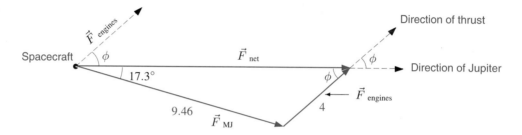

Figure 10.9: The engine-thrust vector is given by $\vec{F}_{\text{engines}} = \vec{F}_{\text{net}} - \vec{F}_{\text{MJ}}$, where $\vec{F}_{\text{net}}$ points directly towards Jupiter and $\vec{F}_{\text{MJ}}$ is as shown in Figure 10.7

We want to find the direction of thrust, labeled ϕ in Figure 10.9. By the Law of Sines, we have

$$\frac{\sin\phi}{9.46} = \frac{\sin 17.3°}{4}$$
$$\sin\phi = 9.46\left(\frac{\sin 17.3°}{4}\right).$$

One solution to this equation is

$$\phi = \arcsin\left(\frac{9.46}{4}\sin 17.3°\right) = 44.7°.$$

The engine thrusters should be aimed 44.7° counterclockwise from the direction of Jupiter.

There is another possible value for ϕ, the direction of the engine thrust in Example 3. This is because there are two solutions to the equation

$$\sin\phi = \frac{9.46}{4}\sin 17.3°.$$

The second solution, $180° - 44.7° = 135.3°$, is shown in Figure 10.10. Both values of ϕ serve to counteract the gravity of Mars, but the 44.7° direction helps push the craft toward Jupiter, whereas the 135.3° direction helps push it away from Jupiter. Thus, the first solution is probably preferable.

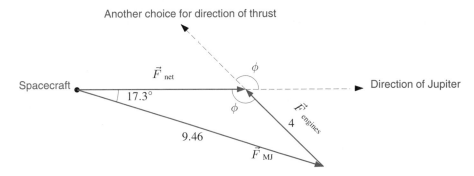

Figure 10.10: There are two possible choices for the direction of thrust. One has the effect of pushing the spacecraft toward Jupiter, the other (shown here) pushes the spacecraft away from Jupiter

Scalar Multiplication

A vector $\vec{u}$ can be added to itself. In Figure 10.11. The resulting vector, $\vec{u} + \vec{u}$, points in the same direction as $\vec{u}$ but is twice as long. The vector $\vec{u} + \vec{u} + \vec{u}$ also points in the same direction as $\vec{u}$, but is three times as long. We write

$$2\vec{u} \quad \text{for} \quad \vec{u} + \vec{u} \qquad \text{and} \qquad 3\vec{u} \quad \text{for} \quad \vec{u} + \vec{u} + \vec{u}.$$

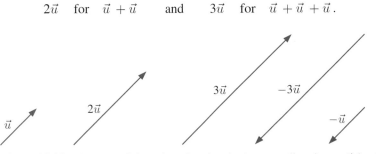

Figure 10.11: The vector $2\vec{u} = \vec{u} + \vec{u}$, points in the same direction as $\vec{u}$ but is twice as long as $\vec{u}$. Similarly, $3\vec{u} = \vec{u} + \vec{u} + \vec{u}$ points in the same direction and is 3 times as long as $\vec{u}$, and $-\vec{u}$ is in the opposite direction as $\vec{u}$

Similarly, $0.5\vec{u}$ points in the same direction as $\vec{u}$ but is only half as long. In general, if $\vec{v}$ is a vector and k a positive number (a positive scalar), then $k\vec{v}$ is a vector that is k times as long as $\vec{v}$ and pointing in the same direction as $\vec{v}$. The vector $k\vec{v}$ is called a *scalar multiple* of $\vec{v}$.

If k is a negative number, the vector $k\vec{v}$ has length $|k|$ times the length of $\vec{v}$, but points on the opposite direction to $\vec{v}$. For example, $-3\vec{u}$ is the vector that is three times as long as $\vec{u}$ but pointing in the opposite direction. See Figure 10.11.

The Zero Vector

The *zero vector*, denoted by the symbol $\vec{0}$, represents no displacement, which is the displacement that leaves you where you started. The following rules summarize scalar multiplication.

- If $k > 0$, then $k\vec{v}$ points in the same direction as $\vec{v}$ and is k times as long.
- If $k < 0$, then $k\vec{v}$ points in the opposite direction as $\vec{v}$ and is $|k|$ times as long.
- If $k = 0$, then $k\vec{v} = \vec{0}$, the zero vector.

Alternate view of Subtraction

Since the vector $-\vec{v}$ points in the opposite direction to $\vec{v}$, the difference $\vec{u} - \vec{v}$ is the same as the sum $\vec{u} + (-\vec{v})$. See Figure 10.12.

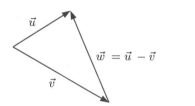
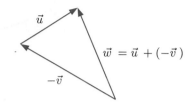

Figure 10.12: Illustration of the fact that $\vec{u} - \vec{v} = \vec{u} + (-\vec{v})$

Properties of Vector Addition and Scalar Multiplication

The following properties hold true for any three vectors $\vec{u}$, $\vec{v}$, $\vec{w}$ and any two scalars a and b:

1.   *Commutativity of addition:* $\vec{u} + \vec{v} = \vec{v} + \vec{u}$
2.   *Associativity of addition:* $(\vec{u} + \vec{v}) + \vec{w} = \vec{u} + (\vec{v} + \vec{w})$
3.   *Associativity of scalar multiplication:* $a(b\vec{v}) = (ab)\vec{v}$
4.   *Distributivity of scalar multiplication:* $(a + b)\vec{v} = a\vec{v} + b\vec{v}$ and $a(\vec{u} + \vec{v}) = a\vec{u} + a\vec{v}$
5.   *Identities:* $\vec{v} + \vec{0} = \vec{v}$ and $1 \cdot \vec{v} = \vec{v}$

These properties are analogous to the corresponding properties for addition and multiplication of numbers. In other words, vector addition and scalar multiplication of vectors behave as expected.

Problems for Section 10.1

In Problems 1–4, say whether the given quantity is a vector or a scalar.

1.   The distance from Seattle to St. Louis.   2.   The population of the US.

3.   The wind velocity at a point on the earth's surface.

4.   The temperature at a point on the earth's surface.

5.   The vectors $\vec{w}$ and $\vec{u}$ are in Figure 10.13. Match the vectors $\vec{p}, \vec{q}, \vec{r}, \vec{s}, \vec{t}$ with five of the following vectors: $\vec{u} + \vec{w}$, $\vec{u} - \vec{w}$, $\vec{w} - \vec{u}$, $2\vec{w} - \vec{u}$, $\vec{u} - 2\vec{w}$, $2\vec{w}$, $-2\vec{w}$, $2\vec{u}$, $-2\vec{u}$, $-\vec{w}$, $-\vec{u}$.

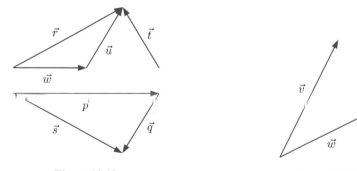

Figure 10.13   *Figure 10.14*

Given the displacement vectors $\vec{v}$ and $\vec{w}$ in Figure 10.14, draw the vectors in Problems 6–10.

6. $\vec{v} + \vec{w}$ 7. $\vec{v} - \vec{w}$ 8. $2\vec{v}$ 9. $2\vec{v} + \vec{w}$ 10. $\vec{v} - 2\vec{w}$

11. A person leaves home and walks 2 miles due west. She then walks 3 miles southwest. How far away from home is she? In what direction must she walk to head directly home? (Your answer should be an angle in degrees and should include a sketch.)

12. The person from Problem 11 next walks 4 miles southeast. How far away from home is she? In what direction must she walk to head directly home?

13. A helicopter is hovering at 3000 meters directly over the eastern perimeter of a secret Air Force installation. The eastern perimeter is 5000 meters from the installation headquarters. The helicopter receives a transmission from headquarters that a UFO has been sighted at 7500 meters directly over the installation's western perimeter, which is 9000 meters from the headquarters. How far must the helicopter travel to intercept the UFO? In what direction must it head? Be specific.

14. Suppose instead the UFO in Problem 13 is sighted directly over the installation's northern perimeter, which is 12,000 meters from headquarters. How far must the helicopter travel to intercept the UFO? In what direction must it head? Be specific.

Use the definitions of addition and scalar multiplication to explain each of the properties in Problems 15–21.

15. $\vec{w} + \vec{v} = \vec{v} + \vec{w}$

16. $(a + b)\vec{v} = a\vec{v} + b\vec{v}$

17. $a(\vec{v} + \vec{w}) = a\vec{v} + a\vec{w}$

18. $(\vec{u} + \vec{v}) + \vec{w} = \vec{u} + (\vec{v} + \vec{w})$

19. $\vec{v} + \vec{0} = \vec{v}$

20. $1\vec{v} = \vec{v}$

21. $\vec{v} + (-1)\vec{w} = \vec{v} - \vec{w}$

10.2 THE COMPONENTS OF A VECTOR

So far, vectors have been described in terms of length and direction. There is another way to think about a vector, and that is in terms of *components*. We begin with an example illustrating the use of components; we define this term precisely later.

Example 1   A ship travels 200 miles in a direction that its compass says is due east. The captain then discovers that the compass is faulty and that the ship has actually been heading in a direction that is 17.4° to the north of due east. How far north is the ship from its intended course?

Solution   In Figure 10.15, a vector $\vec{v}$ has been drawn for the ship's actual displacement. We know that $\|\vec{v}\| = 200$ miles and the direction of $\vec{v}$ is 17.4° to the north of due east. We see that $\vec{v}$ can be considered the sum of two vectors, one pointing due east and the other pointing due north, so

$$\vec{v} = \vec{v}_{\text{north}} + \vec{v}_{\text{east}}.$$

From Figure 10.15, we have

$$\sin 17.4° = \frac{\|\vec{v}_{\text{north}}\|}{200},$$

so

$$\|\vec{v}_{\text{north}}\| = 200 \sin 17.4° = 59.8 \text{ miles.}$$

Therefore, the ship is 59.8 miles due north of its intended course.

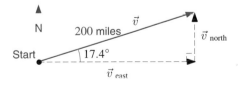

Figure 10.15: How far north is the ship from its intended eastward course?

Unit Vectors

A *unit vector* is a vector of length 1 unit. The unit vectors in the directions of the positive x- and y-axes are called $\vec{i}$ and $\vec{j}$, respectively. See Figure 10.16. These two vectors are important because any vector in the plane can be expressed in terms of $\vec{i}$ and $\vec{j}$.

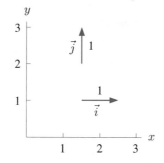

Figure 10.16: The unit vectors $\vec{i}$ and $\vec{j}$

Example 2 The x-axis points east and the y-axis points north. A person walks 3 miles east and then 4 miles north. Express her displacement in terms of the unit vectors $\vec{i}$ and $\vec{j}$.

Solution The unit of distance is miles, so the unit vector $\vec{i}$ represents a displacement of 1 mile east, and the unit vector $\vec{j}$ represents a displacement of 1 mile north. The person's displacement can be written

$$\text{Displacement} = 3 \text{ miles east} + 4 \text{ miles north} = 3\vec{i} + 4\vec{j}.$$

Example 3 The x-axis points east and the y-axis points north. Describe the displacement $5\vec{i} - 7\vec{j}$ in words.

Solution Since $\vec{i}$ represents a 1 mile displacement to the east, $5\vec{i}$ represents a 5 mile displacement to the east. Since $\vec{j}$ represents a 1 mile displacement to the north, $-7\vec{j}$ represents a 7 mile displacement to the south. Thus, $5\vec{i} - 7\vec{j}$ represents a 5 mile displacement to the east followed by a 7 mile displacement to the south.

Resolving a Vector in the Plane into Components

In this book, we focus chiefly on vectors that lie flat in the plane. Such vectors can always be broken (or *resolved*) into sum of two vectors,[2] one parallel to the x-axis and the other to the y-axis. These two vectors are called the *components* parallel to the axes.

For the vector $\vec{v}$ in Figure 10.17, we can write

$$\vec{v} = \vec{v}_1 + \vec{v}_2,$$

where $\vec{v}_1$ is called the *x-component* of $\vec{v}$, and $\vec{v}_2$ is the *y-component* of $\vec{v}$. What can we say about the components $\vec{v}_1$ and $\vec{v}_2$?

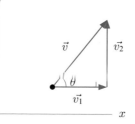

Figure 10.17: A vector $\vec{v}$ can be written as the sum of components, $\vec{v}_1$ and $\vec{v}_2$, parallel to the x- and y-axes

---

[2]However, one or both of the vectors may be of zero length.

In Figure 10.17, the vector $\vec{v}$ has length $||\vec{v}||$ and makes an angle θ with the positive x-axis, with $0 \leq \theta < 2\pi$. Thus, we see that

$$\cos\theta = \frac{||\vec{v}_1||}{||\vec{v}||} \quad \text{and} \quad \sin\theta = \frac{||\vec{v}_2||}{||\vec{v}||}.$$

Therefore

$$||\vec{v}_1|| = ||\vec{v}||\cos\theta \quad \text{and} \quad ||\vec{v}_2|| = ||\vec{v}||\sin\theta.$$

We know that $\vec{v}_1$ is parallel to the x-axis, so it is a scalar multiple of the unit vector $\vec{i}$. Similarly, $\vec{v}_2$ is a scalar multiple of $\vec{j}$. After checking that the signs of $\cos\theta$ and $\sin\theta$ give the correct directions for $\vec{v}_1$ and $\vec{v}_2$, we have

$$\vec{v}_1 = \left(||\vec{v}||\cos\theta\right)\vec{i} \quad \text{and} \quad \vec{v}_2 = \left(||\vec{v}||\sin\theta\right)\vec{j}.$$

These are the x- and y- components of $\vec{v}$. We have the following result:

In the plane, a vector $\vec{v}$ of length $||\vec{v}||$ which makes an angle θ with the positive x-axis can be written in terms of its components:

$$\vec{v} = \left(||\vec{v}||\cos\theta\right)\vec{i} + \left(||\vec{v}||\sin\theta\right)\vec{j}.$$

Example 4   In Example 1, suppose the x-axis points east and the y-axis points north. Resolve the displacement vector $\vec{v}$ into components parallel to the axes.

Solution   The vector $\vec{v}_{\text{north}}$ is the y-component of $\vec{v}$ and $\vec{v}_{\text{east}}$ is the x-component. We already know that $||\vec{v}_{\text{north}}|| = 59.8$ miles. From Figure 10.15, we see that $||\vec{v}_{\text{east}}|| = 200\cos 17.4° = 190.8$ miles. (This last piece of information tells us that although the ship traveled a total of 200 miles, its eastward progress amounted to only 190.8 miles.) The easterly and northerly components are

$$\vec{v}_{\text{east}} = 190.8\vec{i} \qquad \text{and} \qquad \vec{v}_{\text{north}} = 59.8\vec{j}.$$

Thus, the course taken by the ship can be written as follows:

$$\vec{v} = \vec{v}_{\text{east}} + \vec{v}_{\text{north}} = 190.8\vec{i} + 59.8\vec{j}.$$

To check the calculation, notice that

$$||\vec{v}_{\text{east}}||^2 + ||\vec{v}_{\text{north}}||^2 = (190.8)^2 + (59.8)^2 \approx 40,000 = ||\vec{v}||^2.$$

Sums and scalar multiples are easy to compute if vectors are written in terms of components.

Example 5   Let $\vec{u} = 5\vec{i} + 7\vec{j}$, $\vec{v} = \vec{i} + \vec{j}$, and $\vec{w} = 6\vec{j} - 3\vec{i} + 2\vec{j} - 2\vec{i} - 4\vec{j} + 7\vec{i}$. Describe in words the similarities and differences between the following displacements:

$$2\vec{i} + 4\vec{j}, \quad \vec{u} - 3\vec{v}, \quad \vec{w}.$$

Solution   We show that the displacements $2\vec{i} + 4\vec{j}$, $\vec{u} - 3\vec{v}$, and $\vec{w}$ are all the same. Using the properties on page 422, we have

$$\begin{aligned}
\vec{u} - 3\vec{v} &= (5\vec{i} + 7\vec{j}) - 3(\vec{i} + \vec{j}) \\
&= (5\vec{i} + 7\vec{j}) - (3\vec{i} + 3\vec{j}) \\
&= (5\vec{i} - 3\vec{i}) + (7\vec{j} - 3\vec{j}) = 2\vec{i} + 4\vec{j}.
\end{aligned}$$

The vector $\vec{w}$ can be simplified as

$$\vec{w} = (-3\vec{i} - 2\vec{i} + 7\vec{i}) + (6\vec{j} + 2\vec{j} - 4\vec{j}) = 2\vec{i} + 4\vec{j}.$$

Alternatively, we can see that all three displacements are the same by doing head-to-tail addition for each case, with the same starting point. Figure 10.18 shows that all three ending points are the same, although the path followed is different in each case.

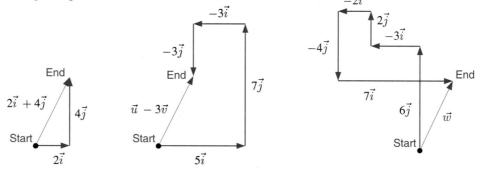

Figure 10.18: These three displacements, $2\vec{i} + 4\vec{j}$, $\vec{u} - 3\vec{v}$, $\vec{w}$, are all the same

Displacement Vectors

Vectors that indicate displacement are often designated by giving two points that specify the initial and final positions. For example, the vector $\vec{v} = \overrightarrow{PQ}$ describes a displacement or motion from point P to point Q. If we know the coordinates of P and Q, we can easily find the components of $\vec{v}$.

Example 6   If P is the point $(-3, 1)$ and Q is the point $(2, 4)$, find the components of the vector $\overrightarrow{PQ}$.

Solution   Figure 10.19 shows the vector $\overrightarrow{PQ}$ and its components $\vec{v}_1$ and $\vec{v}_2$. The length of $\vec{v}_1$ is the horizontal distance from P to Q, so

$$\|\vec{v}_1\| = 2 - (-3) = 5.$$

The length of $\vec{v}_2$ is the vertical distance from P to Q, so

$$\|\vec{v}_2\| = 4 - 1 = 3.$$

Thus, $\vec{v}_1 = 5\vec{i}$ and $\vec{v}_2 = 3\vec{j}$, so $\vec{v} = \overrightarrow{PQ} = 5\vec{i} + 3\vec{j}$.

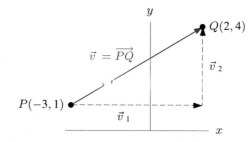

Figure 10.19: Finding components $\vec{v}_1$ and $\vec{v}_2$ from the coordinates of the points P and Q

In general, we have the following result:

If $P = (x_1, y_1)$ and $Q = (x_2, y_2)$, then the vector $\vec{v} = \overrightarrow{PQ}$ from P to Q is given by

$$\vec{v} = (x_2 - x_1)\vec{i} + (y_2 - y_1)\vec{j}.$$

Vectors in n Dimensions

So far we have considered 2-dimensional vectors which have two components. Vectors in space need three components and they are also very useful. In addition, there are *n-dimensional* vectors, which have n components, where n is a positive integer.

Example 7   A balloon rises vertically a distance of 2 miles, floats west a distance of 3 miles, and floats north a distance of 4 miles. The balloon's displacement vector has three components: a vertical component, a westward component, and a northward component. Therefore, this displacement is a 3-dimensional vector. We write this vector as

$$-3\vec{i} + 4\vec{j} + 2\vec{k},$$

where $\vec{k}$ is the unit vector that points vertically upward.

Problems for Section 10.2

For Problems 1–2, perform the indicated computations.

1. $(4\vec{i} + 2\vec{j}) - (3\vec{i} - \vec{j})$
2. $(\vec{i} + 2\vec{j}) + (-3)(2\vec{i} + \vec{j})$

3. On the graph in Figure 10.20, draw the vector $\vec{v} = 4\vec{i} + \vec{j}$ twice, once with its tail at the origin and once with its tail at the point $(3, 2)$.

Resolve the vectors in Problems 4–6 into components.

4. A vector starting at the point $P = (1, 2)$ and ending at the point $Q = (4, 6)$.
5. A vector starting at the point $Q = (4, 6)$ and ending at the point $P = (1, 2)$.
6. The vector shown in Figure 10.21, with components expressed in inches.

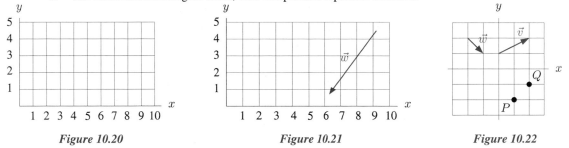

Figure 10.20   *Figure 10.21*   *Figure 10.22*

In Problems 7–9, use the information in Figure 10.22. Each grid square is 1 unit along each side.

7. Write the following vectors in component form.

   (a) $\vec{v}$ (b) $2\vec{w}$ (c) $\vec{v} + \vec{w}$ (d) $\vec{w} - \vec{v}$

   (e) The displacement vector $\overrightarrow{PQ}$.   (f) The vector from the point P to the point $(2, 0)$.
   (g) A vector perpendicular to the y-axis.   (h) A vector perpendicular to the x-axis.

8. What is the angle between the vector $\vec{w}$ and the negative y-axis?

9. What is the angle between the vector $\vec{w}$ and the displacement vector $\overrightarrow{PQ}$?

Find the length of the vectors in Problems 10–13.

10. $\vec{v} = \vec{i} - \vec{j} + 3\vec{k}$

11. $\vec{v} = \vec{i} - \vec{j} + 2\vec{k}$

12. $\vec{v} = 1.2\vec{i} - 3.6\vec{j} + 4.1\vec{k}$

13. $\vec{v} = 7.2\vec{i} - 1.5\vec{j} + 2.1\vec{k}$

14. (a) Find a unit vector from the point $P = (1, 2)$ toward the point $Q = (4, 6)$.
    (b) Find a vector of length 10 pointing in the same direction.

15. Which is traveling faster, a car whose velocity vector is $21\vec{i} + 35\vec{j}$, or a car whose velocity vector is $40\vec{i}$, assuming that the units are the same for both directions?

16. A truck is traveling due north at 30 km/hr toward a crossroad. On a perpendicular road a police car is traveling west toward the intersection at 40 km/hr. Both vehicles will reach the crossroad in exactly one hour. Find the vector currently representing the displacement of the truck with respect to the police car.

17. A car is traveling at a speed of 50 km/hr. The positive y-axis is north and the positive x-axis is east. Resolve the car's velocity vector into two components if the car is traveling in each of the following directions:
    (a) East (b) South (c) Southeast (d) Northwest

18. Shortly after taking off, a plane is climbing northwest through still air at an airspeed of 200 km/hr, and rising at a rate of 300 m/min. Resolve into components its velocity vector in a coordinate system in which the x-axis points east, the y-axis points north, and the z-axis points up.

19. Which of the following vectors are parallel?

$$\vec{u} = 2\vec{i} + 4\vec{j} - 2\vec{k}, \quad \vec{v} = \vec{i} - \vec{j} + 3\vec{k}, \quad \vec{w} = -\vec{i} - 2\vec{j} + \vec{k},$$
$$\vec{p} = \vec{i} + \vec{j} + \vec{k}, \quad \vec{q} = 4\vec{i} - 4\vec{j} + 12\vec{k}, \quad \vec{r} = \vec{i} - \vec{j} + \vec{k}.$$

20. The hour hand and the minute hand of a clock are represented by the vectors $\vec{h}$ and $\vec{m}$, respectively, with $\|\vec{h}\| = 2$ and $\|\vec{m}\| = 3$. The origin is at the center of the clock and the positive x-axis goes through the three o'clock position.

    (a) What are the components of $\vec{h}$ and $\vec{m}$ at the following times? Illustrate with a sketch.
        (i) 12 noon (ii) 3 pm (iii) 1 pm (iv) 1:30 pm
    (b) Sketch the displacement vector from the tip of the hour hand to the tip of the minute hand at 3 pm. What are the components of this displacement vector?
    (c) Sketch the vector representing the sum $\vec{h} + \vec{m}$ at 1:30 pm. What are the components of this sum?

A cat is sitting on the ground at the point $(1, 4, 0)$ watching a squirrel at the top of a tree. The tree is one unit high and its base is at the point $(2, 4, 0)$. Find the displacement vectors in Problems 21–24.

21. From the origin to the cat.

22. From the bottom of the tree to the squirrel.

23. From the bottom of the tree to the cat.

24. From the cat to the squirrel.

10.3 APPLICATION OF VECTORS

Alternate Notation for the Components of a Vector

The vector $v = 3\vec{i} + 4\vec{j}$ is sometimes written $\vec{v} = (3, 4)$. This notation can be confused with the coordinate notation used for points. For instance, it isn't clear whether $(3, 4)$ means the point $x = 3$, $y = 4$ or the vector $3\vec{i} + 4\vec{j}$. Nevertheless, this notation is useful for vectors in n dimensions.

Population Vectors

Vectors in n dimensions are useful for keeping track of n quantities, as illustrated in the next example.

Example 1   The 1998 population of the six New England states (Connecticut, Maine, Massachusetts, New Hampshire, Rhode Island, and Vermont) can be thought of as a 6-dimensional vector $\vec{P}$. The components of $\vec{P}$ are the populations of the six states. Using the alternate notation, we write

$$\vec{P} = (P_{CT}, P_{ME}, P_{MA}, P_{NH}, P_{RI}, P_{VT}),$$

where P_{CT} is the population of Connecticut, P_{ME} is the population of Maine, and so on.

The population vector $\vec{P}$ from the previous example does not have a geometrical interpretation. We can't draw a picture of $\vec{P}$, or interpret its length and direction as we do for displacement vectors. However, the following example shows how it is used.

Example 2   The vectors $\vec{P}$ and $\vec{Q}$ give the populations of the six New England states in 1990 and 1998, respectively. According to the US Census Bureau,[3] these vectors are given, in millions of people, by

$$\vec{P} = (3.29, 1.23, 6.02, 1.11, 1.00, 0.56)$$
$$\vec{Q} = (3.27, 1.24, 6.15, 1.19, 0.99, 0.59)$$

For instance, the population of Connecticut was 3.29 million in 1990 and 3.27 million in 1998.

(a)   Find $\vec{R} = \vec{Q} - \vec{P}$. Explain its significance in terms of the population of New England.

(b)   Let $\vec{S}$ be the estimated population of New England in the year 2006. One expert predicts that $\vec{S} = \vec{Q} + 2\vec{R}$. Find $\vec{S}$ and explain the assumption this expert is making about the New England population.

(c)   Another expert makes a different prediction, claiming that $\vec{T} = 1.08\vec{Q}$ will give the population of New England in 2006. Find $\vec{T}$ and explain the assumption that this expert is making.

Solution   (a)   We have

$$\vec{R} = \vec{Q} - \vec{P}$$
$$= (3.27, 1.24, 6.15, 1.19, 0.99, 0.59) - (3.29, 1.23, 6.02, 1.11, 1.00, 0.56)$$
$$= (3.27 - 3.29, 1.24 - 1.23, 6.15 - 6.02, 1.19 - 1.11, 0.99 - 1.00, 0.59 - 0.56)$$
$$= (-0.02, 0.01, 0.13, 0.08, -0.01, 0.03).$$

The components of $\vec{R}$ give the change in population for each New England state. For instance, the population of Connecticut fell from 3.29 million in 1990 to 3.27 million in 1998, a change of $3.27 - 3.29 = -0.02$ million people, so $R_{CT} = -0.02$.

(b)   The formula $\vec{S} = \vec{Q} + 2\vec{R}$ means that

Population in 2006 = Population in 1998 + 2 · Change between 1990 and 1998.

According to this expert, states whose populations dropped between 1990 and 1998 will see them drop twice as much between 1998 and 2006. Likewise, states whose populations climbed between 1990 and 1998 will see them climb twice as much between 1998 and 2006. Algebraically, we have

$$\vec{S} = \vec{Q} + 2\vec{R}$$
$$= (3.27, 1.24, 6.15, 1.19, 0.99, 0.59) + 2 \cdot (-0.02, 0.01, 0.13, 0.08, -0.01, 0.03)$$
$$= (3.27, 1.24, 6.15, 1.19, 0.99, 0.59) + (-0.04, 0.02, 0.26, 0.16, -0.02, 0.06)$$
$$= (3.23, 1.26, 6.41, 1.35, 0.97, 0.65).$$

---

[3] From: http://www.census.gov

For instance, Connecticut's population dropped by 0.02 million between 1990 and 1998. The predicted population of Connecticut in 2006 is given by

$$S_{CT} = Q_{CT} + 2 \cdot R_{CT} = 3.27 + 2 \cdot (-0.02) = 3.23.$$

The 0.04 million drop in Connecticut's population between 1998 and 2006 is twice as large as the 0.02 million drop between 1990 and 1998.

(c)   The formula $\vec{T} = 1.08\vec{Q}$ means that each component of $\vec{T}$ is 1.08 times as large as the corresponding component of $\vec{Q}$. In other words, the population of each state is predicted to grow by 8%. We have

$$\begin{aligned}
\vec{T} = 1.08\vec{Q} &= 1.08 \cdot (3.27, 1.24, 6.15, 1.19, 0.99, 0.59) \\
&= (1.08 \cdot 3.27, 1.08 \cdot 1.24, 1.08 \cdot 6.15, 1.08 \cdot 1.19, 1.08 \cdot 0.99, 1.08 \cdot 0.59) \\
&= (3.53, 1.34, 6.64, 1.29, 1.07, 0.64).
\end{aligned}$$

For instance, the population of New Hampshire is predicted to grow from $Q_{NH} = 1.19$ million in 1998 to $T_{NH} = 1.08Q_{NH} = 1.08(1.19) = 1.29$ million in the year 2006.

Notice how vector addition and subtraction was used in parts (a) and (b) of the previous example, and scalar multiplication was used in parts (b) and (c).

Economics

The blockbuster movie *Independence Day* grossed $306 million at domestic (US) box offices, far more than any other domestic release in 1996.[4] It made even more money at international box offices, where it grossed $474 million. Rounding out its wild success was grossing $252 million for video rentals, for a grand total of over one billion dollars.

An economist might describe this situation with a *revenue vector* $\vec{r}$, writing $\vec{r} = (306, 474, 252)$ to indicate the domestic, international, and videotape gross revenues, in millions of dollars. This is helpful in bookkeeping since it is easier to write $\vec{r}$ instead of three separate revenue variables.

Example 3   After *Independence Day*, the three next most financially successful US movies in 1996 were, in order, *Twister*, *Mission Impossible*, and *The Rock*. If $\vec{s}$, $\vec{t}$, and $\vec{u}$ are the revenue vectors for these three movies, then

$$\vec{s} = (242, 252, 108) \qquad \vec{t} = (181, 272, 92) \qquad \vec{u} = (134, 197, 55).$$

Find $\vec{R}$, the total revenue vector for all three of these movies.

Solution   The total revenue, $\vec{R}$, is the sum of $\vec{s}$, $\vec{t}$, and $\vec{u}$:

$$\vec{R} = \vec{s} + \vec{t} + \vec{u} = (242 + 181 + 134, 252 + 272 + 197, 108 + 92 + 55) = (557, 721, 255).$$

Price and Consumption Vectors

A car dealership has several different models of cars in its inventory, with different prices for each model. A *price vector*, $\vec{P}$, gives the price of each model:

$$\vec{P} = (P_1, P_2, \ldots, P_n).$$

Here P_1 is the price of car model 1 and P_2 is the price of model 2, and so on. A *consumption vector* $\vec{C}$ gives the number of each model of car purchased (or consumed) during a given month:

$$\vec{C} = (C_1, C_2, \ldots, C_n).$$

---

[4] *Entertainment Weekly*, February 7, 1997.

Example 4   Suppose $\vec{I}$ gives the number of cars of each model a car dealer currently has in inventory. Explain the meaning of the following expressions and statements in terms of the car dealership.

(a) $\vec{I} - \vec{C}$ (b) $\vec{I} - 2\vec{C}$ (c) $\vec{C} = 0.3\vec{I}$

Solution   (a) This expression represents the difference between the number of each model currently in inventory and the number of that model purchased each month. So $\vec{I} - \vec{C}$ represents the numbers of each model that remain on the lot after one month.

(b) This expression represents the number of each model the dealer has left after two months, assuming no new cars are added to inventory and that $\vec{C}$ does not change.

(c) This equation tells us that 30% of the inventory of each model is purchased by consumers each month.

Example 5   Let $\vec{E}$ represent expenses incurred by the dealer for acquisition, insurance, and overhead, for each model. What does the expression $\vec{P} - \vec{E}$ represent?

Solution   This represents the dealer's profit for each model. For instance, a model that the dealer sells for $29,000 may cost $25,000 to acquire from the factory, insure, and maintain. The difference of $4,000 represents profit for the dealer. The vector $\vec{P} - \vec{E}$ keeps track of this profit for each model.

Physics

Newton's law of gravitation states that the magnitude of force exerted on an object of mass m by the earth is given by

$$\|\vec{F}_E\| = G\frac{mM_E}{r_E^2},$$

where M_E is the mass of the earth, r_E is the distance from the object to earth's center, and G is a constant. Similarly, the force exerted on the object by the moon has magnitude

$$\|\vec{F}_L\| = G\frac{mM_L}{r_L^2},$$

where M_L is the mass of the moon and r_L is the distance to the moon's center. The force exerted by the earth is directed toward the earth; the force exerted by the moon is exerted toward the moon.

Example 6   Figure 10.23 shows the position of a 100,000 kg spacecraft relative to the earth and moon. The vector $\vec{r}$ is the position of the moon relative to the earth, $\vec{r}_E$ is the position of the spacecraft relative to the earth, and $\vec{r}_L$ is the position of the spacecraft relative to the moon. If distances are measured in thousands of kilometers, then

$$\vec{r} = 384\vec{i} \quad \text{and} \quad \vec{r}_E = 280\vec{i} + 90\vec{j}.$$

Which force is stronger, the pull of the moon on the spacecraft or the pull of the earth? [Note that $M_E = 5.98 \cdot 10^{24}$ kg, $M_L = 7.34 \cdot 10^{22}$ kg, and $G = 6.67 \cdot 10^{-23}$.]

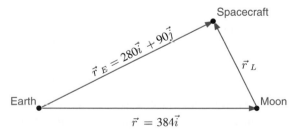

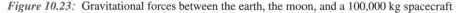

Figure 10.23: Gravitational forces between the earth, the moon, and a 100,000 kg spacecraft

Solution   To use Newton's law of gravitation, we first find $||\vec{r}_E||$ and $||\vec{r}_L||$. Since $\vec{r}_E = 280\vec{i} + 90\vec{j}$,

$$||\vec{r}_E|| = \sqrt{(280)^2 + (90)^2} = \sqrt{86500} = 294.$$

Since $\vec{r}_L = \vec{r}_E - \vec{r}$, we have

$$\vec{r}_L = (280\vec{i} + 90\vec{j}) - 384\vec{i} = -104\vec{i} + 90\vec{j}.$$

Thus,

$$||\vec{r}_L|| = \sqrt{(-104)^2 + (90)^2} = \sqrt{18916} = 138.$$

Finally, using the values of m, M_E, M_L and G given in the problem, we calculate

$$F_E = G\frac{mM_E}{||\vec{r}_E||^2} = 461 \qquad \text{and} \qquad F_L = G\frac{mM_L}{||\vec{r}_L||^2} = 25.9.$$

This means that the pull of the earth on this spacecraft is much stronger than the pull of the moon; in fact, it is over ten times as strong. The earth's pull is stronger, even though the spaceship is closer to the moon, because the earth is much more massive than the moon.

Computer Graphics: Position Vectors

Video games nearly always incorporate computer generated graphics, as do the flight simulators used by commercial and military flight training schools. Hollywood makes increasing use of computer graphics in its films. The film *Terminator 2: Judgement Day* was a watershed event for the use of computer generated special effects. Its famous liquid metal assassin was created almost entirely with computers. So were the toys and tornados in the more recent blockbusters *Toy Story* and *Twister*. Enormous amounts of computation are involved in creating such effects and much of this computation involves vectors.

A computer screens can be thought of as an xy-grid, with the origin at the lower left corner. The position of a point on the screen is specified by a vector pointing from the origin to the point. Such a vector is called a *position vector*. The tail of a position vector is always fixed at the origin.

Example 7   A video game shows two airplanes on the screen at the points $(3, 5)$ and $(7, 2)$. See Figure 10.24. Both airplanes move a distance of 3 units at an angle of $70°$ counterclockwise from the x-axis. What are the new positions of the airplanes?

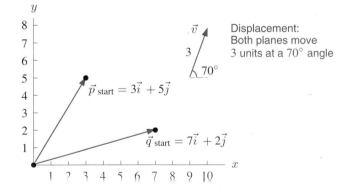

Figure 10.24: Computer screen showing two airplanes moving by a displacement vector $\vec{v}$

Solution   The initial position of the first airplane is given by the position vector $\vec{p}_{\text{start}} = 3\vec{i} + 5\vec{j}$. The plane's displacement can also be thought of as a vector, $\vec{v}$. Resolving $\vec{v}$ into components:

$$\vec{v} = (3\cos 70°)\vec{i} + (3\sin 70°)\vec{j} = 1.03\vec{i} + 2.82\vec{j}.$$

The airplane's final position, $\vec{p}_{end}$, is given by

$$\vec{p}_{end} = \vec{p}_{start} + \vec{v} = 3\vec{i} + 5\vec{j} + 1.03\vec{i} + 2.82\vec{j}$$
$$= 4.03\vec{i} + 7.82\vec{j}.$$

The second airplane's initial position is given by the position vector $\vec{q}_{start} = 7\vec{i} + 2\vec{j}$. The second plane's displacement, however, is exactly the same as the first airplane's, $\vec{v} = 1.03\vec{i} + 2.82\vec{j}$. Therefore, the final position of the second airplane, $\vec{q}_{end}$, is given by

$$\vec{q}_{end} = \vec{q}_{start} + \vec{v} = 7\vec{i} + 2\vec{j} + 1.03\vec{i} + 2.82\vec{j}$$
$$= 8.03\vec{i} + 4.82\vec{j}.$$

Example 7 illustrates the difference between position vectors and other vectors: The tail of a position vector is fixed to the origin, but the tail of an ordinary vector (like the displacement vector $\vec{v}$) can be anywhere, so long as its length and orientation do not change.

Problems for Section 10.3

1. A man walks 5 miles in a direction 30° north of east. He then walks a distance x miles due east. He turns around to look back at his starting point, which is at an angle of 10° south of west.

    (a) Make a sketch. Give vectors in $\vec{i}$ and $\vec{j}$ components, for each part of the man's walk.
    (b) What is x?
    (c) How far is the man from his starting point?

2. Two children are throwing a ball back-and-forth straight across the back seat of a car. Suppose the ball is being thrown at 10 mph relative to the car and the car is going 25 mph down the road.

    (a) Make a sketch showing the relevant velocity vectors.
    (b) If one child does not catch the ball and it goes out an open window, what angle does the ball's horizontal motion make with the road?

3. Figure 10.25 shows a rectangle whose four corners are the points

    $$a = (2, 1), \quad b = (4, 1), \quad c = (4, 2), \quad \text{and} \quad d = (2, 2).$$

    As part of a video game, this rectangle is rotated counterclockwise through an angle of 35° about the origin. See Figure 10.26. What are the new coordinates of the corners of the rectangle?

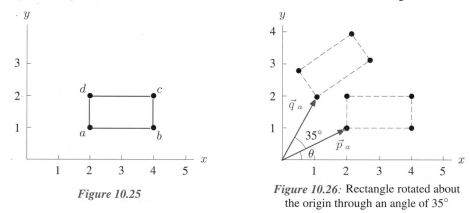

Figure 10.25

Figure 10.26: Rectangle rotated about the origin through an angle of 35°

4. There are five students in a class. Their scores on the midterm (out of 100) are given by the vector $\vec{v} = (73, 80, 91, 65, 84)$. Their scores on the final (out of 100) are given by $\vec{w} = (82, 79, 88, 70, 92)$. The final counts twice as much as the midterm. Find a vector giving the total scores (out of 100) of the students.

5. An airplane is heading northeast at an airspeed of 700 km/hr, but there is a wind blowing from the west at 60 km/hr. In what direction does the plane end up flying? What is its speed relative to the ground? [Hint: Resolve the velocity vectors for the airplane and the wind into components.]

6. An airplane is flying at an airspeed of 600 km/hr in a cross-wind that is blowing from the northeast at a speed of 50 km/hr. In what direction should the plane head to end up going due east?

10.4 THE DOT PRODUCT

We have added and subtracted vectors and we have multiplied vectors by scalars. We now see how to multiply one vector by another vector.

To compute the product of $\vec{u}$ and $\vec{v}$, we multiply each coordinate of $\vec{u}$ by the corresponding coordinate of $\vec{v}$ and add these products. The result is called the *dot product*, written $\vec{u} \cdot \vec{v}$. Notice that the dot product of two vectors is a scalar, not a vector.

Example 1   A car dealer sells five different models of car. The number of each model sold each week is given by the consumption vector $\vec{C} = (22, 14, 8, 12, 19)$. For instance, we see that in one week the dealer sells 22 of the first model, 14 of the second, and so forth.

The price of each model is given by the vector $\vec{P} = (14, 18, 35, 42, 27)$, where the units are $1000s. Thus, the price of the first model is $14,000, the price of the second is $18,000, and so forth. Find the dealer's weekly revenue.

Solution   The total revenue earned by the dealer each week (in 1000s of dollars) is given by

$$\text{Revenue (in \$1000s)} = 22\,\text{cars} \cdot 14\,\text{per car} + 14\,\text{cars} \cdot 18\,\text{per car} + 8\,\text{cars} \cdot 35\,\text{per car}$$
$$+ 12\,\text{cars} \cdot 42\,\text{per car} + 19\,\text{cars} \cdot 27\,\text{per car}$$
$$= 1857.$$

The dealer brings in $1,857,000 each week. Since the revenue is obtained by multiplying the corresponding coordinates of $\vec{C}$ and $\vec{P}$ and adding, we can write revenue as a dot product:

$$\text{Revenue} = \vec{C} \cdot \vec{P}.$$

Note the revenue is a scalar, not a vector, as the dot product always give a number.

In general, we define the dot product of two vectors as follows:

if $\vec{u} = (u_1, u_2, \ldots, u_n)$ and $\vec{v} = (v_1, v_2, \ldots, v_n)$ are two n-dimensional vectors, then the dot product of $\vec{u}$ and $\vec{v}$ is the scalar given by

$$\vec{u} \cdot \vec{v} = u_1 v_1 + u_2 v_2 + \cdots + u_n v_n.$$

Example 2   The population vector $\vec{P}_{\text{NewEngland}} = (3.29, 1.23, 6.02, 1.11, 1.00, 0.56)$ gives the populations (in millions) of the six New England states (CT, ME, MA, NH, RI, VT) in 1990. The vector $\vec{r} = (-0.6\%, 0.8\%, 2.2\%, 7.2\%, -1.0\%, 5.4\%)$ gives the percent change in population between 1990 and 1998 for each state. We see that

$$\text{Change in Connecticut population} = -0.006 \cdot 3.29 = -0.0197,$$
$$\text{Change in Maine population} = 0.008 \cdot 1.23 = 0.0098, \qquad \text{and so on.}$$

If $\Delta P_{\text{New England}}$ represents the overall change in the population of New England, then

$$\Delta P_{\text{New England}} = -0.006 \cdot 3.29 + 0.008 \cdot 1.23 + 0.022 \cdot 6.02$$
$$+ 0.072 \cdot 1.11 - 0.01 \cdot 1.0 + 0.054 \cdot 0.56$$
$$= 0.2227,$$

so the population of New England increased by 0.2227 million people (222,700) between 1990 and 1998. Notice that the total change in the New England population is given by the dot product.

$$\Delta P_{\text{New England}} = \vec{r} \cdot \vec{P}_{\text{NewEngland}}.$$

Properties of the Dot Product

- $\vec{u} \cdot \vec{v} = \vec{v} \cdot \vec{u}$ (*Commutative Law*)
- $\vec{u} \cdot (\vec{v} + \vec{w}) = \vec{u} \cdot \vec{v} + \vec{u} \cdot \vec{w}$ (*Distributive Law*)
- $\vec{v} \cdot \vec{v} = ||\vec{v}||^2$
- $\vec{u} \cdot \vec{v} = ||\vec{u}|| \cdot ||\vec{v}|| \cos\theta$, where θ is the angle between $\vec{u}$ and $\vec{v}$, and $0 \le \theta \le 180$.

The commutative law holds because multiplication of coordinates is commutative. A proof of the distributive law using coordinates is outlined in the Problem 14. For the third property, note that

$$\vec{v} \cdot \vec{v} = (v_1, v_2) \cdot (v_1, v_2) = v_1^2 + v_2^2.$$

By the Pythagorean theorem, $||\vec{v}||^2 = v_1^2 + v_2^2$, so $\vec{v} \cdot \vec{v} = ||v||^2$. A similar argument applies to vectors of higher dimension.

The fourth property provides a useful way to calculate the dot product without using coordinates. We use the Law of Cosines to show why it works for 2-dimensional vectors.

Justification of $\vec{u} \cdot \vec{v} = ||\vec{u}|| \cdot ||\vec{v}|| \cos\theta$ Figure 10.27 shows two vectors, $\vec{u}$

and $\vec{v}$, with an angle θ between them. The vectors $\vec{u}$, $\vec{v}$, and $\vec{w} = \vec{u} - \vec{v}$ form a triangle. Now, we know (from the third property of dot products) that $\vec{w} \cdot \vec{w} = ||\vec{w}||^2$. We can also calculate $\vec{w} \cdot \vec{w}$ using the distributive and commutative laws:

$$\vec{w} \cdot \vec{w} = (\vec{v} - \vec{u}) \cdot (\vec{v} - \vec{u})$$
$$= \vec{v} \cdot \vec{v} - \vec{u} \cdot \vec{v} - \vec{v} \cdot \vec{u} + \vec{u} \cdot \vec{u}$$
$$= ||\vec{v}||^2 + ||\vec{u}||^2 - 2\vec{u} \cdot \vec{v}.$$

Since $\vec{w} \cdot \vec{w} = ||\vec{w}||^2$, we have shown that

$$||\vec{w}||^2 = ||\vec{v}||^2 + ||\vec{u}||^2 - 2\vec{u} \cdot \vec{v}.$$

But applying the Law of Cosines to the triangle in Figure 10.27 gives

$$||\vec{w}||^2 = ||\vec{u}||^2 + ||\vec{v}||^2 - 2||\vec{u}|| \cdot ||\vec{v}|| \cos\theta.$$

Thus setting these two expressions for $||\vec{w}||^2$ equal gives

$$||\vec{u}||^2 + ||\vec{v}||^2 - 2\vec{u} \cdot \vec{v} = ||\vec{u}||^2 + ||\vec{v}||^2 - 2||\vec{u}|| \cdot ||\vec{v}|| \cos\theta,$$

which simplifies to the formula we wanted:

$$\vec{u} \cdot \vec{v} = ||\vec{u}|| \cdot ||\vec{v}|| \cos\theta.$$

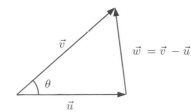

Figure 10.27: Triangle used to justify $\vec{u} \cdot \vec{v} = ||\vec{u}|| \cdot ||\vec{v}|| \cos\theta$

Example 3   (a)   Find $\vec{v} \cdot \vec{w}$ where $\vec{v} = 3\vec{i} + 4\vec{j}$ and $\vec{w} = 2\vec{i} + 5\vec{j}$.

(b)   One person walks 3 miles east and then 4 miles north to point A. Another person walks 2 miles east and then 5 miles north to point B. Both started from the same spot, O. What is the angle of separation of these two people? (See Figure 10.28.)

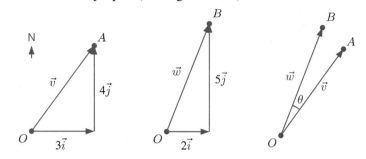

Figure 10.28: What is the angle of separation θ between the people at A and B?

Solution   (a)   We have $\vec{v} \cdot \vec{w} = (3\vec{i} + 4\vec{j}) \cdot (2\vec{i} + 5\vec{j}) = 3 \cdot 2 + 4 \cdot 5 = 26$.

(b)   Assuming $\vec{i}$ points east and $\vec{j}$ points north, we see that $\vec{v}$ gives the first person's position and $\vec{w}$ gives the second person's position. The angle of separation between these two people is labeled θ in Figure 10.28. We use the formula

$$\vec{v} \cdot \vec{w} = ||\vec{v}|| \cdot ||\vec{w}|| \cos \theta.$$

Since $||\vec{v}|| = \sqrt{3^2 + 4^2} = 5$ and $||\vec{w}|| = \sqrt{2^2 + 5^2} = \sqrt{29}$ and, from part (a), $\vec{v} \cdot \vec{w} = 26$, we have

$$26 = 5\sqrt{29} \cos \theta,$$
$$\cos \theta = \frac{26}{5\sqrt{29}},$$
$$\theta = \arccos \frac{26}{5\sqrt{29}} = 15.1°.$$

What Does the Dot Product Mean?

The dot product can be interpreted as a measure of the alignment of two vectors. If two vectors $\vec{u}$ and $\vec{v}$ are perpendicular, then the angle between them is $\theta = 90°$ and they are not at all in alignment. (See Figure 10.29.) In this case, $\cos \theta = \cos 90° = 0$, so

$$\vec{u} \cdot \vec{v} = ||\vec{u}|| \cdot ||\vec{v}|| \cos 90° = 0.$$

A dot product of zero tells us that the two vectors are not aligned at all.

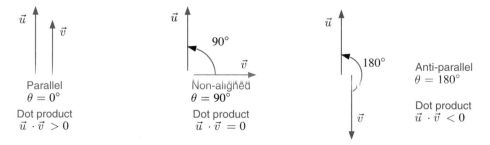

Figure 10.29: Dot product is positive, zero, or negative depending on alignment of vectors

If the two vectors are perfectly aligned, they are parallel, so $\theta = 0°$ and $\cos\theta = \cos 0° = 1$. In this case,

$$\vec{u} \cdot \vec{v} = ||\vec{u}|| \cdot ||\vec{v}|| \cos 0° = ||\vec{u}|| \cdot ||\vec{v}||,$$

so the dot product is positive.

If θ is between $0°$ and $90°$, the vectors are partially aligned and $\cos\theta$ is between 0 and 1. In this case $\vec{u} \cdot \vec{v}$ is positive, but smaller than $||\vec{u}|| \cdot ||\vec{v}||$.

Two vectors are also perfectly aligned if they are pointing in opposite directions. Such vectors are *anti-parallel*. In this case, $\theta = 180°$ and $\cos\theta = -1$, and we have

$$\vec{u} \cdot \vec{v} = ||\vec{u}|| \cdot ||\vec{v}|| \cos 180° = -||\vec{u}|| \cdot ||\vec{v}||,$$

so the dot product is negative.

To summarize:

- Perfect alignment results in the largest possible value for $\vec{u} \cdot \vec{v}$. It occurs if $\vec{u}$ and $\vec{v}$ are parallel, with $\theta = 0°$.

- Perpendicularity results in $\vec{u} \cdot \vec{v} = 0$. It occurs if $\vec{u}$ and $\vec{v}$ are at angle of $\theta = 90°$.

- Perfect alignment in opposite directions results in the most negative value for $\vec{u} \cdot \vec{v}$. It occurs if $\vec{u}$ and $\vec{v}$ are anti-parallel, with $\theta = 180°$.

Work

In physics, the concept of *work* is represented by the dot product. In everyday language, work means effort expended. In physics, the term has a similar, but more precise, meaning.

Suppose you load a heavy refrigerator onto a truck. The refrigerator is on casters and glides with little effort along the floor. However, to lift the refrigerator takes a lot of work. The *work* done, in moving the refrigerator against the force of gravity is defined by

$$\text{Work} = -\vec{F} \cdot \vec{r},$$

where $\vec{F}$ is the weight of the refrigerator and $\vec{r}$ is the displacement. Weight is written as a vector because it has both magnitude (how heavy the refrigerator is) and a direction (downwards, towards the center of the earth). In fact, the refrigerator's weight is an example of a force, which is why we have labeled it $\vec{F}$. If we measure distance in feet and weight in pounds, work is measured in foot-pounds, where 1 foot-pound is the amount of work required to raise 1 pound a distance of 1 foot.

Suppose we push the refrigerator up a ramp that makes a $10°$ angle with the floor. If the ramp is 12 ft long and if the refrigerator weighs 350 lbs, then the weight $\vec{F}$ has a magnitude of 350, the displacement $\vec{r}$ has a magnitude of 12, and the angle θ between them is $10° + 90° = 100°$. (See Figure 10.30.) The work done by the force $\vec{F}$ is

$$\text{Work} = -||\vec{F}|| \cdot ||\vec{r}|| \cos 100° = -350 \cdot 12 \cos 100° = 729.3 \text{ ft-lbs.}$$

To push the refrigerator up the ramp, we do 729.3 ft-lbs of work.

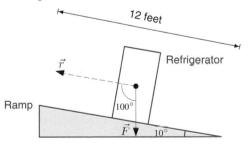

Figure 10.30: Refrigerator being pushed up ramp; the angle between $\vec{F}$ and $\vec{r}$ is $\theta = 100°$

It is informative to consider the two extreme cases: horizontal and vertical ramps. A horizontal ramp leads to a 90° angle between $\vec{F}$ and $\vec{r}$, so

$$\text{Work} = -||\vec{F}|| \cdot ||\vec{r}|| \cos 90° = -350 \cdot 12 \cdot 0 = 0 \text{ ft-lbs}.$$

Since we push the refrigerator in a direction perpendicular to its weight, we don't have to fight the weight at all (assuming frictionless casters). For a vertical ramp, we have

$$\text{Work} = -||\vec{F}|| \cdot ||\vec{r}|| \cos 180° = -350 \cdot 12 \cdot (-1) = 4200 \text{ ft-lbs},$$

In this case, we are pushing (or hoisting) the refrigerator in a direction opposite to that of its weight, so we feel its full force.

Problems for Section 10.4

For Problems 1–6, perform the following operations on the given 3-dimensional vectors.

$$\vec{a} = 2\vec{j} + \vec{k} \qquad \vec{b} = -3\vec{i} + 5\vec{j} + 4\vec{k} \qquad \vec{c} = \vec{i} + 6\vec{j} \qquad \vec{y} = 4\vec{i} - 7\vec{j} \qquad \vec{z} = \vec{i} - 3\vec{j} - \vec{k}$$

1. $\vec{c} \cdot \vec{y}$

2. $\vec{a} \cdot \vec{z}$

3. $\vec{a} \cdot \vec{b}$

4. $(\vec{a} \cdot \vec{b})\vec{a}$

5. $(\vec{a} \cdot \vec{y})(\vec{c} \cdot \vec{z})$

6. $((\vec{c} \cdot \vec{c})\vec{a}) \cdot \vec{a}$

7. How much work is done in pushing a 350 lb refrigerator up a 12 ft ramp which makes a 30° angle with the floor?

8. Which pairs of the vectors $\sqrt{3}\vec{i} + \vec{j}, 3\vec{i} + \sqrt{3}\vec{j}, \vec{i} - \sqrt{3}\vec{j}$ are parallel and which are perpendicular?

9. Compute the angle between the vectors $\vec{i} + \vec{j} + \vec{k}$ and $\vec{i} - \vec{j} - \vec{k}$.

10. For what values of t are $\vec{u} = t\vec{i} - \vec{j} + \vec{k}$ and $\vec{v} = t\vec{i} + t\vec{j} - 2\vec{k}$ perpendicular? Are there values of t for which $\vec{u}$ and $\vec{v}$ are parallel?

11. Suppose $\vec{a}$ and $\vec{b}$ have lengths given by $||\vec{a}|| = 7$ and $||\vec{b}|| = 4$, but that $\vec{a}$ and $\vec{b}$ can point in any direction. What are the maximum and minimum possible lengths for the vectors $\vec{a} + \vec{b}$ and $\vec{a} - \vec{b}$? Illustrate your answers with sketches.

12. A 100-meter dash is run on a track in the direction of the vector $\vec{v} = 2\vec{i} + 6\vec{j}$. The wind velocity $\vec{w}$ is $5\vec{i} + \vec{j}$ km/hr. The rules say that a legal wind speed measured in the direction of the dash must not exceed 5 km/hr. Will the race results be disqualified due to an illegal wind? Justify your answer.

13. Let A, B, C be the points $A = (1, 2); B = (4, 1); C = (2, 4)$. Is triangle $\triangle ABC$ a right triangle?

14. In this problem, you will check the distributive law for 2-vectors. Show that if $\vec{u} = (u_1, u_2), \vec{v} = (v_1, v_2)$, and $\vec{w} = (w_1, w_2)$, then
$$\vec{u} \cdot (\vec{v} + \vec{w}) = \vec{u} \cdot \vec{v} + \vec{u} \cdot \vec{w}.$$

15. Show that the vectors $(\vec{b} \cdot \vec{c})\vec{a} - (\vec{a} \cdot \vec{c})\vec{b}$ and $\vec{c}$ are perpendicular.

16. (a) Bread, eggs, and milk cost $1.50 per loaf, $1.00 per dozen, and $2.00 per gallon, respectively, at Acme Store. Use a price vector $\vec{a}$ and a consumption vector $\vec{c}$ to write a vector equation that describes what may be bought for $20.

    (b) At Beta Mart, where the food is fresher, the price vector is $\vec{b} = (1.60, 0.90, 2.25)$. Explain the meaning of $(\vec{b} - \vec{a}) \cdot \vec{c}$ in practical terms. Is $\vec{b} - \vec{a}$ ever perpendicular to $\vec{c}$?

    (c) Some people think Beta Mart's freshness makes each grocery item at Beta equivalent to 110% of the corresponding Acme item. What does it mean for a consumption vector to satisfy the inequality $(1/1.1)\vec{b} \cdot \vec{c} < \vec{a} \cdot \vec{c}$?

17. Recall that in 2 or 3 dimensions, if θ is the angle between $\vec{v}$ and $\vec{w}$, the dot product is given by

$$\vec{v} \cdot \vec{w} = \|\vec{v}\|\|\vec{w}\| \cos\theta.$$

We use this relationship to define the angle between two vectors in n-dimensions. If $\vec{v}$, $\vec{w}$ are n-vectors, then the dot product, $\vec{v} \cdot \vec{w} = v_1 w_1 + v_2 w_2 + \cdots + v_n w_n$, is used to define the angle θ by

$$\cos\theta = \frac{\vec{v} \cdot \vec{w}}{\|\vec{v}\|\|\vec{w}\|} \qquad \text{provided } \|\vec{v}\|, \|\vec{w}\| \neq 0.$$

We now use this idea of angle to measure how close two populations are to one another genetically. Table 10.1 shows the relative frequencies of four alleles (variants of a gene) in four populations.

TABLE 10.1

Allele	Eskimo	Bantu	English	Korean
A_1	0.29	0.10	0.20	0.22
A_2	0.00	0.08	0.06	0.00
B	0.03	0.12	0.06	0.20
O	0.67	0.69	0.66	0.57

Let $\vec{a}_1, \vec{a}_2, \vec{a}_3, \vec{a}_4$ be the 4-vectors showing the relative frequencies in the Eskimo, Bantu, English, Korean populations, respectively. The genetic distance between two populations is defined as the angle between the corresponding vectors. Using this definition, is the English population closer genetically to the Bantus or to the Koreans? Explain.[5]

18. Let S be the triangle with vertices $A = (2, 2, 2)$, $B = (4, 2, 1)$, and $C = (2, 3, 1)$.

   (a) Find the length of the shortest side of S.
   (b) Find the cosine of the angle BAC at vertex A.

19. A basketball gymnasium is 25 meters high, 80 meters wide and 200 meters long. For a half time stunt, the cheerleaders want to run two strings, one from each of the two corners of the gym above one basket to the diagonally opposite corners of the gym floor. What is the angle made by the strings as they cross?

REVIEW PROBLEMS FOR CHAPTER TEN

For Problems 1–2, perform the indicated computations.

1. $-4(\vec{i} - 2\vec{j}) - 0.5(\vec{i} - \vec{k})$

2. $2(0.45\vec{i} - 0.9\vec{j} - 0.01\vec{k}) - 0.5(1.2\vec{i} - 0.1\vec{k})$

Resolve the vectors in Problems 3–4 into components.

3.

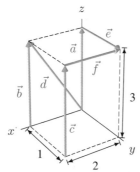

4.

5. Find the length of the vectors $\vec{u}$ and $\vec{v}$ in Problem 4.

---

[5] Adapted from Cavalli-Sforza and Edwards, "Models and Estimation Procedures," Am J. Hum. Genet., Vol. 19 (1967), pp. 223-57.

6. Find a vector that points in the same direction as $\vec{i} - \vec{j} + 2\vec{k}$, but has length 2.

7. (a) Are the vectors $4\vec{i} + a\vec{j} + 6\vec{k}$ and $a\vec{i} + (a-1)\vec{j} + 3\vec{k}$ parallel for any values of the constant a?
   (b) Are these vectors ever perpendicular?

8. A point P is on the rim of a moving bicycle wheel of radius 1 ft.. The bicycle is moving forward at 6π ft./second.

   (a) Sketch the velocity of P relative to the wheel's axle.
   (b) Sketch the velocity of the axle relative to the ground.
   (c) Sketch the velocity of P relative to the ground.
   (d) Does P ever stop, relative to the ground? What is the fastest speed that P moves, relative to the ground?

9. A plane is heading due east and climbing at the rate of 80 km/hr. If its airspeed is 480 km/hr and there is a wind blowing 100 km/hr to the northeast, what is the ground speed of the plane?

10. A particle moving with speed v hits a barrier at an angle of 60° and bounces off at an angle of 60° in the opposite direction with speed reduced by 20 percent, as shown in Figure 10.31. Find the velocity vector of the object after impact.

11. Figure 10.32 shows a molecule with four atoms at O, A, B and C. Show that every atom in the molecule is 2 units away from every other atom.

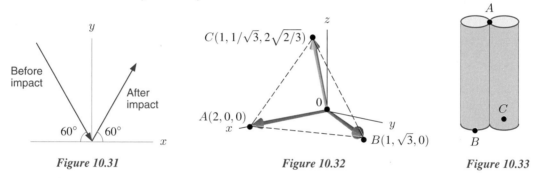

Figure 10.31   Figure 10.32   Figure 10.33

12. Two cylindrical cans of radius 2 and height 7 are shown in Figure 10.33. The cans touch down the side. Let A be the point at the top rims where they touch. Let B be the front point on the bottom rim of the left can, and C be the back point on the bottom rim of the right can. The origin is at A and the z-axis points upward; the x-axis points forward (out of the paper) and the y-axis points to the right.

   (a) Write vectors in component form for $\overrightarrow{AB}$ and $\overrightarrow{AC}$.
   (b) What is the angle between $\overrightarrow{AB}$ and $\overrightarrow{AC}$?

In Problems 13–15, a 5-pound block sits on a plank of wood. If one end of the plank is raised, the block slides down the plank. However, friction between the block and the plank prevents it from sliding until the plank has been raised a certain height. It turns out that the *sliding force* exerted by gravity on the block is proportional to the sine of the angle made by the plank with the ground.

13. Find a formula for $F = g(\theta)$, the sliding force (in lbs) exerted on a block if the plank makes an angle of θ with the ground. [Hint: What is sliding force if the plank is horizontal? Vertical?]

14. One end of the plank is lifted at a constant rate of 2 ft per second, while the other end rests on the ground.

   (a) Find a formula for $F = h(t)$, the sliding force exerted on the block as a function of time.
   (b) Suppose the block begins to slide once the sliding force equals 3 lbs. At what time will the block begin to slide?

15. The 5-lb force exerted on the block by gravity can be resolved into two components, the sliding force $\vec{F}_s$ parallel to the ramp and the normal force $\vec{F}_N$ perpendicular to the ramp. Use this information to show that your formula in Problem 13 is correct.

16. A consumption vector of three goods is given by $\vec{x} = (x_1, x_2, x_3)$, where x_1, x_2 and x_3 are the quantities consumed of the three goods. Consider a budget constraint represented by the equation $\vec{p} \cdot \vec{x} = k$, where $\vec{p}$ is the price vector of the three goods and k is a constant. Show that the difference between two consumption vectors corresponding to points satisfying the same budget constraint is perpendicular to the price vector $\vec{p}$.

17. (a) Using the fact that $\vec{u} \cdot \vec{v} = \|\vec{u}\| \cdot \|\vec{v}\| \cos\theta$, show that

$$\vec{u} \cdot (-\vec{v}) = -(\vec{u} \cdot \vec{v}).$$

[Hint: What happens to the angle when you multiply $\vec{v}$ by -1?]

(b) Using the fact that $\vec{u} \cdot \vec{v} = \|\vec{u}\| \cdot \|\vec{v}\| \cos\theta$, show that for any negative scalar λ

$$\vec{u} \cdot (\lambda\vec{v}) = \lambda(\vec{u} \cdot \vec{v})$$
$$(\lambda\vec{u}) \cdot \vec{v} = \lambda(\vec{u} \cdot \vec{v}).$$

18. Consider the grid in Figure 10.34. Write expressions for $\overrightarrow{AB}$ and $\overrightarrow{CD}$ in terms of $\vec{i}$ and $\vec{j}$.

19. Consider the grid of equilateral triangles in Figure 10.35. Find expressions for $\overrightarrow{AB}, \overrightarrow{BC}, \overrightarrow{AC}$, and $\overrightarrow{AD}$ in terms of $\vec{u}$ and $\vec{v}$.

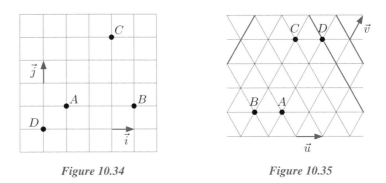

Figure 10.34   *Figure 10.35*

20. Consider the regular hexagon in Figure 10.36. Express the six sides and all three diameters in terms of $\vec{m}$ and $\vec{n}$.

21. Consider the grid of regular hexagons in Figure 10.37. Express $\overrightarrow{AC}, \overrightarrow{AB}, \overrightarrow{AD}$ and $\overrightarrow{BD}$ in terms of $\vec{m}$ and $\vec{n}$.

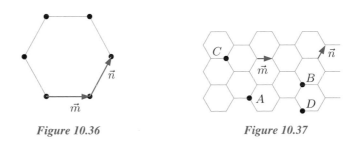

Figure 10.36   *Figure 10.37*

CHAPTER ELEVEN

OTHER WAYS OF DEFINING FUNCTIONS

Up until now, we have focused on four primary ways to define and analyze functions: verbally, graphically, numerically, and algebraically. In this chapter we explore some new ways to define functions and relationships. In many cases, these new approaches build on the four approaches we already know.

11.1 DEFINING FUNCTIONS USING SUMS

Landscape timbers are large beams of wood used to landscape gardens. To make the terrace in Figure 11.1, one timber is set into the slope, followed by a stack of two, then a stack of three, then a stack of four. The stacks are separated by earth.

The total number of timbers used is the sum of the number of timbers in each stack:

$$\text{Number of timbers} = 1 + 2 + 3 + 4 = 10.$$

For a larger terrace using 5 stacks of timbers, the total number is given by

$$\begin{array}{c}\text{Number of timbers}\\ \text{in 5 stacks}\end{array} = 1 + 2 + 3 + 4 + 5 = 15.$$

For an even larger terrace using 6 stacks, the total number is given by

$$\begin{array}{c}\text{Number of timbers}\\ \text{in 6 stacks}\end{array} = 1 + 2 + 3 + 4 + 5 + 6 = 21.$$

In general, for a terrace made from n stacks of landscape timbers, we see that

$$\begin{array}{c}\text{Number of timbers}\\ \text{in } n \text{ stacks}\end{array} = 1 + 2 + \cdots + n.$$

The symbol $\cdots$ means that all the integers from 1 to n are included in the sum.

Letting T be the total number of timbers in a terrace made using n stacks of timbers, we see that T is a function of n, so

$$T = f(n) = 1 + 2 + \cdots + n.$$

In this section we see how to evaluate functions defined by sums.

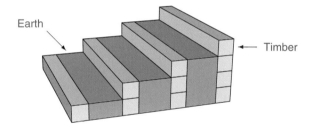

Figure 11.1: A slope terraced for planting using landscape timbers

Arithmetic Sequences

Each stack of landscape timbers contains one more timber than the previous one. The list, or *sequence*, of numbers

$$1, 2, 3, 4, 5, \ldots$$

is an example of an *arithmetic sequence*.

> ### Arithmetic Sequence
>
> A **sequence** is a list of numbers. In an **arithmetic sequence**, the terms increase (or decrease) by adding (or subtracting) a fixed quantity.

The nth Term in an Arithmetic Sequence

The terms in an arithmetic sequence increase (or decrease) by a fixed number, d. If a_1 is the first term, a_2 the second, a_3 the third, we have

$$a_2 = a_1 + d$$
$$a_3 = a_2 + d = (a_1 + d) + d = a_1 + 2d$$
$$a_4 = a_3 + d = (a_1 + 2d) + d = a_1 + 3d,$$

and so on. In general:

> A formula for a_n, the nth **term of an arithmetic sequence** with first term a_1 and difference d between consecutive terms, is given by
>
> $$a_n = a_1 + \underbrace{d + d + \cdots + d}_{n-1 \text{ times}} = a_1 + (n-1)d.$$

Arithmetic Series

To find the total number of timbers in n stacks, we add the terms in our arithmetic sequence. Such a sum is called an *arithmetic series*.

> ### Arithmetic Series
>
> A **series** is a sum of the numbers in a list or sequence. An **arithmetic series** is the sum of terms in an arithmetic sequence.

The total number of landscape timbers is the arithmetic series consisting of the sum of the first n positive integers, written

$$S_n = 1 + 2 + \cdots + n.$$

A famous story concerning this series[1] is told about the great mathematician Carl Friedrich Gauss (1777–1855), who as a young boy was asked by his teacher to add the numbers from 1 to 100. He did so almost immediately:

$$S_{100} = 1 + 2 + 3 + \cdots + 100 = 5050.$$

Of course, no one really knows how Gauss accomplished this, but he probably did not perform the calculation directly, by adding 100 terms. He might have noticed that the terms in the sum can be regrouped into pairs, as follows:

$$S_{100} = 1 + 2 + \cdots + 99 + 100 = \underbrace{(1 + 100) + (2 + 99) + \cdots + (50 + 51)}_{50 \text{ pairs}}$$
$$= \underbrace{101 + 101 + \cdots + 101}_{50 \text{ terms}} \quad \text{Each pair adds to 101}$$
$$= 50 \cdot 101 = 5050.$$

---

[1] As told by E.T. Bell in *The Men of Mathematics*, p.221 (New York: Simon and Schuster, 1937), the series involved was arithmetic, but more complicated than this one.

The approach of pairing numbers works for the sum from 1 to n, no matter how large n is. Provided n is an even number, we can write

$$S_n = 1 + 2 + \cdots + (n-1) + n = \underbrace{(1+n) + (2 + (n-1)) + (3 + (n-2)) + \cdots}_{\frac{1}{2}n \text{ pairs}}$$

$$= \underbrace{(1+n) + (1+n) + \cdots}_{\frac{1}{2}n \text{ pairs}} \quad \text{Each pair adds to } 1+n$$

so we have the formula:

$$\boxed{S_n = \frac{1}{2}n(n+1).}$$

To check this formula, let $n = 100$:

$$S_{100} = 1 + 2 + \cdots + 100 = \frac{1}{2}(100)(100 + 1) = 50(101) = 5050.$$

A similar derivation shows that this formula for S_n also holds for odd values of n.

The Sum of an Arithmetic Series

To find a formula for the sum of a general arithmetic series, we first assume that n is even and write

$$S_n = a_1 + a_2 + \cdots + a_n$$
$$= \underbrace{(a_1 + a_n) + (a_2 + a_{n-1}) + (a_3 + a_{n-2}) + \cdots}_{\frac{1}{2}n \text{ pairs}}.$$

We pair off the first term with the last term, the second term with the next to last term, and so on, just as Gauss may have done. Each pair of terms adds up to the same value, just as Gauss's pairs added to 101. Using the formula for the terms of an arithmetic sequence, $a_n = a_1 + (n-1)d$, the first pair, $a_1 + a_n$, can be written as

$$a_1 + a_n = a_1 + \underbrace{a_1 + (n-1)d}_{a_n} = 2a_1 + (n-1)d,$$

and the second pair, $a_2 + a_{n-1}$, can be written as

$$a_2 + a_{n-1} = \underbrace{a_1 + d}_{a_2} + \underbrace{a_1 + (n-2)d}_{a_{n-1}} = 2a_1 + (n-1)d.$$

Both the first two pairs have the same sum: $2a_1 + (n-1)d$. The remaining pairs also all have the same sum, so

$$S_n = \underbrace{(a_1 + a_n) + (a_2 + a_{n-1}) + (a_3 + a_{n-2}) + \cdots}_{\frac{1}{2}n \text{ pairs}} = \frac{1}{2}n\left(2a_1 + (n-1)d\right).$$

The sum, S_n, of the first n terms of the **arithmetic series** with $a_n = a_1 + (n-1)d$ is

$$S_n = \frac{1}{2}n(2a_1 + (n-1)d).$$

The same formula gives the sum of the series when n is odd. See Problem 18 on page 450.

Example 1   Calculate the sum $1 + 2 + \cdots + 100$ using the formula for S_n.

Solution   Here, $a_1 = 1$, $n = 100$, and $d = 1$. We get the same answer as before:

$$S_{100} = \frac{1}{2}n\left(2a_1 + (n-1)d\right) = \frac{1}{2}(100)\left(2(1) + (100-1)(1)\right) = 50(101) = 5050,$$

Summation Notation

The symbol Σ is used to represent a sum such as $a_1 + a_2 + \cdots + a_n$. This symbol, pronounced *sigma*, is the Greek capital letter for *S*, which stands for sum. Using this notation, we write

$$\sum_{i=1}^{n} a_i$$

to stand for the sum $a_1 + a_2 + \cdots + a_n$. The Σ tells us we are adding some numbers. The a_i tells us that the numbers we are adding are called a_1, a_2, and so on. The sum begins with a_1 and ends with a_n because the subscript i starts at $i = 1$ (at the bottom of the Σ sign) and ends at $i = n$ (at the top of the Σ sign):

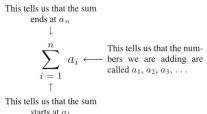

Example 2   Using sigma notation, write a formula for $T = f(n)$, the total number of landscape timbers in n stacks.

Solution   We have $T = f(n) = 1 + 2 + \cdots + n$. We are adding n numbers, starting at $i = 1$ and ending at $i = n$. This sum can be written

$$T = f(n) = \sum_{i=1}^{n} i.$$

Example 3   Use sigma notation to write the sum of the first 20 positive odd numbers. Evaluate this sum.

Solution   The odd numbers form an arithmetic sequence: $1, 3, 5, 7, \ldots$ with $a_1 = 1$ and $d = 2$. The ith odd number is

$$a_i = 1 + (i-1)2 = 2i - 1.$$

The sum of the first 20 odd numbers is given by

$$\text{Sum} = \sum_{i=1}^{20} a_i = \sum_{i=1}^{20}(2i - 1).$$

We evaluate the sum using $n = 20$, $a_1 = 1$, and $d = 2$:

$$\text{Sum} = \frac{1}{2}n\left(2a_1 + (n-1)d\right) = \frac{1}{2}(20)\left(2(1) + (20-1)(2)\right) = 400.$$

Falling Objects

If air resistance is neglected, every falling object travels 16 ft during the first second, 48 ft during the next, 80 ft during the next, and so on. These distances form the arithmetic sequence $16, 48, 80, \ldots$. In this sequence, $a_1 = 16$ and $d = 32$. The nth term is $a_n = a_1 + (n-1)d = 16 + 32(n-1) = 32n - 16$. Thus, the fifth term is $a_5 = 32(5) - 16 = 144$. The value of $a_{10} = 32(10) - 16 = 304$.

Example 4   Calculate S_1, S_2, and S_3, the distances an object falls in 1, 2, and 3 seconds, respectively.

Solution   Since $a_1 = 16$ and $d = 32$,

$$S_1 = a_1 = 16 \text{ feet}$$
$$S_2 = a_1 + a_2 = a_1 + (a_1 + d) = 16 + 48 = 64 \text{ feet}$$
$$S_3 = a_1 + a_2 + a_3 = 16 + 48 + 80 = 144 \text{ feet}$$

Alternatively, the values of S_n can be calculated using the formula $S_n = \frac{1}{2}n\left(2a_1 + (n-1)d\right)$:

$$S_1 = \frac{1}{2} \cdot 1(2 \cdot 16 + 0 \cdot 32) = 16, \quad S_2 = \frac{1}{2} \cdot 2(2 \cdot 16 + 1 \cdot 32) = 64, \quad S_3 = \frac{1}{2} \cdot 3(2 \cdot 16 + 2 \cdot 32) = 144.$$

Example 5   An object falls from 1000 feet starting at time $t = 0$ seconds. What is its height, h, in feet above the ground at $t = 1, 2, 3$ seconds? Show these values on a graph of height against time.

Solution   At time $t = 0$, the height $h = 1000$ feet. At time $t = 1$, the object has fallen 16 feet, so

$$h = 1000 - 16 = 984 \text{ feet.}$$

At time $t = 2$, the object has fallen a total distance of $S_2 = 64$ feet, so

$$h = 1000 - 64 = 936 \text{ feet.}$$

At time $t = 3$, the object has fallen a total distance of 144 feet, so

$$h = 1000 - 144 = 856 \text{ feet.}$$

These heights are marked on the graph in Figure 11.2.

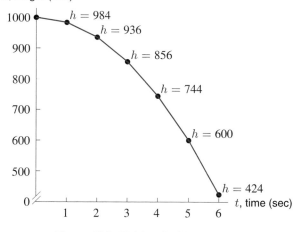

Figure 11.2: Height of a falling object

Arithmetic Sequences and Linear Functions

You may have noticed that the arithmetic sequence for a falling object is like a linear function. The formula for the nth term, $a_n = 32n - 16$, is the same as the formula for a linear function with slope $m = 32$ and initial value $b = -16$. However, for a sequence we consider only nonnegative integer values of n, whereas a linear function is defined for all values of n. We can think of an arithmetic sequence as a linear function whose domain has been restricted to the nonnegative integers.

Problems for Section 11.1

In Problems 1–4, write each of the sums using sigma notation.

1. $3 + 6 + 9 + 12 + 15 + 18 + 21$
2. $10 + 13 + 16 + 19 + 22$
3. $1/2 + 1 + 3/2 + 2 + 5/2 + 3 + 7/2 + 4$
4. $30 + 25 + 20 + 15 + 10 + 5$

5. Find the sum of the first nine terms of the series: $7 + 14 + 21 + \cdots$.

6. Find the thirtieth positive multiple of 5 and the sum of the first thirty positive multiples of 5.

Without using a calculator, find the sum of the series in Problems 7–10.

7. $\displaystyle\sum_{i=1}^{50} 3i$
8. $\displaystyle\sum_{i=1}^{30} (5i + 10)$
9. $\displaystyle\sum_{n=0}^{15} \left(2 + \frac{1}{2}n\right)$
10. $\displaystyle\sum_{n=0}^{10} (8 - 4n)$

Problems 11–14 refer to the falling object of Example 4 on page 448.

11. Calculate the distance, S_{10}, the object falls in 10 seconds.

12. (a) Find the total distance that the object falls in 4, 5, 6 seconds.
    (b) The object falls from 1000 feet at time $t = 0$. Calculate its height at $t = 4$, $t = 5$, $t = 6$ seconds. Explain how you can check your answer using Figure 11.2.

13. Find a formula for $f(n)$, the distance fallen by the object in n seconds.

14. If the object falls from 1000 feet, how long does it take to hit the ground?

15. A boy is dividing M&Ms between himself and his sister. He gives one to his sister and takes one for himself. He gives another to his sister and takes two for himself. He gives a third to his sister and takes three for himself, and so on.

    (a) On the nth round, how many M&Ms does the boy give his sister? How many does he take himself?
    (b) After n rounds, how many M&Ms does his sister have? How many does the boy have?

16. An auditorium has 30 seats in the first row, 34 seats in the second row, 38 seats in the third row, and so on. If there are twenty rows in the auditorium, how many seats are there in the last row? How many seats are there in the auditorium?

17. In a workshop, it costs $300 to make one piece of furniture. The second piece costs a bit less, $280. The third costs even less, $263, and the fourth costs only $249. The cost for each additional piece of furniture is called the *marginal cost* of production. These costs are recorded in Table 11.1. In most cases, including this problem, as the quantity produced increases, the marginal cost first decreases and then increases again. The table also shows the difference between each marginal cost—for instance, $280 - 300 = -20$. Notice that although the sequence of numbers 300, 280, 263, ..., is not arithmetic, the sequence of differences, $-20, -17, -14, \ldots$, is arithmetic and the terms eventually become positive.

TABLE 11.1

n, Number pieces furniture	1		2		3		4
c, Marginal cost ($)	300		280		263		249
Change in marginal cost ($)		-20		-17		-14	

(a) Assume that the arithmetic sequence $-20, -17, -14, \ldots$, continues. Complete the table for $n = 5, 6, \ldots, 12$.

(b) Find a formula for c_n, the marginal cost for producing the nth piece of furniture. Use the fact that c_n is found by adding the terms in an arithmetic sequence. Using your formula, find the cost for producing the 12th piece and the 50th piece of furniture.

(c) A piece of furniture can be sold at a profit provided it costs less than $400 to make. How many pieces of furniture should the workshop make each day? Discuss.

18.   In the text we showed how to calculate the sum of an arithmetic series with an even number of terms. Consider the arithmetic series

$$5 + 12 + 19 + 26 + 33 + 40 + 47 + 54 + 61.$$

Here, there are $n = 9$ terms, and the difference between each term is $d = 7$. Adding these terms directly, we find that their sum is 297. In this problem we find the sum of this arithmetic sequence in two different ways. We then use our results to obtain a general formula for the sum of an arithmetic series with an odd number of terms.

(a)   The sum of the first and last terms is $5 + 61 = 66$, the sum of the second and next-to-last terms is $12 + 54 = 66$, and so on. Find the sum of this arithmetic series by pairing off terms in this way. Notice that since the number of terms is odd, one of them will be unpaired.

(b)   This arithmetic series can be thought of as a series of eight terms $(5 + \cdots + 54)$ plus an additional term (61). Use the formula we found for the sum of an arithmetic series containing an even number of terms to find the sum of the given arithmetic series.

(c)   Find a formula for the sum of an arithmetic series with n terms where n is odd. Let a_1 be the first term in the series, and let d be the difference between consecutive terms. Show that the two approaches used in parts (a) and (b) give the same result, and show that your formula is the same as the formula given for even values of n.

11.2   GEOMETRIC SERIES

In the previous section, we studied *arithmetic series*. An arithmetic series is the sum of terms in a sequence in which each term can be obtained by adding a constant to the preceeding term. In this section, we study another type of sum, a *geometric series*. In a geometric series, each term is a constant multiple of the preceeding term.

Saving Money

A person deposits \$50 every year in an account that pays 6% interest per year, compounded annually. After the first deposit (but before any interest has been earned), the balance of this account is

$$B_1 = 50.$$

After 1 year has passed, the first deposit has earned interest, so the balance becomes $50(1.06)$. Then the second deposit is made and the balance becomes

$$B_2 = \underbrace{2^{\text{nd}} \text{ deposit}}_{50} + \underbrace{1^{\text{st}} \text{ deposit with interest}}_{50(1.06)}$$

$$= 50 + 50(1.06).$$

After 2 years have passed, the third deposit is made, and the balance is

$$B_3 = \underbrace{3^{\text{rd}} \text{ deposit}}_{50} + \underbrace{2^{\text{nd}} \text{ dep. with 1 year interest}}_{50(1.06)} + \underbrace{1^{\text{st}} \text{ deposit with 2 years interest}}_{50(1.06)^2}$$

$$= 50 + 50(1.06) + 50(1.06)^2.$$

Continuing in this fashion, we see that

After 4 deposits	$B_4 = 50 + 50(1.06) + 50(1.06)^2 + 50(1.06)^3$
After 5 deposits	$B_5 = 50 + 50(1.06) + 50(1.06)^2 + 50(1.06)^3 + 50(1.06)^4$

$$\vdots$$

After n deposits	$B_n = 50 + 50(1.06) + 50(1.06)^2 + \cdots + 50(1.06)^{n-1}.$

Example 1   How much money is in this account after 5 years? After 25 years?

Solution   After 5 years, we have made 6 deposits. A calculator gives
$$B_6 = 50 + 50(1.06) + 50(1.06)^2 + \cdots + 50(1.06)^5$$
$$= \$348.77.$$

After 25 years, we have made 26 deposits. Even using a calculator, it would be tedious to evaluate B_{26} by adding 26 terms. Fortunately, there is a shortcut. We begin with the formula for B_{26}:
$$B_{26} = 50 + 50(1.06) + 50(1.06)^2 + \cdots + 50(1.06)^{25}.$$

Multiplying both sides of this equation by 1.06 and adding 50 gives
$$1.06B_{26} + 50 = 1.06 \left(50 + 50(1.06) + 50(1.06)^2 + \cdots + 50(1.06)^{25}\right) + 50.$$

We can simplify the right-hand side to give
$$1.06B_{26} + 50 = 50(1.06) + 50(1.06)^2 + \cdots + 50(1.06)^{25} + 50(1.06)^{26} + 50.$$

Notice that the right-hand side of this equation and the formula for B_{26} have almost every term in common. We can rewrite this equation as
$$1.06B_{26} + 50 = \underbrace{50 + 50(1.06) + 50(1.06)^2 + \cdots + 50(1.06)^{25}}_{B_{26}} + 50(1.06)^{26}$$
$$= B_{26} + 50(1.06)^{26}.$$

Solving for B_{26} gives
$$1.06B_{26} - B_{26} = 50(1.06)^{26} - 50$$
$$0.06B_{26} = 50(1.06)^{26} - 50$$
$$B_{26} = \frac{50(1.06)^{26} - 50}{0.06}.$$

Using a calculator to evaluate this expression for B_{26}, we find that $B_{26} = \$2957.82$.

Geometric Series

The formula for the bank balance,
$$B_{26} = 50 + 50(1.06) + 50(1.06)^2 + \cdots + 50(1.06)^{25},$$
is an example of a *geometric series*. In general, a geometric series is a sum in which each term is a constant multiple of the preceeding term.

> A **geometric series** is a sum of the form
> $$S_n = a + ax + ax^2 + \cdots + ax^{n-1}.$$

The reason we stop at an exponent of $n-1$ in the definition of S_n is so that there is a total of n terms (including the first term, which is $a = ax^0$). For instance, there are 26 terms in the series
$$50 + 50(1.06) + 50(1.06)^2 + \cdots + 50(1.06)^{25},$$
so $n = 26$. For this series, $x = 1.06$ and $a = 50$.

Another Geometric Series: Drug Levels in The Body

Geometric series arise naturally in many different contexts. The following example illustrates a geometric series with decreasing terms.

Example 2　A patient is given a 20 mg injection of a therapeutic drug. Each day, the patient's body metabolizes 50% of the drug present, so that after 1 day only one-half of the original amount remains, after 2 days only one-fourth remains, and so on. The patient is given a 20 mg injection of the drug every day at the same time. Write a geometric series that gives the drug level in this patient's body after n days.

Solution　Immediately after the 1st injection, the drug level in the body is given by

$$Q_1 = 20.$$

One day later, the original 20 mg has fallen to $20 \cdot \frac{1}{2} = 10$ mg and the second 20 mg injection is given. Right after the second injection, the drug level is given by

$$Q_2 = \underbrace{2^{nd} \text{ injection}}_{20} + \underbrace{\text{Residue of 1}^{st} \text{ injection}}_{\frac{1}{2} \cdot 20}$$

$$= 20 + 20 \left(\frac{1}{2} \right).$$

Two days later, the original 20 mg has fallen to $(20 \cdot \frac{1}{2}) \cdot \frac{1}{2} = 20(\frac{1}{2})^2 = 5$ mg, the second 20 mg injection has fallen to $20 \cdot \frac{1}{2} = 10$ mg, and the third 20 mg injection is given. Right after the third injection, the drug level is given by

$$Q_3 = \underbrace{3^{rd} \text{ injection}}_{20} + \underbrace{\text{Residue of 2}^{nd} \text{ injection}}_{20 \cdot \frac{1}{2}} + \underbrace{\text{Residue of 1}^{st} \text{ injection}}_{20 \cdot \frac{1}{2} \cdot \frac{1}{2}}$$

$$= 20 + 20 \left(\frac{1}{2} \right) + 20 \left(\frac{1}{2} \right)^2.$$

Continuing, we see that

After 4th injection　$Q_4 = 20 + 20 \left(\frac{1}{2} \right) + 20 \left(\frac{1}{2} \right)^2 + 20 \left(\frac{1}{2} \right)^3$

After 5th injection　$Q_5 = 20 + 20 \left(\frac{1}{2} \right) + 20 \left(\frac{1}{2} \right)^2 + \cdots + 20 \left(\frac{1}{2} \right)^4$

$$\vdots$$

After n^{th} injection　$Q_n = 20 + 20 \left(\frac{1}{2} \right) + 20 \left(\frac{1}{2} \right)^2 + \cdots + 20 \left(\frac{1}{2} \right)^{n-1}.$

This is another example of a geometric series. Here, $a = 20$ and $x = 1/2$ in the geometric series formula

$$Q_n = a + ax + ax^2 + \cdots + ax^{n-1}.$$

To calculate the drug level for large values of n, we use the same shortcut as in Example 1.

Example 3　What quantity of the drug remains in the patient's body after the 10th injection?

Solution　After the 10th injection, the drug level in the patient's body is given by

$$Q_{10} = 20 + 20 \left(\frac{1}{2} \right) + 20 \left(\frac{1}{2} \right)^2 + \cdots + 20 \left(\frac{1}{2} \right)^9.$$

Multiply both sides of this formula by $1/2$ and add 20, giving

$$\frac{1}{2}Q_{10} + 20 = \frac{1}{2} \cdot \left(20 + 20\left(\frac{1}{2}\right) + 20\left(\frac{1}{2}\right)^2 + \cdots + 20\left(\frac{1}{2}\right)^9\right) + 20$$

$$= \underbrace{20 + 20\left(\frac{1}{2}\right) + 20\left(\frac{1}{2}\right)^2 + \cdots + 20\left(\frac{1}{2}\right)^9}_{Q_{10}} + 20\left(\frac{1}{2}\right)^{10}$$

$$= Q_{10} + 20\left(\frac{1}{2}\right)^{10}.$$

We now solve for Q_{10}. We have

$$\frac{1}{2}Q_{10} + 20 = Q_{10} + 20\left(\frac{1}{2}\right)^{10}$$

$$-\frac{1}{2}Q_{10} = 20\left(\frac{1}{2}\right)^{10} - 20$$

$$Q_{10} = -2\left(20\left(\frac{1}{2}\right)^{10} - 20\right)$$

$$\approx 39.96 \text{ mg}.$$

The Sum of a Geometric Series

The shortcut from Examples 1 and 3 can be used to find the sum of a general geometric series. Let S_n be the sum of a geometric series of n terms, so that

$$S_n = a + ax + ax^2 + \cdots + ax^{n-1}.$$

Multiply both sides of this equation by x and add a, giving

$$xS_n + a = x\left(a + ax + ax^2 + \cdots + ax^{n-1}\right) + a$$

$$= \left(ax + ax^2 + ax^3 + \cdots + ax^{n-1} + ax^n\right) + a.$$

The right-hand side can be rewritten as

$$xS_n + a = \underbrace{a + ax + ax^2 + \cdots + ax^{n-1}}_{S_n} + ax^n$$

$$= S_n + ax^n.$$

Solving the equation $xS_n + a = S_n + ax^n$ for S_n gives

$$xS_n - S_n = ax^n - a$$

$$S_n(x-1) = ax^n - a \qquad \text{factoring out } S_n$$

$$S_n = \frac{ax^n - a}{x-1}$$

$$= \frac{a(x^n - 1)}{x-1}.$$

By multiplying the numerator and denominator by -1, this formula can be rewritten as follows:

The sum of a **geometric series of n terms** is given by

$$S_n = a + ax + ax^2 + \cdots + ax^{n-1} = \frac{a(1-x^n)}{1-x}.$$

Example 4   Check the formula for the sum of a geometric series by using it to solve Examples 1 and 3.

Solution   For Example 1, we need to find B_6 and B_{26} where

$$B_n = 50 + 50(1.06) + 50(1.06)^2 + \cdots + 50(1.06)^{n-1}.$$

Using the formula for S_n with $a = 50$ and $x = 1.06$, we get the same answers as before:

$$B_6 = \frac{50(1 - (1.06)^6)}{1 - 1.06} = 348.77,$$

$$B_{26} = \frac{50(1 - (1.06)^{26})}{1 - 1.06} = 2957.82.$$

For Example 3, we need to find Q_{10} where

$$Q_n = 20 + 20\left(\frac{1}{2}\right) + 20\left(\frac{1}{2}\right)^2 + \cdots + 20\left(\frac{1}{2}\right)^{n-1}.$$

Using the formula for S_n with $a = 20$ and $x = 1/2$, we get the same quantity as before:

$$Q_{10} = \frac{20(1 - (\frac{1}{2})^{10})}{1 - \frac{1}{2}} = 39.96.$$

What Happens to the Drug Level Over Time?

Suppose the patient from Example 2 receives injections over a long period of time. What happens to the drug level in the patient's body? To find out, we calculate the drug level after 10, 15, 20, and 25 injections:

$$Q_{10} = \frac{20(1 - (\frac{1}{2})^{10})}{1 - \frac{1}{2}} = 39.960938 \text{ mg},$$

$$Q_{15} = \frac{20(1 - (\frac{1}{2})^{15})}{1 - \frac{1}{2}} = 39.998779 \text{ mg},$$

$$Q_{20} = \frac{20(1 - (\frac{1}{2})^{20})}{1 - \frac{1}{2}} = 39.999962 \text{ mg},$$

$$Q_{25} = \frac{20(1 - (\frac{1}{2})^{25})}{1 - \frac{1}{2}} = 39.999999 \text{ mg}.$$

The drug level appears to approach 40 mg. To see why this happens, notice that if there is exactly 40 mg of drug in the body, then half of this amount is metabolized in one day, leaving 20 mg. At the next 20 mg injection, the level returns to 40 mg. We say that the *equilibrium* drug level is 40 mg.

Initially the patient has less than 40 mg of the drug in the body. Then the amount metabolized in one day is less than 20 mg. Thus, after the next 20 mg injection, the drug level is higher than it was before. For instance, if there are currently 30 mg, then after one day, half of this has been metabolized, leaving 15 mg. At the next injection the level rises to 35 mg, or 5 mg higher than where it started. Eventually, the quantity levels off to 40 mg.

Infinite Geometric Series

Another way to think about the patient's drug level over time is to consider an *infinite geometric series*. We know that after n injections, the drug level is given by

$$Q_n = 20 + 20\left(\frac{1}{2}\right) + 20\left(\frac{1}{2}\right)^2 + \cdots + 20\left(\frac{1}{2}\right)^{n-1} = \frac{20(1 - (\frac{1}{2})^n)}{1 - \frac{1}{2}}.$$

What happens to the value of this sum as the number of terms approaches infinity? It doesn't seem possible to add up an infinite number of terms. However, we can look at the *partial sums*, Q_n, to see what happens for large values of n. For large values of n, we see that $(1/2)^n$ is very small, so that

$$Q_n = \frac{20\,(1 - \text{Small number})}{1 - \frac{1}{2}}$$
$$= \frac{20\,(1 - \text{Small number})}{1/2}.$$

As n becomes arbitrarily large, that is, as $n \to \infty$, we know that $(1/2)^n \to 0$, so

$$Q_n \to \frac{20(1 - 0)}{1/2} = \frac{20}{1/2} = 40 \text{ mg}.$$

The Sum of an Infinite Geometric Series

Consider the geometric series $S_n = a + ax + ax^2 + \cdots + ax^{n-1}$. In general, if $|x| < 1$, then $x^n \to 0$ as $n \to \infty$, and

$$S_n = \frac{a(1 - x^n)}{1 - x} \to \frac{a(1 - 0)}{1 - x} = \frac{a}{1 - x} \text{ as } n \to \infty.$$

Thus, if $|x| < 1$, the partial sums S_n approach a finite value, S, as $n \to \infty$; we say that the series *converges* to S.

For $|x| < 1$, the **sum of the infinite geometric series** is given by

$$S = a + ax + ax^2 + \cdots + ax^n + \cdots = \frac{a}{1 - x}.$$

On the other hand, if $|x| > 1$, then we say that the series does not converge. The terms in the series get larger and larger as $n \to \infty$, so adding infinitely many of them does not give a finite sum.

What happens when $x = \pm 1$? The formula for S_n does not apply in these cases, and in fact the infinite geometric series does not converge. (To see why, consider what happens to 1^n and $(-1)^n$ as n increases.)

Present Value of a Series of Payments

When basketball player Patrick Ewing was signed by the New York Knicks, he was given a contract for $30 million: $3 million a year for ten years. Of course, since much of the money was to be paid in the future, the team's owners did not have to have all $30 million available on the day of the signing. How much money would the owners have to deposit in a bank account on the day of the signing in order to cover all the future payments? Assuming the account was earning interest, the owners would have to deposit much less than $30 million. This smaller amount is called the *present value* of $30 million. We will calculate the present value of Ewing's contract on the day he signed.

Definition of Present Value

Let's consider a simplified version of this problem, with only one future payment: How much money would we need to deposit in a bank account today in order to have $3 million in one year? At an annual interest rate of 5%, compounded annually, the deposit would grow by a factor of 1.05. Thus,

$$\text{Required deposit} \times 1.05 = \$3 \text{ million},$$
$$\text{Required deposit} = \frac{3,000,000}{1.05} = 2,857,142.86.$$

We would need to deposit \$2,857,142.86. We say that this is the *present value* of the \$3 million. Similarly, if we need \$3 million in 2 years, the amount we would need to deposit is given by

$$\text{Required deposit} \times 1.05^2 = \$3 \text{ million},$$

$$\text{Required deposit} = \frac{3{,}000{,}000}{1.05^2} = 2{,}721{,}088.44.$$

The \$2,721,088.44 is the present value of \$3 million payable two years from today. In general,

> The **present value**, \$$P$, of a future payment, \$$B$, is the amount which would have to be deposited (at some interest rate, r) in a bank account today to have exactly \$$B$ in the account at the relevant time in the future.

If r is the annual interest rate (compounded annually) and if n is the number of years, then

$$B = P\left(1 + r\right)^n, \quad \text{or equivalently,} \quad P = \frac{B}{(1 + r)^n}.$$

Calculating the Present Value of Ewing's Contract

The present value of Ewing's contract represents what it was worth on the day it was signed. Suppose that he receives his money in 10 payments of \$3 million each, the first payment to be made on the day the contract was signed. We calculate the present values of all 10 payments assuming that interest is compounded annually at a rate of 5% per year. Since the first payment is made the day the contract is signed, we have:

$$\text{Present value of first payment, in millions of dollars} = 3.$$

Since the second payment is made a year in the future, in millions of dollars we have:

$$\text{Present value of second payment} = \frac{3}{(1.05)^1} = \frac{3}{1.05}.$$

The third payment is made two years in the future, so in millions of dollars:

$$\text{Present value of third payment} = \frac{3}{(1.05)^2},$$

and so on. Similarly, in millions of dollars:

$$\text{Present value of tenth payment} = \frac{3}{(1.05)^9}.$$

Thus, in millions of dollars,

$$\text{Total present value} = 3 + \frac{3}{1.05} + \frac{3}{(1.05)^2} + \cdots + \frac{3}{(1.05)^9}.$$

Rewriting this expression, we see that it is a finite geometric series with $a = 3$ and $x = 1/1.05$:

$$\text{Total present value} = 3 + 3\left(\frac{1}{1.05}\right) + 3\left(\frac{1}{1.05}\right)^2 + \cdots + 3\left(\frac{1}{1.05}\right)^9.$$

The formula for the sum of a finite geometric series gives

$$\text{Total present value of contract in millions of dollars} = \frac{3\left(1 - (\frac{1}{1.05})^{10}\right)}{1 - \frac{1}{1.05}} \approx 24.3.$$

Thus, the total present value of the contract is about \$24.3 million dollars.

Example 5   Suppose Patrick Ewing's contract with the Knicks guaranteed him and his heirs an annual payment of \$3 million *forever*. How much would the owners need to deposit in an account today in order to provide these payments?

Solution   At 5%, the total present value of an infinite series of payments is given by

$$\text{Total present value} = 3 + \frac{3}{1.05} + \frac{3}{(1.05)^2} + \cdots$$

$$= 3 + 3\left(\frac{1}{1.05}\right) + 3\left(\frac{1}{1.05}\right)^2 + \cdots.$$

The sum of this infinite geometric series can be found using the formula:

$$\text{Total present value} = \frac{3}{1 - \frac{1}{1.05}} = 63 \text{ million dollars.}$$

To see that this answer is reasonable, suppose that \$63 million is deposited in an account today, and that a \$3 million payment is immediately made to Patrick Ewing. Over the course of a year, the remaining \$60 million earns 5% interest, which works out to \$3 million, so the next year the account again has \$63 million. Thus, it would have cost the New York Knicks only about \$40 million more (than the \$24.3 million) to pay Ewing and his heirs \$3 million a year forever.

Summation Notation

In the last section, we saw how to write an arithmetic series using summation notation. We can also use summation notation for geometric series. For example, we write

$$\sum_{i=1}^{10} x^i = x^1 + x^2 + \cdots + x^{10}.$$

The Σ tells us we are adding some numbers. The x^i tells us that the quantities being added are x^1, x^2, and so on, up to x^{10}. The value of i starts at 1 because of the $i = 1$ at the bottom of the Σ sign; the value of i stops at 10 because of the 10 at the top of the Σ sign.

Example 6   Evaluate $\displaystyle\sum_{i=0}^{5} 7\left(\frac{1}{2}\right)^i$.

Solution   The value of i starts at 0 and ends at 5. The quantities being added are $7(1/2)^0$, $7(1/2)^1$, ..., up to $7(1/2)^5$. Thus,

$$\sum_{i=0}^{5} 7\left(\frac{1}{2}\right)^i = 7 + 7\left(\frac{1}{2}\right) + 7\left(\frac{1}{2}\right)^2 + \cdots + 7\left(\frac{1}{2}\right)^5.$$

This is a geometric series of 6 terms with $a = 7$ and $x = 1/2$. Therefore, we have

$$\sum_{i=0}^{5} 7\left(\frac{1}{2}\right)^i = \frac{7(1 - (\frac{1}{2})^6)}{1 - \frac{1}{2}} \approx 13.78.$$

Summation notation provides a compact way of writing geometric series. Using this notation,

The general formula for a geometric series of n terms can be written

$$\sum_{i=0}^{n-1} ax^i = a + ax + ax^2 + \cdots + ax^{n-1}.$$

The general formula for an infinite geometric series can be written

$$\sum_{i=0}^{\infty} ax^2 = a + ax + ax^2 + \cdots.$$

Example 7   Represent the present value of Patrick Ewing's contract using summation notation.

Solution   We have

$$\text{Total present value} = 3 + \frac{3}{1.05} + \frac{3}{(1.05)^2} + \cdots + \frac{3}{(1.05)^9} = \sum_{i=0}^{9} 3\left(\frac{1}{1.05}\right)^i.$$

Example 8   Find the sum of the geometric series $\sum_{i=0}^{17} 7(-z)^i$ and $\sum_{i=0}^{\infty} (-z)^i$ provided $|z| < 1$.

Solution   We have

$$\sum_{i=0}^{17} 7(-z)^i = 7(-z)^0 + 7(-z)^1 + 7(-z)^2 + \cdots + 7(-z)^{17}.$$

Since $(-z)^0 = 1$, this is a geometric series with $a = 7$ and $x = -z$. There are 18 terms, thus we have

$$\sum_{i=0}^{17} 7(-z)^i = 7\frac{1 - (-z)^{18}}{1 - (-z)} = 7\frac{1 - z^{18}}{1 + z}.$$

As for $\sum_{i=0}^{\infty} (-z)^i$, this is an infinite geometric series. Since $|z| < 1$, we have

$$\sum_{i=0}^{\infty} (-z)^i = \frac{1}{1 - (-z)} = \frac{1}{1 + z}.$$

Problems for Section 11.2

In Problems 1–8, decide which of the following are geometric series. For those which are, give the first term and the ratio between successive terms. For those which are not, explain why not.

1. $2 + 1 + \dfrac{1}{2} + \dfrac{1}{4} + \dfrac{1}{8} + \cdots$

2. $1 - \dfrac{1}{2} + \dfrac{1}{4} - \dfrac{1}{8} + \dfrac{1}{16} + \cdots$

3. $1 + \dfrac{1}{2} + \dfrac{1}{3} + \dfrac{1}{4} + \dfrac{1}{5} + \cdots$

4. $5 - 10 + 20 - 40 + 80 - \cdots$

5. $1 - x + x^2 - x^3 + x^4 - \cdots$

6. $1 + x + 2x^2 + 3x^3 + 4x^4 + \cdots$

7. $y^2 + y^3 + y^4 + y^5 + \cdots$

8. $3 + 3z + 6z^2 + 9z^3 + 12z^4 + \cdots$

9. Find the sum of the series in Problem 5.   10. Find the sum of the series in Problem 7.

Write each of the sums in Problems 11–14 in sigma notation.

11. $1 + 4 + 16 + 64 + 256$

12. $3 - 9 + 27 - 81 + 243 - 729$

13. $2 + 10 + 50 + 250 + 1250 + 6250 + 31250$

14. $32 - 16 + 8 - 4 + 2 - 1$

15. Find the sum of the first seven terms of the series: $7 + 14 + 28 + \cdots$

16. Find the sum of the first ten terms of the series: $5/9 + 5/3 + 5 + \cdots$

Find the sum of the series in Problems 17–24.

17. $3 + \dfrac{3}{2} + \dfrac{3}{4} + \dfrac{3}{8} + \cdots + \dfrac{3}{2^{10}} \approx 5.997.$

18. $-2 + 1 - \dfrac{1}{2} + \dfrac{1}{4} - \dfrac{1}{8} + \dfrac{1}{16} - \cdots$

19. $5 + 15 + 45 + 135 + \cdots + 5(3^{12})$

20. $1/125 + 1/25 + 1/5 + \cdots + 625$

21. $\displaystyle\sum_{i=4}^{\infty} \left(\dfrac{1}{3}\right)^{i}$

22. $\displaystyle\sum_{i=0}^{\infty} \dfrac{3^{i} + 5}{4^{i}}$

23. $\displaystyle\sum_{n=1}^{10} 4(2^{n})$

24. $\displaystyle\sum_{k=0}^{7} 2\left(\dfrac{3}{4}\right)^{k}$

25. A repeating decimal can always be expressed as a fraction. This problem shows how writing a repeating decimal as a geometric series enables you to find the fraction. Consider the decimal $0.232323\ldots$.

   (a) Use the fact that $0.232323\ldots = 0.23 + 0.0023 + 0.000023 + \cdots$ to write $0.232323\ldots$ as a geometric series.

   (b) Use the formula for the sum of a geometric series to show that $0.232323\ldots = 23/99$.

In Problems 26–31, use the method of Problem 25 to write each of the decimals as fractions.

26. $0.235235235\ldots$

27. $6.19191919\ldots$

28. $0.12222222\ldots$

29. $0.4788888\ldots$

30. $0.85656565656\ldots$

31. $0.7638383838\ldots$

32. Figure 11.3 shows the quantity of the drug atenolol in the blood as a function of time, with the first dose at time $t = 0$. Atenolol is taken in 50 mg doses once a day to lower blood pressure.

   (a) If the half-life of atenolol in the blood is 6.3 hours, what percentage of the atenolol present at the start of a 24-hour period is still there at the end?

   (b) Find expressions for the quantities Q_0, Q_1, Q_2, Q_3, $\ldots$, and Q_n shown in Figure 11.3. Write the expression for Q_n in closed-form.

   (c) Find expressions for the quantities P_1, P_2, P_3, $\ldots$, and P_n shown in Figure 11.3. Write the expression for P_n in closed-form.

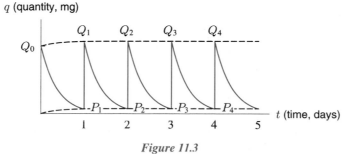

Figure 11.3

33. You have an ear infection and are told to take a 250 mg tablet of ampicillin (a common antibiotic) four times a day (every six hours). It is known that at the end of six hours, about 4% of the drug is still in the body. What quantity of the drug is in the body right after the third tablet? The fortieth? Assuming you continue taking tablets, what happens to the drug level in the long run?

34. In Problem 33 we found the quantity Q_n, the amount (in mg) of ampicillin left in the body right after the nth tablet is taken.

    (a) Make a similar calculation for P_n, the quantity of ampicillin (in mg) in the body right *before* the nth tablet is taken.
    (b) Find a simplified formula for P_n.
    (c) What happens to P_n in the long run? Is this the same as what happens to Q_n? Explain in practical terms why your answer makes sense.

35. Draw a graph like that in Figure 11.3 for 250 mg of ampicillin taken every 6 hours, starting at time $t = 0$. Put on the graph the values of $Q_1, Q_2, Q_3, \ldots$ calculated in Problem 33 and the values of $P_1, P_2, P_3, \ldots$ calculated in Problem 34.

36. Consider Patrick Ewing's contract described on page 456. Determine the present value of the contract if the interest rate is 7% per year, compounded continuously, for the entire 10-year period of the contract.

37. One way of valuing a company is to calculate the present value of all its future earnings. Suppose a farm expects to sell $1000 worth of Christmas trees once a year forever, with the first sale in the immediate future. What is the present value of this Christmas tree business? Assume that the interest rate is 4% per year, compounded continuously.

Problems 38–40 are about *bonds*, which are issued by a government to raise money. An individual who buys a $1000 bond gives the government $1000 and in return receives a fixed sum of money, called the *coupon*, every six months or every year for the life of the bond. At the time of the last coupon, the individual also gets the $1000, or *principal*, back.

38. What is the present value of a $1000 bond which pays $50 a year for 10 years, starting one year from now? Assume interest rate is 6% per year, compounded annually.

39. What is the present value of a $1000 bond which pays $50 a year for 10 years, starting one year from now? Assume the interest rate is 4% per year, compounded annually.

40. (a) What is the present value of a $1000 bond which pays $50 a year for 10 years, starting one year from now? Assume the interest rate is 5% per year, compounded annually.
    (b) Since $50 is 5% of $1000, this bond is often called a 5% bond. What does your answer to part (a) tell you about the relationship between the principal and the present value of this bond when the interest rate is 5%?
    (c) If the interest rate is more than 5% per year, compounded annually, which one is larger: the principal or the value of the bond? Why do you think the bond is then described as *trading at discount*?
    (d) If the interest rate is less than 5% per year, compounded annually, why is the bond described as *trading at a premium*?

11.3 PARAMETRIC EQUATIONS

The Mars Pathfinder

On July 4, 1997, the Mars Pathfinder bounced down onto the surface of the red planet, its impact cushioned by airbags. The next day, the Sojourner—a small, six-wheeled robot—rolled out of the spacecraft and began a rambling exploration of the surrounding terrain. Figure 11.4 shows a photograph of the Sojourner and a diagram of the path it took before radio contact was lost.

In Figure 11.4, Sojourner's path is labeled according to the elapsed number of Martian days or *sols* (short for solar periods). The robot's progress was slow because instructions for every movement had to be calculated by NASA engineers and sent from earth.

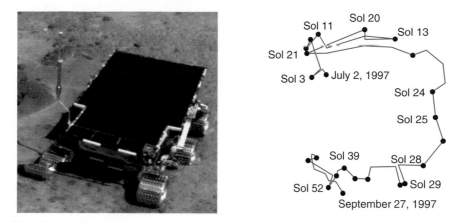

Figure 11.4: The Pathfinder Sojourner robot and the path it followed on the surface of the planet Mars. Image from NASA/Jet Propulsion Laboratory. Diagram adapted from *The New York Times*, July 21, 1998

The Path of the Sojourner Robot

How can Sojourner's path be represented? How can we tell a robot where it should go, how fast, and when? The path in Figure 11.4 is not a function in the ordinary sense—for instance, it crosses over itself in several different places. However, positions on the path can be determined by knowing t, the elapsed time in sols (Martian days). In this section, we describe a path by giving the x- and y-coordinates of points on the path as functions of a *parameter* such as t.

Programming a Robot Using Coordinates

One way to program a robot's motion is to send it coordinates to tell it where to go. Imagine that a robot like the Sojourner is moving around the xy-plane.[2] If we choose the origin $(0,0)$ to be the spot where the robot's spacecraft lands, we can direct its motion by giving it (x,y) coordinates of the points to which it should move.

We select the positive y-axis so that it points north (that is, toward the northern pole of Mars) and the positive x-axis so that it points east. Our units of measurement are meters, so that, the coordinates $(2,4)$ indicate a point that is 2 meters to the east and 4 meters to the north of the landing site. We program the robot to move to the following points:

$$(0,0) \quad \rightarrow \quad (1,1) \quad \rightarrow \quad (2,2) \quad \rightarrow \quad (2,3) \quad \rightarrow \quad (2,4).$$

These points have been plotted in Figure 11.5.

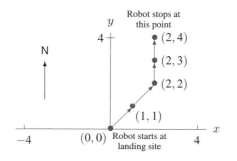

Figure 11.5: The robot is programmed to move along this path

---

[2]The surface of Mars is not as flat as the xy-plane, but this is a useful first approximation.

Programming the Robot Using a Parameter

In order to represent the robot's path using a parameter, we need two functions: one for the robot's x-coordinate, $x = f(t)$, and one for its y-coordinate, $y = g(t)$. The function for x describes the robot's east-west motion, and the function for y describes its north-south motion. Together, they are called *parametric equations* for the robot's path.

Example 1   If t is time in minutes, describe the path followed by a robot given by

$$x = 2t, \qquad y = t \quad \text{for} \quad 0 \le t \le 5.$$

Solution   At time $t = 0$, the robot's position is given by

$$x = 2 \cdot 0 = 0, \qquad y = 0, \qquad \text{so it starts at the point } (0, 0).$$

One minute later, at $t = 1$, its position is given by

$$x = 2 \cdot 1 = 2, \qquad y = 1, \qquad \text{so it has moved to the point } (2, 1).$$

At time $t = 2$, its position is given by

$$x = 2 \cdot 2 = 4, \qquad y = 2, \qquad \text{so it has moved to the point } (4, 2).$$

The path followed by the robot is given by

$$(0, 0) \quad \rightarrow \quad (2, 1) \quad \rightarrow \quad (4, 2) \quad \rightarrow \quad (6, 3) \quad \rightarrow \quad (8, 4) \quad \rightarrow \quad (10, 5).$$

At time $t = 5$, the robot stops at the point $(10, 5)$ because we have restricted the values of t to the interval $0 \le t \le 5$. In Figure 11.6 we see the path followed by the robot; it is a straight line.

In the previous example, we can use substitution to rewrite the formula for x in terms of y. Since $x = 2t$ and $t = y$, we have $x = 2y$. Thus, the path followed by the robot has the equation

$$y = \frac{1}{2}x.$$

Since the parameter t can be easily eliminated from our equations, you may wonder why we use it. One reason is that it is useful to know when the robot gets to each point. The values of x and y tell us where the robot is, while the parameter t tells us when it gets there. In addition, for some pairs of parametric equations, the parameter t cannot be so easily eliminated.

So far, we have assumed that a robot moves in a straight line between two points. We now imagine that a robot can be continuously redirected along a curved path.

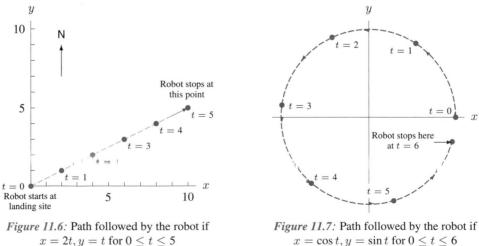

Figure 11.6: Path followed by the robot if $x = 2t, y = t$ for $0 \le t \le 5$

Figure 11.7: Path followed by the robot if $x = \cos t, y = \sin t$ for $0 \le t \le 6$

Example 2   A robot begins at the point $(1, 0)$ and follows the path given by the equations

$$x = \cos t, \qquad y = \sin t \qquad \text{where } t \text{ is in minutes,} \quad 0 \leq t \leq 6.$$

(a)   Describe the path followed by the robot.

(b)   What happens when you try to eliminate the parameter t from these equations?

Solution   (a)   At time $t = 0$, the robot's position is given by

$$x = \cos 0 = 1, \qquad y = \sin 0 = 0$$

so it starts at the point $(1, 0)$, as required. At time $t = 1$, its position is given by

$$x = \cos 1 = 0.54, \qquad y = \sin 1 = 0.84.$$

Thus, the robot has moved west and north. At time $t = 2$, its position is given by

$$x = \cos 2 = -0.42, \qquad y = \sin 2 = 0.91.$$

Now it is farther west and slightly farther north. Continuing, we see that the path followed by the robot is given by

$$(1, 0) \rightarrow (0.54, 0.84) \quad \rightarrow \quad (-0.42, 0.91) \quad \rightarrow \quad (-0.99, 0.14)$$
$$\rightarrow (-0.65, -0.76) \quad \rightarrow \quad (0.28, -0.96) \quad \rightarrow \quad (0.96, -0.28).$$

In Figure 11.7 we see the path followed by the robot; it is circular with a radius of one meter. At the end of 6 minutes the robot has not quite returned to its starting point at $(1, 0)$.

(b)   One way to eliminate t from this pair of equations is to use the Pythagorean identity,

$$\cos^2 t + \sin^2 t = 1.$$

Since $x = \cos t$ and $y = \sin t$, we can substitute x and y into this equation:

$$\underbrace{(\cos t)}_{x}{}^2 + \underbrace{(\sin t)}_{y}{}^2 = 1,$$

giving

$$x^2 + y^2 = 1.$$

This is the equation of a circle of radius 1 centered at $(0, 0)$. Attempting to solve for y in terms of x, we have

$$x^2 + y^2 = 1$$
$$y^2 = 1 - x^2$$
$$y = +\sqrt{1 - x^2} \quad \text{or} \quad y = -\sqrt{1 - x^2}.$$

Thus, we obtain two different equations for y in terms of x. The first, $y = +\sqrt{1 - x^2}$, returns positive values for y (as well as 0), while the second returns negative values for y (as well as 0). The first equation gives the top half of the circle, while the second gives the bottom half.

In the previous example, y is not a function of x, because for all x-values (except 1 and -1) there are two possible y values. This confirms what we already knew, because the graph in Figure 11.7 fails the vertical line test.

Different Motions Along the Same Path

It is possible to parameterize the same curve in more than one way, as the following example illustrates.

Example 3   Describe the motion of the robot if it follows the path given by

$$x = \cos \frac{1}{2}t, \qquad y = \sin \frac{1}{2}t \qquad \text{for} \quad 0 \le t \le 6.$$

Solution   At time $t = 0$, the robot's position is given by

$$x = \cos\left(\frac{1}{2} \cdot 0\right) = 1, \qquad y = \sin\left(\frac{1}{2} \cdot 0\right) = 0$$

so it starts at the point $(1, 0)$, as before. After one minute, that is, at time $t = 1$, its position (rounded to thousandths) is given by

$$x = \cos\left(\frac{1}{2} \cdot 1\right) = 0.878, \qquad y = \sin\left(\frac{1}{2} \cdot 1\right) = 0.479.$$

At time $t = 2$ its position is given by

$$x = \cos\left(\frac{1}{2} \cdot 2\right) = 0.540, \qquad y = \sin\left(\frac{1}{2} \cdot 2\right) = 0.841.$$

Continuing, we see that the path followed by the robot is given by

$$(1, 0) \to (0.878, 0.479) \quad \to \quad (0.540, 0.841) \quad \to \quad (0.071, 0.997)$$
$$\to (-0.416, 0.909) \quad \to \quad (-0.801, 0.598) \quad \to \quad (-0.990, 0.141).$$

Figure 11.8 shows the path; it is again circular with a radius of one meter. However, at the end of 6 minutes the robot has not even made it half way around the circle. Because we have multiplied the parameter t by a factor of $1/2$, the robot moves at half its original rate.

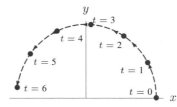

Figure 11.8: Path followed by the robot if $x = \cos \frac{1}{2}t, y = \sin \frac{1}{2}t$ for $0 \le t \le 6$

Example 4   Describe the motion of the robot if it follows the path given by:

(a) $x = \cos(-t), y = \sin(-t)$ for $0 \le t \le 6$ (b) $x = \cos t, y = \sin t$ for $0 \le t \le 10$

Solution   (a)   Figure 11.9 shows the robot's path. The robot travels around the circle in the clockwise direction, opposite to that in Examples 2 and 3.

(b)   See Figure 11.10. The robot travels around the circle more than once but less than twice, coming to a stop roughly southwest of the landing site.

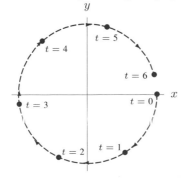

Figure 11.9: Path given by $x = \cos(-t), y = \sin(-t)$ for $0 \le t \le 6$

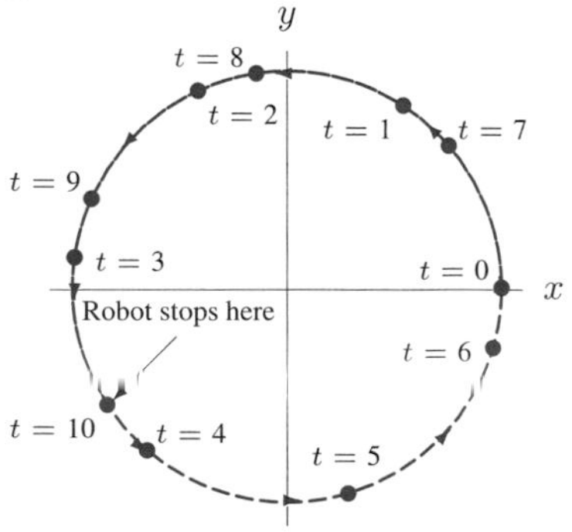

Figure 11.10: Path given by $x = \cos t, y = \sin t$ for $0 \le t \le 10$

Other Parametric Curves

Parametric equations can be used to describe extremely complicated motions. For instance, suppose the robot follows the rambling path described by the parametric equations

$$x = 20\cos t + 4\cos(4\sqrt{5}t) \qquad y = 20\sin t + 4\sin(4\sqrt{8}t).$$

See Figure 11.11. The path is roughly circular, and the robot moves in a more or less counterclockwise direction. The dashed circle in Figure 11.11 is given by the parametric equations

$$x = 20\cos t, \qquad y = 20\sin t \qquad \text{for } 0 \le t \le 2\pi.$$

The other terms, $4\cos(4\sqrt{5}t)$ and $4\sin(4\sqrt{8}t)$, are responsible for the robot's rambling motion.

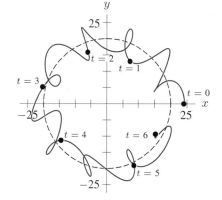

Figure 11.11: Rambling path, parametrically defined

The Archimedean Spiral

In polar coordinates, the Archimedean spiral is the graph of the equation introduced on page 325:

$$r = \theta.$$

Since the relationship between polar coordinates and Cartesian coordinates is

$$x = r\cos\theta \qquad \text{and} \qquad y = r\sin\theta,$$

we can write the Archimedean spiral $r = \theta$ as

$$x = \theta\cos\theta \qquad \text{and} \qquad y = \theta\sin\theta.$$

Replacing θ (an angle) by t (a time), we obtain the parametric equations for the spiral in Figure 11.12:

$$x = t\cos t \qquad y = t\sin t.$$

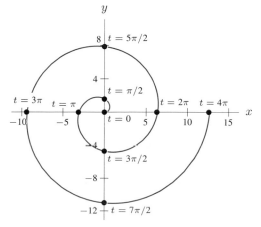

Figure 11.12: Archimedean spiral given by
$x = t\cos t, y = t\sin t, 0 \le t \le 4\pi$

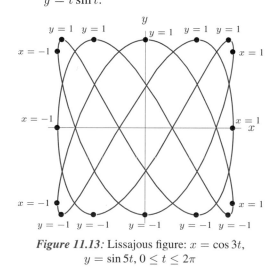

Figure 11.13: Lissajous figure: $x = \cos 3t$,
$y = \sin 5t, 0 \le t \le 2\pi$

Lissajous Figures

The beautiful curve in Figure 11.13 is called a *Lissajous figure*. Its equations are

$$x = \cos 3t, \qquad y = \sin 5t \qquad \text{for } 0 \le t \le 2\pi.$$

To explain the shape of this Lissajous figure, notice that $y = \sin 5t$ completes five full oscillations on the interval $0 \le t \le 2\pi$. Thus, since the amplitude of $y = \sin 5t$ is 1, the value of y reaches a maximum of 1 at five different values of t. Similarly, the value of y is -1 at another five different values of t. Figure 11.13 shows the curve climbs to a high point of $y = 1$ five times and falls to a low of $y = -1$ five times.

Meanwhile, since $x = \cos 3t$, the value of x oscillates between 1 and -1 a total of 3 times on the interval $0 \le t \le 2\pi$. Figure 11.13 shows that the curve moves to its right boundary, $x = 1$, three times and to its left boundary, $x = -1$, three times.

Foxes and Rabbits

In a park, a population of foxes preys on a population of rabbits. Suppose F is the number of foxes, R is the number of rabbits, t is the number of months since January 1, and that

$$R = 1000 - 500 \sin\left(\frac{\pi}{6}t\right) \qquad \text{and} \qquad F = 150 + 50 \cos\left(\frac{\pi}{6}t\right).$$

These equations are a parameterization of the curve in Figure 11.14. Since this curve fails the vertical line test, F is not a function of R.

We can use the curve to analyze the relationship between the rabbit population and the fox population. In January, there are 1000 rabbits but they are dying off; in April only 500 rabbits remain. By July the population has rebounded to 1000 rabbits, and by October it has soared to 1500 rabbits. Then the rabbit population begins to fall again; the following April there are once more only 500 rabbits.

The fox population also rises and falls, though not at the same time of year as the rabbits. In January, the fox population numbers 200. By April, it has fallen to 150 foxes. The fox population continues to fall, perhaps because there are so few rabbits to eat at that time. By July, only 100 foxes remain. The fox population then begins to increase again. By October, when the rabbit population is largest, the fox population is growing rapidly, so that by January it has returned to its maximum size of 200 foxes. By this point, though, the rabbit population is already dropping, perhaps because there are so many hungry foxes around. Then the cycle repeats.

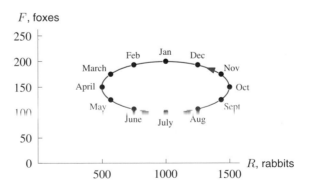

Figure 11.14: The relationship between F, the number of foxes, and R, the number of rabbits

Using Graphs to Parameterize a Curve

Example 5   Figure 11.15 shows the graphs of two functions, $f(t)$ and $g(t)$. Describe the motion of the particle whose coordinates at time t are given by $x = f(t), y = g(t)$.

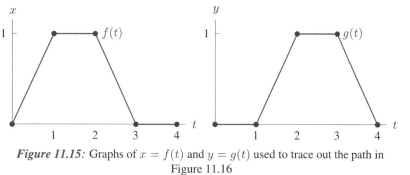
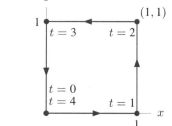

Figure 11.15: Graphs of $x = f(t)$ and $y = g(t)$ used to trace out the path in Figure 11.16

Figure 11.16: Square parameterized by $x = f(t), y = g(t)$ from Figure 11.15

Solution   Between times $t = 0$ and $t = 1$, the x-coordinate increases from 0 to 1, while the y-coordinate stays fixed at 0. The particle moves along the x-axis from $(0, 0)$ to $(1, 0)$. Between times $t = 1$ and $t = 2$, the x-coordinate stays fixed at $x = 1$, while the y-coordinate increases from 0 to 1. Thus, the particle moves along the vertical line from $(1, 0)$ to $(1, 1)$. Between times $t = 2$ and $t = 3$, it moves horizontally backward to $(0, 1)$, and between times $t = 3$ and $t = 4$ it moves down the y-axis to $(0, 0)$. Thus, it traces out the square in Figure 11.16.

Problems for Section 11.3

Sketch the curve represented by the parametric equations in Problems 1–12. Assume that the parameter is restricted to values for which the functions are defined. Indicate the direction of the curve. Eliminate the parameter, t, to obtain an equation for y as a function of x. Check that your sketches are correct by graphing y as a function of x.

1. $x = t + 1, \quad y = 3t - 2$ 2. $x = 5 - 2t, \quad y = 1 + 4t$ 3. $x = \sqrt{t}, \quad y = 2t + 1$

4. $x = \sqrt{t}, \quad y = t^2 + 4$ 5. $x = t - 3, \quad y = t^2 + 2t + 1$ 6. $x = e^t, \quad y = e^{2t}$

7. $x = e^{2t}, \quad y = e^{3t}$ 8. $x = \ln t, \quad y = t^2$ 9. $x = t^3, \quad y = 2 \ln t$

10. $x = 2 \cos t, \quad y = 2 \sin t$ 11. $x = 3 + \sin t, \quad y = 2 + \cos t$ 12. $x = 2 \cos t, \quad y = 3 \sin t$

For Problems 13–16, describe the motion of a particle whose position at time t is given by $x = f(t), y = g(t)$.

13.

14.

15.

16.

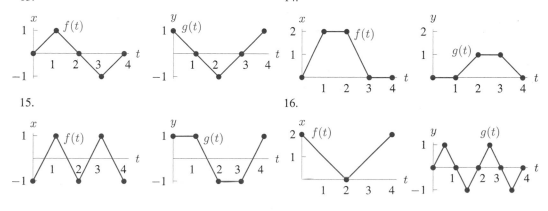

17. Write two different parameterizations for each of the curves in the xy-plane.
    (a) A parabola whose equation is $y = x^2$.
    (b) The parabola from part (a) shifted to the left 2 units and up 1 unit.

Problems 18–21 give parameterizations of the unit circle or a part of it. In each case, describe in words how the circle is traced out, including when and where the particle is moving clockwise and when and where the particle is moving counterclockwise.

18. $x = \cos t$, $y = -\sin t$
19. $x = \sin t$, $y = \cos t$
20. $x = \cos(t^2)$, $y = \sin(t^2)$
21. $x = \cos(\ln t)$, $y = \sin(\ln t)$

22. Describe the similarities and differences among the motions in the plane given by the following three pairs of parametric equations:
    (a) $x = t$, $y = t^2$ 　　(b) $x = t^2$, $y = t^4$ 　　(c) $x = t^3$, $y = t^6$.

23. As t varies, the following parametric equations trace out a line in the plane
    $$x = 2 + 3t, \quad y = 4 + 7t.$$
    (a) What part of the line is obtained by restricting t to nonnegative numbers?
    (b) What part of the line is obtained if t is restricted to $-1 \le t \le 0$?
    (c) How should t be restricted to give the part of the line to the left of the y-axis?

24. Suppose $a, b, c, d, m, n, p, q > 0$. Match each of the following pairs of parametric equations with one of the lines l_1, l_2, l_3, l_4 in Figure 11.17.

    I. $\begin{cases} x = a + ct, \\ y = -b + dt. \end{cases}$ 　　II. $\begin{cases} x = m + pt, \\ y = n - qt. \end{cases}$

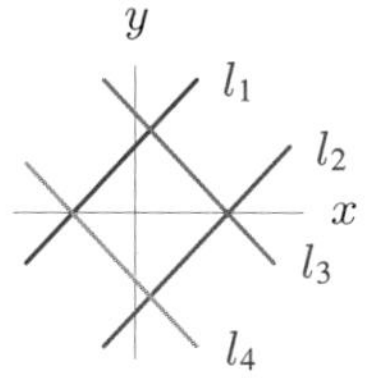

    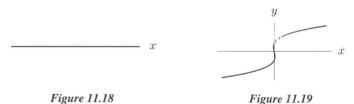

    Figure 11.17

25. A bug is crawling around the Cartesian plane. Let $x = f(t)$ be the function denoting the x-coordinate of the bug's position as a function of time, t, and let $y = g(t)$ be the y-coordinate of the bug's position.
    (a) Suppose $f(t) = t$ and $g(t) = t$. What path does the bug follow?
    (b) Now let $f(t) = \cos t$ and $g(t) = \sin t$. What path does the bug follow? What is its starting point? (That is, where is the bug when $t = 0$?) When does the bug get back to its starting point?
    (c) Now let $f(t) = \cos t$ and $g(t) = 2 \sin t$. What path does the bug follow? What is its starting point? When does the bug get back to its starting point?

Write a parameterization for the lines in the xy-plane in Problems 26–27.
26. A vertical line through the point $(-2, -3)$. 　　27. The line through the points $(2, -1)$ and $(1, 3)$.

Graph the Lissajous figures in Problems 28–31 using a calculator or computer.
28. $x = \cos 2t$, $y = \sin 5t$
29. $x = \cos 3t$, $y = \sin 7t$
30. $x = \cos 2t$, $y = \sin 4t$
31. $x = \cos 2t$, $y = \sin \sqrt{3}t$

32. Motion along a straight line is given by a single equation, say, $x = t^3 - t$ where x is distance along the line. It is difficult to see the motion from a plot; it just traces out the x-line, as in Figure 11.18. To visualize the motion, we introduce a y-coordinate and let it slowly increase, giving Figure 11.19. Try the following on a calculator or computer. Let $y = t$. Now plot the parametric equations $x = t^3 - t, y = t$ for, say, $-3 \le t \le 3$. What does the plot in Figure 11.19 tell you about the particle's motion?

Figure 11.18 　　　　　　　　　*Figure 11.19*

For Problems 33–34, plot the motion along the x-line by introducing a y-coordinate, as in Problem 32. What does the plot tell you about the particle's motion?

33. $x = \cos t,$ $-10 \le t \le 10$

34. $x = t^4 - 2t^2 + 3t - 7,$ $-3 \le t \le 2$

35.   A ball is thrown vertically into the air at time $t = 0$. Its height $d(t)$, in feet, above the ground at time t, in seconds, is given by:
$$d(t) = -16t^2 + 48t + 6.$$

   (a)   Write a parameterization for the curve.
   (b)   Sketch a graph of the curve.
   (c)   What is the height of the ball at $t = 0$? Explain how this makes sense.
   (d)   Does the ball ever reach this height again? If so when? Why?
   (e)   When does the ball reach its maximum height? What is the maximum height?

11.4   IMPLICITLY DEFINED CURVES AND CONIC SECTIONS

In the previous section, we saw that the parametrically defined curve, $x = \cos t$, $y = \sin t$, is the circle of radius 1 centered at the origin. Eliminating the parameter t gives the equation

$$x^2 + y^2 = 1.$$

The circle described by the equation $x^2 + y^2 = 1$ is an example of an *implicitly defined* curve. To be explicit is to state a fact outright; to be implicit is to make a statement in a round-about way. If a curve has an equation of the form $y = f(x)$, then we say that y is an *explicit* function of x. However, the equation $x^2 + y^2 = 1$ does not explicitly state that y depends on x. Instead, this dependence is implied by the equation. If we try to solve for y in terms of x, we do not obtain a function. Rather, we obtain *two* functions, one for the top half of the unit circle and one for the bottom half:

$$y = \begin{cases} +\sqrt{1 - x^2} \\ -\sqrt{1 - x^2} \end{cases}$$

Example 1   Sketch a graph of $|y| = |x|$. Can you find an explicit formula for y in terms of x?

Solution   The equation $|y| = |x|$ is implicit because it does not tell us what y equals, instead, it tells us what $|y|$ equals. Suppose $x = 2$. What values can y equal? Since

$$|y| = |2| = 2, \qquad \text{we have} \qquad y = 2 \text{ or } -2,$$

because for either of these y-values $|y| = 2$. What if x is negative? If $x = -3$, we have

$$|y| = |-3| = 3, \qquad \text{then} \qquad y = 3 \text{ or } -3,$$

because these are the solutions to $|y| = 3$.

For almost all x-values, there are two y-values, one given by $y = +|x|$ and one by $y = -|x|$. The only exception is at $x = 0$, because the only solution to the equation $|y| = 0$ is $y = 0$. Thus, the graph of $|y| = |x|$ has two parts, given by

$$y = \begin{cases} +|x| \\ -|x| \end{cases}$$

The graph of $y = -|x|$ is an upside-down version of the graph of $y = |x|$. Thus, when both parts of the graph of $|y| = |x|$ are plotted together, the resulting graph looks like an X. (See Figure 11.20.)

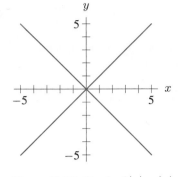

Figure 11.20: Graph of $|y| = |x|$

Conic Sections

The curves known as *conic sections* include two we are already familiar with, circles and parabolas. They also include *ellipses* and *hyperbolas*. An ellipse is a "squashed" circle; an example is given by Figure 11.14 on page 466, which shows the relationship between a population of foxes and a population of rabbits. The function $y = 1/x$ is a special type of hyperbola known as a *rectangular hyperbola*.

Conic sections are so-called because, as was demonstrated by the Greeks, they can be constructed by slicing, or sectioning, a cone. We will not consider this geometrical aspect of these curves, but will study them in terms of parametric and implicit equations. Conic sections arise naturally in physics, since the path of a body orbiting the sun is a conic section. As we have already studied parabolas, we now focus on circles, ellipses and hyperbolas.

Circles

Example 2   Graph the parametric equations

$$x = 5 + 2\cos t \qquad \text{and} \qquad y = 3 + 2\sin t.$$

Solution   We see that x varies between 3 and 7 with a midline of 5 while y varies between 1 and 5 with a midline of 3. Figure 11.21 gives a graph of this function. It appears to be a circle of radius 2 centered at the point $(5, 3)$.

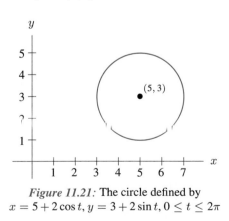

Figure 11.21: The circle defined by
$x = 5 + 2\cos t, y = 3 + 2\sin t, 0 \le t \le 2\pi$

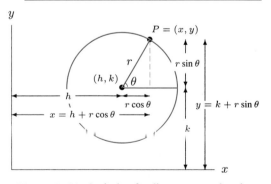

Figure 11.22: A circle of radius r centered at the point (h, k)

Compare Figure 11.21 with Figure 11.22 which shows a circle of radius r centered at the point (h, k). The coordinates of a point $P = (x, y)$, corresponding to an angle θ, are given by

$$x = h + r \cos \theta \qquad \text{and} \qquad y = k + r \sin \theta.$$

Writing t instead of θ, these are parametric equations of the parametric curve in Figure 11.21. This curve is a circle,[3] because the point P is always a fixed distance, r, from the point (h, k).

For $r > 0$, the parametric equations of a **circle** of radius r centered at the point (h, k) are:

$$x = h + r \cos t \qquad y = k + r \sin t \qquad 0 \le t \le 2\pi.$$

Eliminating the Parameter t in the Equations for a Circle

To eliminate the parameter t from the parametric equations for a circle, we rewrite the equations as

$$x - h = r \cos t \qquad \text{so} \qquad (x - h)^2 = r^2 \cos^2 t$$

and

$$y - k = r \sin t \qquad \text{so} \qquad (y - k)^2 = r^2 \sin^2 t.$$

Adding these two equations and applying the Pythagorean identity, $\cos^2 t + \sin^2 t = 1$, gives

$$\begin{aligned}
(x - h)^2 + (y - k)^2 &= r^2 \cos^2 t + r^2 \sin^2 t \\
&= r^2 (\cos^2 t + \sin^2 t) \\
&= r^2.
\end{aligned}$$

An implicit equation for the **circle** of radius r centered at the point (h, k) is:

$$(x - h)^2 + (y - k)^2 = r^2.$$

This is called the **standard form** of the equation of a circle.

For the unit circle, we have $h = 0$, $k = 0$, and $r = 1$. This gives

$$x^2 + y^2 = 1.$$

For the circle in Example 2, we have $h = 5$, $k = 3$, and $r = 2$. This gives

$$(x - 5)^2 + (y - 3)^2 = 2^2 = 4.$$

If we expand the equation for the circle in Example 2, we get a quadratic equation in two variables:

$$x^2 - 10x + 25 + y^2 - 6y + 9 = 4,$$

or

$$x^2 + y^2 - 10x - 6y + 30 = 0,$$

Note that the coefficients of x^2 and y^2 (in this case, 1) are equal. Such equations often describe circles; we put them into standard form, $(x - h)^2 + (y - k)^2 = r^2$, by completing the square.

---

[3] If $r < 0$, the curve is a circle of radius $|r|$.

Example 3   Describe in words the curve defined by the equation
$$x^2 + 10x + 20 = 4y - y^2.$$

Solution   Rearranging terms gives
$$x^2 + 10x + y^2 - 4y = -20.$$

We complete the square for the terms involving x and (separately) for the terms involving y:

$$\underbrace{(x^2 + 10x + 25)}_{(x+5)^2} + \underbrace{(y^2 - 4y + 4)}_{(y-2)^2} \quad \underbrace{-25 - 4}_{\substack{\text{Compensating} \\ \text{terms}}} \quad = -20$$

$$(x + 5)^2 + (y - 2)^2 - 29 = -20$$
$$(x + 5)^2 + (y - 2)^2 = 9.$$

This equation is a circle of radius $r = 3$ with center $(h, k) = (-5, 2)$.

Ellipses

In this section we consider the graph of an ellipse, which is a "squashed" circle.

> The parametric equations of an **ellipse** centered at (h, k) are:
>
> $$x = h + a \cos t \qquad y = k + b \sin t \qquad 0 \le t \le 2\pi.$$

In the special case where $a = b$, these equations[4] give a circle of radius $r = a$. Thus, a circle is a special kind of ellipse. The parametric equations of an ellipse are transformations of the parametric equations of the unit circle, $x = \cos t$, $y = \sin t$. These transformations have the effect of shifting and stretching the unit circle into an ellipse.

Example 4   Graph the ellipse given by
$$x = 7 + 5 \cos t, \qquad y = 4 + 2 \sin t, \qquad 0 \le t \le 2\pi.$$

Solution   See Figure 11.23. The center of the ellipse is $(h, k) = (7, 4)$. The value $a = 5$ determines the horizontal "radius" of the ellipse, while $b = 2$ determines the vertical "radius". The value of x varies from a maximum of 12 to a minimum of 2 about the vertical midline $x = 7$. Similarly, the value of y varies from a maximum of 6 to a minimum of 2 about the horizontal midline $y = 4$. The ellipse is symmetric about the midlines $x = 7$ and $y = 4$.

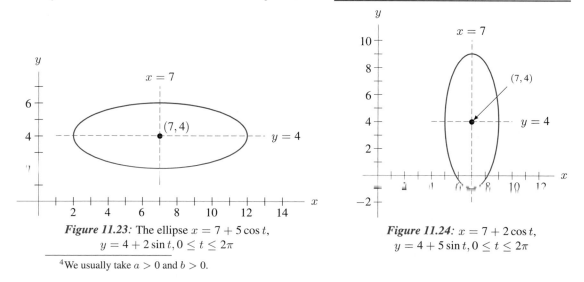

Figure 11.23: The ellipse $x = 7 + 5 \cos t$, $y = 4 + 2 \sin t, 0 \le t \le 2\pi$

Figure 11.24: $x = 7 + 2 \cos t$, $y = 4 + 5 \sin t, 0 \le t \le 2\pi$

[4]We usually take $a > 0$ and $b > 0$.

In general, the horizontal axis of an ellipse has length $2a$, and the vertical axis has length $2b$. This is similar to a circle, which has a diameter $2r$. The difference here is that the diameter of a circle is the same in every direction, whereas the "diameter" of an ellipse depends on the direction.

The longer axis of an ellipse is called the *major* axis and the shorter axis is called the *minor* axis. Either the horizontal or the vertical axis can be the longer axis.

Example 5   Graph the ellipse given by

$$x = 7 + 2\cos t, \qquad y = 4 + 5\sin t, \qquad 0 \le t \le 2\pi.$$

How is this ellipse similar to the one in Example 4? How is it different?

Solution   This ellipse has the same center $(h, k) = (7, 4)$ as the ellipse in Example 4. However, as we see in Figure 11.24, the vertical axis of this ellipse is the longer one and the horizontal axis is the shorter one. This is the opposite of the situation in Example 4.

Eliminating the Parameter t in the Equations for an Ellipse

As with circles, we can eliminate t from the parametric equations for an ellipse. We first rewrite the equation for x as follows:

$$x = h + a\cos t$$
$$x - h = a\cos t$$
$$\frac{x - h}{a} = \cos t$$
$$\left(\frac{x - h}{a}\right)^2 = \cos^2 t.$$

Similarly, we rewrite the equation for y as

$$\left(\frac{y - k}{b}\right)^2 = \sin^2 t.$$

Adding these two equations gives

$$\left(\frac{x - h}{a}\right)^2 + \left(\frac{y - k}{b}\right)^2 = \cos^2 t + \sin^2 t = 1.$$

The implicit equation for an **ellipse** centered at (h, k) and with horizontal axis $2a$ and vertical axis $2b$ is:

$$\frac{(x - h)^2}{a^2} + \frac{(y - k)^2}{b^2} = 1.$$

Hyperbolas

We now use another form of the Pythagorean identity. With $\tan t = \sin t / \cos t$ and $\sec t = 1/\cos t$, on page 271 we showed that

$$\sec^2 t - \tan^2 t = 1.$$

We use this identity to investigate the curve described by the parametric equations

$$x = \sec t \qquad \text{and} \qquad y = \tan t.$$

By eliminating the parameter t, we get

$$x^2 - y^2 = 1.$$

This implicit equation looks similar to the equation for a unit circle $x^2 + y^2 = 1$. However, the curve it describes is very different. Solving for y, we find that

$$y^2 = x^2 - 1.$$

For large values of x, the values of y^2 and x^2 are very nearly equal. For instance, when $x = 100$, we see that $x^2 = 10,000$ and $y^2 = x^2 - 1 = 9999$. Thus,

$$y^2 \approx x^2 \qquad \text{for large values of } x.$$

So, for large values of x, we have

$$y \approx \pm\sqrt{x^2}$$
$$y = \pm|x| \qquad \text{Because } \sqrt{x^2} = |x|.$$

The graph of $y = \pm|x|$ is in Figure 11.20 on page 470; it looks like an X.

What happens for smaller values of x? Writing the equation as $y^2 = x^2 - 1$, shows us that x^2 cannot be less than 1, for otherwise y^2 would be negative. See Table 11.2 and Figure 11.25.

TABLE 11.2   *Points on the graph of the hyperbola,* $x^2 - y^2 = 1$

x	y	Points	x	y	Points
-3	± 2.83	$(-3, 2.83)$ and $(-3, -2.83)$	3	± 2.83	$(3, 2.83)$ and $(3, -2.83)$
-2	± 1.73	$(-2, 1.73)$ and $(-2, -1.73)$	2	± 1.73	$(2, 1.73)$ and $(2, -1.73)$
-1	± 0	$(-1, 0)$ and $(-1, -0)$	1	± 0	$(1, 0)$ and $(1, -0)$
0	Undefined	None			

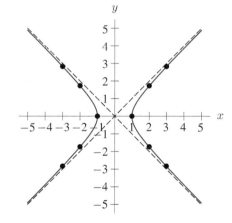

Figure 11.25: A graph of $x^2 - y^2 = 1$. The X-shaped graph of $y^2 = x^2$ has been dashed in

This graph is an example of a *hyperbola*. By analogy to the unit circle, it is called the *unit hyperbola*. It has two branches, one to the right of the y-axis and one to the left. As x grows large in magnitude, either towards $+\infty$ or $-\infty$, the graph approaches the asymptotes shown by the dashed lines, $y^2 = x^2$.

A General Formula for Hyperbolas

We can shift and stretch the graphs of the parametric equations for the unit hyperbola by writing them as $x = h + a \sec t$ and $y = k + b \tan t$. This has the effect of shifting and stretching the unit hyperbola so that it is centered at (h, k).

> The parametric equations for a **hyperbola** centered at (h, k) are
>
> $$x = h + a \sec t \qquad y = k + b \tan t \qquad 0 \le t \le 2\pi.$$

We can eliminate the parameter, t, from the parametric equations to obtain the implicit equation[5]

$$\frac{(x - h)^2}{a^2} - \frac{(y - k)^2}{b^2} = 1.$$

Notice that this equation is similar to the general equation for an ellipse.

Example 6   Graph the equation $\dfrac{(x - 4)^2}{9} - \dfrac{(y - 7)^2}{25} = 1$.

Solution   We see that $h = 4$ and $k = 7$, so we have shifted the unit hyperbola 4 units to the right and 7 units up. We have $a^2 = 9$ and $b^2 = 25$, so $a = 3$ and $b = 5$. Thus, we have stretched the unit hyperbola horizontally by a factor of 3 and vertically by a factor of 5.

To make it easier to draw hyperbolas, it is useful to imagine a "unit square" centered at the origin. This unit square helps us locate the vertices and asymptotes of the hyperbola $x^2 - y^2 = 1$. See Figure 11.26. Stretching and shifting the hyperbola causes this square to be transformed into a rectangle centered at the point $(h, k) = (4, 7)$ with width $2a = 6$ and height $2b = 10$. See Figure 11.27. The rectangle enables us to draw the X-shaped asymptotes as diagonal lines through the corners. The vertices of the transformed hyperbola are located at the midpoints of the vertical sides of the rectangle.

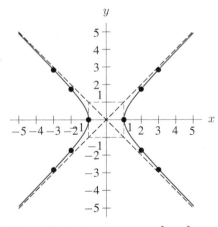

Figure 11.26: The hyperbola $x^2 - y^2 = 1$, with the unit square dashed in

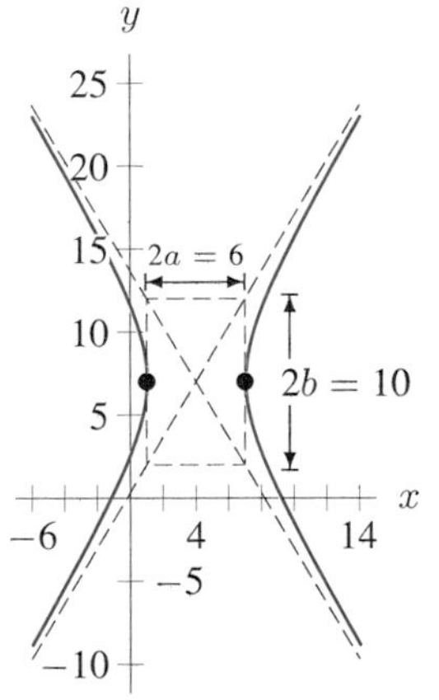

Figure 11.27: A graph of $\frac{(x-4)^2}{9} - \frac{(y-7)^2}{25} = 1$. A rectangle of width $2a = 6$ height $2b = 10$ has been dashed in, centered at the point $(h, k) = (4, 7)$

Note that the ellipse $\frac{(x-4)^2}{9} + \frac{(y-7)^2}{25} = 1$ would exactly fit inside the rectangle in Figure 11.27. In this sense, a hyperbola is an ellipse turned "inside-out."

---

[5] Because of the symmetry of the graph, it turns out not to matter if a and b are positive or negative.

Example 7   Graph the equation $\dfrac{(y-7)^2}{25} - \dfrac{(x-4)^2}{9} = 1.$

Solution   Figure 11.28 shows a graph of this equation. This graph is similar to the one in Figure 11.27 except that it opens up and down instead of right and left.

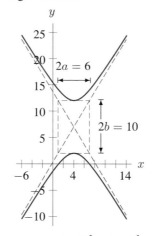

Figure 11.28: A graph of $\frac{(y-7)^2}{25} - \frac{(x-4)^2}{9} = 1$. This graph is similar to the one in Figure 11.27 except that here the hyperbola opens up and down instead of right and left

In summary:

The implicit equation for a **hyperbola**,

$$\frac{(x-h)^2}{a^2} - \frac{(y-k)^2}{b^2} = 1,$$

describes a hyperbola that opens left and right. Its asymptotes are diagonal lines through the corners of a rectangle of width $2a$ and height $2b$ centered at the point (h, k). The graph with equation

$$\frac{(y-k)^2}{b^2} - \frac{(x-h)^2}{a^2} = 1$$

has a similar shape, except that it opens up and down.

Problems for Section 11.4

1. Identify the center and radius for each of the following circles.
   (a) $(x-2)^2 + (y+4)^2 = 20$ (b) $2x^2 + 2y^2 + 4x - 8y = 12$

2. For each of the following ellipses, find the center and lengths of the major and minor axes. Sketch a graph of each ellipse.
   (a) $\dfrac{(x+1)^2}{4} + \dfrac{(y-3)^2}{6} = 1$ (b) $2x^2 + 3y^2 - 6x + 6y = 12$

3. For each of the following hyperbolas, find the center, vertices, and asymptotes. Sketch a graph of each hyperbola.

    (a) $\dfrac{(x+5)^2}{6} - \dfrac{(y-2)^2}{4} = 1$

    (b) $x^2 - y^2 + 2x = 4y + 17$

In Problems 4–8. identify the equation as a circle, ellipse, or hyperbola. In each case, find the center. For a circle, find the length of the radius; for an ellipse, find the lengths of the horizontal and vertical axes; for a hyperbola, find the width and length of the rectangle centered at the center of the hyperbola.

4. $x^2 + y^2 - x + 3y = 4$

5. $x^2 + 4y^2 + 2x - 8y = 11$

6. $3x^2 - 3y^2 - 6x + 12y = 36$

7. $y^2 - 9x^2 - 18x + 6y = 9$

8. $25x^2 + 4y^2 - 50x - 16y + 1 = 60$

In Problems 9–13, write an implicit equation for the indicated conic section.

9. Circle with center $(2, 3)$ and radius 6

10. Ellipse, center at $(5, -3)$; horizontal axis of length 12; vertical axis of length 10.

11. Hyperbola, center at $(-2, 4)$; width and length of rectangle are 8; hyperbola opens up and down.

12. Hyperbola, center at $(-6, -5)$; vertical width of rectangle is 6, horizontal length of rectangle is 12; hyperbola opens left and right.

13. Ellipse, center at $(-4, -2)$; horizontal axis of length 4, vertical axis of length 8.

Write a parameterization for each of the curves in the xy-plane in Problems 14–19.

14. A circle of radius 3 centered at the origin and traced clockwise.

15. A circle of radius 5 centered at the point $(2, 1)$ and traced counterclockwise.

16. A circle of radius 2 centered at the origin traced clockwise starting from $(-2, 0)$ when $t = 0$.

17. The circle of radius 4 centered at the point $(4, 4)$ starting on the x-axis when $t = 0$.

18. An ellipse centered at the origin and crossing the x-axis at ± 5 and the y-axis at ± 7.

19. An ellipse centered at the origin, crossing the x-axis at ± 3 and the y-axis at ± 7. Start at the point $(-3, 0)$ and trace out the ellipse counterclockwise.

What curves do the parametric equations in Problems 20–22 trace out? Find an implicit or explicit equation for each curve.

20. $x = 2 + \cos t, \; y = 2 - \sin t$ 21. $x = 2 + \cos t, \; y = \cos^2 t$ 22. $x = 2 + \cos t, \; y = 2 - \cos t$

State whether the equations in Problems 23–25 represent a curve parametrically, implicitly, or explicitly. Give the two other types of representations for the same curve.

23. $xy = 1$ for $x > 0$

24. $x = e^t, \quad y = e^{2t}$ for all t 25. $x^2 - 2x + y^2 = 0$ for $y < 0$

26. An ant, starting at the origin, moves at 2 units/sec along the x-axis to the point $(1, 0)$. The ant then moves counterclockwise along the unit circle to $(0, 1)$ at a speed of $3\pi/2$ units/sec, then straight down to the origin at a speed of 2 units/sec along the y-axis.

    (a) Express the ant's coordinates as a function of time, t, in secs.
    (b) Express the reverse path as a function of time.

27.   What can you say about the values of a, b and k if the equations

$$x = a + k \cos t, \quad y = b + k \sin t, \qquad 0 \le t \le 2\pi,$$

trace out each of the circles in Figure 11.29?   (a) C_1 (b) C_2 (c) C_3

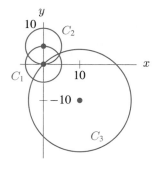

Figure 11.29

11.5 COMPLEX NUMBERS AND POLAR COORDINATES

Extending the Definition of Functions

Some of the functions we have studied, such as x^2, e^x, and $\sin x$, are defined for all real numbers, whereas other functions, such as $\sqrt{x}$, are defined for some real numbers but not for others. The function $\sqrt{x}$ is undefined for $x < 0$.

In this section, we expand our idea of number to include *complex numbers*. These numbers make it possible to interpret expressions like $\sqrt{-4}$, in a meaningful way.

Using Complex Numbers to Solve Equations

The quadratic equation

$$x^2 - 2x + 2 = 0$$

is not satisfied by any real number x. Applying the quadratic formula gives

$$x = \frac{2 \pm \sqrt{4 - 8}}{2} = 1 \pm \frac{\sqrt{-4}}{2}.$$

But -4 doesn't have a square root which is a real number. To overcome this problem, we define the imaginary number i such that:

$$i^2 = -1.$$

Using i, we see that $(2i)^2 = -4$. Now returning to our solution of the quadratic:

$$x = 1 \pm \frac{\sqrt{-4}}{2} = 1 \pm \frac{2i}{2} = 1 \pm i.$$

This solves our quadratic equation. The numbers $1 + i$ and $1 - i$ are examples of complex numbers.

A **complex number** is defined as any number that can be written in the form

$$z = a + bi,$$

where a and b are real numbers and $i = \sqrt{-1}$.
The *real part* of z is the number a; the *imaginary part* is the number b.

Calling the number i imaginary makes it sound like i doesn't exist in the same way as real numbers. In some cases, it is useful to make such a distinction between real and imaginary numbers. If we measure mass or position, we want our answers to be real. But imaginary numbers are just as legitimate mathematically as real numbers. For example, complex numbers are used in studying wave motion in electric circuits.

Algebra of Complex Numbers

Numbers such as 0, 1, $\frac{1}{2}$, π, e and $\sqrt{2}$ are called *purely real*, because their imaginary part is zero. Numbers such as i, $2i$, and $\sqrt{2}i$ are called *purely imaginary*, because they contain only the number i multiplied by a nonzero real coefficient.

Two complex numbers are called *conjugates* if their real parts are equal and if their imaginary parts differ only in sign. The complex conjugate of the complex number $z = a + bi$ is denoted $\bar{z}$, so

$$\bar{z} = a - bi.$$

(Note that z is real if and only if $z = \bar{z}$.)

- Two complex numbers, $z = a + bi$ and $w = c + di$, are equal only if $a = c$ and $b = d$.
- Adding two complex numbers is done by adding real and imaginary parts separately:
$$(a + bi) + (c + di) = (a + c) + (b + d)i.$$
- Subtracting is similar: $(a + bi) - (c + di) = (a - c) + (b - d)i.$
- Multiplication works just like for polynomials, using $i^2 = -1$ to simplify:
$$(a + bi)(c + di) = a(c + di) + bi(c + di) = ac + adi + bci + bdi^2$$
$$= ac + adi + bci - bd = (ac - bd) + (ad + bc)i.$$
- Powers of i: We know that $i^2 = -1$; then, $i^3 = i \cdot i^2 = -i$, and $i^4 = (i^2)^2 = (-1)^2 = 1$. Then $i^5 = i \cdot i^4 = i$, and so on. Thus we have

$$i^n = \begin{cases} i & \text{for } n = 1, 5, 9, 13, \ldots \\ -1 & \text{for } n = 2, 6, 10, 14, \ldots \\ -i & \text{for } n = 3, 7, 11, 15, \ldots \\ 1 & \text{for } n = 4, 8, 12, 16, \ldots \end{cases}$$

- The product of a number and its conjugate is always real and nonnegative:
$$z \cdot \bar{z} = (a + bi)(a - bi) = a^2 - abi + abi - b^2i^2 = a^2 + b^2.$$
- Dividing is done by multiplying the numerator and denominator by the conjugate of the denominator, thereby making the denominator real:
$$\frac{a + bi}{c + di} = \frac{a + bi}{c + di} \cdot \frac{c - di}{c - di} = \frac{ac - adi + bci - bdi^2}{c^2 + d^2} = \frac{ac + bd}{c^2 + d^2} + \frac{bc - ad}{c^2 + d^2}i.$$

Example 1   Write $(2 + 7i)(4 - 6i) - i$ as a single complex number.

Solution $(2 + 7i)(4 - 6i) - i = 8 + 28i - 12i - 42i^2 - i = 8 + 15i + 42 = 50 + 15i.$

Example 2   Compute $\dfrac{2 + 7i}{4 - 6i}$.

Solution $\dfrac{2 + 7i}{4 - 6i} = \dfrac{2 + 7i}{4 - 6i} \cdot \dfrac{4 + 6i}{4 + 6i} = \dfrac{8 + 12i + 28i + 42i^2}{4^2 + 6^2} = \dfrac{-34 + 40i}{52} = \dfrac{-17}{26} + \dfrac{10}{13}i.$

The Complex Plane and Polar Coordinates

It is often useful to picture a complex number $z = x + iy$ in the plane, with x along the horizontal axis and y along the vertical. The xy-plane is then called the *complex plane*. Figure 11.30 shows the complex numbers $3i$, $-2 + 3i$, -3, $-2i$, 2, and $1 + i$.

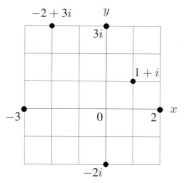

Figure 11.30: Points in the complex plane

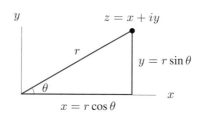

Figure 11.31: The point $z = x + iy$ in the complex plane, showing polar coordinates

The triangle in Figure 11.31 shows that a complex number can be written using polar coordinates as follows:

$$z = x + iy = r\cos\theta + ir\sin\theta.$$

Example 3 Express $z = -2i$ and $z = -2 + 3i$ using polar coordinates. (See Figure 11.30.)

Solution For $z = -2i$, the distance of z from the origin is 2. Thus $r = 2$. Also, one value for θ is $\theta = 3\pi/2$. Thus, using polar coordinates,

$$z = -2i = 2\cos(3\pi/2) + i\,2(\sin 3\pi/2).$$

For $z = -2 + 3i$, we have $x = -2$, $y = 3$, so $r = \sqrt{(-2)^2 + 3^2} \approx 3.61$ and $\tan\theta = 3/(-2)$, so, for example, $\theta \approx 2.16$. Thus using polar coordinates,

$$z = -2 + 3i \approx 3.61\cos(2.16) + i\,3.61\sin(2.16).$$

Example 4 Consider the point with polar coordinates $r = 5$ and $\theta = 3\pi/4$. What complex number does this point represent?

Solution Since $x = r\cos\theta$ and $y = r\sin\theta$ we see that $x = 5\cos 3\pi/4 = -5/\sqrt{2}$, and $y = 5\sin 3\pi/4 = 5/\sqrt{2}$, so $z = -5/\sqrt{2} + i\,5/\sqrt{2}$.

Euler's Formula

Based on what we have seen so far, there is no reason to expect a connection between the exponential function e^x and the trigonometric functions $\sin x$ and $\cos x$. However, in the eighteenth century, the Swiss mathematician Leonhard Euler discovered a surprising connection between these functions that involves complex numbers. This result, called *Euler's formula*, states that, for real θ in radians,

$$e^{i\theta} = \cos\theta + i\sin\theta$$

Example 5   Evaluate $e^{i\pi}$.

Solution   Using Euler's formula,
$$e^{i\pi} = \cos \pi + i \sin \pi = -1 + i \cdot 0 = -1.$$

This statement, known as *Euler's identity*, is sometimes written $e^{i\pi} + 1 = 0$. It is famous because it relates five of the most fundamental constants in mathematics: 0, 1, e, π, and i.

You may be wondering what it means to raise a number to an imaginary power. Euler's formula doesn't tell us what complex powers mean, but it does provide a consistent way to evaluate complex powers. We will use Euler's formula without proof; we also assume that complex exponentials, such as 4^i and $e^{i\theta}$, obey the usual rules of manipulation.

Example 6   Convert 4^i to the form $a + bi$.

Solution   Using Euler's formula and the exponent rules, we can evaluate the expression 4^i. Since $4 = e^{\ln 4}$,
$$\begin{aligned}
4^i &= \left(e^{\ln 4}\right)^i \\
&= e^{i(\ln 4)} \qquad \text{Using an exponent rule} \\
&= \cos(\ln 4) + i \sin(\ln 4) \qquad \text{Using Euler's formula} \\
&= 0.183 + 0.983i \qquad \text{Using a calculator.}
\end{aligned}$$

Example 7   Is the expression $\sqrt{i}$ defined? If so, find a possible value for it.

Solution   From Example 5, we know that
$$e^{i\pi} = -1.$$

Taking the square root of both sides gives
$$\begin{aligned}
\sqrt{e^{i\pi}} &= \sqrt{-1} \\
\left(e^{i\pi}\right)^{1/2} &= i \\
e^{i\pi/2} &= i \qquad \text{Using an exponent rule}
\end{aligned}$$

We can now find $\sqrt{i}$. Taking the square root of both sides of this equation and using an exponent rule, we have
$$\begin{aligned}
\sqrt{e^{i\pi/2}} &= \sqrt{i} \\
\left(e^{i\pi/2}\right)^{1/2} &= \sqrt{i} \\
e^{i\pi/4} &= \sqrt{i}.
\end{aligned}$$

Thus, $\sqrt{i} = e^{i\pi/4}$. We can write this number in the form $a + bi$ by using Euler's formula:
$$\begin{aligned}
e^{i\pi/4} &= \cos \frac{\pi}{4} + i \sin \frac{\pi}{4} \\
&= \frac{\sqrt{2}}{2} + i\frac{\sqrt{2}}{2}.
\end{aligned}$$

Thus, we see that
$$\sqrt{i} = \frac{\sqrt{2}}{2} + i\frac{\sqrt{2}}{2}.$$

To see that this value for $\sqrt{i}$ is correct, we square it to see that we get i:

$$
\left(\frac{\sqrt{2}}{2} + i\frac{\sqrt{2}}{2} \right)^2 = \left(\frac{\sqrt{2}}{2} + i\frac{\sqrt{2}}{2} \right) \left(\frac{\sqrt{2}}{2} + i\frac{\sqrt{2}}{2} \right)
$$

$$
= \left(\frac{\sqrt{2}}{2} \right)^2 + \frac{\sqrt{2}}{2}\left(i\frac{\sqrt{2}}{2} \right) + i\frac{\sqrt{2}}{2}\left(\frac{\sqrt{2}}{2} \right) + \left(i\frac{\sqrt{2}}{2} \right)^2
$$

$$
= \frac{1}{2} + \frac{1}{2}i + \frac{1}{2}i - \frac{1}{2}
$$

$$
= i.
$$

We have not shown that this is the only number whose square is i. In fact, as you can check, $z = -\frac{\sqrt{2}}{2} - i\frac{\sqrt{2}}{2}$ is another square root of i.

The previous example is significant for the following reason: Having introduced i as $\sqrt{-1}$, it would be inconvenient if we needed another new symbol for $\sqrt{i}$, and another for $\sqrt{\sqrt{i}}$, and so on. However, it turns out that no new numbers other than i are needed. By including i, the system of complex numbers is closed in the sense that no algebraic operation (other than division by 0) leads to numbers outside the system.

Polar Form of a Complex Number

Euler's formula allows us to write the complex number represented by the point with polar coordinates (r, θ) in the following form:

$$
z = r(\cos\theta + i\sin\theta) = re^{i\theta}.
$$

The expression $r^{i\theta}$ is called the **polar form** of the complex number z.

Similarly, since $\cos(-\theta) = \cos\theta$ and $\sin(-\theta) = -\sin\theta$, we have

$$
re^{-i\theta} = r\left(\cos(-\theta) + i\sin(-\theta) \right) = r(\cos\theta - i\sin\theta).
$$

Thus, $re^{-i\theta}$ is the conjugate of $re^{i\theta}$.

Example 8   Express the complex number represented by the point with polar coordinates $r = 8$ and $\theta = 3\pi/4$, in Cartesian form, $a + bi$, and in polar form, $z = re^{i\theta}$.

Solution   Using Cartesian coordinates, the complex number is

$$
z = 8\left(\cos\left(\frac{3\pi}{4} \right) + i\sin\left(\frac{3\pi}{4} \right) \right) = \frac{-8}{\sqrt{2}} + i\frac{8}{\sqrt{2}}.
$$

Knowing the polar coordinates we write the polar form as

$$
z = 8e^{i\,3\pi/4}.
$$

Among its many benefits, the polar form of complex numbers makes finding powers and roots of complex numbers much easier. Using the polar form, $z = re^{i\theta}$, for a complex number, we may find any power of z as follows:

$$
z^p = (re^{i\theta})^p = r^p e^{ip\theta}.
$$

To find roots, let p be a fraction, as in the following example.

Example 9   Find a cube root of the complex number represented by the point with polar coordinates $(8, 3\pi/4)$.

Solution   In Example 8, we saw that this complex number could be written as $z = 8e^{i3\pi/4}$. Thus,

$$\sqrt[3]{z} = \left(8e^{i\,3\pi/4}\right)^{1/3} = 8^{1/3}e^{i(3\pi/4)\cdot(1/3)} = 2e^{\pi i/4} = 2\left(\cos\left(\frac{\pi}{4}\right) + i\sin\left(\frac{\pi}{4}\right)\right)$$

$$= 2\left(\frac{1}{\sqrt{2}} + \frac{i}{\sqrt{2}}\right) = \frac{2}{\sqrt{2}} + \frac{2i}{\sqrt{2}}.$$

Euler's Formula and Trigonometric Identities

We can use Euler's formula to derive trigonometric identities.

Example 10   Derive the Pythagorean identity, $\cos^2\theta + \sin^2\theta = 1$, using Euler's formula.

Solution   According to the exponent rules, we have

$$e^{i\theta} \cdot e^{-i\theta} = e^{i\theta - i\theta} = e^0 = 1.$$

Using Euler's formula, we rewrite this as

$$\underbrace{(\cos\theta + i\sin\theta)}_{e^{i\theta}}\underbrace{(\cos\theta - i\sin\theta)}_{e^{-i\theta}} = 1.$$

Multiplying out the left-hand side gives

$$(\cos\theta + i\sin\theta)(\cos\theta - i\sin\theta) = \cos^2\theta - i\sin\theta\cos\theta + i\sin\theta\cos\theta - i^2\sin^2\theta$$

$$= \cos^2\theta + \sin^2\theta,$$

and so $\cos^2\theta + \sin^2\theta = 1$.

Example 11   Derive the identities for $\cos(\theta + \phi)$ and $\sin(\theta + \phi)$.

Solution   Using the exponent rules, we see from Euler's formula that

$$e^{i(\theta+\phi)} = e^{i\theta} \cdot e^{i\phi}$$

$$= (\cos\theta + i\sin\theta)(\cos\phi + i\sin\phi)$$

$$= \cos\theta\cos\phi + \underbrace{i\cos\theta\sin\phi + i\sin\theta\cos\phi}_{i(\cos\theta\sin\phi + \sin\theta\cos\phi)} + \underbrace{i^2\sin\theta\sin\phi}_{-\sin\theta\sin\phi}$$

$$= \underbrace{\cos\theta\cos\phi - \sin\theta\sin\phi}_{\text{Real part}} + i\underbrace{(\cos\theta\sin\phi + \sin\theta\cos\phi)}_{\text{Imaginary part}}.$$

But it is also true that

$$e^{i(\theta+\phi)} = \underbrace{\cos(\theta + \phi)}_{\text{Real part}} + i\underbrace{\sin(\theta + \phi)}_{\text{Imaginary part}}.$$

Two complex numbers are equal only if their real and imaginary parts are equal. Setting real parts equal gives

$$\cos(\theta + \phi) = \cos\theta\cos\phi - \sin\theta\sin\phi.$$

Setting imaginary parts equal gives

$$\sin(\theta + \phi) = \cos\theta\sin\phi + \sin\theta\cos\phi.$$

This last identity is usually written in the following way:

$$\sin(\theta + \phi) = \sin\theta\cos\phi + \sin\phi\cos\theta.$$

Problems for Section 11.5 ━━━━━━━

For Problems 1–6, express the given complex number in polar form, $z = re^{i\theta}$.

   1. $2i$ 2. -5 3. $-3 - 4i$ 4. 0 5. $-i$ 6. $-1 + 3i$

For Problems 7–14, perform the indicated calculations. Give your answer in Cartesian form, $z = x + iy$.

   7. $(2 + 3i) + (-5 - 7i)$ 8. $(2 + 3i)(5 + 7i)$ 9. $(2 + 3i)^2$ 10. $(0.5 - i)(1 - i/4)$

  11. $(2i)^3 - (2i)^2 + 2i - 1$ 12. $(e^{i\pi/3})^2$ 13. $\sqrt{e^{i\pi/3}}$ 14. $\sqrt[4]{10e^{i\pi/2}}$

By writing the complex numbers in polar form, $z = re^{i\theta}$, find a value for the quantities in Problems 15–22. Give your answer in Cartesian form, $z = x + iy$.

  15. $\sqrt{4i}$ 16. $\sqrt{-i}$ 17. $\sqrt[3]{i}$ 18. $\sqrt{7i}$

  19. $(1 + i)^{2/3}$ 20. $(\sqrt{3} + i)^{1/2}$ 21. $(\sqrt{3} + i)^{-1/2}$ 22. $(\sqrt{5} + 2i)^{\sqrt{2}}$

Solve the simultaneous equations in Problems 23–24 for the complex numbers A_1 and A_2.

  23. $A_1 + A_2 = 2$
       $(1 - i)A_1 + (1 + i)A_2 = 0$

  24. $A_1 + A_2 = i$
       $iA_1 - A_2 = 3$

  25. Let $z_1 = -3 - i\sqrt{3}$ and $z_2 = -1 + i\sqrt{3}$.

     (a) Find $z_1 z_2$ and z_1/z_2. Give your answer in Cartesian form, $z = x + iy$.
     (b) Put z_1 and z_2 into polar form, $z = re^{i\theta}$. Find $z_1 z_2$ and z_1/z_2 using the polar form, and verify that you get the same answer as in part (a).

  26. If the roots of the equation $x^2 + 2bx + c = 0$ are the complex numbers $p \pm iq$, find expressions for p and q in terms of a and b.

Are the statements in Problems 27–32 true or false? Explain your answer.

  27. Every nonnegative real number has a real square root.

  28. For any complex number z, the product $z \cdot \bar{z}$ is a real number.

  29. The square of any complex number is a real number.

  30. If f is a polynomial, and $f(z) = i$, then $f(\bar{z}) = i$.

  31. Every nonzero complex number z can be written in the form $z = e^w$, where w is another complex number.

  32. If $z = x + iy$, where x and y are positive, then $z^2 = a + ib$ has a and b positive.

For Problems 33–34, use Euler's formula to derive the following relationships. (Note that if a, b, c, d are real numbers, $a + bi = c + di$ means that $a = c$ and $b = d$.)

  33. $\sin 2\theta = 2 \sin \theta \cos \theta$ 34. $\cos 2\theta = \cos^2 \theta - \sin^2 \theta$

11.6 HYPERBOLIC FUNCTIONS

There are two combinations of the functions e^x and e^{-x} used so often in engineering that they are given their own names. They are the hyperbolic sine, abbreviated sinh (pronounced "cinch") and hyperbolic cosine, abbreviated cosh (rhymes with "gosh"). They are defined as follows:

Hyperbolic Sine and Hyperbolic Cosine

$$\cosh x = \frac{e^x + e^{-x}}{2} \qquad \sinh x = \frac{e^x - e^{-x}}{2}$$

Properties of Hyperbolic Functions

The graphs of the $\cosh x$ and $\sinh x$ are given in Figures 11.32 and 11.33 together with the graphs of related multiples of e^x and e^{-x}.

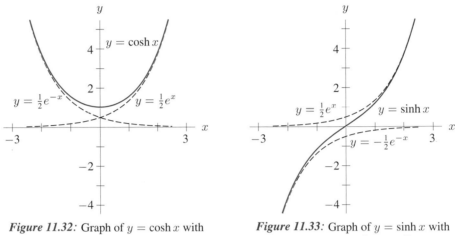

Figure 11.32: Graph of $y = \cosh x$ with multiples of e^x and e^{-x}

Figure 11.33: Graph of $y = \sinh x$ with multiples of e^x and e^{-x}

The graphs suggest that $\cosh x$ is even and $\sinh x$ is odd and that $\cosh x$ has a y-intercept of 1, whereas $\sinh x$ has a y-intercept of 0:

$$\cosh 0 = 1 \qquad \sinh 0 = 0$$
$$\cosh(-x) = \cosh x \qquad \sinh(-x) = -\sinh x$$

To show that the hyperbolic functions really do have these properties, we use their formulas.

Example 1   Show that   (a) $\cosh(0) = 1$ (b) $\cosh(-x) = \cosh x$

Solution   (a)   Substituting $x = 0$ into the formula for $\cosh x$ gives the y-intercept:

$$\cosh 0 = \frac{e^0 + e^{-0}}{2} = \frac{1 + 1}{2} = 1.$$

(b)   Substituting $-x$ for x gives

$$\cosh(-x) = \frac{e^{-x} + e^{-(-x)}}{2} = \frac{e^{-x} + e^x}{2} = \cosh x.$$

Thus, $\cosh x$ is an even function.

Example 2   Describe and explain the behavior of $\cosh x$ as $x \to \infty$ and then as $x \to -\infty$.

Solution   From Figure 11.32, we see that as $x \to \infty$, the graph of $\cosh x$ resembles the graph of $\frac{1}{2}e^x$. Similarly, as $x \to -\infty$, the graph of $\cosh x$ resembles the graph of $\frac{1}{2}e^{-x}$. Using the formula for $\cosh x$, and the facts that $e^{-x} \to 0$ as $x \to \infty$ and $e^x \to 0$ as $x \to -\infty$, we can predict these results algebraically:

$$\text{As } x \to \infty, \qquad \cosh x = \frac{e^x + e^{-x}}{2} \to \frac{1}{2}e^x.$$

$$\text{As } x \to -\infty, \qquad \cosh x = \frac{e^x + e^{-x}}{2} \to \frac{1}{2}e^{-x}.$$

Figures 11.32 and 11.33 suggest that the graph of $\cosh x$ is always above the graph of $\sinh x$. In the following example, we use the formulas to confirm this.

Example 3   Show that $\cosh x > \sinh x$ for all x.

Solution   The difference between $\cosh x$ and $\sinh x$ is given by

$$\cosh x - \sinh x = e^{-x}.$$

Since e^{-x} is positive for all x, we see that

$$\cosh x > \sinh x \quad \text{for all } x.$$

Identities Involving cosh x and sinh x

The hyperbolic functions have names that remind us of the trigonometric functions because they have some properties that are similar to those of the trigonometric functions. Consider the expression

$$(\cosh x)^2 - (\sinh x)^2.$$

This can be simplified using the formulas and the fact that $e^x \cdot e^{-x} = 1$:

$$(\cosh x)^2 - (\sinh x)^2 = \left(\frac{e^x + e^{-x}}{2}\right)^2 - \left(\frac{e^x - e^{-x}}{2}\right)^2$$

$$= \frac{e^{2x} + 2e^x \cdot e^{-x} + e^{-2x}}{4} - \frac{e^{2x} - 2e^x \cdot e^{-x} + e^{-2x}}{4}$$

$$= \frac{e^{2x} + 2 + e^{-2x} - e^{2x} + 2 - e^{-2x}}{4}$$

$$= 1.$$

Thus, writing $\cosh^2 x$ for $(\cosh x)^2$ and $\sinh^2 x$ for $(\sinh x)^2$, we have the identity

$$\boxed{\cosh^2 x - \sinh^2 x = 1}$$

This identity is reminiscent of the Pythagorean identity $\cos^2 x + \sin^2 x = 1$. Extending the analogy, we define

Hyperbolic Tangent

$$\tanh x = \frac{\sinh x}{\cosh x}$$

Parameterizing the Hyperbola Using Hyperbolic Functions

Consider the curve parameterized by the equations

$$x = \cosh t, \quad y = \sinh t, \quad -\infty < t < \infty.$$

The parameter, t, can be eliminated from these equations using the identity $\cosh^2 t - \sinh^2 t = 1$, This gives

$$x^2 - y^2 = \underbrace{\cosh^2 t}_{x^2} - \underbrace{\sinh^2 t}_{y^2} = 1,$$

which is the implicit equation for the hyperbola in Figure 11.34:

$$\boxed{x^2 - y^2 = 1.}$$

Notice that since $\cosh t > 0$, the parameterization gives only the right branch of the hyperbola. Solving the equation $x^2 - y^2 = 1$ for x, we obtain two values of x for every value of y:

$$x^2 - y^2 = 1$$
$$x^2 = 1 + y^2$$
$$x = \pm\sqrt{1 + y^2}.$$

The graph of $x = \sqrt{1 + y^2}$ appears in Figure 11.35. The equation $x = -\sqrt{1 + y^2}$ is a reflection across the y-axis of the graph of $x = \sqrt{1 + y^2}$. The two branches make up a single hyperbola.

This parameterization $x = \cosh t, y = \sinh t$ gives us information about the graph of the hyperbola. At $t = 0$, we have $(x, y) = (\cosh 0, \sinh 0) = (1, 0)$, so the curve passes through the point $(1, 0)$. As $t \to \infty$, we know that both $\cosh t$ and $\sinh t$ approach $(1/2)e^t$, so for large values of t we see that $\cosh t \approx \sinh t$. Thus, as t increases the curve draws close to the diagonal line $y = x$. Since $\sinh x < \cosh x$ for all x, we know that $y < x$, so the curve approaches this asymptote from below. See Figure 11.35.

Similarly, as $t \to -\infty$, we know that $x \to e^{-t}/2$ and $y \to -e^{-t}/2$, which means that $y \approx -x$. Thus, as $t \to -\infty$, the curve draws close to the asymptote $y = -x$. In this case, though, the curve approaches the asymptote from above, because the y-coordinate is slightly larger (less negative) than the x-coordinate.

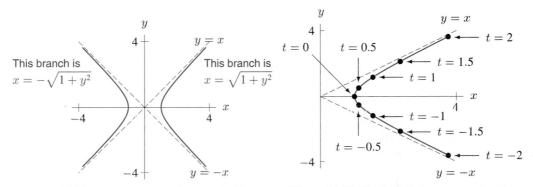

Figure 11.34: The hyperbola defined implicitly by $x^2 - y^2 = 1$

Figure 11.35: The hyperbola parameterized by $x = \cosh t, y = \sinh t, -\infty < t < \infty$

Relating Trigonometric and Hyperbolic Functions

Provided x is real, we know from Euler's formula that

$$e^{ix} = \cos x + i \sin x.$$

Since $\cos(-x) = \cos x$ and $\sin(-x) = -\sin x$, we have

$$e^{-ix} = e^{i(-x)} = \cos(-x) + i \sin(-x) = \cos x - i \sin x.$$

Then

$$e^{ix} + e^{-ix} = (\cos x + i \sin x) + (\cos x - i \sin x)$$
$$= 2\cos x,$$

so

$$\cos x = \frac{e^{ix} + e^{-ix}}{2}.$$

The expression on the right looks like the formula for the hyperbolic cosine function, except it involves imaginary exponents. Extending the definition of $\cosh x$ to imaginary values of the input, we can write

$$\cosh(ix) = \frac{e^{ix} + e^{-ix}}{2}.$$

Thus, we have the following relationship between the cosine function and the hyperbolic cosine function:

$$\boxed{\cosh(ix) = \cos x}$$

In Problem 7, we derive the analogous formula

$$\boxed{\sinh(ix) = i \sin x.}$$

Problems for Section 11.6

1. Show that $\sinh 0 = 0$.

2. Show that $\sinh(-x) = -\sinh(x)$.

3. Using the method of Example 2 on page 486, describe and explain the behavior of $\sinh x$ as $x \to \infty$ and as $x \to -\infty$.

4. Is there an identity analogous to $\sin 2x = 2 \sin x \cos x$ for the hyperbolic functions? Explain.

5. Is there an identity analogous to $\cos 2x = \cos^2 x - \sin^2 x$ for the hyperbolic functions? Explain.

6. Consider the family of functions $y = a\cosh(x/a)$ for $a > 0$. Sketch graphs for $a = 1, 2, 3$. Describe in words the effect of increasing a.

7. Show that $\sinh(ix) = i \sin x$.

8. Find an expression for $\cos(ix)$ in terms of $\cosh x$.

9. Find an expression for $\sin(ix)$ in terms of $\sinh x$.

REVIEW PROBLEMS FOR CHAPTER ELEVEN

1. Write the following using sigma notation $100 + 90 + 80 + 70 + \cdots + 0$.

2. (a) Write the sum $\sum_{n=1}^{5}(4n - 3)$ in expanded form.

   (b) Compute the sum.

3. Find the sum of the first eighteen terms of the series: $8 + 11 + 14 + \cdots$. What is the eighteenth term?

In Problems 4–5, decide which of the following are geometric series. For those which are, give the first term and the ratio between successive terms. For those which are not, explain why not.

4. $1 - y^2 + y^4 - y^6 + \cdots$

5. $1 + 2z + (2z)^2 + (2z)^3 + \cdots$

6. Find the sum of the series in Problem 4.

7. Find the sum of the series in Problem 5.

8. A store clerk has 108 cans to stack. He can fit 24 cans on the bottom row and can stack the cans 8 rows high. Use arithmetic series to determine how he can stack the cans so that each row contains fewer cans than the row beneath it and that the number of cans in each row decreases at a constant rate.

9. A ball is dropped from a height of 10 feet and bounces. Each bounce is $3/4$ of the height of the bounce before. Thus after the ball hits the floor for the first time, the ball rises to a height of $10(3/4) = 7.5$ feet, and after it hits the floor for the second time, it rises to a height of $7.5(3/4) = 10(3/4)^2 = 5.625$ feet.

   (a) Find an expression for the height to which the ball rises after it hits the floor for the nth time.

   (b) Find an expression for the total vertical distance the ball has traveled when it hits the floor for the first, second, third, and fourth times.

   (c) Find an expression for the total vertical distance the ball has traveled when it hits the floor for the nth time. Express your answer in closed-form.

10. You might think that the ball in Problem 9 keeps bouncing forever since it takes infinitely many bounces. This is not true! It can be shown that a ball dropped from a height of h feet reaches the ground in $\frac{1}{4}\sqrt{h}$ seconds. It is also true that it takes a bouncing ball the same amount of time to rise h feet. Use these facts to show that the ball in Problem 9 stops bouncing after

$$\frac{1}{4}\sqrt{10} + \frac{1}{2}\sqrt{10}\sqrt{\frac{3}{4}}\left(\frac{1}{1 - \sqrt{3/4}}\right)$$

seconds, or approximately 11 seconds.

11. This problem illustrates how banks create credit and can thereby lend out more money than has been deposited. Suppose that initially \$100 is deposited in a bank. Experience has shown bankers that on average only 8% of the money deposited is withdrawn by the owner at any time. Consequently, bankers feel free to lend out 92% of their deposits. Thus \$92 of the original \$100 is loaned out to other customers (to start a business, for example). This \$92 will become someone else's income and, sooner or later, will be redeposited in the bank. Then 92% of \$92, or $\$92(0.92) = \$84.64$, is loaned out again and eventually redeposited. Of the \$84.64, the bank again loans out 92%, and so on.

   (a) Find the total amount of money deposited in the bank.

   (b) The total amount of money deposited divided by the original deposit is called the *credit multiplier*. Calculate the credit multiplier for this example and explain what this number tells us.

12. Let x be the position of a particle moving along a horizontal line. Plot the motion $x = t \ln t, 0.01 \le t \le 10$ by introducing a y-coordinate. What does the plot tell you about the particle's motion?

Write a parameterization for each of the curves in Problems 13–15.

13. The horizontal line through the point $(0, 5)$.

14. The circle of radius 1 in the xy-plane centered at the origin, traversed counterclockwise when viewed from above.

15. The circle of radius 2 centered at the origin starting at the point $(0, 2)$ when $t = 0$.

16. On a graphing calculator or a computer, plot $x = 2t/(t^2 + 1)$, $y = (t^2 - 1)/(t^2 + 1)$, first for $-50 \leq t \leq 50$, and then for $-5 \leq t \leq 5$. Explain what you see. Is the curve really a circle?

17. Plot the Lissajous figure given by $x = \cos 2t$, $y = \sin t$ using a graphing calculator or computer. Explain why it looks like part of a parabola. [Hint: Use a double angle identity.]

18. A planet P in the xy-plane orbits the star S counterclockwise in a circle of radius 10 units, completing one orbit in 2π units of time. In addition, a moon M orbits the planet P counterclockwise in a circle of radius 3 units, completing one orbit in $2\pi/8$ units of time. The star S is fixed at the origin $x = 0$, $y = 0$, and at time $t = 0$ the planet P is at the point $(10, 0)$ and the moon M is at the point $(13, 0)$.

    (a) Find parametric equations for the x- and y-coordinates of the planet at time t.
    (b) Find parametric equations for the x- and y-coordinates of the moon at time t. [Hint: For the moon's position at time t, take a vector from the sun to the planet at time t and add a vector from the planet to the moon].
    (c) Plot the path of the moon in the xy-plane, using a graphing calculator or computer.

19. A wheel of radius 1 meter rests on the x-axis with its center on the y-axis. There is a spot on the rim at the point $(1, 1)$. See Figure 11.36. At time $t = 0$ the wheel starts rolling on the x-axis in the direction shown at a rate of 1 radian per second.

    (a) Find parametric equations describing the motion of the center of the wheel.
    (b) Find parametric equations describing the motion of the spot on the rim. Plot its path.

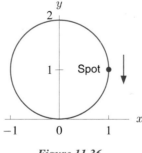

Figure 11.36

For Problems 20–21, express the given complex number in polar form, $z = re^{i\theta}$.

20. $1 + i$

21. $5 - 12i$

For Problems 22–23, perform the indicated calculations. Give your answer in Cartesian form, $z = x + iy$.

22. $(1 + i)^2 + (1 + i)$

23. $(5e^{i7\pi/6})^3$

By writing the complex numbers in polar form, $z = re^{i\theta}$, find a value for the quantities in Problems 24–25. Give your answer in Cartesian form, $z = x + iy$.

24. $(1 + i)^{100}$

25. $(-4 + 4i)^{2/3}$

APPENDICES

Many of the skills you need for this book were covered in previous courses. However, it may have been a long time since you systematically practiced these skills. This appendix is designed as a review of some basic rules and definitions. As you work through the problems of the appendix, try to make both the vocabulary and the manipulations second nature so that you can use them quickly and appropriately.

A EXPONENTS

Positive Integer Exponents

Repeated addition leads to multiplication. For example,

$$\underbrace{2 + 2 + 2 + 2 + 2}_{5 \text{ terms in sum}} = 5 \times 2.$$

Similarly, repeated multiplication leads to *exponentiation*. For example,

$$\underbrace{2 \times 2 \times 2 \times 2 \times 2}_{5 \text{ factors in product}} = 2^5.$$

Here, 5 is the *exponent* of 2, and 2 is called the *base*. Notice that 2^5 is not the same as 5^2, because $2^5 = 32$ and $5^2 = 25$.

In general, if a is a real number and n is a positive integer, then we define exponentiation as an abbreviation for multiplication:

$$\underbrace{a \cdot a \cdot a \cdots a}_{n \text{ factors}} = a^n.$$

It's worth noticing that $a^1 = a$, because here we have only 1 factor of a.

Exponent Rules

There are five basic facts, called exponent rules, which follow as a direct consequence of the definition of exponents. These rules are summarized in the following box.

Exponent Rules

If a and b are real numbers and m and n are positive integers,

1. $a^n \cdot a^m = a^{n+m}$ 2. $\dfrac{a^n}{a^m} = a^{n-m}$ 3. $(a^m)^n = a^{m \cdot n}$

4. $(ab)^n = a^n b^n$ 5. $\left(\dfrac{a}{b}\right)^n = \dfrac{a^n}{b^n}$

The explanation of all five rules are very straightforward. If you forget one of the rules, you should be able to immediately rederive it using the following logic.

For exponent rule 1, we have

$$a^n \cdot a^m = \underbrace{a \cdot a \cdot a \cdots a}_{n \text{ factors}} \underbrace{a \cdot a \cdot a \cdots a}_{m \text{ factors}} = \underbrace{a \cdot a \cdot a \cdots a}_{n + m \text{ factors}} = a^{n+m}.$$

For example, $2^3 \cdot 2^5 = (2 \cdot 2 \cdot 2) \cdot (2 \cdot 2 \cdot 2 \cdot 2 \cdot 2) = 2^8$, and $8 = 3 + 5$.

For exponent rule 2, suppose n and m are positive integers and $n > m$. Then the m factors of a in the denominator cancel with some of the n factors of a in the numerator, leaving only $n - m$ factors of a:

$$\frac{a^n}{a^m} = \frac{\overbrace{a \cdot a \cdot a \cdot a \cdots a}^{n \text{ factors of } a}}{\underbrace{a \cdot a \cdots a}_{m \text{ factors of } a}} = \frac{\overbrace{\cancel{a} \cdot \cancel{a} \cdots \cancel{a} \cdot}^{m \text{ factors of } a \text{ cancel}} \overbrace{a \cdots a}^{n-m \text{ factors of } a \text{ are left after cancelling}}}{\underbrace{\cancel{a} \cdot \cancel{a} \cdots \cancel{a}}_{\substack{m \text{ factors} \\ \text{of } a \text{ cancel}}}} = \underbrace{a \cdot a \cdot a \cdots a}_{n - m \text{ factors}} = a^{n-m}$$

For example,

$$\frac{4^5}{4^3} = \frac{4 \cdot 4 \cdot 4 \cdot 4 \cdot 4}{4 \cdot 4 \cdot 4} = \frac{\cancel{4} \cdot \cancel{4} \cdot \cancel{4} \cdot 4 \cdot 4}{\cancel{4} \cdot \cancel{4} \cdot \cancel{4}} = 4 \cdot 4 = 4^2 = 4^{5-3},$$

For exponent rule 3,

$$(a^m)^n = \underbrace{(a \cdot a \cdot a \cdots a)}_{m \text{ factors of } a}{}^n = \overbrace{\underbrace{(a \cdot a \cdot a \cdots a)}_{m \text{ factors of } a}\underbrace{(a \cdot a \cdot a \cdots a)}_{m \text{ factors of } a} \cdots \underbrace{(a \cdot a \cdot a \cdots a)}_{m \text{ factors of } a}}^{\substack{\text{The } m \text{ factors of } a \text{ are multiplied } n \text{ times,} \\ \text{giving a total of } m \cdot n \text{ factors of } a}} = a^{m \cdot n}.$$

For example,

$$(a^2)^3 = (\underbrace{a \cdot a}_{2 \text{ factors of } a})^3 = \underbrace{(a \cdot a)(a \cdot a)(a \cdot a)}_{3 \text{ times 2 factors of } a} = a^6.$$

An alternative explanation for rule 3 uses rule 1:

$$(a^m)^n = \underbrace{a^m \cdot a^m \cdots a^m}_{n \text{ factors of } a^m} = a^{\overbrace{m + m + \cdots + m}^{n \text{ terms in sum}}} = a^{m \cdot n}.$$

For example, $(2^5)^3 = 2^5 \cdot 2^5 \cdot 2^5 = 2^{5+5+5} = 2^{15}$, and $15 = 5 \cdot 3$.

To justify exponent rule 4, we use the commutative property of multiplication, which states that the order in which numbers are multiplied together does not affect the result. For example, $2 \cdot 9 \cdot 13 = 13 \cdot 2 \cdot 9 = 9 \cdot 13 \cdot 2 = 234$.

$$(a \cdot b)^n = \underbrace{(a \cdot b)(a \cdot b)(a \cdot b) \cdots (a \cdot b)}_{n \text{ factors of } (a \cdot b)} = \underbrace{\overbrace{(a \cdot a \cdot a \cdots a)}^{n \text{ factors of } a} \cdot \overbrace{(b \cdot b \cdot b \cdots b)}^{n \text{ factors of } b}}_{\substack{\text{Since we can rearrange the order using the} \\ \text{commutative property of multiplication}}} = a^n \cdot b^n.$$

For example, $(5 \cdot 8)^2 = (5 \cdot 8)(5 \cdot 8) = (5 \cdot 5)(8 \cdot 8) = 5^2 \cdot 8^2 = 1600$.

For exponent rule 5,

$$\left(\frac{a}{b}\right)^n = \underbrace{\left(\frac{a}{b}\right) \cdot \left(\frac{a}{b}\right) \cdot \left(\frac{a}{b}\right) \cdots \left(\frac{a}{b}\right)}_{n \text{ factors of } a/b} = \frac{\overbrace{a \cdot a \cdot a \cdots a}^{n \text{ factors of } a}}{\underbrace{b \cdot b \cdot b \cdots b}_{n \text{ factors of } b}} = \frac{a^n}{b^n}.$$

For example, $\left(\dfrac{2}{7}\right)^3 = \left(\dfrac{2}{7}\right) \cdot \left(\dfrac{2}{7}\right) \cdot \left(\dfrac{2}{7}\right) = \dfrac{2 \cdot 2 \cdot 2}{7 \cdot 7 \cdot 7} = \dfrac{2^3}{7^3} = \dfrac{8}{343}$.

Be aware of the following notational conventions:

$$ab^n = a(b^n), \qquad \text{but } ab^n \neq (ab)^n,$$
$$-b^n = -(b^n), \qquad \text{but } -b^n \neq (-b)^n,$$
$$-ab^n = (-a)(b^n).$$

For example, $-2^4 = -(2^4) = -16$, but $(-2)^4 = (-2)(-2)(-2)(-2) = +16$. Also, be sure to realize that for values of n other than 0 and 1,

$$(a + b)^n \neq a^n + b^n \qquad \text{Power of a sum} \neq \text{Sum of powers}.$$

Rational Exponents

The natural definition for exponentiation as an abbreviation for multiplication holds for positive integers only. For example, 4^5 means 4 multiplied times itself 5 times, but we cannot use the same definition for 4^0 or 4^{-1} or $4^{1/2}$. We must choose a definition for exponents which are not positive integers.

The five exponent rules follow from the definition of positive integer exponents. Therefore, it makes sense to choose a definition for exponents like 0, -1, $1/2$ which is consistent with the five exponent rules. The definitions are summarized in the following box. After the box, explanations are given for these choices of definition.

Definitions For Exponentiation

If a is a real number and m and n are positive integers:[1]

- $a^0 = 1$ (for $a \neq 0$)
- $a^{-n} = \frac{1}{a^n}$ (for $a \neq 0$)
- $a^{1/n} = \sqrt[n]{a}$ (if n is even, then we assume $a \geq 0$.)
- $a^{m/n} = \sqrt[n]{a^m} = (\sqrt[n]{a})^m$ (if n is even, then we assume $a^m \geq 0$.)

To understand the definition of a^0, consider the following example. If we apply exponent rule 1 when one of the exponents is $m = 0$, and $a \neq 0$, then

$$a^n \cdot a^m = a^n \cdot a^0 = a^{n+0} = a^n,$$

so

$$a^n \cdot a^0 = a^n.$$

Notice that if we divide both sides of the equation by a^n, we get $a^0 = 1$. Therefore, if we want exponent rule 1 to hold for any exponent, including $m = 0$, we must define $a^0 = 1$, for $a \neq 0$. (We leave 0^0 undefined.)

Next we define a^{-n}, where n is a positive integer. Consider the following example. By the definition of a^2 and a^5,

$$\frac{a^2}{a^5} = \frac{a \cdot a}{a \cdot a \cdot a \cdot a \cdot a} = \frac{\cancel{a} \cdot \cancel{a}}{\cancel{a} \cdot \cancel{a} \cdot a \cdot a \cdot a} = \frac{1}{a^3}.$$

Alternatively, if we apply exponent rule 2, we see that the result is a negative exponent:

$$\frac{a^2}{a^5} = a^{2-5} = a^{-3}.$$

Therefore, if we want exponent rule 2 to hold for any exponent, including negative numbers, we must define $a^{-3} = \frac{1}{a^3}$. In general, $a^{-n} = \frac{1}{a^n}$. Notice that, in particular, $a^{-1} = \frac{1}{a}$.

Example 1   Use the rules of exponents to simplify the following:

(a) $3b(2b)^3(b^{-1})$

(b) $\dfrac{y^{-1}(x^{-3}y^{-1})^2}{2x^{-1}}$

(c) $\left(\dfrac{2^{-3}}{L}\right)^{-2}$

(d) $\dfrac{5(2s+1)^4(s+3)^{-2}}{2s+1}$

(e) $-2^2p^0(-3q)^4$

(f) $(-z)^{-n}$

---

[1] When we write a fractional power, we assume that the base is restricted to the values for which the power is defined.

Solution   (a) $3b(2b)^3 \left(b^{-1}\right) = 3b^1(2^3b^3)b^{-1} = 3 \cdot 8 \cdot b^1 b^3 b^{-1} = 3 \cdot 8 \cdot b^{3+(-1)+1} = 24b^3$

(b) $\dfrac{y^4 \left(x^3 y^{-2}\right)^2}{2x^{-1}} = \dfrac{y^4 x^6 y^{-4}}{2x^{-1}} = \dfrac{y^{(4-4)} x^{(6-(-1))}}{2} = \dfrac{y^0 x^7}{2} = \dfrac{x^7}{2}$

(c) $\left(\dfrac{2^{-3}}{L}\right)^{-2} = \dfrac{2^{(-3)(-2)}}{L^{-2}} = \dfrac{2^6}{L^{-2}} = 64L^2$ (Note that $\dfrac{1}{L^{-2}} = \dfrac{L^0}{L^{-2}} = L^{0-(-2)} = L^2$.)

(d) $\dfrac{5(2s+1)^4(s+3)^{-2}}{(2s+1)} = \dfrac{5(2s+1)^{4-1}}{(s+3)^2} = \dfrac{5(2s+1)^3}{(s+3)^2}$

(e) $-2^2 p^0(-3q)^4 = -4 \cdot 1(-3)^4 q^4 = -4(81)q^4 = -324q^4.$

(f) $(-z)^{-n} = \dfrac{1}{(-z)^n} = \dfrac{(-1)^n}{z^n}.$

Before explaining the definitions for fractional exponents, we review some notation:

Let n be an integer greater than 1:
- For $a \geq 0$,

  $\sqrt{a}$ is the positive number whose square is a.

  $\sqrt[n]{a}$ is the positive number whose n^{th} power is a.

- For $a < 0$,

  If n is even, $\sqrt[n]{a}$ is not a real number.

  If n is odd, $\sqrt[n]{a}$ is the negative number whose n^{th} power is a.

We say that $\sqrt[n]{a}$ is the nth **root of** a.

For example, $\sqrt{49} = 7$ because $7^2 = 49$. Likewise, $\sqrt[3]{125} = 5$ because $5^3 = 125$, and $\sqrt[5]{32} = 2$ because $2^5 = 32$. Similarly, $\sqrt[3]{-27} = -3$ because $(-3)^3 = -27$, and $\sqrt{-9}$ is not a real number, because the square of no real number is negative. When $\sqrt[n]{a}$ and $\sqrt[n]{b}$ are real numbers with $b \neq 0$, the rules for radicals are as follows:

$$\sqrt[n]{ab} = \sqrt[n]{a}\,\sqrt[n]{b}, \qquad \sqrt[n]{\frac{a}{b}} = \frac{\sqrt[n]{a}}{\sqrt[n]{b}}$$

Example 2   Simplify the following radicals:

(a) $\sqrt{36}$

(b) $\sqrt[3]{-8x^6}$

(c) $\sqrt{\dfrac{16R^8}{25}}$

(d) $\sqrt[4]{16u^8 w^{12}}$

Solution   (a) $\sqrt{36} = +6$ (Notice that the answer is the positive root, not ±6.)

(b) $\sqrt[3]{-8x^6} = \sqrt[3]{-8} \cdot \sqrt[3]{x^6} = -2x^2$

(c) $\sqrt{\dfrac{16R^8}{25}} = \dfrac{\sqrt{16R^8}}{\sqrt{25}} = \dfrac{\sqrt{16}\sqrt{R^8}}{\sqrt{25}} = \dfrac{4R^4}{5}$

(d) $\sqrt[4]{16u^8 w^{12}} = \sqrt[4]{16}\sqrt[4]{u^8}\sqrt[4]{w^{12}} = 2u^2 w^3$, because $2^4 = 16$, $(u^2)^4 = u^8$, and $(w^3)^4 = w^{12}$.

Special case: n-th root of a^n
If n is odd,
$$\sqrt[n]{a^n} = a \qquad \text{by definition.}$$

If n is even,
$$\sqrt[n]{a^n} = |a| = \begin{cases} a & \text{if } a \geq 0 \\ -a & \text{if } a < 0 \end{cases}.$$

Example 3   Simplify:

(a) $\sqrt[4]{(-1)^4}$

(b) $\sqrt{9x^2}$

Solution   (a) $\sqrt[4]{(-1)^4} = \sqrt[4]{1} = 1$ (Notice that in this case $\sqrt[n]{a^n} \neq a$.)

(b) $\sqrt{9x^2} = 3|x|$ (since x can be either positive or negative).

To see why fractional exponents are defined in the way they are, we again consider the consequences of the exponent rules. Suppose we evaluate an expression with a fractional exponent, such as $(a^{1/2})^2$, for $a \geq 0$. If we apply exponent rule 3, we get
$$(a^{1/2})^2 = a^{(1/2) \cdot 2} = a^1 = a, \quad \text{so} \quad (a^{1/2})^2 = a.$$

Taking the square root of both sides results in $a^{1/2} = \sqrt{a}$. Therefore, if we want exponent rule 3 to hold for any exponent, including fractions, we must define $a^{1/2} = \sqrt{a}$. In general, $a^{1/n} = \sqrt[n]{a}$. Note that when n is even, then a must be non-negative, and $a^{1/n} \geq 0$.

We can extend this explanation to an expression such as $(a^{1/3})^2$. Since $a^{1/3} = \sqrt[3]{a}$, we must have $(a^{1/3})^2 = (\sqrt[3]{a})^2$. Exponent rule 3 gives
$$(a^{1/3})^2 = a^{(1/3) \cdot 2} = a^{2/3}.$$

Therefore, if we want exponent rule 3 to hold for any exponent, including fractions, we must define $a^{2/3} = (\sqrt[3]{a})^2$. In general, $a^{m/n} = (\sqrt[n]{a})^m$. Because multiplication is commutative, we have
$$a^{m/n} = a^{m \cdot (1/n)} = a^{(1/n) \cdot m}$$
$$= (a^m)^{1/n} = (a^{1/n})^m$$
$$\text{so} \quad a^{m/n} = \sqrt[n]{a^m} = (\sqrt[n]{a})^m.$$

We have chosen these definitions so that exponent rules hold for any rational exponent.

Example 4   Find $(27)^{2/3}$.

Solution $(27)^{2/3} = \sqrt[3]{27^2} = \sqrt[3]{729} = 9$, or, equivalently, $(27)^{2/3} = \left(27^{\frac{1}{3}}\right)^2 = \left(\sqrt[3]{27}\right)^2 = 3^2 = 9.$

Example 5   Simplify:

(a) $\left(\dfrac{M^{1/5}}{3N^{-1/2}}\right)^2$

(b) $\dfrac{3u^2\sqrt{uw}}{w^{1/3}}$

Solution   (a) $\left(\dfrac{M^{1/5}}{3N^{-1/2}}\right)^2 = \dfrac{\left(M^{1/5}\right)^2}{\left(3N^{-1/2}\right)^2} = \dfrac{M^{2/5}}{3^2 N^{-1}} = \dfrac{M^{2/5}N}{9}$

(b) $\dfrac{3u^2\sqrt{uw}}{w^{1/3}} = \dfrac{3u^2 u^{1/2} w}{w^{1/3}} = 3u^{(2+1/2)} w^{(1-1/3)} = 3u^{5/2} w^{2/3}$

Calculator Note: Some calculators will not compute $a^{m/n}$ for $m \neq 1$ when a is negative, even if n is odd. For example, though $(-1)^{2/3}$ is well defined, a calculator may display "error."

Example 6   Evaluate, if possible:

(a) $(-2197)^{2/3}$

(b) $(-256)^{3/4}$

Solution   (a) To find $(-2197)^{2/3}$ on a calculator, we can first evaluate $(-2197)^{1/3}$, and then square the result. This gives 169.

(b) $(-256)^{3/4}$ is not a real number since $(-256)^{1/4}$ is not real.

Irrational and Variable Exponents

The rules for exponents given for integer and fractional exponents apply also when the exponent is an irrational number, as in $x^{\sqrt{3}}$, or a variable, as in 5^x.

Example 7   Simplify:

(a) $p^{\sqrt{2}} \cdot p^{\sqrt{8}}$

(b) $\sqrt{\dfrac{x^{3\pi}}{x^{\pi}}}$

Solution   (a) $p^{\sqrt{2}} \cdot p^{\sqrt{8}} = p^{(\sqrt{2}+\sqrt{8})} = p^{(\sqrt{2}+2\sqrt{2})} = p^{3\sqrt{2}}$

(b) $\sqrt{\dfrac{x^{3\pi}}{x^{\pi}}} = \left(\dfrac{x^{3\pi}}{x^{\pi}}\right)^{1/2} = \left(x^{3\pi-\pi}\right)^{1/2} = \left(x^{2\pi}\right)^{1/2} = x^{2\pi(1/2)} = x^{\pi}$

Example 8   Simplify:

(a) $-3^x \cdot 3^{-x}$

(b) $\dfrac{4^p}{2^p}$

(c) $\dfrac{a^{2/3}(a^k)(a^k)}{a}$

(d) $\dfrac{f^{2y}g^y}{f^{3y}}$

Solution   (a) $-3^x \cdot 3^{-x} = -\left(3^x\right)\left(3^{-x}\right) = -\left(3^{x-x}\right) = -\left(3^0\right) = -1$

(b) $\dfrac{4^p}{2^p} = \left(\dfrac{4}{2}\right)^p = 2^p$

(c) $\dfrac{a^{2/3}\left(a^k\right)\left(a^k\right)}{a} = a^{2/3}\left(a^k\right)\left(a^k\right)\left(a^{-1}\right) = a^{((2/3)+k+k-1)} = a^{2k-1/3}$

(d) $\dfrac{f^{2y}g^y}{f^{3y}} = \left(\dfrac{f^2 g}{f^3}\right)^y = \left(\dfrac{g}{f}\right)^y$

Problems for Section A

For Problems 1–16, evaluate without a calculator.

1. 4^3

2. $(-5)^2$

3. 11^2

4. 10^4

5. $(-1)^{12}$

6. $(-1)^{13}$

7. $\dfrac{5^3}{5^2}$

8. $\dfrac{5^3}{5}$

9. $\dfrac{10^8}{10^5}$

10. $\dfrac{6^4}{6^4}$

11. 8^0

12. $\sqrt{4}$

13. $\sqrt{4^2}$

14. $\sqrt{4^3}$

15. $\sqrt{4^4}$

16. $\sqrt{(-4)^2}$

For Problems 17–44, simplify the following expressions.

17. $\sqrt{x^4}$

18. $\sqrt{y^8}$

19. $\sqrt{w^8 z^4}$

20. $\sqrt{x^5 y^4}$

21. $\sqrt{16x^3}$

22. $\sqrt{49w^9}$

23. $\sqrt{25x^3 z^4}$

24. $\sqrt{r^2}$

25. $\sqrt{r^3}$

26. $\sqrt{r^4}$

27. $\sqrt{36t^2}$

28. $\sqrt{64s^7}$

29. $\sqrt{50x^4 y^6}$

30. $\sqrt{48u^{10} v^{12} y^5}$

31. $\sqrt{8m}\sqrt{2m^3}$

32. $\sqrt{6s^2 t^3 v^5}\sqrt{6st^5 v^3}$

33. $(32)^{1/5}$

34. $(16)^{1/2}$

35. $16^{1/4}$

36. $16^{3/4}$

37. $16^{5/4}$

38. $16^{5/2}$

39. 3^{-1}

40. 3^{-2}

41. $3^{-(3/2)}$

42. 25^{-1}

43. 25^{-2}

44. $25^{-(3/2)}$

Evaluate the quantities in Problems 45–59 mentally.

45. $\dfrac{1}{7^{-2}}$

46. $\dfrac{2^7}{2^3}$

47. $(-1)^{445}$

48. -11^2

49. $(-2)\left(3^2\right)$

50. $\left(5^0\right)^3$

51. $2.1\left(10^3\right)$

52. $\sqrt[3]{-125}$

53. $\sqrt{(-4)^2}$

54. $(-1)^3 \sqrt{36}$

55. $(0.04)^{1/2}$

56. $(-8)^{2/3}$

57. $(1/27)^{-1/3}$

58. $(0.125)^{1/3}$

59. $100^{5/2}$

Simplify the expressions in Problems 60–77 and leave without radicals.

60. $(0.1)^2 \left(4xy^2\right)^2$

61. $3\left(3^{x/2}\right)^2$

62. $\left(4L^{2/3}P\right)^{3/2}(P)^{-3/2}$

63. $7\left(5w^{1/2}\right)\left(2w^{1/3}\right)$

64. $\left(S\sqrt{16xt^2}\right)^2$

65. $\sqrt{e^{2x}}$

66. $(3AB)^{-1}\left(A^2 B^{-1}\right)^2$

67. $e^{kt} \cdot e^3 \cdot e$

68. $\sqrt{M+2}\,(2+M)^{3/2}$

69. $\left(3x\sqrt{x^3}\right)^2$

70. $x^e \left(x^e\right)^2$

71. $\left(y^{-2} e^y\right)^2$

72. $\dfrac{4x^{(3\pi+1)}}{x^2}$

73. $\dfrac{4A^{-3}}{(2A)^{-4}}$

74. $\dfrac{a^{n+1} 3^{n+1}}{a^n 3^n}$

75. $\dfrac{12u^3}{3\left(uv^2 w^4\right)^{-1}}$

76. $\left(a^{-1} + b^{-1}\right)^{-1}$

77. $\left(\dfrac{35(2b+1)^9}{7(2b+1)^{-1}}\right)^2$

(Do not expand $(2b+1)^9$.)

If possible, evaluate the quantities in Problems 78–86. Check your answers with a calculator.

78. $(-32)^{3/5}$

79. $-32^{3/5}$

80. $-625^{3/4}$

81. $(-625)^{3/4}$

82. $(-1728)^{4/3}$

83. $64^{-3/2}$

84. $-64^{1/2}$

85. $(-64)^{2/3}$

86. $81^{3/4}$

Determine whether the statements in Problems 87–96 are true or false.

87. $t^3 t^4 = t^{12}$

88. $(u+v)^{-1} = \dfrac{1}{u} + \dfrac{1}{v}$

89. $-4w^2 - 3w^3 = -w^2(4+3w)$

90. $5z^{-4} = \dfrac{1}{5z^4}$

91. $x^2 y^5 = (xy)^{10}$

92. $(p^3)^8 = p^{11}$

93. $\sqrt[3]{-64b^3 c^6} = -4bc^2$

94. $\dfrac{m^8}{2m^2} = \dfrac{1}{2}m^4$

95. $5u^2 + 5u^3 = 10u^5$

96. $(3r)^2 9s^2 = 81r^2 s^2$

B MULTIPLYING ALGEBRAIC EXPRESSIONS

The *distributive property* for real numbers a, b, and c tells us that

$$a(b + c) = ab + ac,$$

and

$$(b + c)a = ba + ca.$$

We use the distributive property and the rules of exponents to multiply algebraic expressions involving parentheses. This process is sometimes referred to as *expanding* the expression.

Example 1   Multiply the following expressions and simplify.

(a) $3x^2\left(x + \dfrac{1}{6}x^{-3}\right)$

(b) $\left((2t)^2 - 5\right)\sqrt{t}$

(c) $2^x(3^x + 2^{x-1})$

Solution   (a) $3x^2\left(x + \dfrac{1}{6}x^{-3}\right) = (3x^2)(x) + (3x^2)\left(\dfrac{1}{6}x^{-3}\right) = 3x^3 + \dfrac{1}{2}x^{-1}$

(b) $\left((2t)^2 - 5\right)\sqrt{t} = (2t)^2(\sqrt{t}) - 5\sqrt{t} = (4t^2)\left(t^{1/2}\right) - 5t^{1/2} = 4t^{5/2} - 5t^{1/2}$

(c) $2^x\left(3^x + 2^{x-1}\right) = (2^x)(3^x) + (2^x)\left(2^{x-1}\right) = (2 \cdot 3)^x + 2^{x+x-1} = 6^x + 2^{2x-1}$

If there are two terms in each factor, then there are four terms in the product:

$$(a + b)(c + d) = a(c + d) + b(c + d) = ac + ad + bc + bd.$$

The following special cases of the above product occur frequently. Learning to recognize their forms aids in factoring.

$$(a + b)(a - b) = a^2 - b^2$$
$$(a + b)^2 = a^2 + 2ab + b^2$$
$$(a - b)^2 = a^2 - 2ab + b^2$$

Example 2   Expand the following and simplify by gathering like terms.

(a) $(5x^2 + 2)(x - 4)$

(b) $(2\sqrt{r} + 2)(4\sqrt{r} - 3)$

(c) $(e^x + 1)(2x + e^{-x})$

(d) $\left(3 - \dfrac{1}{2}x\right)^2$

Solution   (a) $(5x^2 + 2)(x - 4) = (5x^2)(x) + (5x^2)(-4) + (2)(x) + (2)(-4) = 5x^3 - 20x^2 + 2x - 8$

(b) $(2\sqrt{r} + 2)(4\sqrt{r} - 3) = (2)(4)(\sqrt{r})^2 + (2)(-3)(\sqrt{r}) + (2)(4)(\sqrt{r}) + (2)(-3) = 8r + 2\sqrt{r} - 6$

(c) $(e^x + 1)(2x + e^{-x}) = (e^x)(2x) + (e^x)(e^{-x}) + (1)(2x) + (1)(e^{-x})$

$$= 2xe^x + e^{x-x} + 2x + e^{-x}$$
$$= 2xe^x + 1 + 2x + e^{-x}$$

(d) $\left(3 - \dfrac{1}{2}x\right)^2 = 3^2 - 2(3)\left(\dfrac{1}{2}x\right) + \left(-\dfrac{1}{2}x\right)^2 = 9 - 3x + \dfrac{1}{4}x^2$

Problems for Section B

Simplify the expressions in Problems 1–2.

1. $-(x - 3) - 2(5 - x)$ 2. $(x - 5)6 - 5(1 - (2 - x))$

For Problems 3–20, expand each of the following products.

3. $3(x + 2)$	4. $5(x - 3)$	5. $2(3x - 7)$
6. $-4(y + 6)$	7. $12(x + y)$	8. $-7(5x - 8y)$
9. $x(2x + 5)$	10. $3z(2x - 9z)$	11. $-10r(5r + 6rs)$
12. $x(3x - 8) + 2(3x - 8)$	13. $5z(x - 2) - 3(x - 2)$	14. $(x + 1)(x + 3)$
15. $(x - 2)(x + 6)$	16. $(5x - 1)(2x - 3)$	17. $(x + 2)(3x - 8)$
18. $(y + 1)(z + 3)$	19. $(12y - 5)(8w + 7)$	20. $(5z - 3)(x - 2)$

Multiply and write the expressions in Problems 21–30 without parentheses. Gather like terms.

21. $\left(3x - 2x^2\right)(4) + (5 + 4x)(3x - 4)$ 22. $\left(t^2 + 1\right)(50t) - \left(25t^2 + 125\right)(2t)$

23. $P(p - 3q)^2$ 24. $\left(A^2 - B^2\right)^2$ 25. $4(x - 3)^2 + 7$ 26. $-\left(\sqrt{2x} + 1\right)^2$

27. $u\left(u^{-1} + 2^u\right)2^u$ 28. $K(R - r)r^2$ 29. $(x + 3)\left(\dfrac{24}{x} + 2\right)$ 30. $\left(\dfrac{e^x + e^{-x}}{2}\right)^2$

C   FACTORING ALGEBRAIC EXPRESSIONS

To write an expanded expression in factored form, we "un-multiply" the expression. Some techniques for factoring are given in this section. We can check factoring by remultiplying.

Removing a Common Factor

It is sometimes useful to factor out the same factor from each of the terms in an expression. This is basically the distributive law in reverse:

$$ab + ac = a(b + c).$$

One special case is removing a factor of -1, which gives

$$\boxed{-a - b = -(a + b)}$$

Another special case is

$$\boxed{(a - b) = -(b - a)}$$

Example 1   Factor the following:

(a) $\dfrac{2}{3}x^2y + \dfrac{4}{3}xy$ (b) $e^{2x} + xe^x$ (c) $(2p + 1)p^3 - 3p(2p + 1)$ (d) $-\dfrac{s^2t}{8w} - \dfrac{st^2}{16w}$

Solution   (a) $\dfrac{2}{3}x^2y + \dfrac{4}{3}xy = \dfrac{2}{3}xy(x + 2)$

(b) $e^{2x} + xe^x = e^x \cdot e^x + xe^x = e^x(e^x + x)$

(c) $(2p + 1)p^3 - 3p(2p + 1) = (p^3 - 3p)(2p + 1) = p(p^2 - 3)(2p + 1)$

     (Note that the expression $(2p + 1)$ was one of the factors common to both terms.)

(d) $-\dfrac{s^2t}{8w} - \dfrac{st^2}{16w} = -\dfrac{st}{8w}\left(s + \dfrac{t}{2}\right).$

Grouping Terms

Even though all the terms may not have a common factor, we can sometimes factor by first grouping the terms and then removing a common factor.

Example 2   Factor $x^2 - hx - x + h$.

Solution $x^2 - hx - x + h = \left(x^2 - hx\right) - (x - h) = x(x - h) - (x - h) = (x - h)(x - 1)$

Factoring Quadratics

One way to factor quadratics is to mentally multiply out the possibilities.

Example 3   Factor $t^2 - 4t - 12$.

Solution   If the quadratic factors, it will be of the form
$$t^2 - 4t - 12 = (t + ?)(t + ?).$$

We are looking for two numbers whose product is -12 and whose sum is -4. By trying combinations, we find
$$t^2 - 4t - 12 = (t - 6)(t + 2).$$

Example 4   Factor $4 - 2M - 6M^2$.

Solution $4 - 2M - 6M^2 = (2 - 3M)(2 + 2M)$

Perfect Squares and the Difference of Squares

Recognition of the special products $(x + y)^2$, $(x - y)^2$ and $(x + y)(x - y)$ in expanded form is useful in factoring. Reversing the results in the last section, we have
$$a^2 + 2ab + b^2 = (a + b)^2,$$
$$a^2 - 2ab + b^2 = (a - b)^2,$$
$$a^2 - b^2 = (a - b)(a + b).$$

When we can see that terms in an expression we want to factor are squares, it often makes sense to look for one of these forms. The difference of squares identity (the third one listed above) is especially useful.

Example 5   Factor: (a) $16y^2 - 24y + 9$ (b) $25S^2R^4 - T^6$ (b) $x^2(x - 2) + 16(2 - x)$

Solution   (a) $16y^2 - 24y + 9 = (4y - 3)^2$
(b) $25S^2R^4 - T^6 = \left(5SR^2\right)^2 - \left(T^3\right)^2 = \left(5SR^2 - T^3\right)\left(5SR^2 + T^3\right)$
(c) $x^2(x - 2) + 16(2 - x) = x^2(x - 2) - 16(x - 2) = (x - 2)\left(x^2 - 16\right) = (x - 2)(x - 4)(x + 4)$

Example 6   Factor $z^{2/3} - z^{1/3} - 6$.

Solution   Notice that $z^{2/3} - z^{1/3} - 6$ is a quadratic. It is helpful to substitute $u = z^{1/3}$. Then
$$z^{2/3} - z^{1/3} - 6 = u^2 - u - 6$$
$$= (u - 3)(u + 2).$$

Now undo the substitution, that is, let $u = z^{1/3}$. Therefore,
$$z^{2/3} - z^{1/3} - 6 = (z^{1/3} - 3)(z^{1/3} + 2).$$

Problems for Section C

For Problems 1–42, factor completely if possible.

1. $2x + 6$	2. $3y + 15$	3. $5z - 30$
4. $4t - 6$	5. $10w - 25$	6. $u^2 - 2u$
7. $3u^4 - 4u^3$	8. $3u^7 + 12u^2$	9. $12x^3y^2 - 18x$
10. $14r^4s^2 - 21rst$	11. $x^2 + 3x + 2$	12. $x^2 + 3x - 2$
13. $x^2 - 3x + 2$	14. $x^2 - 3x - 2$	15. $x^2 + 2x + 3$
16. $x^2 - 2x - 3$	17. $x^2 - 2x + 3$	18. $x^2 + 2x - 3$
19. $2x^2 + 5x + 2$	20. $3x^2 - x - 4$	21. $2x^2 - 10x + 12$
22. $x^2 + 3x - 28$	23. $x^3 - 2x^2 - 3x$	24. $x^3 + 2x^2 - 3x$
25. $x^2 - 1.4x - 3.92$	26. $a^2x^2 - b^2$	27. $\pi r^2 + 2\pi rh$
28. $B^2 - 10B + 24$	29. $c^2 + x^2 - 2cx$	30. $x^2 + y^2$
31. $a^4 - a^2 - 12$	32. $(t + 3)^2 - 16$	33. $hx^2 + 12 - 4hx - 3x$
34. $r(r - s) - 2(s - r)$	35. $y^2 - 3xy + 2x^2$	36. $x^2e^{-3x} + 2xe^{-3x}$
37. $t^2e^{5t} + 3te^{5t} + 2e^{5t}$	38. $(s + 2t)^2 - 4p^2$	39. $P(1 + r)^2 + P(1 + r)^2r$
40. $x \sin x - \sin x$	41. $\cos^2 x - 2\cos x + 1$	42. $e^{2x} + 2e^x + 1$

D WORKING WITH FRACTIONS

Algebraic fractions are combined in the same manner as numeric fractions–that is, according to the following rules:

$$\frac{a}{c} + \frac{b}{c} = \frac{a + b}{c} \qquad \text{add numerators when denominators are equal, or}$$

$$\frac{a}{b} + \frac{c}{d} = \frac{a(d)}{b(d)} + \frac{(b)c}{(b)d} = \frac{ad + bc}{bd} \qquad \text{find a common denominator}$$

$$\frac{a}{b} \cdot \frac{c}{d} = \frac{ac}{bd} \qquad \text{multiply numerators and denominators for a product}$$

$$\frac{a/b}{c/d} = \frac{a}{b} \cdot \frac{d}{c} = \frac{ad}{bc} \qquad \text{to divide by a fraction, multiply by its reciprocal}$$

(We assume that no denominators are zero.)

To expand the last case, where either the numerator or denominator of a fraction is itself a fraction, remember that

$$\frac{\frac{a}{b}}{c} = \frac{\frac{a}{b}}{\frac{c}{1}} = \frac{a}{b} \cdot \frac{1}{c} = \frac{a}{bc} \qquad \text{and} \qquad \frac{a}{\frac{b}{c}} = \frac{\frac{a}{1}}{\frac{b}{c}} = \frac{a}{1} \cdot \frac{c}{b} = \frac{ac}{b}.$$

In all cases, we cannot divide by zero; that is, $(a/0)$ is not defined. Also the sign of a fraction is changed by changing the sign of the numerator or the denominator (but not both):

$$-\frac{a}{b} = \frac{-a}{b} = \frac{a}{-b}.$$

Example 1   Perform the indicated operations and express the answers as a single fraction.

(a) $\dfrac{4}{x^2+1} - \dfrac{1-x}{x^2+1}$

(b) $\dfrac{M}{M^2 - 2M - 3} + \dfrac{1}{M^2 - 2M - 3}$

(c) $\dfrac{-H^2 P}{17} \cdot \dfrac{\left(PH^{1/3}\right)^2}{K^{-1}}$

(d) $\dfrac{2z/w}{w(w-3z)}$

Solution   (a) $\dfrac{4}{x^2+1} - \dfrac{1-x}{x^2+1} = \dfrac{4 - (1-x)}{x^2+1} = \dfrac{3+x}{x^2+1}$

(b) $\dfrac{M}{M^2 - 2M - 3} + \dfrac{1}{M^2 - 2M - 3} = \dfrac{M+1}{(M^2 - 2M - 3)} = \dfrac{M+1}{(M+1)(M-3)} = \dfrac{1}{M-3}$

(c) $\dfrac{-H^2 P}{17} \cdot \dfrac{\left(PH^{1/3}\right)^2}{K^{-1}} = \dfrac{-H^2 P \left(P^2 H^{2/3}\right)}{17 K^{-1}} = -\dfrac{H^{8/3} P^3 K}{17}$

(d) $\dfrac{2z/w}{w(w-3z)} = \dfrac{2z}{w} \cdot \dfrac{1}{w(w-3z)} = \dfrac{2z}{w^2(w-3z)}$

Example 2   Simplify the following expressions:

(a) $2x^{-1/2} + \dfrac{\sqrt{x}}{3}$

(b) $2\sqrt{t+3} + \dfrac{1-2t}{\sqrt{t+3}}$

Solution   (a) $2x^{-1/2} + \dfrac{\sqrt{x}}{3} = \dfrac{2}{\sqrt{x}} + \dfrac{\sqrt{x}}{3} = \dfrac{2\cdot 3 + \sqrt{x}\sqrt{x}}{3\sqrt{x}} = \dfrac{6+x}{3\sqrt{x}} = \dfrac{6+x}{3x^{1/2}}$

(b) $2\sqrt{t+3} + \dfrac{1-2t}{\sqrt{t+3}} = \dfrac{2\sqrt{t+3}}{1} + \dfrac{1-2t}{\sqrt{t+3}}$

$= \dfrac{2\sqrt{t+3}\sqrt{t+3} + 1 - 2t}{\sqrt{t+3}}$

$= \dfrac{2(t+3) + 1 - 2t}{\sqrt{t+3}}$

$= \dfrac{7}{\sqrt{t+3}} = \dfrac{7}{(t+3)^{1/2}}$

Finding a Common Denominator

We can multiply (or divide) both the numerator and denominator of a fraction by the same non-zero number without changing the fraction's value. This is equivalent to multiplying by a factor of $+1$. We are using this rule when we add or subtract fractions with different denominators. For example, to add $\dfrac{x}{3a} + \dfrac{1}{a}$, we multiply $\dfrac{1}{a} \cdot \dfrac{3}{3} = \dfrac{3}{3a}$. Then

$$\frac{x}{3a} + \frac{1}{a} = \frac{x}{3a} + \frac{3}{3a} = \frac{x+3}{3a}.$$

Example 3   Perform the indicated operations:

(a) $3 - \dfrac{1}{x-1}$

(b) $\dfrac{2}{x^2+x} + \dfrac{x}{x+1}$

Solution   (a) $3 - \dfrac{1}{x-1} = 3\dfrac{(x-1)}{(x-1)} - \dfrac{1}{x-1} = \dfrac{3(x-1)-1}{x-1} = \dfrac{3x-3-1}{x-1} = \dfrac{3x-4}{x-1}$

(b) $\dfrac{2}{x^2+x} + \dfrac{x}{x+1} = \dfrac{2}{x(x+1)} + \dfrac{x}{x+1} = \dfrac{2}{x(x+1)} + \dfrac{x(x)}{(x+1)(x)} = \dfrac{2+x^2}{x(x+1)}$

Note: We can multiply (or divide) the numerator and denominator by the same non-zero number because this is the same as multiplying by a factor of $+1$, and multiplying by a factor of 1 does not change the value of the expression. However, we cannot perform any other operation that would change the value of the expression. For example, we cannot *add* the same number to the numerator and denominator of a fraction *nor* can we square both, take the logarithm of both, etc., without changing the fraction.

Reducing Fractions: Canceling

We can reduce a fraction when we have the same (non-zero) factor in both the numerator and the denominator. For example,

$$\frac{ac}{bc} = \frac{a}{b} \cdot \frac{c}{c} = \frac{a}{b} \cdot 1 = \frac{a}{b}.$$

Example 4   Reduce the following fractions (if possible).

(a) $\dfrac{2x}{4y}$

(b) $\dfrac{2+x}{2+y}$

(c) $\dfrac{5n-5}{1-n}$

(d) $\dfrac{x^2(4-2x) - (4x-x^2)(2x)}{x^4}$

Solution   (a) $\dfrac{2x}{4y} = \dfrac{2}{2} \cdot \dfrac{x}{2y} = \dfrac{x}{2y}$

(b) $\dfrac{2+x}{2+y}$ cannot be reduced further.

(c) $\dfrac{5n-5}{1-n} = \dfrac{5(n-1)}{(-1)(n-1)} = -5$

(d) $\dfrac{x^2(4-2x) - \left(4x-x^2\right)(2x)}{x^4} = \dfrac{x^2(4-2x) - (4-x)(2x^2)}{x^4}$

$$= \dfrac{[(4-2x) - 2(4-x)]}{x^2}\left(\dfrac{x^2}{x^2}\right)$$

$$= \dfrac{4-2x-8+2x}{x^2} = \dfrac{-4}{x^2}$$

Complex Fractions

A *complex fraction* is a fraction whose numerator or denominator (or both) contains one or more fractions. To simplify a complex fraction, we change the numerator and denominator to single fractions and then divide.

Example 5   Write the following as simple fractions in reduced form.

(a) $\dfrac{\dfrac{1}{x+h} - \dfrac{1}{x}}{h}$

(b) $\dfrac{a+b}{a^{-2} - b^{-2}}$

Solution (a) $\dfrac{\dfrac{1}{x+h}-\dfrac{1}{x}}{h}=\dfrac{\dfrac{x-(x+h)}{x(x+h)}}{h}=\dfrac{\dfrac{-h}{x(x+h)}}{\dfrac{h}{1}}=\dfrac{-h}{x(x+h)}\cdot\dfrac{1}{h}=\dfrac{-1}{x(x+h)}\dfrac{(h)}{(h)}=\dfrac{-1}{x(x+h)}$

(b) $\dfrac{a+b}{a^{-2}-b^{-2}}=\dfrac{a+b}{\dfrac{1}{a^2}-\dfrac{1}{b^2}}=\dfrac{a+b}{\dfrac{b^2-a^2}{a^2b^2}}=\dfrac{a+b}{1}\cdot\dfrac{a^2b^2}{b^2-a^2}=\dfrac{(a+b)(a^2b^2)}{(b+a)(b-a)}=\dfrac{a^2b^2}{b-a}$

Splitting Expressions

We can reverse the rule for adding fractions to split up an expression into two fractions,

$$\frac{a+b}{c}=\frac{a}{c}+\frac{b}{c}.$$

Example 6 Split $\dfrac{3x^2+2}{x^3}$ into two reduced fractions.

Solution $\dfrac{3x^2+2}{x^3}=\dfrac{3x^2}{x^3}+\dfrac{2}{x^3}=\dfrac{3}{x}+\dfrac{2}{x^3}$

Sometimes we can alter the form of the fraction even further if we can create a duplicate of the denominator within the numerator. This technique is useful when graphing some rational functions. For example, we may rewrite the fraction $\dfrac{x+3}{x-1}$ by creating a factor of $(x-1)$ within the numerator. To do this, we write

$$\frac{x+3}{x-1}=\frac{x-1+1+3}{x-1}$$

which can be written as

$$\frac{(x-1)+4}{x-1}.$$

Then, splitting this fraction, we have

$$\frac{x+3}{x-1}=\frac{x-1}{x-1}+\frac{4}{x-1}=1+\frac{4}{x-1}.$$

Problems for Section D

For Problems 1–23, perform the following operations. Express answers in reduced form.

1. $\dfrac{3}{5}+\dfrac{4}{7}$ 2. $\dfrac{7}{10}-\dfrac{2}{15}$ 3. $\dfrac{1}{2x}-\dfrac{2}{3}$ 4. $\dfrac{6}{7y}+\dfrac{9}{y}$

5. $\dfrac{-2}{yz}+\dfrac{4}{z}$ 6. $\dfrac{-2z}{y}+\dfrac{4}{y}$ 7. $\dfrac{2}{x^2}-\dfrac{3}{x}$ 8. $\dfrac{6}{y}+\dfrac{7}{y^3}$

9. $\dfrac{\frac{3}{4}}{\frac{7}{20}}$ 10. $\dfrac{\frac{5}{6}}{15}$ 11. $\dfrac{\frac{3}{x}}{\frac{x^2}{6}}$ 12. $\dfrac{\frac{3}{x}}{\frac{6}{x^2}}$

13. $\dfrac{13}{x-1}+\dfrac{14}{2x-2}$ 14. $\dfrac{14}{x-1}+\dfrac{13}{2x-2}$ 15. $\dfrac{4z}{x^2y}-\dfrac{3w}{xy^4}$

16. $\dfrac{10}{y-2}+\dfrac{3}{2-y}$ 17. $\dfrac{8y}{y-4}+\dfrac{32}{y-4}$ 18. $\dfrac{8y}{y-4}+\dfrac{32}{4-y}$

19. $\dfrac{\dfrac{1}{x} - \dfrac{2}{x^2}}{\dfrac{2x-4}{x^5}}$

20. $\dfrac{9}{x^2 + 5x + 6} + \dfrac{12}{x+3}$

21. $\dfrac{8}{3x^2 - x - 4} - \dfrac{9}{x+1}$

22. $\dfrac{5}{(x-2)^2(x+1)} - \dfrac{18}{(x-2)}$

23. $\dfrac{15}{(x-3)^2(x+5)} + \dfrac{7}{(x-3)(x+5)^2}$

In Problems 24–35, perform the specified operations. Express answers in reduced form.

24. $\dfrac{3}{x-4} - \dfrac{2}{x+4}$

25. $\dfrac{x^2}{x-1} - \dfrac{1}{1-x}$

26. $\dfrac{1}{2r+3} + \dfrac{3}{4r^2 + 6r}$

27. $u + a + \dfrac{u}{u+a}$

28. $\dfrac{1}{\sqrt{x}} - \dfrac{1}{(\sqrt{x})^3}$

29. $\dfrac{1}{e^{2x}} + \dfrac{1}{e^x}$

30. $\dfrac{a+b}{2} \cdot \dfrac{8x+2}{b^2 - a^2}$

31. $\dfrac{0.07}{M} + \dfrac{3}{4}M^2$

32. $\dfrac{1}{r_1} + \dfrac{1}{r_2} + \dfrac{1}{r_3}$

33. $\dfrac{8y}{y-4} - \dfrac{32}{y-4}$

34. $\dfrac{a}{a^2 - 9} + \dfrac{1}{a-3}$

35. $\dfrac{x^3}{x-4} \Big/ \dfrac{x^2}{x^2 - 2x - 8}$

In Problems 36–49, simplify, if possible.

36. $\dfrac{\dfrac{1}{x+y}}{x+y}$

37. $\dfrac{\dfrac{w+2}{2}}{w+2}$

38. $\dfrac{\dfrac{1}{(x+h)^2} - \dfrac{1}{x^2}}{h}$

39. $\dfrac{a^{-2} + b^{-2}}{a^2 + b^2}$

40. $\dfrac{a^2 - b^2}{a^2 + b^2}$

41. $\dfrac{[4 - (x+h)^2] - [4 - x^2]}{h}$

42. $\dfrac{b^{-1}(b - b^{-1})}{b+1}$.

43. $\dfrac{1 - a^{-2}}{1 + a^{-1}}$.

44. $\dfrac{x^{-1} + x^{-2}}{1 - x^{-2}}$.

45. $p - \dfrac{q}{\dfrac{p}{q} + \dfrac{q}{p}}$

46. $\dfrac{\dfrac{3}{xy} - \dfrac{5}{x^2 y}}{\dfrac{6x^2 - 7x - 5}{x^4 y^2}}$

47. $\dfrac{\frac{1}{x}(3x^2) - (\ln x)(6x)}{(3x^2)^2}$

48. $\dfrac{2x(x^3 + 1)^2 - x^2(2)(x^3 + 1)(3x^2)}{[(x^3 + 1)^2]^2}$

49. $\dfrac{\frac{1}{2}(2x - 1)^{-1/2}(2) - (2x-1)^{1/2}(2x)}{(x^2)^2}$

In Problems 50–55, split into a sum or difference of reduced fractions.

50. $\dfrac{26x + 1}{2x^3}$

51. $\dfrac{\sqrt{x} + 3}{3\sqrt{x}}$

52. $\dfrac{6l^2 + 3l - 4}{3l^4}$

53. $\dfrac{7 + p}{p^2 + 11}$

54. $\dfrac{\frac{1}{3}x - \frac{1}{2}}{2x}$

55. $\dfrac{t^{-1/2} + t^{1/2}}{t^2}$

In Problems 56–61, rewrite in the form $1 + (A/B)$.

56. $\dfrac{x-2}{x+5}$

57. $\dfrac{q-1}{q-4}$

58. $\dfrac{R+1}{R}$

59. $\dfrac{3 + 2u}{2u + 1}$

60. $\dfrac{\cos x + \sin x}{\cos x}$

61. $\dfrac{1 + e^x}{e^x}$

Determine whether the statements in Problems 62–67 are true or false.

62. $\dfrac{a+c}{a} = 1 + c$

63. $\dfrac{rs - s}{s} = r - 1$

64. $\dfrac{y}{y+z} = 1 + \dfrac{y}{z}$

65. $\dfrac{2u^2 - w}{u^2 - w} = 2$

66. $\dfrac{x^2 yz}{2x^2 y} = \dfrac{z}{2}$

67. $x^{5/3} - 3x^{2/3} = \dfrac{x^2 - 3x}{x^{1/3}}$

E CHANGING THE FORM OF EXPRESSIONS

Rearranging Coefficients and Exponents

Changing the form of an expression can often be useful. Manipulations like the following occur frequently:

- $\dfrac{x}{2} = \left(\dfrac{1}{2}\right) x$

- $\dfrac{3}{4(2r + 1)^{10}} = \dfrac{3}{4}(2r + 1)^{-10}$

- $2^{-n} = \left(\dfrac{1}{2}\right)^{n}$

- $2^{x+3} = 2^{x} \cdot 2^{3} = 8(2^{x})$

- $\dfrac{3x + \sqrt{2x}}{\sqrt{x}} = \dfrac{3x}{\sqrt{x}} + \dfrac{\sqrt{2x}}{\sqrt{x}} = \dfrac{3x}{\sqrt{x}} + \dfrac{\sqrt{2}\sqrt{x}}{\sqrt{x}} = 3x^{(1-1/2)} + \sqrt{2} = 3x^{1/2} + \sqrt{2}$

Completing the Square

Another example of changing the form of an expression is the conversion of $ax^2 + bx + c$ into the form $a(x - h)^2 + k$. We make this conversion by *completing the square*, a method for producing a perfect square within a quadratic expression. A perfect square is an expression of the form:

$$(x + n)^2 = x^2 + 2nx + n^2$$

for some number n.

In order to complete the square in a given expression, we must find that number n. Observe that when a perfect square is multiplied out, the coefficient of x is 2 times the number n. Therefore, we can find n by dividing the coefficient of the x term by 2. Once we know n, then we know that the constant term in the perfect square must be n^2. In summary:

> **To complete the square** in the expression $x^2 + bx + c$, divide the coefficient of x by 2, giving $b/2$. Then add and subtract $(b/2)^2 = b^2/4$ and factor the perfect square:
>
> $$x^2 + bx + c = \left(x + \frac{b}{2}\right)^2 - \frac{b^2}{4} + c.$$
>
> To complete the square in the expression $ax^2 + bx + c$, factor out a first.

Example 1   Rewrite $x^2 - 10x + 4$ in the form $a(x - h)^2 + k$.

Solution   Notice that half of the coefficient of x is $\frac{1}{2}(-10) = -5$. Squaring -5 gives 25. We have

$$x^2 - 10x + 4 = x^2 - 10x + (25 - 25) + 4$$
$$= (x^2 - 10x + 25) - 25 + 4$$
$$= (x - 5)^2 - 21.$$

Thus, $a = +1$, $h = +5$, and $k = -21$.

Example 2   Complete the square in the formula $h(x) = 5x^2 + 30x - 10$.

Solution   We first factor out 5:

$$h(x) = 5(x^2 + 6x - 2).$$

Now we complete the square in the expression $x^2 + 6x - 2$.

Step 1: Divide the coefficient of x by 2, giving 3.

Step 2: Square the result: $3^2 = 9$.

Step 3: Add the result after the x term, then subtract it:

$$h(x) = 5(\underbrace{x^2 + 6x + 9}_{\text{Perfect square}} - 9 - 2).$$

Step 4: Factor the perfect square and simplify the rest:

$$h(x) = 5\left((x + 3)^2 - 11\right).$$

Now that we have completed the square, we can multiply by the 5:

$$h(x) = 5(x + 3)^2 - 55.$$

The Quadratic Formula

We derive a general formula for the zeros of $q(x) = ax^2 + bx + c$, with $a \neq 0$, by completing the square. To find the zeros, set $q(x) = 0$:

$$ax^2 + bx + c = 0.$$

Before we complete the square, we factor out the coefficient of x^2:

$$a\left(x^2 + \frac{b}{a}x + \frac{c}{a}\right) = 0.$$

Since $a \neq 0$, we can divide both sides by a:

$$x^2 + \frac{b}{a}x + \frac{c}{a} = 0.$$

To complete the square, we add and then subtract $\left((b/a)/2\right)^2 = b^2/(4a^2)$:

$$\underbrace{x^2 + \frac{b}{a}x + \frac{b^2}{4a^2}}_{\text{Perfect square}} - \frac{b^2}{4a^2} + \frac{c}{a} = 0.$$

We factor the perfect square and simplify the constant term, giving:

$$\left(x + \frac{b}{2a}\right)^2 - \left(\frac{b^2 - 4ac}{4a^2}\right) = 0 \qquad \text{since } \frac{-b^2}{4a^2} + \frac{c}{a} = \frac{-b^2}{4a^2} + \frac{4ac}{4a^2} = -\left(\frac{b^2 - 4ac}{4a^2}\right)$$

$$\left(x + \frac{b}{2a}\right)^2 = \frac{b^2 - 4ac}{4a^2} \qquad \text{adding } \frac{b^2 - 4ac}{4a^2} \text{ to both sides}$$

$$x + \frac{b}{2a} = \pm\sqrt{\frac{b^2 - 4ac}{4a^2}} = \frac{\pm\sqrt{b^2 - 4ac}}{2a} \qquad \text{taking the square root}$$

$$x = \frac{-b}{2a} \pm \frac{\sqrt{b^2 - 4ac}}{2a} \qquad \text{subtracting } b/2a$$

$$x = \frac{-b \pm \sqrt{b^2 - 4ac}}{2a}.$$

Problems for Section E

In Problems 1–12, rewrite each expression as a sum of powers of the variable.

1. $3x^2 \left(x^{-1}\right) + \dfrac{1}{2x} + x^2 + \dfrac{1}{5}$

2. $10 \left(3q^2 - 1\right) (6q)$

3. $\left(y - 3y^{-2}\right)^2$

4. $x(x + x^{-1})^2$

5. $2P^2(P) + (9P)^{1/2}$

6. $\dfrac{(1 + 3\sqrt{t})^2}{2}$

7. $\dfrac{18 + x^2 - 3x}{-6}$

8. $\left(\dfrac{1}{N} - N\right)^2$

9. $\dfrac{-3(4x - x^2)}{7x}$

10. $\dfrac{x^4 + 2x + 1}{2\sqrt{x}}$

11. $\dfrac{ax^2 + bx + c}{x}$

12. $\sqrt{1 + p^2} \cdot (1 + p^2)^{3/2}$

In Problems 13–18, rewrite each expression in the form $a(bx + c)^n$.

13. $\dfrac{12}{\sqrt{3x + 1}}$

14. $\dfrac{250\sqrt[3]{10 - s}}{0.25}$

15. $0.7(x - 1)^3(1 - x)$

16. $\dfrac{1}{2(x^2 + 1)^3}$

17. $4(6R + 2)^3(6)$

18. $\sqrt{\dfrac{28x^2 - 4\pi x}{x}}$

In Problems 19–30, rewrite each expression in the form ab^x or ab^t.

19. $\dfrac{1^x}{2^x}$

20. $\dfrac{1}{2^x}$

21. $10{,}000(1 - 0.24)^t$

22. e^{2x+1}

23. $2 \cdot 3^{-x}$

24. $2^x \cdot 3^{x-1}$

25. $16^{t/2}$

26. $\dfrac{e^3}{e^{-x+4}}$

27. $\dfrac{5^x}{-3^x}$

28. $\dfrac{e \cdot e^x}{0.2}$

29. $\left(10e^t\right)^2$

30. $\dfrac{e^{kt}}{1/A}$

For Problems 31–36, rewrite each expression as a sum of positive powers of the variable.

31. $x^2 + x^{-3}$

32. $\dfrac{5}{x^{-2}} + x + 1$

33. $(2y^2 - 5)3y$

34. $y^3(y + 1)^2$

35. $z^4(1 + z^{-2})^2$

36. $z^5(z^{-3}) - 2z^{-1} + 6$

For Problems 37–42, complete the square for each expression.

37. $x^2 + 8x$

38. $y^2 - 12y$

39. $w^2 + 7w$

40. $2r^2 + 20r$

41. $s^2 + 6s - 8$

42. $3t^2 + 24t - 13$

In Problems 43–46, rewrite each expression in the form $a(x - h)^2 + k$.

43. $x^2 - 2x - 3$ 44. $10 - 6x + x^2$ 45. $-x^2 + 6x - 2$ 46. $3x^2 - 12x + 13$

In Problems 47–52, simplify and rewrite using only positive exponents .

47. $-3 \left(x^2 + 7\right)^{-4} (2x)$

48. $-2 (1 + 3^x)^{-3} (\ln 2) (2^x)$

49. $-(\sin(\pi t))^{-1}(- \cos(\pi t))\pi$

50. $-(\tan z)^{-2} \left(\dfrac{1}{\cos^2 z}\right)$

51. $\dfrac{-e^x \left(x^2\right) - e^{-x}2x}{\left(x^2\right)^2}$

52. $-x^{-2}(\ln x) + x^{-1} \left(\dfrac{1}{x}\right)$

In Problems 53–60, simplify and rewrite in radical form.

53. $(5x)^{1/2}$

54. $(3x - 2)^{-(1/2)}$

55. $6y(4z - 5)^{1/3}$

56. $(27z)^{5/3}(3w - 1)^{-(1/2)}$

57. $\frac{1}{2}(x^2 + 16)^{-1/2}(2x)$

58. $\frac{1}{2}(x^2 + 10x + 1)^{-1/2}(2x + 10)$

59. $\frac{1}{2}(\sin(2x))^{-1/2}(2)\cos(2x)$

60. $\frac{2}{3}\left(x^2 - e^{3x}\right)^{-5/3}\left(3x^2 - e^{3x}(3)\right)$

F SOLVING EQUATIONS

Solving in Your Head

When we first look at an equation, we see if we can guess the answer by mentally trying numbers. Consider the following equations and mental solutions.

$\sqrt{x} - 4 = 0$	*"I'm looking for a number whose square root is 4."*	$\sqrt{16} - 4 = 0$
$2x - 3 = 0$	*"What value can I use for x that will give $2x = 3$?"*	$2\left(\frac{3}{2}\right) - 3 = 0$
$\frac{3}{x} + 1 = 0$	*"Three divided by what number gives -1?"*	$\frac{3}{(-3)} + 1 = 0$
$e^x = 1$	*"What exponent can I use with the base e to get 1?"*	$e^{(0)} = 1$
$x^2(x + 2) = 0$	*"What number makes each factor zero?"*	$0^2 = 0$ and $(-2) + 2 = 0$
$\frac{(x + 1)(3 - x)}{(1 - x)^2} = 0$	*"What numbers make the numerator equal to 0?"*	$(-1) + 1 = 0$ and $3 - (3) = 0$
$1 - \sin x = 0$	*"What numbers make the sine value equal 1?"*	$1 - \sin\left(\frac{\pi}{2}\right) = 0$

Notice that the last equation, $\sin x = 1$, has many other solutions as well.

Solving Exactly Versus Solving Approximately

Some equations can be solved exactly, often by using algebra. For example, the equation $7x - 1 = 0$ has the exact solution $x = 1/7$. Other equations can be hard or even impossible to solve exactly. However, it is often possible, and sometimes easier, to find an approximate solution to an equation by using a graph or a numerical method on a calculator. The equation $7x - 1 = 0$ has the approximate solution $x \approx 0.14$ (since $1/7 = 0.142857\ldots$). We use the sign $\approx$, meaning approximately equal, when we want to emphasize that we are making an approximation.

Example 1   Give exact and approximate solutions to $x^2 = 3$.

Solution   The exact solutions are $x = \pm\sqrt{3}$; approximate ones are $x \approx \pm 1.73$, or $x \approx \pm 1.732$, or $x \approx \pm 1.73205$. (since $\sqrt{3} = 1.732050808\ldots$). Notice that the equation $x^2 = 3$ has only two exact solutions, but many possible approximate solutions, depending on how much accuracy is required.

Operations on Equations

For more complicated equations, additional steps may be needed in order to find a solution.

Linear Equations

To solve a linear equation, we clear any parentheses and then isolate the variable.

Example 2   Solve $3 - [5.4 + 2(4.3 - x)] = 2 - (0.3x - 0.8)$ for x.

Solution   We begin by clearing the innermost parentheses on each side. This gives
$$3 - [5.4 + 8.6 - 2x] = 2 - 0.3x + 0.8.$$

Then
$$3 - 14 + 2x = 2 - 0.3x + 0.8$$
$$2.3x = 13.8,$$
$$x = 6.$$

Example 3   Solve for q if $p^2 q + r(-q - 1) = 4(p + r)$.

Solution

$$p^2 q - rq - r = 4p + 4r$$
$$p^2 q - rq = 4p + 5r$$
$$q(p^2 - r) = 4p + 5r$$
$$q = \frac{4p + 5r}{p^2 - r}$$

Solving by Factoring

Some equations can be put into factored form such that the product of the factors is zero. Then we solve by using the fact that if $a \cdot b = 0$, then either a or b (or both) is zero.

Example 4   Solve $(x + 1)(x + 3) = 15$ for x.

Solution   Do not make the mistake of setting $x + 1 = 15$ and $x + 3 = 15$. It is not true that $a \cdot b = 15$ means that $a = 15$ or $b = 15$ (or both). (Although it is true that if $a \cdot b = 0$, then $a = 0$ or $b = 0$, or both.) So, we must expand the left-hand side and set the equation equal to zero:
$$x^2 + 4x + 3 = 15,$$
$$x^2 + 4x - 12 = 0.$$

Then, factoring gives
$$(x - 2)(x + 6) = 0.$$

Thus $x = 2$ and $x = -6$ are solutions.

Example 5   Solve $2(x + 3)^2 = 5(x + 3)$.

Solution   You might be tempted to divide both sides by $(x + 3)$. However, if you do this you will overlook one of the solutions. Instead, write
$$2(x + 3)^2 - 5(x + 3) = 0$$
$$(x + 3)\left(2(x + 3) - 5\right) = 0$$
$$(x + 3)(2x + 6 - 5) = 0$$
$$(x + 3)(2x + 1) = 0.$$

Thus, $x = -\dfrac{1}{2}$ and $x = -3$ are solutions.

Example 6   Solve $e^x + xe^x = 0$.

Solution   Factoring gives $e^x(1+x) = 0$. Since e^x is never zero, $x = -1$ is the only solution.

Using the Quadratic Formula

If an equation is in the form $ax^2 + bx + c = 0$, we can use the quadratic formula to find the solutions,

$$x = \frac{-b + \sqrt{b^2 - 4ac}}{2a} \quad \text{or} \quad x = \frac{-b - \sqrt{b^2 - 4ac}}{2a}.$$

provided that $\sqrt{b^2 - 4ac}$ is a real number.

Example 7   Solve $11 + 2x = x^2$.

Solution   The equation is
$$-x^2 + 2x + 11 = 0.$$

The expression on the left does not factor using integers, so we use

$$x = \frac{-2 + \sqrt{4 - 4(-1)(11)}}{2(-1)} = \frac{-2 + \sqrt{48}}{-2} = \frac{-2 + \sqrt{16 \cdot 3}}{-2} = \frac{-2 + 4\sqrt{3}}{-2} = 1 - 2\sqrt{3},$$

$$x = \frac{-2 - \sqrt{4 - 4(-1)(11)}}{2(-1)} = \frac{-2 - \sqrt{48}}{-2} = \frac{-2 - \sqrt{16 \cdot 3}}{-2} = \frac{-2 - 4\sqrt{3}}{-2} = 1 + 2\sqrt{3}.$$

The exact solutions are $x = 1 - 2\sqrt{3}$ and $x = 1 + 2\sqrt{3}$.

The decimal approximations to these numbers $x = 1 - 2\sqrt{3} = -2.46$ and $x = 1 + 2\sqrt{3} = 4.46$ are approximate solutions to this equation. The approximate solutions could be found directly from a graph or calculator.

Fractional Equations

If an equation involves fractions, we can eliminate the fractions by multiplying both sides of the equation by the least common denominator and then solving as before. However, we must check for extraneous solutions at the end.

Example 8   Solve $\dfrac{2x}{x+1} - 3 = \dfrac{2}{x^2 + x}$ for x.

Solution   To look for a common denominator, factor $x^2 + x$. Then we have
$$\frac{2x}{x+1} - 3 = \frac{2}{x(x+1)}.$$

Multiplying both sides by $x(x+1)$ gives

$$x(x+1)\left(\frac{2x}{x+1} - 3\right) = 2$$
$$2x^2 - 3x(x+1) = 2$$
$$2x^2 - 3x^2 - 3x = 2$$
$$-x^2 - 3x - 2 = 0,$$

or
$$x^2 + 3x + 2 = 0.$$

Factoring, we have
$$(x+2)(x+1) = 0,$$

so $x = -2$ and $x = -1$ are potential solutions. However, the original equation is not defined for $x = -1$, so $x = -1$ is extraneous. The only solution is $x = -2$.

Example 9   Solve for P_2 if $\dfrac{1}{P_1} + \dfrac{1}{P_2} = \dfrac{1}{P_3}$.

Solution   Multiplying by $P_1 P_2 P_3$ gives

$$P_2 P_3 + P_1 P_3 = P_1 P_2.$$

We are solving for P_2, so we move all P_2 terms to one side

$$P_2 P_3 - P_1 P_2 = -P_1 P_3.$$

Factoring out P_2 gives

$$P_2(P_3 - P_1) = -P_1 P_3$$

$$P_2 = \frac{-P_1 P_3}{P_3 - P_1} = \frac{P_1 P_3}{P_1 - P_3}.$$

Note that we assume that none of P_1, P_2, or $P_3 = 0$. Also, we assume $P_3 \neq P_1$, or else the denominator of the solution is 0.

Radical Equations

We solve radical equations by raising both sides of the equation to the same power. The principle is that if $a = b$, then $a^r = b^r$. Again, we must check for extraneous solutions.

Example 10   Solve $2\sqrt{x} = x - 3$.

Solution   Squaring both sides gives

$$(2\sqrt{x})^2 = (x - 3)^2$$

$$4x = x^2 - 6x + 9.$$

Then

$$x^2 - 10x + 9 = 0$$

$$(x - 1)(x - 9) = 0,$$

so $x = 1$ and $x = 9$ are potential solutions. Since $x = 1$ is not a solution of the original equation, the only solution is $x = 9$.

Example 11   Solve $4 = x^{-1/2}$.

Solution   Taking both sides to the -2 power:

$$(4)^{-2} = (x^{-1/2})^{-2}$$

$$(4)^{-2} = x,$$

$$x = \frac{1}{16}.$$

Exponential Equations

When the variable we want to solve for is in the exponent, we again "do the same thing" to both sides of the equation. This time we take logarithms using the property that if $a = b$ then $\log a = \log b$, (provided $a, b > 0$). Note that we could also use the natural logarithm. The logarithm rule that $\log(m^x) = x \log m$ enables us to solve the equation.

Example 12   Solve $10^{2x+1} = 3$ for x.

Solution   Taking logs,

$$\log 10^{2x+1} = \log 3$$
$$2x + 1 = \log 3, \qquad \text{Since } \log 10^P = P$$
$$x = \frac{(\log 3) - 1}{2}.$$

The solution $x = ((\log 3) - 1)/2$ is exact; converting the solution to $x \approx -0.2614$ gives an approximate solution.

Example 13   Solve $2e^x = 12$.

Solution   We have

$$e^x = 6 \qquad \text{Dividing both sides by 2}$$
$$\ln e^x = \ln 6 \qquad \text{Taking the natural log of both sides}$$
$$x = \ln 6 \qquad \text{Since } \ln e^P = P.$$

Example 14   Solve $1.07 = 4^{-x}$.

Solution   Taking natural logs of both sides

$$\ln 1.07 = \ln(4^{-x})$$
$$\ln 1.07 = (-x)(\ln 4)$$
$$x = -\frac{\ln 1.07}{\ln 4}$$

When applying the logarithm function to both sides of an equation, we may use log base 10 or log base e (the natural log). In Example 12, log base 10 is most convenient, as is the natural log for Example 13. The answers $x = ((\log 3) - 1)/2$ and $x = \ln 6$ are *exact* solutions. In order to compare answers in different forms or use these solutions in a numerical computation, we use a calculator to find a decimal approximation.

Example 15   Solve for x in Examples 13 and 14 by taking the logarithm base 10 (rather than the natural logarithm) of both sides of the equations. Determine decimal approximations (to the accuracy of your calculator) for the solutions you find and for the solutions given in Examples 13 and 14. Do the answers agree?

Solution   In Example 13, we have

$$2e^x = 12$$
$$e^x = 6$$
$$\log e^x = \log 6$$
$$x \log e = \log 6$$
$$x = \frac{\log 6}{\log e} \approx 1.791759469.$$

The exact answer $x = \ln 6$ given in Example 13 is in a simpler form, but $\ln 6 \approx 1.791759469$, an approximate solution. Yes, the answers agree (to at least nine decimal places)

In Example 14,

$$1.07 - 4^{-x}$$
$$\log 1.07 = \log 4^{-x}$$
$$\log 1.07 = -x \log 4$$
$$x = -\frac{\log 1.07}{\log 4} \approx -0.0488053983.$$

This agrees with the exact solution $x = -\dfrac{\ln 1.07}{\ln 4} \approx -0.0488053983$.

The solutions $-\log 1.07/\log 4$ and $-\ln 1.07/\ln 4$ are two different ways of writing the exact solution, while -0.0488053983 is an approximate solution.

Example 16 Solve for t in the equation $P = P_0 e^{kt}$.

Solution Dividing both sides by P_0, we have

$$\frac{P}{P_0} = e^{kt},$$

so

$$\ln\left(\frac{P}{P_0}\right) = \ln e^{kt},$$

$$\ln\left(\frac{P}{P_0}\right) = kt.$$

Thus, $t = \dfrac{1}{k}\ln\left(\dfrac{P}{P_0}\right)$.

Problems for Section F

For Problems 1–23, solve each of the following equations for the variable.

1. $3x = 15$
2. $-2y = 12$
3. $4z = 22$
4. $x + 3 = 10$
5. $y - 5 = 21$
6. $w - 23 = -34$
7. $2x - 5 = 13$
8. $7 - 3y = -14$
9. $13t + 2 = 47$
10. $2x - 5 = 4x - 9$
11. $17 - 28y = 13y + 24$
12. $x^2 + 7x + 6 = 0$
13. $y^2 - 5y - 6 = 0$
14. $2w^2 + w - 10 = 0$
15. $4s^2 + 3s - 15 = 0$
16. $\dfrac{2}{x} + \dfrac{3}{2x} = 8$
17. $\dfrac{3}{x-1} + 1 = 5$
18. $\sqrt{y-1} = 13$
19. $\sqrt{5y+3} = 7$
20. $\sqrt{2x-1} + 3 = 9$
21. $\dfrac{21}{z-5} - \dfrac{13}{z^2 - 5z} = 3$
22. $-16t^2 + 96t + 12 = 60$
23. $2(r+5) - 3 = 3(r-8) + 21$

For Problems 24–35, solve each of the following equations for the specified variable.

24. $A = l \cdot w$, for l.
25. $C = 2\pi r$, for r.
26. $I = Prt$, for P.
27. $C = \dfrac{5}{9}(F - 32)$, for F.
28. $e^{kt} = 1.0573$, for k.
29. $0.079 = \ln B$, for B.
30. $3xy + 1 = 2y - 5x$, for y.
31. $\dfrac{At - B}{C - B(1 - 2t)} = 3$, for t.
32. $\dfrac{a - cx}{b + dx} + a = 0$, for x.
33. $Ab^5 = C$, for b.
34. $|2x + 1| = 7$, for x.
35. $P = Qe^{kt} - P$, for t.

Solve the equations in Problems 36–58.

36. $B - 4[B - 3(1 - B)] = 42$

37. $1.06s - 0.01(248.4 - s) = 22.67s$

38. $\dfrac{5}{3}(y + 2) = \dfrac{1}{2} - y$

39. $3t - \dfrac{2(t - 1)}{3} = 4$

40. $8 + 2x - 3x^2 = 0$

41. $2p^3 + p^2 - 18p - 9 = 0$

42. $N^2 - 2N - 3 = 2N(N - 3)$

43. $\dfrac{1}{64}t^3 = t$

44. $x^2 - 1 = 2x$

45. $4x^2 - 13x - 12 = 0$

46. $60 = -16t^2 + 96t + 12$

47. $y^2 + 4y - 2 = 0$

48. $\dfrac{2}{z - 3} + \dfrac{7}{z^2 - 3z} = 0$

49. $\dfrac{x^2 + 1 - 2x^2}{(x^2 + 1)^2} = 0$

50. $4 - \dfrac{1}{L^2} = 0$

51. $2 + \dfrac{1}{q + 1} - \dfrac{1}{q - 1} = 0$

52. $\sqrt{r^2 + 24} = 7$

53. $\dfrac{1}{\sqrt[3]{x}} = -2$

54. $3\sqrt{x} = \dfrac{1}{2}x$

55. $10 = \sqrt{\dfrac{v}{7\pi}}$

56. $\dfrac{(3x + 4)(x - 2)}{(x - 5)(x - 1)} = 0$

57. $5^{2x} - 5^x - 6 = 0$

58. $2 \cdot 16^x - 5 \cdot 4^x = 12$

For Problems 59–62, express answers in exact form and give a decimal approximation (to two decimal places).

59. $5000 = 2500(0.97)^t$

60. $280 = 40 + 30e^{2t}$

61. $\dfrac{1}{2}(2^x) = 16$

62. $1 + 10^{-x} = 4.3$

In Problems 63–76, solve for the indicated variable.

63. $\left(\dfrac{1}{2}\right)^{t/1000} = e^{kt}$.   Solve for k (to six decimal places).

64. $\dfrac{1}{2}P_0 = P_0(0.8)^x$.   Solve for x (to three decimal places).

65. $T = 2\pi\sqrt{\dfrac{l}{g}}$. Solve for l.

66. $y'y^2 + 2xyy' = 4y$. Solve for y'.

67. $l = l_0 + \dfrac{k}{2}w$. Solve for w.

68. $2x - (xy' + yy') + 2yy' = 0$. Solve for y'.

69. $by - d = ay + c$. Solve for y.

70. $u(v + 2) + w(v - 3) = z(v - 1)$. Solve for v.

71. $S = \dfrac{rL - a}{r - 1}$. Solve for r.

72. $h = v_0t + \dfrac{1}{2}at^2$. Solve for a.

73. $K = \dfrac{MA(y - z)}{N}$. Solve for z.

74. $R = \dfrac{eb + de}{e + f}$. Solve for e.

75. $x + z = \dfrac{y + z}{y} + w$. Solve for z.

76. $\dfrac{3s - t}{2s} = 4 - \dfrac{v}{w}$. Solve for s.

77. Solve for x by completing the square: $8x^2 - 1 = 2x$.

78. Solve for x:

$$\dfrac{x^2 - 5mx + 4m^2}{x - m} = 0.$$

G   SYSTEMS OF EQUATIONS

To solve for two unknowns, we must have two equations— that is, two relationships between the unknowns. Similarly, three unknowns require three equations, and n unknowns (n an integer) require n equations. The group of equations is known as a *system* of equations. To solve the system, we find the *simultaneous* solutions to all equations in the system.

Example 1   Solve for x and y in the following system of equations.

$$\begin{cases} y + \dfrac{x}{2} = 3 \\ 2(x + y) = 1 - y \end{cases}$$

Solution   Solving the first equation for y, we write $y = 3 - x/2$. Substituting for y in the second equation gives

$$2\left(x + \left(3 - \frac{x}{2}\right)\right) = 1 - \left(3 - \frac{x}{2}\right).$$

Then

$$2x + 6 - x = -2 + \frac{x}{2}$$

$$x + 6 = -2 + \frac{x}{2}$$

$$2x + 12 = -4 + x$$

$$x = -16.$$

Using $x = -16$ in the first equation to find the corresponding y, we have

$$y - \frac{16}{2} = 3$$

$$y = 3 + 8 = 11.$$

Thus, the solution that simultaneously solves both equations is $x = -16$, $y = 11$.

Example 2   Solve for x and y in the following system.

$$\begin{cases} y = x - 1 \\ x^2 + y^2 = 5 \end{cases}$$

Solution   We substitute the expression $(x - 1)$ for y in the second equation. Then
$$x^2 + (x - 1)^2 = 5$$
$$x^2 + x^2 - 2x + 1 = 5$$
$$2x^2 - 2x - 4 = 0$$
$$x^2 - x - 2 = 0.$$

Factoring gives $(x - 2)(x + 1) = 0$, so $x = 2$ or $x = -1$.

We then find the y-values which correspond to each x. If $x = 2$, then $y = (2) - 1 = 1$. If $x = -1$, then $y = (-1) - 1 = -2$. The solutions to the system are $x = 2$ and $y = 1$ or $x = -1$ and $y = -2$. Notice that $x = 2$ and $y = -2$ is *not* a solution; nor is $x = -1$ and $y = 1$.

Example 3   Solve for Q_0 and a in the system

$$\begin{cases} 90.7 = Q_0 a^{10} \\ 91 = Q_0 a^{13}. \end{cases}$$

Solution   Taking ratios,

$$\frac{Q_0 a^{13}}{Q_0 a^{10}} = \frac{91}{90.7}$$

$$a^3 = \frac{91}{90.7},$$

$$a = \sqrt[3]{\frac{91}{90.7}} \approx 1.0011.$$

To find Q_0,

$$90.7 = Q_0(1.0011)^{10},$$

$$Q_0 = 89.7083.$$

The simultaneous solution is $a \approx 1.0011$ and $Q_0 \approx 89.7083$.

Graphically, the simultaneous solutions to a system of equations give us the coordinates of the point (or points) of intersection for the graphs of the equations in the system. For example, the solutions to Example 2 give the coordinates of the points where the line $y = x - 1$ intersects the circle centered at the origin with radius $\sqrt{5}$.

Example 4   Find the points of intersection for the graphs in Figure G.1.

Solution   We can solve $y = x^2 - 1$ and $y = x + 1$ simultaneously by setting the y-values equal to one another. Then

$$x^2 - 1 = x + 1$$
$$x^2 - x - 2 = 0$$
$$(x + 1)(x - 2) = 0,$$

so $x = -1$ and $x = 2$ are solutions. To get the corresponding y-values, we use either equation to find $x = -1$ gives $y = 0$, and $x = 2$ gives $y = 3$. The graphs intersect at the points $(-1, 0)$ and $(2, 3)$.

For some systems of equations, it is impossible to find the simultaneous solution(s) using algebra. In this case, the only choice is to find the solution(s) by approximating the point(s) of intersection graphically.

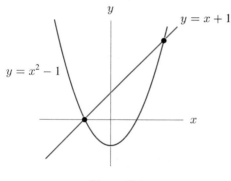

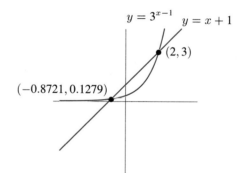

Figure G.1   *Figure G.2:* The graphs reveal two intersection points

---

Example 5   Solve the system:

$$\begin{cases} y = x + 1 \\ y = 3^{x-1} \end{cases}$$

Solution   We are looking for the point or points of intersection of the line $y = x + 1$ and the exponential equation $y = 3^{x-1}$. Setting the expressions for y equal to one another gives

$$x + 1 = 3^{x-1}.$$

We cannot use the algebraic techniques of the previous section to solve this equation. We might try to guess a solution. Note that

$$2 + 1 = 3^{2-1}$$
$$3 = 3,$$

so $x = 2$, $y = 3$ is a solution to the system. If you were not able to guess this solution, you would be able to approximate it from the graph.

However, the graphs of the equations in Figure G.2 reveal that the system has two solutions. There is also a value of x such that $-1 < x < 0$ where the two graphs intersect. By tracing along the graph, we find

$$x \approx -0.8721,\ y \approx 0.1279.$$

Thus, the solutions to the system are $x = 2$ and $y = 3$ or $x \approx -0.8721$ and $y \approx 0.1279$.

Simultaneous solutions to a system of equations in two variables can always be approximated by graphing, whereas it is only sometimes possible to find exact solutions using algebra.

Problems for Section G

Solve the systems of equations in Problems 1–12 for x and y.

1. $\begin{cases} x + y = 3 \\ x - y = 5 \end{cases}$

2. $\begin{cases} 2x - y = 10 \\ x + 2y = 15 \end{cases}$

3. $\begin{cases} 3x - 2y = 6 \\ y = 2x - 5 \end{cases}$

4. $\begin{cases} x = 7y - 9 \\ 4x - 15y = 26 \end{cases}$

5. $\begin{cases} x^2 + y^2 = 36 \\ y = x - 3 \end{cases}$

6. $\begin{cases} 2x + 3y = 7 \\ y = -\frac{3}{5}x + 6 \end{cases}$

7. $\begin{cases} y = 2x - x^2 \\ y = -3 \end{cases}$

8. $\begin{cases} y = 4 - x^2 \\ y - 2x = 1 \end{cases}$

9. $\begin{cases} y = 1/x \\ y = 4x \end{cases}$

10. $\begin{cases} 2(x + y) = 3 \\ x = y + 3(x - 5) \end{cases}$

11. $\begin{cases} y = x^3 - 1 \\ y = e^x \end{cases}$

12. $\begin{cases} ax + y = 2a \\ x + ay = 1 + a^2 \end{cases}$

13. Let ℓ be the line of slope 3 passing through the origin. Find the points of intersection of the line ℓ and the parabola whose equation is $y = x^2$. Sketch the line and the parabola, and label the points of intersection.

Determine the points of intersection for Problems 14–16.

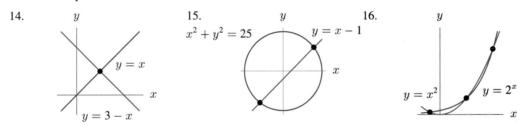

14.   15.   16.

H INEQUALITIES

Just as we can solve some simple equations mentally, we can solve some inequalities mentally. Consider the following examples.

To solve:	We think:	The solution is:
$x - 1 > 0$	"If x is larger than +1, the left hand side is *positive*."	$x > 1$
$x + 4 < 10$	"As long as x stays smaller than +6, the left hand side will be less than the right hand side."	$x < 6$
$3 - x < 0$	"When x gets larger than +3, the left hand side will be negative."	$x > 3$
$x - 2 \geq 0$	"If x = 2 then x − 2 is 0. When x is larger than 2, then x − 2 is positive."	$x \geq 2$
$x^2(x + 5) > 0$	"The value of x^2 is always positive, so x + 5 needs to be positive."	$x > -5$

We solve inequalities using some of the techniques used in solving equations. However, there are some important differences in the application of these techniques to inequalities.

For a, b, and c real numbers, if $a < b$, then

$$a + c < b + c$$
$$a \cdot c < b \cdot c \qquad \text{if } c > 0$$
$$a \cdot c > b \cdot c \qquad \text{if } c < 0 \qquad \text{(Inequality reverses direction).}$$

For a and b positive, when $a < b$, then

$$a^2 < b^2 \quad , \text{and} \quad \frac{1}{a} > \frac{1}{b} \qquad \text{(Inequality reverses direction).}$$

Linear Inequalities

We solve linear inequalities with the techniques we use to solve linear equations; however, the inequality is reversed when we multiply or divide both sides of the inequality by a negative number.

Example 1 Solve for x:

$$1 - \frac{3}{2}x \leq 16.$$

Solution We have

$$1 - \frac{3}{2}x - 1 \leq 16 - 1$$
$$-\frac{3}{2}x \leq 15$$
$$\left(-\frac{2}{3}\right)\left(-\frac{3}{2}x\right) \geq \left(-\frac{2}{3}\right)15$$
$$x \geq -10.$$

Example 2 Solve for x:

$$-6 < 4x - 7 < 5.$$

Solution The solutions to this double inequality is all values of x such that $4x - 7$ is between -6 and 5. We can operate on all three sections of the inequality at once (if we're careful). We start by adding 7 to each part which gives

$$-6 + 7 < 4x - 7 + 7 < 5 + 7.$$
$$1 < 4x < 12$$
$$\frac{1}{4} < x < 3.$$

The solution is all numbers between 1/4 and 3.

Checking the Sign of an Expression

When an expression is a product or a quotient, we can determine the sign of the expression on an interval by looking at the sign of each factor over the interval.

Example 3   Determine the values of x which make the expression $(4 - x)^2 e^{-x}$ positive, negative, and zero respectively.

Solution   The expression e^{-x} is always positive, and $(4 - x)^2$ is positive for any value of x except $x = 4$, where it equals zero. Therefore, $(4 - x)^2 e^{-x} > 0$ for $x \neq 4$, and $(4 - x)^2 e^{-x} = 0$ if $x = 4$.

Example 4   Determine the values of x which make the expression $(x + 1)(x - 7)$ positive, negative, and zero respectively.

Solution   The expression is zero if $x = -1$ or $x = +7$. The product can change sign at $x = -1$ or $x = +7$. We check the sign of the expression by looking at the sign of each factor over the intervals established by dividing the number line at $x = -1$ and $x = +7$. The expression is positive when both factors have the same sign.

$$
\begin{array}{ccccc}
(-)(-) & & (+)(-) & & (+)(+) \\
\hline
\text{Positive} & -1 & \text{Negative} & 7 & \text{Positive}
\end{array} \quad x
$$

Thus,

$$(x + 1)(x - 7) = 0 \quad \text{for } x = -1 \text{ or } x = 7,$$
$$(x + 1)(x - 7) < 0 \quad \text{for } -1 < x < 7,$$
$$(x + 1)(x - 7) > 0 \quad \text{for } x < -1 \text{ or } x > 7.$$

Example 5   Determine the values of r for which the expression $\dfrac{2r + 5}{(r - 1)(r - 3)}$ is zero, positive, negative, or undefined.

Solution   A fraction is equal to 0 if the numerator is 0, so this expression equals 0 if $r = -\frac{5}{2}$. The values $r = 1$ and $r = 3$ make the expression undefined because the denominator is zero there. We divide the number line by marking $-\frac{5}{2}$, 1, and 3, and we check the sign of each each factor on each interval. This gives

$$
\begin{array}{ccccccc}
\dfrac{(-)}{(-)(-)} & & \dfrac{(+)}{(-)(-)} & & \dfrac{(+)}{(+)(-)} & & \dfrac{(+)}{(+)(+)} \\
\hline
\text{Negative} & -\frac{5}{2} & \text{Positive} & 1 & \text{Negative} & 3 & \text{Positive}
\end{array} \quad r
$$

Therefore,

$$\frac{2r + 5}{(r - 1)(r - 3)} \quad \text{is not defined for } r = 1 \text{ or } r = 3.$$

$$\frac{2r + 5}{(r - 1)(r - 3)} - 0 \quad \text{for } r = \frac{5}{2},$$

$$\frac{2r + 5}{(r - 1)(r - 3)} > 0 \quad \text{for } -\frac{5}{2} < r < 1 \text{ or } r > 3,$$

$$\frac{2r + 5}{(r - 1)(r - 3)} < 0 \quad \text{for } r < -\frac{5}{2} \text{ or } 1 < r < 3,$$

Example 6   Determine y-values which make the expression $\dfrac{\sqrt{y+2}}{y^3}$ positive, negative, zero, or undefined.

Solution   The expression is zero if $y = -2$. It is not defined if y is less than -2 because we cannot take the square root of a negative number. Furthermore, the quotient is not defined if $y = 0$ because that puts a zero in the denominator. The radical is always positive (where defined), so the numerator is positive. Because the cube of a negative number is negative, we can visualize signs in each interval.

So

$$\frac{\sqrt{y+2}}{y^3} \quad \text{is undefined for } y < -2 \text{ or } y = 0,$$

$$\frac{\sqrt{y+2}}{y^3} = 0 \quad \text{for } y = -2,$$

$$\frac{\sqrt{y+2}}{y^3} > 0 \quad \text{for } y > 0.$$

$$\frac{\sqrt{y+2}}{y^3} < 0 \quad \text{for } -2 < y < 0,$$

Solving Nonlinear Inequalities by Factoring and Checking Signs

We can often solve polynomial and rational inequalities by starting out the same way we would with an equation and then using a number line to find the intervals on which the inequality holds.

Example 7   Solve for x if

$$2x^2 + x^3 \le 3x.$$

Solution   We first set the inequality so that zero is on one side. Then

$$x^3 + 2x^2 - 3x \le 0,$$
$$x(x^2 + 2x - 3) \le 0,$$
$$x(x + 3)(x - 1) \le 0.$$

The left hand side equals 0 for $x = 0$, $x = -3$, and $x = 1$. To solve the inequality we want to select the x-values for which the left hand side is negative. We check signs over the four intervals created by marking $x = 0, -3, +1$.

So $x^3 + 2x^2 - 3x < 0$ for $x < -3$ or $0 < x < 1$ and $x^3 + 2x^2 - 3x = 0$ at $x = 0$, $x = -3$, and $x = 1$. Therefore, $2x^2 + x^3 \le 3x$ for $x \le -3$ or $0 \le x \le 1$.

Example 8   Solve for q: $\dfrac{1}{2 - q} > q.$

Solution First, put 0 on the right, then combine the fractions on the left.

$$\frac{1}{2-q} - q > 0,$$

$$\frac{1}{2-q} - q\frac{(2-q)}{(2-q)} > 0,$$

$$\frac{1 - q(2-q)}{2-q} > 0,$$

$$\frac{1 - 2q + q^2}{2-q} > 0.$$

We want this quotient to be positive, so the numerator and denominator must have the same sign. Notice that $q \neq 2$. Factoring the numerator, we have $\frac{(1-q)^2}{2-q} > 0$. The numerator is never negative, and the denominator is positive if $2 - q > 0$ which gives $q < 2$. However, $(1-q)^2 = 0$ if $q = 1$, so $q = 1$ must be excluded in order to preserve the inequality. Therefore, $\frac{1}{2-q} > q$ for $q < 2$ and $q \neq 1$.

Radical inequalities

We can eliminate radicals by raising both sides of an equation to the same power. However, we must check our answers for extraneous roots.

Example 9 Solve for x: $\sqrt{x-6} < 2$.

Solution The expression $\sqrt{x-6}$ is not defined unless $x \geq 6$, so we can restrict the x-values we consider to $x \geq 6$. Also, $\sqrt{x-6} \geq 0$, so we can square both sides giving

$$(\sqrt{x-6})^2 < 2^2,$$

$$x - 6 < 4,$$

$$x < 10.$$

However, we are only considering x-values which are 6 or larger, so $\sqrt{x-6} < 4$ for $6 \leq x < 10$.

Example 10 Solve for x: $\sqrt{2-x} > 3$.

Solution Squaring both sides gives

$$(\sqrt{2-x})^2 > 3^2,$$

$$2 - x > 9,$$

$$-x > 7,$$

$$x < -7.$$

In this case, the fact that the radical is only defined for $x \leq 2$ does not affect our solution because any number which is less than -7 is also less than 2.

Inequalities and Interval Notation

Inequalities can be used to describe sets of numbers. For instance $x \leq 5$ represents the set of all real numbers less than or equal to 5. Likewise, $x \geq -2$ represents the set of all real numbers greater than or equal to -2. Graphically, we represent the set of real numbers less than or equal to 5 as

5

The dot at 5 shows that the number 5 is included. Alternatively, this set of real numbers can be represented using interval notation. For example, $x \leq 5$ can be written as $(-\infty, 5]$, where $-\infty$ means that we consider arbitrarily large negative numbers.

Example 11 Represent the intervals (a) $-2 \leq y \leq 5$ (b) $-2 < y < 5$ graphically and using interval notation.

Solution (a)

and $[-2, 5]$.

(b)

and $(-2, 5)$.

Notice that the endpoint of an interval is included if the dot on a number line is filled in, or if square brackets, [,], are used in interval notation. If the dot is not filled in or parentheses, (,), are used in the interval notation, the endpoint of the intervals is not included.

Example 12 Use interval notation to represent the following intervals of real numbers:

(a) $-2 \leq x < 3$ (b) $r < 0$ (c) $10 < t \leq 17.5$

(d) $\dfrac{1}{2} \leq w \leq \dfrac{9}{2}$ (e) $-12 < r$ (f) $|x| < 9$

Solution (a) $[-2, 3)$ (b) $(-\infty, 0)$ (c) $(10, 17.5]$ (d) $[1/2, 9/2]$ (e) $(-12, \infty)$ (f) $(-9, 9)$

Problems for Section H

Solve the inequalities in Problems 1–15 mentally.

1. $2(x - 7) \geq 0$ 2. $\sqrt{x} > 4$ 3. $x^2 < 25$

4. $x - 3 > 2$ 5. $1 + \sqrt{x + 4} > 0$ 6. $|t| < 5$

7. $x^2 \geq 16$ 8. $1 + x^2 > 0$ 9. $5 - x < 0$

10. $2^{-x} > 0$ 11. $|\theta| - 1 \leq 7$ 12. $2x^2 + 1 < 0$

13. $\dfrac{x - 5}{x^2 + 1} > 0$ 14. $e^t(t + 3) < 0$ 15. $\dfrac{2 + z}{5 + z^4} < 0$

Write the statements in Problem 16–21 as inequalities.

16. The x-values which are less than 0.001   17. The y-values between -1 and 1

18. All p-values except 5   19. All the positive values of k

20. All the r-values which are not negative   21. The t-values during or after the year 1995

For Problems 22–35, solve the following inequalities and graph the solutions on a number line.

22. $x - 3 > 1$ 23. $3x - 5 > 1$ 24. $-2x > 8$ 25. $-4x + 7 < 13$

26. $x^2 - 5x + 6 > 0$ 27. $x^2 - 7x < 8$ 28. $\dfrac{1}{x - 3} > 0$ 29. $\dfrac{2}{x - 5} > 3$

30. $|x| > 2$ 31. $|y - 2| > 4$ 32. $|3w - 2| - 4 < 0$ 33. $|2z - 7| < 8$

34. $|r| \geq 0$ 35. $|r| < 0$

In Problems 36–47, determine the real number values of the variable (if any) which will make each expression (a) undefined, (b) zero, (c) positive, (d) negative.

36. $3x^2 + 6x$

37. $(2x)e^x + x^2 e^x$

38. 2^{-x}

39. $6t^2 - 30t + 36$

40. $\frac{1}{3}x^{-2/3}$

41. $\frac{-24}{p^3}$

42. $\frac{1}{2\sqrt{x^2 + 1}}(2x)$

43. $\frac{\ln x}{x}$

44. $\frac{t - 3}{t^2 + 10}$

45. $\frac{x - 1}{x - 2}$

46. $\frac{1 - 3u^2}{(u^2 + 1)^3}$

47. $-\frac{2x - 1}{\left(x(x - 1)\right)^2}$

Solve the inequalities in Problems 48–59.

48. $4 - x^2 > 0$

49. $-1 \le 4x - 3 \le 1$

50. $0 \le \frac{1}{2} - n < 11$

51. $\sqrt{3l} - \frac{1}{4} > 0$

52. $t^2 - 3t - 4 \ge 0$

53. $2(x-1)(x+4)+(x-1)^2 > 0$

54. $\frac{5x}{3} - \frac{3}{2} \le \frac{8x}{5}$

55. $\frac{1}{x} < \frac{1}{2}$

56. $2 + \frac{r}{r - 3} > 0$

57. $\frac{1}{x} > \frac{1}{x + 1}$

58. $\frac{2x^2 - (2x + 1)(2x)}{x^4} < 0$

59. $\frac{3(x + 2)^2 - 6x(x + 2)}{(x + 2)^4} > 0$

ANSWERS TO ODD NUMBERED PROBLEMS

Section 1.1

1 (a) 69°F

(b) July 17 and 20

(c) Yes

(d) No

5 Independent: A
Dependent: n

7 (a) 40

(b) 2

9 (a) (I), (III), (IV), (V)
(VII), (VIII)

(b) (i) (V) and (VI)

(ii) (VIII)

(c) (III) and (IV)

15 (a) One

(b) Yes

19 $S = 6\pi r^2$

21 (a) No

(b) Yes

23 (b) $C = 2 + (0.5)l$

25 (a) Yes

(b) No

Section 1.2

1 $k = 3/2; y = (3x)/2; x = 5.33$

3 $k = 5; c = 5d; c = 125$

5 Direct; $k = -1, p = 1$

7 Direct; $k = 1, p = 1$

9 Direct; $k = 0.17, p = 1$

11 Inverse; $k = 2/3, p = -1$

13 (a) $C(x) = kx$

(b) $k = 9.5; C(x) = 9.5x$

(d) $52.25

15 $c = 81.67x; 326.68$

17 $h = 192.5/v; 64.167$ mph

19 (a) 830

(b) 104

(c) 2880

(d) 277,778

21 Yes; $k = 32, p = 5$

23 Yes; $k = 3, p = -5$

25 Yes; $k = -5, p = 4$

27 Yes; $k = 1/6$ and $p = -7$

29 No

31 No

33 (a) $A = \pi d^2/4$
$V = \pi d^3/6$

(b) Yes, d^2
Yes, d^3

35 (a) $d = 1.7, 3.4, 20.4, 102$
$d = 0.34t$

(b) 9.8 mins

(c) $A = 9.1, 36.3, 1307, 32685$
$A = 0.363t^2$

(d) $P = 11.25t^2$

(e) 298 sec, or 5 min

37 (b) $C = 36.96d^2$

(c) $C = (400\pi/34)d^2$

Section 1.3

1 (a) 2.9, -19.5

(b) 48.5, -26.5

3 Between 3000 and
4000 years ago

5 No

7 (a) 300, -150 hundred/yr

(b) 167, -83 hundred/yr

(c) 235, -118 hundred/yr

9 (a) (i) -2

(ii) -6

(iii) -4

(b) $2 \leq x \leq 4$

11 Yes; No

13 (b) Between 80 and 90 meters; no

15 (a) E, III

(b) G, I

(c) F, II

Chapter 1 Review

1 (a) No

(b) No

(c) Yes

3 (a) 11 cubic feet

(c) 12 cubic feet

(d) $S = n + 6$

7 G is function of N
N not necessarily function of G

9 $0.945P$

11 Yes, 5

13 No

15 Yes, 1/9

17 No

19 Yes, 42

21 $n = 1.33s$; 32 cookies

23 (a) $N = kp$

(b) Doubles

(c) Increases by $1000k$

25 (a) 1200 miles

(b) $T = f(R) = 1200$ miles/R

(c) $f(300) = 4$ hrs.

27 (a) $r = f(V) = 0.6204\sqrt[3]{V}$

(c) 12.60 cm

31 (b) July

(c) Increasing: Jan–July
Decreasing: July–Dec

33 (a) 4.45%

(b) 4.45%; $281 billion

(c) $28.1 billion

35 No

Section 2.1

3 (c) 22 million

(d) 0.3 million people/year

5 (a) Side length and perimeter

(c) 4

7 (a) $r = f(v)$ is linear

9 $P = 18,310 + 58t$

11 (a) $T = 1900

(b) $C = 7$

(d) Twelve credits

(e) Fixed costs that do not depend
on the number of credits taken

13 (a) $y = (1/250)x + 300$

(b) $y = 400; y = 500$

(c) $100,000

15 No

19 (a) (i) -1.246

(ii) -1.246

(iii) -1.246

(b) $f(x) = -0.367 - 1.246x$

Section 2.2

1 $y = 4 - 3/5x$

3 $y = -0.3 + 5x$

5 $y = 2/3 + 5/3x$

7 $y = -40/3 - 2/3x$

9 $y = 5 + 0x$

11 $y = 3 - 2x$

13 $y = -5 + 5x/3$

15 $y = 0.03 + 0.1x$

19 $g(x) = 80 - 0.05x$

21 (a) Membership fee: $55;
Fixed price per meal: $3.25

(b) $C = 55 + 3.25n$

(c) $217.50

(d) $n = (C - 55)/3.25$

(e) 75 meals

23 (b) $v = 40 - 32t$

25 $y = -4 + 4x$

27 $y = \frac{16+5\sqrt{7}}{2+\sqrt{7}} - \frac{3}{2+\sqrt{7}}x$ or
$y = (1 + 2\sqrt{7}) + (2 - \sqrt{7})x$

29 (a) $q = 120 - 50p$

Section 2.3

1   $x = 5, y = -2$

5   (d)   Steepness appears to decrease as vertical height of window increases

7   A, h
     B, f
     C, g
     D, j
     E, k

9   (a)   Fixed cost= \$8000, unit cost= \$200
     (b)   Fixed cost= \$5000, unit cost= \$200
     (c)   Fixed cost= \$10,000, unit cost= \$100
     (d)   Fixed cost= \$0; unit cost= \$50

11   (a) $P = (a, 0)$
     (b) $A = (0, b), B = (-c, 0)$
         $C = (a + c, b), D = (a, 0)$

13   (a)   (i) $y = 9 - \frac{2}{3}x$
         (ii) $y = -4 + \frac{3}{2}x$

15   $(1, 0)$

17   $(3, 2.5)$

19   (a)   Company A: $20 + 0.2x$
         Company B: $35 + 0.1x$
         Company C: 70
     (c)   Slope: mileage rate
         Vertical intercept: fixed cost/day
     (d)   A for $x < 150$
         B for $150 < x < 350$
         C for $x > 350$

21   $y = 13/x + 2/3.$

23   (a) $m_1 = m_2$ and $b_1 \neq b_2$
     (b) $m_1 = m_2$ and $b_1 = b_2$
     (c) $m_1 \neq m_2$
     (d)   Not possible

Section 2.4

1   (a) $r = 1$
     (b) $r = 0.7$
     (c) $r = 0$
     (d) $r = -0.98$
     (e) $r = -0.25$
     (f) $r = -0.5$

3   (c) $y = 15x - 80$
     (e)   Yes, a strong positive correlation

5   (b) $r \approx -1/2$
     (c) $y = 27.5139 - 0.1674x$
         $r = -0.5389$

7   (b)   Estimates will vary, e.g. $H = 37t - 37$
     (c) $H = 37.26t - 39.85$, $r = 0.9995$, $r = 1$. The correlation between the data set and the regression line is very good.
     (d)   Slope gives the average number of home-runs per year, about 37

Chapter 2 Review

1   $y = 6 - 3/5x$

3   Parallel line:
     $y = -4x + 9$
     Perpendicular line:
     $y = 0.25x + 4.75$

5   (a)   (V)
     (b)   (VI)
     (c)   (I)
     (d)   (IV)
     (e)   (III)
     (f)   (II)

9   $d = 60 + 50t$

11   $c = 4000 + 80r$

13   (a)   Increasing function of x
     (b) $F(x) = 1.242x$
     (c) ≈ 3.7 pounds

15   \$10,500

17   $d(0) \approx 0.2968$ inches
     $d(12.5) \approx 0.1016$ inches
     $g < 19$

19   (a) $s = (\sqrt{\pi})r$
     (b)   Linear
     (c) $r = 0$

21   (a)   Demand decreases to 100
     (b) $D = 1100 - 200p$
     (c)   \$5.25

25   (b) $0 \leq t \leq 4$

29   l_1

Section 3.1

1   100

3   (a) $t = 6$
     (b) $t = 1, t = 2$

5   (a) -4
     (b) ± 2

7   (a) $-1/2$
     (b) -1

9   (b)   3
     (c) -8
     (d)   16
     (e)   6

11   (a) $P = (b, a)$
         $Q = (d, e)$
     (b) $f(b) = a$
     (c) $x = d$
     (d) $x = h$
     (e) $a = -e$

13   (a)   48 feet for both
     (b)   4 sec, 64 ft

15   (a) $s(2) = 146$
     (b)   Solve $v(t) = 65$
     (c)   At 3 hours

17   (c)   5050

Section 3.2

1   (a)   (i)   248
         (ii)   142
         (iii)   4
         (iv)   12
         (v)   378
         (vi) -18
         (vii)   248
         (viii)   570
         (ix)   13
     (b)   (i) $x = 2$
         (ii) $x = 8$
         (iii) $x = 7$
     (c) $x = 1, 4$

3   (a) -25
     (b) -4
     (d)   7
     (e)   3, 5
     (f) $2, 2\sqrt[3]{3} \approx 2.88$

5   (a) $(5, 3)$
     (b) $(21, 5)$
     (c) $\sqrt{7} - \sqrt{6}$

7   (a) $g(-3x) = 9x^2 - 3x$
     (b) $g(1 - x) = x^2 - 3x + 2$
     (c) $g(x + \pi) = x^2 + (2\pi + 1)x + \pi^2 + \pi$
     (d) $g(\sqrt{x}) = x + \sqrt{x}$
     (e) $g(1/(x + 1)) = (x + 2)/(x + 1)^2$
     (f) $g(x^2) = x^4 + x^2$

9   (a)   (iii)
     (b)   (i)
     (c)   (ii)

13   I is (b)
     II is (d)
     III is (c)
     IV is (h)

15   (a)   2
     (b) -3
     (c) -3
     (d)   2

17   (a) $C(3,5) = \$6.25$
     (b) $C^{-1}(\$3.5) \approx 1.67$

19   (a) $A = f(s) = 4s$
     (b) $f(s + 4) = 4(s + 4) = 4s + 16$
     (c) $f(s + 4) = 4s + 4$
     (d)   Meters

Section 3.3

1   $t = 1, 2, 3, \ldots, 12$
5   (a)   Domain $-2 \le x \le 2$

   (c)   Range $0 \le y \le 2$

7   Domain: $x \ge 3$
   Range: $y \ge 0$
9   Domain: $x, x \ne 0$
   Range: $y > 0$
11 Domain: $x, x \ne -1$
   Range: $y < 0$
13 Domain: all real x
   Range: $y \ge -4$
15 Domain: all real x
   Range: all real y
17 Domain: all real x
   Range: all real y
19 $y \le -(1/2)$ or $y \ge (1/2)$
21 $0 \le y \le 3$
23 (a)   Domain: all real numbers
      Range: all real numbers

   (b)   Domain: all real numbers
      Range: $n(x) \le 9$

   (c)   Domain: $x \ge 3$ and $x \le -3$
      Range: $q(x) \ge 0$

25 (a)   8

   (b)   Domain: $1, 2, \ldots, N$,
      where $N = $ number of listings

   (c)   Range: $1, 2, \ldots, 9$

27 (a)   Domain: $x \ge -3$
      Range: $0 \le y < 1$

   (b) $f(x) = (3+x)/(9+x)$

   (c)   Domain: $x \ne -9$
      Range: $y \ne 1$

Section 3.4

1   (a)   Yes

   (b)   No

   (c) $y = 1, 2, 3, 4$
3   (c)   Domain: all $x, x \ne 0$
      Range: -1 and 1

   (d)   False, $u(0)$ is undefined
9   (a) $n(L) = 2L + 10$

   (b)   Domain: $L \ge 5$
      Range: $n(L) = $
      $20, 21, 22, 23, \ldots$

   (c)   Domain: All real numbers
      Range: All real numbers
11 (b)   Integers from 1 to 50
      Even integers from 40 to 120
13 (a) $1.12
      $1.26

   (b)   Domain: All positive integers
      Range: All positive multiples of
      0.14
15 (a)   First and third

   (b) > 45 mph or < 4 mph

   (c) ≥ 46 mph

Chapter 3 Review

1   $(0, 2)$
3   Intersect at $x = 2$
5   (a)   11,000

   (b)   16.064

   (c)   2,541,000
7   (c)   Domain is $-\infty < x < \infty$
      Range is $0 \le y < \infty$

   (d)   Increasing: $0 < x < \infty$
      Decreasing: $-\infty < x < 0$
9   (c)   Domain is all real numbers
      except 0
      Range is all real numbers
      except 0

   (d)   Decreasing: $-\infty < x < 0$ and
      $0 < x < \infty$
11 (c)   Domain is $0 \le x < \infty$
      Range is $0 \le y < \infty$

   (d)   Increasing: $0 \le x < \infty$
15 Domain: all real numbers
   Range: $h(x) \ge -16$
17 Domain: $4 \le x \le 20$
   Range: $0 \le r(x) \le 2$
19 $a - a^2$
21 (a)   7000

   (b)   8500; 4 weeks after the begin-
      ning of the epidemic

   (c) $w = 1, w = 10$

   (d) $1.5 \le w \le 7.5$
23 (a) $f(1) = 2$
      $g(3) = 4$

   (c) $f(5) = 14$
      $f(-2) = -7$
      $g(5) = 16$
      $g(-2) = 9$

   (d) $f(x) = -1 + 3x$
      $g(x) = (x - 1)^2$
25 (a)   2 kg cost $1.50

   (b)   0.1 kg costs 75¢

   (c) $3 buys 4 kg

   (d) $1.50 buys 2 kg
29 (a)   (i)   6

      (ii) 5

      (iii) Not defined

   (b)   (i) $50 \le s \le 75$

      (ii) $76 \le s \le 125$
31 (a)   Domain: $1 \le n \le 19$,
      n an integer
      Range: $0 \le p(n) \le 700$,
      $p(n)$ an integer

   (c)   Nondecreasing

Section 4.1

1   20%; 2%.
3   $109,272.70
5   $f(n) = P_0(0.8)^n$
7   (a) $C = 100(0.84)^t$

   (b)   41.8 mg

9   18.55%
11 (h)   Unpaid balance: $1772.76
      Paid off: $227.24
      Interest paid: $340.84
13 $b(b^4 - 1)$
15 (a)   25 mg

   (b)   15%

   (c)   4.92 mg

   (d)   20 hours
17 (a) $N(r) = 64(1/2)^r$

   (b)   6
19 (a)   13.4%

   (b)   6.49%

Section 4.2

3   $x < -1.7$ and $x > 2$
5   $f(x) = -3(1/2)^x$
7   $g(x) = 5(0.707)^x$
9   $y = 10^x$
11 $y = \frac{1}{5}(3)^x$
13 $y = 5(2/3)^x$
15 (a) $P = 1154.23(1.2011)^t$

   (b) $1154.23

   (c)   20.1%
17 (a) $g(x)$ is linear

   (b) $g(x) = 2x$
19 (a) $i(x)$ is linear

   (b) $i(x) = 18 - 4x$
21 $f(x) = 16,384(1/4)^x$
23 (a)   (v)

   (b)   (iii)

   (c)   (iv)

   (d)   (ii)

   (e)   (i)
25 (a) $f(0) = 1000$,
      $f(10) \approx 1480$

   (b) $0 \le t \le 10$,
      $0 \le P \le 1500$;
      $0 \le t \le 50$,
      $0 \le P \le 8000$

   (c) $t \approx 23.4$
27 (a) $P \approx 0.538$ millibars

   (b) $h \approx 0.784$ km
29 (a) $V = (39375)(0.9157)^t$

   (b) $V = 39375 - 2587.5t$

   (c)   Linear

Section 4.3

5   $f(x)$:bottom $g(x)$;middle,$h(x)$:top
7   $f(x) = e^{-x}$
   $g(x) = e^x$
   $h(x) = -e^x$
9   (a)   Concave up

   (b)   Concave down

   (c)   Concave down

   (d)   Neither
11 (a)   All

   (b) b

   (c) b, a, c, p

(d) $a = c$

(e) d and q

15 Domain: all t
Range: $Q > 0$

17 (a) Graphs agree

(c) Models 1 and 3 agree (approx); Model 2 predicts larger populations

19 $f(x) = 5e^x$

21 $y = 4e^x$

23 $y = 2e^x$

25 (a) \$24,102.64

(b) About 124 years

27 $4 < f(5) < 7$

29 (a) Decreasing

(b) Concave down then up

(c) About 1992

Section 4.4

1 3.47712

3 (a) 0

(b) 0

(c) 5

(d) 1/2

(e) 2

(f) $-1/2$

5 (a) $\log(A \cdot B) = \log A + \log B$;
$\log \frac{A}{B} = \log A - \log B$;
$\log A^B = B \log A$

(b) $p(\log A + \log B - \log C)$
$p \log \left(\dfrac{AB}{C} \right)$
$p(\log AB - \log C)$

7 (a) True

(b) False

(c) False

(d) True

(e) True

(f) False

9 (a) $2x$

(b) x^3

(c) $-3x$

11 (a) 10^p

(b) 10^{3q}

(c) $p + 3q$

(d) $p/2$

13 $\log(7/4)/\log(1.171/1.088)$

15 $\log(17/3)/\log 2$

17 $(\ln 7 - 5)/(1 - \ln 2)$

19 $-2, 1/3, -1/3$

21 $(\log c - \log d)/\log b$

23 $(\ln Q - \ln P)/k$

25 $(\log(14/3))/(\log 1.081) \approx 19.8$

27 $(\log(8/5))/(\log(1.15/1.07)) \approx 6.5$

29 $(\ln(88/121))/(-0.112) \approx 2.84$

31 $100 \ln(17/18) \approx -5.72$

33 2.3

Section 4.5

1 $y = 0, y = 0, x = 0$

3 (a) 0

(b) $-\infty$

9 (a) $t(x)$

(b) $r(x)$

(c) $s(x)$

11 $A: y = 3^x$, $B: y = 2^x$, $C: y = \ln x$
$D: y = \log x$, $E: y = e^{-x}$

13 $x \neq 0$

15 $x > 1$

17 $y = \log x$

19 $y = 2b^x, b > 1$

21 $y = 0.1b^x, 0 < b < 1$

23 (a) Domain: all x
Range: $y > 0$
Asymptote: $y = 0$

(b) Domain: all $x > 0$
Range: all y
Asymptote: $x = 0$

Section 4.6

1 $(\log(2/17))/\log(9.261/4.5) \approx 0.225178$

3 $-6.2, 61.9$

5 $(1 - \ln 8)/3 \approx -0.360$

7 In 1969

9 (a) $P = 3(3.22)^t$

(b) 0.939 hours

11 (a) 15.9%

(b) 7.5 years

13 (a) 300; 600

(b) 34.7 years

15 (a) 11 years

(b) $b = 0.9389$

17 $t \approx 3.466$

19 (a) $o(t) = 245 \cdot 1.03^t$

(b) $h(t) = 63 \cdot 2^{t/10} \approx 63 \cdot (1.07)^t$

(c) 34.2 years

21 (a) $f(x) = \frac{1}{2}(4)^x$
$g(x) = 4 \left(\frac{1}{3} \right)^x$
$h(x) = x + 2$

(b) $x = \log 8 / \log 12$

(c) $x = 1.38$ or $x = -1.97$

23 (a) $1.01 \cdot 10^{-5}$ moles/liter

(b) 5

25 (a) $D_2 - D_1 = 10 \log(I_2/I_1)$

(b) 3.01 decibels louder

Section 4.7

1 $A_0 = 5$
$k = \ln 2$

3 (a) 1412 bacteria

(b) 10.0 hours

(c) 1 hour

5 (a) $P(t) = 25{,}000 e^{0.075t}$

(b) 7.79%

7 (a) 2.03 volts

(b) after 5.36 seconds

(c) 25.9%

9 (a) $a = 30$; $b = 1.0414$; b is annual growth factor

(b) $k = 0.0405$; k is continuous growth rate per year

11 (a) 5.73%

(b) 5.57%

13 (a) $\approx 33.5\%$

(b) $k \approx 4.1\%$, continuous hourly decay rate

15 3.053%/yr

17 12,146 years old

19 18.6 hours

21 39.8 years

23 45%

25 (a) 9.15%, 8.32%

(b) 576 months

Section 4.8

1 (a) \$1270.24

(b) \$1271.01

(c) \$1271.22

(d) \$1271.25

3 45.65 years

5 32.5 years

7 (a) 27.5 years

(b) 28 years

9 (a) $V(t) = 1000(1.005)^{12t}$

(b) 5.99%

11 Eff. yield: 20.9%
Cont. rate: 19%

13 (a) 8.300%

(b) 8.322%

(c) 8.328%

(d) 8.329%

15 7.35%

17 (a) A = \$1146.16, B = \$1143.39, C = 1148.69

(b) From best to worst: A, B, C

19 \$27,399.14

Section 4.9

1 $10^{-3.65}$ million years ago

Section 4.10

1 (b) Linear function

(c) $f(x)$ is exponential; $g(x)$ is linear

(d) It is linear; yes

3 (b) Exponential

(c) Linear

5 (a) $y = 3603 + 221 e^{\mu}$; $r \approx 0.7016$

(b) $y = 4.8(1.22)^x$; $r \approx 0.9998$

(c) $y = 0.007 x^{3.68}$; $r \approx 0.9751$

7 (a) $y = 2237 + 2570x$

(b) $\ln y = 8.227 + 0.2766x$

(c) $y = 3741 e^{0.2766x}$

9 (a) Negative

(b) $y = b + px$

(c) $(0, 244)$

(d)   No

11   (a) $a \approx -7.786, b \approx 86.28$

(c) $69,954$ minutes ≈ 45 days
0.17 minutes ≈ 10 seconds

13   (a) $N = -14t^4 + 433t^3 - 2255t^2 + 5634t - 4397$

(b)   Good fit

(c)   Model fails for $t > 20$

Chapter 4 Review

1   (a)   \$8811.71; \$15529.24

(b)   6.12 years; 12.2 years

3   (a)   60%

(b) $(0.70)^n$

(c)   7 times

5   (a)   (ii)

(b)   (i)

(c)   (iii)

7   $Q = 0.7746 \cdot (0.3873)^t$

9   $y = 2(3/2)^x$

11   $(\ln(18.5/16.3))/(\ln 1.072) \approx 1.821$

13   No solution

15   $3 + M$

17   $f(x) = 0.0124 + 0.093x$
$g(x) = 3.2(1.2)^x$

19   $-0.587 < x < 4.91$

21   (a)   Exponential

(b) $P(t) = 20,000(1.0414)^t$.

23   (a) $15.27(1.12)^t$

(b)   104,844

25   35.97%

27   (a) $\ln 8 - 3 \approx -0.9206$

(b) $\log 1.25 / \log 1.12 \approx 1.9690$

(c) $-\dfrac{\ln 4}{0.13} \approx -10.6638$

(d)   105

(e) $\frac{1}{3}e^{3/2} \approx 1.4939$

(f) $e^{1/2}/(e^{1/2} - 1) \approx 2.5415$

(g) -1.599 or 2.534

(h)   2.478 or 3

(i)   0.653

29   (a) $P(t) = 0.755\,(1.194)^t$

(b)   20 years; 2135

31   (a)   3%, 2.96%

(b)   3.05%, 3%

(c)   Decay: 6%, 6.19%

(d)   Decay: 4.97%, 5.1%

(e)   3.93%, 3.85%

(f)   Decay: 2.73%, 2.77%

33   (a) $a = 5, k = 0.1142,$
$f(x) = 5e^{0.1142x}$

(b) $a = 17, b = 1.0986,$
$g(x) = 17(1.0986)^x$

(c) $a = 22, b = 1.0473,$
$h(x) = 22(1.0473)^x$
$a = 22, k = 0.0462,$
$h(x) = 22e^{0.0462x}$

35   9.71%

37   (a)   177 microns

(b)   1770 microns

(c)   Differs by 1; differs by factor of 10

39   (b) $W_F(t) = 11.591e^{0.058(t - 1980)}$
(women)
$W_M(t) = 19.173e^{0.045(t - 1980)}$
(men)

(e)   Yes, in about 2020

41   (a)   African-American infants

(b)   Suggests Caucasian declined faster

(d)   Neither declined exponentially

43   (a) $f(x) = p_0(2,087,372,982)^x$

(b)   8.55

(c)   BAC of 0.051

45   (b)   0.69

(c)   1.10

(d) $e \approx 2.72$

Section 5.1

1   (a) $-3, 0, 2, 1, -1$
One unit right

(b) $-3, 0, 2, 1, -1$
One unit left

(c) $0, 3, 5, 4, 2$
Up three units

(d) $0, 3, 5, 4, 2$
One right and three up

3   $f(p) : 0, -3, -4, -3, 0, 5, 12$
$g(p); -4, -3, 0, 5, 12, 21, 32$
$h(p) : 12, 5, 0, -3, -4, -3, 0$
g is f shifted left two
h is f shifted right two

5   (a) $3^w - 3$

(b) 3^{w-3}

(c) $3^w + 1.8$

(d) $3^{w + \sqrt{5}}$

(e) $3^{w+2.1} - 1.3$

(f) $3^{w-1.5} - 0.9$

7   (a)   (vi)

(b)   (iii)

(c)   (ii)

(d)   (v)

(e)   (i)

(f)   (iv)

11   (a) $T(d) = S(d) + 1$

(b) $P(d) = S(d - 1)$

13   (a) $h(x) = f(x) - 2$

(b) $g(x) = f(x + 1)$

(c) $i(x) = f(x + 1) - 2$

17   Vertical shifts

19   (a) $y = k \cdot 2^x, k > 0$ arbitrary constant

(b) $y = 2^x + b$, b arbitrary constant

(c) $y = k \cdot 2^x + b, k > 0$ and b arbitrary constants

23   (a) $t(x) = 5 + 3x$ for $x > 0$

(b) $n(x) = t(x) + 1$ (vertical shift)

(c) $p(x) = 10 + 3(x - 2)$ for $x > 2$, or
$p(x) = t(x - 2) + 5$ for $x > 2$

25   (d) $I(d + 1000)$

(e) $I(d) + 200$

(f)   \$19,385

Section 5.2

3   Reflected across y-axis;
$f(-x) = 4^{-x}$

7   Reflections across x-axis

11   (a) $f(-x) = \sqrt{4 - x^2}$

(c)   Even

15   (b) $d(t)$ reflected in t-axis, raised 142

17   (a) $y = 2^{-x} - 3$

(b) $y = 2^{-x} - 3$

(c)   Yes

19   (a)   Even

(b)   Odd

(c)   Neither

(d)   Neither

21   (a)   Symmetric about the y-axis; even

(b)   Symmetric about the origin; odd

(c)   Not symmetric

(d)   Not symmetric

29   If $f(x)$ is odd, then $f(0) = 0$

33   (a)   Even if and only if $m = 0$

(b)   Odd if and only if $b = 0$

(c)   Both if and only if $m = 0$ and $b = 0$

35   Yes, $f(x) = 0$

Section 5.3

3   (d)   All three

9   (i)   i

(ii)   c

(iii)   b

(iv)   g

(v)   d

11   (a) $f(x)$

(b)   None

(c) $g(x)$

(d)   None

(e) $j(x)$

13   (a) $y = f(t) + 2$
Asymptote: $y = 7$

(b) $y = f(t + 1)$
Asymptote: $y = 5$

(c) $y = f(t-2) - 3$
Asymptote: $y = 2$

15 (a) $y = k \cdot 2^x$, k arbitrary constant

(b) $y = k \cdot 2^x$, k arbitrary constant

Section 5.4

3 (b) Yes

(c) No

7 (b) $-0.5 \le x \le 0.5$

(c) $-8 \le x \le 8$

13 (a) $+1$

(b) $\log(10x) = 1 + \log x$

(c) $k = \log a$

15 Not true

17 $r(t)$: half the level
$s(t)$: half the rate

Section 5.5

1 $y = (1/3)x^2 + 2x + 5$

3 (a) $a = -2, b = 4, c = 16$
Axis of symmetry: $x = 1$
Vertex: $(1, 18)$
Zeros: $x = -2, 4$
y-intercept: $y = 16$

(b) $a = 1, b = 0, c = 3$
Axis of symmetry: y-axis
Vertex: $(0, 3)$
No zeros
y-intercept: $y = 3$

7 $y = -2(x+4)(x-5)$

9 $y = -(x-2)^2$

11 $y = -\frac{1}{9}(x+6)^2 + 9$

13 Shift $y = x^2$ right by 5 units to get
$y = (x-5)^2 = x^2 - 10x + 25$

15 Yes. $f(x) = -(x-1)^2 + 4$

19 No

21 Vertex is $(6, 8)$
Axis of symmetry is $x = 6$

23 Vertex is $(-5, 106)$
Axis of symmetry is $x = 5$

29 $y = -(x-3)^2$

31 -2.4% in 2004

33 (a) $y = 0.01x^2 + 2x + 1$

(b) $y = 2.03x + 0.98$

(c) 0.02

(d) 23.52

(e) $-0.791 \le x \le 3.791$

35 (a) $\$1631.25$; $\$1.25$

(b) $P(x-2)$; $\$3.25$

(c) $P(x) = 50,015x - 0.75$

37 Conjecture: $x^2 + 1 \ge 2x$

39 (a) $f(x) = 27 - 3x$, $g(x) = 17 - x$

(b) 3

(c) $1.87, 7.13$

(e) 2.5

41 (b) Maximum height: $t = T/2$

Chapter 5 Review

1 (a) 4

(b) 1

(c) 5

(d) -2

3 (a) $(6, 5)$

(b) $(2, 1)$

(c) $(1/2, 5)$

(d) $(2, 20)$

5 (a) No

(b) $e^x + 1$

7 (a) $y = |x-1| + 2$

(b) $y = (-x-2)/(x+1)$

(c) $y = -(x+1)^3 + 1$

17 0

19 (a) $x = -c\sqrt{c}$ or $x = c\sqrt{c}$

21 $p \approx 15$, $q \approx 7190$

25 (a) $y = 3h(x)$

(b) $y = -h(x-1)$

(c) $y = -h(2-2x)$

29 (a) $L(x) = 12x$

(b) Short by 6 cm; short by 72 cm

(c) $L(x+2)$

(d) $S(x) = 6x^2$

(e) $V(x) = x^3$

(f) $S(x+10)$

(g) $V(x+10)$

(h) $L(x+10)$

(i) $2L(x+10)$

(j) $L(1.2x) = 1.2L(x)$

31 1 second

Section 6.1

7 3 (or 9) o'clock; descending;
5 minutes; 40 meters; 0 meters;
11.25 minutes

9 3 (or 9) o'clock; rising; 4 minutes;
30 meters; 5 meters; 11 minutes

11 (a) Periodic

(b) Not periodic

(c) Not periodic

(d) Not periodic

(e) Periodic

(f) Not periodic

(g) Periodic

13 Midline: $y = 10$;
Period: 1;
Amplitude: 4;
Minimum: 6 cm;
Maximum: 14 cm

15 Graph is same except starts at a peak

17 (b) Period: 5 hours;
Amplitude: 40°;
Midline: $T = 70°$

19 Midline: $h = 2$;
Amplitude: 1;
Period: 1

21 (b) Peak every twelve years

(c) Midline: $y = 15$;
Amplitude: 3;
Period: 12

Section 6.2

1 (a) $(-0.174, 0.985)$

(b) $(-0.940, -0.342)$

(c) $(-0.940, 0.342)$

(d) $(0.707, -0.707)$

(e) $(0.174, -0.985)$

(f) $(1, 0)$

3 $A = (0.87, 0.5)$, $B = (-0.71, 0.71)$,
$C = (0.87, -0.5)$

5 $P = (-1, 0)$, $Q = (-1, 0)$, $R = (0, 1)$

7 $A = (2.60, 1.5)$
$B = (-2.12, 2.12)$
$C = (2.60, -1.5)$

9 (a) $\sin 135° = 0.71$; $\cos 135° = -0.71$

(b) $\sin 285° = -0.97$; $\cos 285° = 0.26$

11 (a) 307°

(b) 127°

13 $2\sqrt{2}$ meters

15 (a) 72°

(b) 180°

(c) 216°

17 (a) a

(b) $-a$

(c) a

(d) a

(e) $-a$

(f) $-a$

19 466.5 feet

21 $(60, 0)$, $(7.5, 0)$
$(60\cos\theta, 60\sin\theta)$
$(7.5\cos\theta, 7.5\sin\theta)$

Section 6.3

1 $\pi/4$

3 0.2967 radians

5 (a) $\pi/6$ or 0.52

(b) $2\pi/3$ or 2.09

(c) $10\pi/9$ or 3.49

(d) $7\pi/4$ or 5.50

7 (a) Yes

(b) Just over 6

9 (a) $-2\pi/3 < 2/3 < 2\pi/3 < 2.3$

(b) $\cos 2.3 < \cos(-2\pi/3) = \cos(2\pi/3) < \cos(2/3)$

11 $\pi/9$ radians or 20°

13 5π feet

15 $\sin\theta = 0.6$;
$\cos\theta = -0.8$

17 $(-0.99, 0.14)$

21 (a) 72.77 cm

(b) 0.2 radians

23   7π inches

25   0.1345 radians

29   $t \approx 0.74$

Section 6.4

3   $A = 1, B = \pi/2, C = 2,$
    $D = \pi, E = 4, F = 3\pi/2,$
    $G = 5$

5   $g(x) = 2\sin x, a = \pi, b = 2$

7   $f(x) = (\sin x) + 1$
    $g(x) = (\sin x) - 1$

11   They are equal

13   (a)   (i) p

          (ii) s

          (iii) q

          (iv) r

15   (a) $1/\sqrt{2}$

     (b) $1/2$

     (c) $-\sqrt{3}/2$

     (d) $-1/2$

     (e) $1/\sqrt{2}$

19   One; three

21   (a) $(\sin(a+h) - \sin a)/h$

     (b) -0.15

23   (b) $g(x) + h(x) = 1$

Section 6.5

1   (a)   Periodic with period 2π

     (b)   Periodic with period 2

     (c)   Not periodic

     (d)   Periodic with period 4π

7   Amplitude: 20
    Period: 1/2
    Phase shift: 0
    Horizontal shift: 0

9   Amplitude: 1;
    Period: 8π;
    Phase Shift: $\pi/4$;
    Horizontal Shift: π (right)

11   Amplitude: 10;
    Period: 20π;
    Phase Shift: $-\pi$;
    Horizontal Shift: -10π (left)

13   $g(t) = -2\cos(t/2) + 2$

15   $y = -2\sin(\pi\theta/6) + 2$

17   $y = \cos(2\pi\theta/13) - 4$

19   $g(\theta) = -3\cos(\pi\theta/2) + 3$

21   $f(x) = \sin x, a = \pi/2, b = \pi,$
    $c = 3\pi/2, d = 2\pi, e = 1$

23   Amplitude: 20
    Period: 3/4 seconds

25   $f(t) = 14 + 10\sin(\pi t + \pi/2)$

27   $h = 20 + 20\sin((2\pi/5)t - \pi)$

29   $f(t) = 20 + 15\sin((\pi/2)t + \pi/2)$

31   (a) $18°$/min

     (b) $\theta = (18t - 90)°$

     (c) $f(t) =$
        $250 + 250\sin(18t - 90)°$

     (d)   Amp = Midline = 250 feet
        Period = 20 min

33   $(\sin x)^2 = -\frac{1}{2}\cos(2x) + \frac{1}{2}$

35   (a) $P = f(t) =$
        $-450\cos(\pi t/6) + 1750$

     (c) $t_1 \approx 1.9; t_2 \approx 10.1$

37   Both f and g have periods of 1, amplitudes of 1, and midlines $y = 0$

45   $y = -2g(x/3)$

47   $y = 2g(x) - 3$

49   (b) $P = f(t) =$
        $-70\cos(\pi t/6) + 160$

     (c) $t \approx 1.67$ min

51   (b) $23.2°$; 12 months

     (c) $T = f(t) =$
        $-23.2\cos((\pi/6)t) + 58.6$

     (d) $T = f(9) \approx 58.6°$

53   $f(t) = 3\sin((\pi/6)(t - 74)) + 15$

Section 6.6

1   0, 1, 0

3   0

5   -1

7   -1

9   $\sqrt{3}$

11   1

13   $1/\sqrt{3}$

15   (a) $\sin\alpha = -\sqrt{22}/5,$
        $\tan\alpha = \sqrt{22/3}$

     (b) $\sin\beta = -4/5,$
        $\cos\beta = -3/5$

17   $\cos\theta = \sqrt{1 - y^2}$

19   $\sin\theta = \sqrt{x^2 - 16}/x,$
    $\tan\theta = \sqrt{x^2 - 16}/4$

21   $\cos\theta = 9/\sqrt{x^2 + 81},$
    $\sin\theta = x/\sqrt{x^2 + 81}$

23   $y = y_0 + (\tan\theta)(x - x_0)$

25   (a) $\ldots, -3\pi/2, -\pi/2, \pi/2, 3\pi/2,$
        $\ldots$; It has t-intercepts.

     (b) $\ldots, -2\pi, -\pi, 0, \pi, 2\pi, \ldots$; It
        has t-intercepts.

29   $u = -5\cos 2$
    $v = 5\sin 2$
    $w = 5\sqrt{2(1 - \cos 2)}$

Section 6.7

1   (a)   1.88, 4.41

     (b)   1.88, 4.41

5   (a) $-1/2$

     (b) $\sqrt{2}/2$

     (c) $-\sqrt{2}/2$

     (d) $-\sqrt{3}/2$

7   (a) $\sqrt{3}/2$

     (b) $-1/\sqrt{2}$

     (c)   1

     (d) $\sqrt{3}/2$

9   $t = 1.8, 4.5$

11   (a) $\theta \approx 0.7, 2.4$

     (b) $t \approx 1.8, 4.5$

     (c) $x \approx 1.2, 4.4$

13   $t = \pi/4, 3\pi/4,$
    $5\pi/4$, or $7\pi/4$

15   $t = \pi/3, 4\pi/3,$
    $5\pi/3$, or $2\pi/3$

17   $\theta = -\pi/4 + 2\pi k, 5\pi/4 + 2\pi k, k$ an integer

19   $\theta = -\pi/6 + \pi k, k$ an integer

21   (b) $d = f(t) = 11.4 +$
        $5.8\cos(\frac{\pi}{6.2}t)$

     (c)   4:20 pm

23   $\theta \approx 0.848$ and $\theta \approx 2.294$

25   $3\pi/4, 7\pi/4$

27   (a) $\pi/3$

     (b) π

     (c) ≈ 0.1

29   (a) $t_1 \approx 0.161$ and $t_2 \approx 0.625$.

     (b) $t_1 = \arcsin(3/5)/4$ and
        $t_2 = \pi/4 - \arcsin(3/5)/4$

31   (a) $\pi/2$

     (b) $\pi/4$

     (c)   0.84

     (d)   1.56

     (e)   Undefined

     (f)   1.11

     (g)   0.91

     (h) -2.19

33   (a) $\pi/3$

     (b) $2\pi/3$

     (c) $1/2$

     (d) $\pi/3$

35   Statement II is always true; statement I isn't always true

37   (a) $d = \sqrt{2rx + x^2}$

     (b) $d = 25,239$ meters

Chapter 6 Review

1   True

3   False

5   True

7   True

9   False

11   False

13   (a)   II

     (b)   III

     (c)   IV

     (d)   I

     (e)   III

17   (a) $\pi/3 = 1.047197551$

     (b) $\sqrt{2}/2 = 0.7071067812$

     (c) $\pi/6 = 0.5235987756$

     (d) $\pi/4 = 0.7853981634$

     (e) $\sqrt{3}/2 = 0.8660254038$

     (f) $\pi/2 = 1.570796327$

19   $\theta \approx 1.42, 1.72, 3.51, 3.82, 5.6, 5.9$

21   $\theta = \pi/4, 3\pi/4, 5\pi/4, 7\pi/4$

23    $\alpha = \pi/3, \, 5\pi/3$

25    $\alpha = \pi/4, \, 5\pi/4$

27   0.6155, 2.5261, 3.7571, 5.6677

29   1.2310, 5.0522, π

31   5.8π inches

33   (a) $(2 - \sqrt{2})/2$ meters

       (b) $3/2$ meters

35    $0.52°$

37   Outer edge: 3770 cm/min;
         Inner edge: 471 cm/min

39   (a)   I

       (b)   I and III

       (c)   II and IV

       (d)   IV

       (e)   III

41   (a) $C(t)$

       (b) $D(t)$

       (c) $A(t)$

       (d) $B(t)$

43   (a) $4\sin(2\pi t) + 10$

       (b)   1/12, 5/12, 13/12, 17/12 sec

45    $f(t) = 150 + 150\sin((\pi/10)t - \pi/2)$

47    $f(t) = -900\cos((\pi/4)t) + 2100$

51    $(1 + \arcsin(-1/3) \pm 2\pi k)/\pi$ and
       $(1 + \pi - \arcsin(-1/3) \pm 2\pi k)/\pi$
       for $k = 0, 1, 2, \ldots$

53   (a)   3 min; 4 min

       (b)   9:15 am

       (c)   12 min

55   (a)   False

       (b)   True

       (c)   True

       (d)   False

57    $6\sin(\pi(t-3)/6) + 25$

Section 7.1

1   (a) $\tan\theta = 2$

      (b) $\sin\theta = 2/\sqrt{5}$

      (c) $\cos\theta = 1/\sqrt{5}$

3   (a) $5/\sqrt{125}$

      (b) $10/\sqrt{125}$

      (c) $10/\sqrt{125}$

      (d) $5/\sqrt{125}$

      (e) $1/2$

      (f)   2

5   0.979

7   184.7 feet

9    $\approx 39.8°$

11   63.4°

13    $d = 3\cos\alpha$ meters

15    $d = 3\tan\phi$ miles

17   (a) $h \approx 88.39$ feet

      (b) $h = 62.5$ feet

      (c) $c \approx 88.39$ feet,
           $d \approx 108.25$ feet

Section 7.2

1    $x \approx 19.12$

3    $b \approx 5.12, c \approx 6.50; \beta = 52°$

5    $a \approx 11.82, b \approx 2.08; \theta = 80°$

7    $a \approx 10.46; \theta \approx 16.6°, \psi \approx 143.4°$

9    $a_1 \approx 487.07$ ft; $\alpha_1 \approx 70.1°, \beta_1 \approx$
      $79.9°$ or
      $a_2 \approx 396.23$ ft; $\alpha_2 \approx 49.9°, \beta_2 \approx$
      $100.1°$

11    $b \approx 7.9$ m; $\alpha \approx 62.4°, \gamma \approx 95.6°$

13   (a) ≈ 4.95 and 4.24

       (b) ≈ 17

15   (a) $\sin\theta = 0.282$

       (b) $\theta \approx 16.37°$

       (c) 12.08 cm^2

19   Length of arc ≈ 2.617994 ft
       Length of chord ≈ 2.588190 ft

21   B closer by 23.87 feet

23   984 feet

25   125.04 feet

27   159 feet

Section 7.3

3    $(\cos(2\theta))^2 + (\sin(2\theta))^2 = 1$

9   a and i; b and l;
      c and d and f; e and g;
      h and j

11   Not an identity

13   Not an identity

15   Not an identity

17   Identity

19   Not an identity

23   (a) $\theta = 60°, \, 180°,$ and $300°$

       (b) $\theta = \frac{7\pi}{6}, \frac{3\pi}{2}, \frac{11\pi}{6}$

25   (a) $\cos\theta = x$

       (b) $\cos(\pi/2 - \theta) = \sqrt{1 - x^2}$

       (c) $\tan^2\theta = (1 - x^2)/(x^2)$

       (d) $\sin 2\theta = 2\sqrt{1 - x^2}(x)$

       (e) $\cos 4\theta = 1 - 8x^2(1 - x^2)$

       (f) $\sin(\cos^{-1} x) = \sqrt{1 - x^2}$

27    $\sin 2\theta = -24/25;$
      $\cos 2\theta = 7/25;$
      $\tan 2\theta = -24/7$

29    $\cos 2\theta = 1 - 2(x + 1)^2/25$

31    $2(2\cos^2\theta - 1)^2 - 1$

33   (a) $f(x) = 3|\cos(\sin^{-1}(x/3))|$

       (b) $g(x) = \sqrt{5}|\cos(\sin^{-1}(x/\sqrt{5}))|$

Section 7.4

1   (a)   0.839

      (b) -0.454

      (c)   0.545

      (d)   0.891

3    $\sin 15° = \cos 75° = (\sqrt{6} - \sqrt{2})/4$
      $\cos 15° = \sin 75° = (\sqrt{6} + \sqrt{2})/4$

5    $10\sin(t - 0.64)$

7    $\sqrt{29}\sin(3t + 1.95)$

15    $\cos 3\theta = 4\cos^3\theta - 3\cos\theta$

17    $\sin(\ln(xy)) \approx 0.515160$

21   (c) $+, -, -$

       (d) $-, -, +$

       (e) $-, +, -$

Section 7.5

1   Maximum: 2;
      Minimum: ≈ -0.94

3   (a) $f(t) = 65 - 25\cos(\pi t/12)$

       (b)   Period: 24;
           Amplitude: 30

       (c) $\approx$4 am and 1 pm

       (d)   184 mw at 3:21 pm

       (e) $h(t) = 145 - \sqrt{1525}\sin\left(\frac{\pi}{12}t + 0.6947\right);$
           $\max = 145 + \sqrt{1525}$

5   (a) $m = 2.5; b = 20; A = 10$

       (b)   Roughly in January and December

       (c)   Roughly between May and September

7   (a) $y = 1$

       (b) f oscillates faster and faster between -1 and 1 as t increases.

       (c) ≈ 0.54

       (d) $t_1 = \ln(\pi/2)$

       (e) $t_2 = \ln(3\pi/2)$

9   (b)   No; no, but possibly for 1979-1989

       (c)   Multiply by an exponential function:
           $f(t) = (e^{0.05t})(-1.4\cos(\frac{2\pi}{6})t + 1.6)$

       (e)   4.6 billion dollars

11   (a)   1 meter

       (b)   2

       (d) $t = k/2$ for any integer k

Section 7.6

1    $H : x = 0, y = 3; r = 3, \theta = \pi/2$
      $M : x = 0, y = 4; r = 4, \theta = \pi/2$

3    $H : x = -3, y = 0; r = 3, \theta = \pi$
      $M : x = 0, y = 4; r = 4, \theta = \pi/2$

5    $H : x = 3\sqrt{2}/2, y = 3\sqrt{2}/2; r = 3, \theta = \pi/4$
      $M : x = 0, y = -4; r = 4, \theta = 3\pi/2$

7    $H : x \approx 2.90, y \approx -0.78; r = 3, \theta = 23\pi/12$
      $M : x = 0, y = 4; r = 4, \theta = 3\pi/2$

9    $\sqrt{8} \leq r \leq \sqrt{18}$ and $\pi/4 \leq \theta \leq \pi/2$

11    $0 \leq \theta \leq \pi/2$ and $1 \leq r \leq 2/\cos\theta$

13   n loops

15   An inner and outer loop

17    $0 \leq \theta \leq 2\pi$ and $3/16 \leq r \leq 1/2$

19   (a) $0 \leq \theta \leq \pi/4$ and $0 \leq r \leq 1$

(h) Two pieces:
$0 \le x \le \sqrt{2}/2$ and $0 \le y \le x$;
$\sqrt{2}/2 \le x \le 1$ and $0 \le y \le \sqrt{1-x^2}$

Chapter 7 Review

1 $\phi \approx 53.1°; \theta = 36.9°$
3 $\theta \approx 22.6°$
5 $5; 67.38°, 22.62°$
7 Both are right
9 $\cos\theta = 3/\sqrt{14}$
11 $7\pi/6, 11\pi/6, \pi/2$
13 Appear to be same
15 $\alpha = 11.31°; \beta = 16.70°$
19 4 rolls
23 (a) True
   (b) True
   (c) True
   (d) False
25 (a) OE
   (b) OA
   (c) DB
   (d) OF
   (e) OC
   (f) GH

Section 8.1

1 $r(0) = 4, r(1) = 5, r(2) = 2, r(3) = 0, r(4) = 3, r(5)$ undefined
3 $f : 2, 2, 0$
  $g : -2, -2, 0$
  $h : 0, -2, -2, 0, -2, -2, 0$
7 $9x^2 - 15x + 4$
9 (a) $-3(x^2 + x)$
  (b) $2 - x$
  (c) $x^2 + x + \pi$
  (d) $\sqrt{(x^2 + x)}$
  (e) $2/(x + 1)$
  (f) $(x^2 + x)^2$
11 $1/(x^2 - 2)$
13 $\sqrt{x^2 + 1}$
15 $1/(x - 2)$
17 $m(k(x)) = 1/(x^2 - 1)$
19 $n(k(x)) = 2x^4/(x^2 + 1)$
21 $n(m(x)) = 2/(x(x - 1))$
23 $\left(m(x)\right)^2 = 1/(x^2 - 2x + 1)$
25 $n(n(x)) = 8x^4/((x + 1)(8x^4 + x + 1))$
27 $2x + h + 1$
29 $-1/(x(x + h))$
31 13
33 $u(x) = \sqrt{x}, v(x) = 3 - 5x$
35 $u(x) = \sqrt{x}, v(x) = x + 8$
37 $u(x) = 1 - x, v(x) = \sqrt{x}$
39 $u(x) = 1/x, v(x) = x^2$
41 $u(x) = 1/x, v(x) = 1 - x$
43 $u(x) = 3^x, v(x) = 2x - 1$
45 $u(x) = \sqrt{x}, v(x) = 1 - 4x^2$
47 $u(x) = \sqrt{x}$
   $v(x) = 1 - x$
   $w(x) = x^2$
49 $u(x) = 1 - x$
   $v(x) = \sqrt{x}$
   $w(x) = x - 1$
51 $u(x) = x^2$
   $v(x) = 1 + x$
   $w(x) = 1/x$
53 $h(x) = x^3$
55 $f(x) = 1/x$
57 $f(x) = x + 1$ or $f(x) = -(x + 1)$
59 $j(x) = (x^2 + 3)^2 + 3$
61 $v(x) = x + 1/x$
63 (a) $h(x) = 3x^2$
   (b) $j(x) = 2x + 1$
65 (a) $f(x) = 84.62x$
       $g(x) = 1.0908x$
       $h(x) = 0.0129x$
   (b) \$1000 trades for yen, then for 1091.598 euros

Section 8.2

1 (a) $f^{-1}(R) = (1/5)R - 30$
3 Invertible
5 Invertible
7 Not invertible
9 $f^{-1}(x) = x^{1/3}$
11 (a) 2
   (b) Unknown
   (c) 5
13 $f^{-1}(3) < f(3) < 0 < f(0) < f^{-1}(0) < 3$
19 $g^{-1}(x) = (x - 5)/2$
21 $h^{-1}(x) = \sqrt[3]{x/12}$
23 $k^{-1}(x) = (2x + 2)/(x - 1)$
25 $f^{-1}(x) = \log x$
27 $k^{-1}(x) = \frac{1}{2}\ln(x/3)$
29 $n^{-1}(x) = 10^x + 3$
31 $j^{-1}(x) = 1 + 10^{(x-2)}$
33 $f^{-1}(x) = (2x - 3)/(5x + 2)$
35 $f^{-1}(x) = (9 - 1/x)^2 + 4$
37 $m^{-1}(x) = 1/(x^2 - 1)$
39 $q^{-1}(x) = (3 + 5e^x)/(e^x - 1)$
43 $x = \frac{1}{3}\arcsin\left(\frac{2}{7}\right) \pm k\left(\frac{2\pi}{3}\right)$,
   $x = \pi - \frac{1}{3}\arcsin\left(\frac{2}{7}\right) \pm k\left(\frac{2\pi}{3}\right)$ for integer k
45 $x = 1.09^{1/1.05}$
47 $x = -7/2$
49 (a) $A = \pi r^2$
   (c) $r \ge 0$
   (d) $f^{-1}(A) = \sqrt{A/\pi}$
   (f) Yes
51 (a) $P(t) = 0.75t + 14.25$
   (b) $P(20) = 29.25$;
       $P(-10) = 6.75$

(c) Slope is 0.75; $(0, 14.25)$, $(-19, 0)$
(d) $P^{-1}(y) = \frac{4}{3}y - 19$
(e) $P^{-1}(20) = 23/3$,
    $P^{-1}(5) = -37/3$
53 (b) There will be approximately 325,500 people after 50 years
   (c) $f^{-1}(P) = (\log P - \log 37.8)/\log 1.044$
   (d) $f^{-1}(50) = 6.5$
55 (a) $f(t) = 7.11(1.08998)^t$
   (b) $f^{-1}(P) = (\log(P/7.11))/(\log 1.08998)$
   (c) $f(25) = 61.3$
       $f^{-1}(25) = 14.59$
57 (a) $C(0) = 99\%$
   (b) $C(x) = (99 - x)/(100 - x)$
   (c) $C^{-1}(y) = (99 - 100y)/(1 - y)$
59 (b) Domain: all $x \ge 0$;
       Range: $f(x) \ge 0$
   (d) Yes
61 $f^{-1}(x) = h^{-1}(g^{-1}(x))$
63 $1 - t^2$

Section 8.3

5 (a) $h(x) = x^2 + x, h(3) = 12$
  (b) $j(x) = x^2 - 2x - 3, j(3) = 0$
  (c) $k(x) = x^3 + x^2 - x - 1,$
      $k(3) = 32$
  (d) $m(x) = x - 1$ for $x \ne -1$,
      $m(3) = 2$
  (e) $n(x) = 2x + 2, n(3) = 8$
7 $g(x) = (1/x) - 1$
9 $j(x) = 2x^2 - x$
11 $l(x) = 3x - (1/x) - 2$
13 (a) $P(t) = 4{,}017{,}857{,}143 \cdot (1.023)^t$
   (b) Population increases by 2.3%
   (c) $N(t) = 30{,}000 - 900t$
   (d) $f(0) = 0.00000747$
       $f(5) = 0.00000567$
       $f(10) = 0.00000416$
       $f(15) = 0.00000290$
   (e) Neither
   (f) Per capita number of warheads
17 (a) Yes
   (b) The function has no zeros
21 False
25 40

Chapter 8 Review

1 (a) $f(2x) = 4x^2 + 2x$
  (b) $g(x^2) = 2x^2 - 3$
  (c) $h(1 - x) = (1 - x)/x$
  (d) $(f(x))^2 = (x^2 + x)^2$
  (e) $g^{-1}(x) = (x + 3)/2$
  (f) $(h(x))^{-1} = (1 - x)/x$
  (g) $f(x)g(x) = (x^2 + x)(2x - 3)$

(h) $h(f(x)) = (x^2 + x)/(1 - x^2 - x)$

3 Horizontal line $y = 4$

5 (a) $f(t) = 800 - 14t$ gals

   (b) (i) 800 gals

       (ii) 57.1 days

       (iii) 28.6 days

       (iv) $14t$

7 (a) $P(t) = P_0(3)^{t/7}$

   (b) 17%

   (c) $P^{-1}(x) = (7\log(x/P_0))/\log 3$

   (d) 4.42 years

31 $x/(1 + x) + x + 1$

33 (b) Not linear

   (c) Yes

35 $g(2000) = 100$, the dollar cost per square foot for building 2000 square feet of office space

37 $g(q) < g(p) < f(p) < f(q)$

39 $j(x) = \dfrac{x}{h(x)}$

41 $g(x) = x + 1$, $h(x) = 2x$

43 $g(x) = \sqrt{x}$, $h(x) = 1 + \sqrt{x}$

45 $g(x) = \dfrac{1}{x^2}$, $h(x) = x + 4$

47 $h(x) = \sqrt{x} + 3$

49 $h(x) = \dfrac{x}{1 + \sqrt{x}}$

51 $f^{-1}(x) = (x + 7)/3$

53 $j^{-1}(x) = (x^2 - 1)^2$

55 $k^{-1}(x) = (3 - 2x)^2/(x + 1)^2$

57 $f^{-1}(x) = \left(\log\left(\frac{1-3x}{3x-3}\right)\right)/\log 2$

59 $h^{-1}(x) = (5 + 4 \cdot 10^x)/(10^x - 1)$

61 $g^{-1}(x) = \arcsin(\ln x/\ln 2)$

63 $j^{-1}(x) = \arcsin(2x/(x + 1))$

65 False

67 False

69 False

71 Increasing

73 Increasing

75 Can't tell

77 (a) $f(8) = 2$, $f(17) = 2$, $f(29) = 2$, $f(99) = 0$

   (b) $f(3x) = 0$

   (c) No

   (d) $f(f(x)) = f(x)$

   (e) No

Section 9.1

1 (a) A - (iii)

   (b) B - (ii)

   (c) C - (iv)

   (d) D - (i)

5 (a) $x^{-10} \to +\infty$, $-x^{10} \to 0$

   (b) $x^{-10} \to 0$, $-x^{10} \to -\infty$

   (c) $x^{-10} \to 0$, $-x^{10} \to -\infty$

9 (a) $f(x) = x^{1/n}$
       $g(x) = x^n$

(b) $(1, 1)$

11 19.276 cm

13 (a) $h(x) = -2x^2$

   (b) $j(x) = \frac{1}{4}x^2$

15 (a) $p < 0$, $x \neq 0$

   (b) $p > 0$, $y \geq 0$;
       $p < 0$, $y > 0$

   (c) $p > 0$, y is any real;
       $p < 0$, $y \neq 0$

   (d) p even: y-axis symmetry;
       p odd: origin symmetry

17 4320 earth days

Section 9.2

1 (a) Neither

   (b) $j(x) = 3 \cdot 9^x$

   (c) Neither

   (d) $n(x) = 6 \cdot 8^x$

   (e) $p(x) = 25^x$

   (f) Neither

   (g) $r(x) = 2(\frac{1}{9})^x$

   (h) $s(x) = \frac{4}{5}x^3$

3 3^{-x} approaches zero faster

5 (a) C, $f(x)$;
       A, $g(x)$;
       B, $h(x)$

   (b) Yes

   (c) No

7 (a) $v = 40$

   (b) $r(x) > t(x)$

   (c) $t(x) > r(x)$

9 $j(x) = 2x^3$

11 $g(x) = -\frac{1}{6}x^3$

13 $f(x) = (3/2) \cdot x^{-2}$

15 $g(x) = 3x^{-2}$

17 (a) $f(x) = 112x - 96$

   (b) $f(x) = 2(8)^x$

   (c) $f(x) = 16x^3$

19 $f(x) = 2\sin(\frac{\pi}{2}x) + 4$ (trigonometric);
   $g(x) = -\frac{5}{2}x^3$ (power function);
   $h(x) = \frac{1}{3}(\frac{1}{2})^x$ (exponential)

21 Formula not unique

23 ≈ 129 Earth days

25 $f(d) = bd^{p/q}$, $p < q$;
   $g(d) = ad^{p/q}$, $p > q$

Section 9.3

1 3rd degree

3 Degree: 3; Terms: 3;
   $x \to -\infty$: $y \to -\infty$;
   $x \to +\infty$: $y \to +\infty$

5 Degree: 4; Terms: 3;
   $x \to \pm\infty$: $y \to -\infty$

9 -16.54

11 (a) $\frac{1}{2}x$

   (c) $x > 158$ or $x < -158$.

13 $-1.764 < x < 0.875$, or $x > 3.889$

15 (b) $1000 per unit

   (c) Profit: $1000 < x < 2000$;
       Break even: $x = 1000$ or
       $x = 2000$;
       Loss: $x > 2000$ or
       $0 < x < 1000$

17 (c) $T \approx 3.96°C$

21 (a) False

   (b) False

   (c) True

   (d) False

23 (a) $p_5(r) = 1000[(1 + r)^5 + (1 + r)^4 + (1+r)^3 + (1+r)^2 + (1 + r) + 1]$;
      $p_{10}(r) = 1000[(1+r)^{10} + (1 + r)^9 + (1+r)^8 + (1+r)^7 + (1 + r)^6 + (1 + r)^5 + (1 + r)^4 + (1 + r)^3 + (1 + r)^2 + (1 + r) + 1]$

   (b) Approximately 20%

Section 9.4

1 C

5 $f(x) = x(x + 2)^2(x - 3)$

7 $y = -\frac{3}{2}(x + 4)(x + 2)(x - 2)$

9 $y = k(x + 3)(x + 1)(x - 2)^2$, for $k > 0$

11 $y = -\frac{1}{3}(x + 2)^2(x)(x - 2)^2$

13 $g(x) = -\frac{1}{4}(x + 2)^2(x - 2)(x - 3)$

15 $g(x) = -\frac{1}{3}(x + 2)(x - 2)x^2$

17 $j(x) = (x + 3)(x + 2)(x + 1) + 4$

19 $f(x) = 4x(2x + 5)(x - 3)$;
   Zeros: 0, $-5/2$, 3

21 $f(x) = x^2 - 2x$

23 $f(x) = -\frac{1}{2}(x + 3)(x - 1)(x - 4)$

25 $p(x) = x^2 + 2x - 3$

27 $x = -2$ or $x = -3$

29 $x = \pm\frac{1}{2}$

31 $x = (3 \pm \sqrt{33})/4$

33 (a) $V(x) = x(6 - 2x)(8 - 2x)$

   (b) $0 < x < 3$

   (d) ≈ 24.26 in^3

35 7.82 by 5.32 by 1.59 inches

37 (b) No

   (c) 3; $-6 \leq x \leq 3$, $-3 \leq y \leq 3$

   (d) $f(x) = (x + 5)(x - 1)(x - 2)^2$

   (e) 3; No

39 (a) Never true

   (b) Sometimes true

   (c) Sometimes true

   (d) Never true

   (e) True

   (f) Never true

Section 9.5

1 (c)   Horizontal: $y = 0$
            Vertical: $x = 3$
3 (c)   Horizontal: $y = 1$
            Vertical: $x = 0$
5   As $x \to \pm\infty$, $f(x) \to 1$, $g(x) \to x$, and $h(x) \to 0$
7   $y = 1$
9   $y = 4$
11 $f^{-1}(x) = (4x + 4)/(5x + 3)$
13 (a) $C(x) = (1 + x)/(2 + x)$
     (b) $C(0.5) = 60\%$;
            $C(-0.5) = 33\%$
15 (a) $C(n_0)/n_0$
     (b)   Slope is average cost for n_0 units
17 (a) $C(x) = 30000 + 3x$
     (b) $a(x) = 3 + 30000/x$
     (f) $a^{-1}(y) = 30000/(y - 3)$
     (g)   15,000
19 (a) $h(x) = (2x^3 + 2)/(3x^2)$

Section 9.6

1   Zero: $x = -3$;
     Asymptote: $x = -5$;
     $y \to 1$ as $x \to \pm\infty$
3   Zeros: $x = 4$;
     Asymptote: $x = \pm 3$;
     $y \to 0$ as $x \to \pm\infty$
7   (a)   (iii)
     (b)   (i)
     (c)   (ii)
     (d)   (iv)
     (e)   (vi)
     (f)   (v)
9   (b) $f(x) = -x(x - 2)$
11 (a) $y = (1/x) - 3$
     (b) $y = (-3x + 1)/x$
     (c) $(1/3, 0)$
13 (a) $y = 1/(x - 1) + 2$
     (b) $y = (2x - 1)/(x - 1)$
     (c) $(1/2, 0)$ and $(0, 1)$
15 (a) $y = 1/(x - 2)^2 - 1$
     (b) $y = (-x^2 + 4x - 3)/(x^2 - 4x + 4)$
     (c) $(0, -3/4)$, $(1, 0)$ and $(3, 0)$
17 $p = 1$, $(0, 11/3)$, $(11/4, 0)$
     $x = 3$, $y = 4$
19 $p = 1$, $(0, 3)$, $(3/2, 0)$
     $x - 1$, $y = 2$
21 (a) $1/x$
     (b) $y = (1/x) + 2$
23 (a) $y = 1/x$
     (b) $y = x/(2x - 4)$
25 $y = -(x + 1)/(x - 2)$
27 $y = x/((x + 2)(x - 3))$
29 $y = -(x - 3)(x + 2)/((x + 1)(x - 2))$
31 $y = (x - 2)/((x + 1)(x - 1))$
33 $g(x) = (x - 5)/((x + 2)(x - 3))$

35 7.12 by 4.62 by 1.94 inches
37 $y = x^2 + 1$; $(-5, 26)$

Chapter 9 Review

1   $y = 2\cos(\pi x) + 1$
3   $y = 2(x^2 - 4)(x - 1)$
5   $y = \frac{1}{2}(5^x)$
7   Graph (i): J;
     Graph (ii): L;
     Graph (iii): O;
     Graph (iv): H
9   $g(x) = -\frac{1}{3}(x^2)(x + 2)(x - 2)$
11 $f(x) = \frac{1}{36}(x + 4)^2(x - 3)^2$
13 $f(x) = kx^3(x + 1)(x - 2)$ for $k > 0$
17 27.4% of earth's radius
21 (a) $f^{-1}(x) = 5x/(1 - x)$
     (b) $f^{-1}(0.2) = 1.25$
     (c) $x = 0$
     (d) $y = -5$
23 $C < A < 0 < B < D$
25 $A < 0 < C < D < B$
27 $p(x) = \frac{7}{1080}(x + 3)(x - 2)(x - 5)(x - 6)^2$
29 $f(x) = -3(x - 2)(x - 3)/(x - 5)^2$
31 (b)   Second
     (c)   Horizontal: $y = \pi R^2$;
            Vertical: $v = 0$
33 (a) $-3/r$ is attractive;
            $1/r^2$ is repulsive
     (c) $r = 0.67$
     (d) $y = 0$
     (e) $r = 0$
35 (a)   2.25
     (b)   Circle of radius r_0
37 (a)   1.2 meters/sec
     (b)   21% longer than the existing ship

Section 10.1

1   Scalar
3   Vector
5   $\vec{p} = 2\vec{w}$
     $\vec{q} = -\vec{u}$
     $\vec{r} = \vec{u} + \vec{w}$
     $\vec{s} = 2\vec{w} - \vec{u}$
     $\vec{t} = \vec{u} - \vec{w}$
11 4.64 miles; 27.32° north of east
13 14,710 meters;
     angle of 17.82° from horizontal

Section 10.2

1   $\vec{i} + 3\vec{j}$
5   $-3\vec{i} - 4\vec{j}$
7   (a) $2\vec{i} + \vec{j}$
     (b) $2\vec{i} - 2\vec{j}$
     (c) $3\vec{i}$
     (d) $-\vec{i} - 2\vec{j}$
     (e) $\vec{i} + \vec{j}$
     (f) $\vec{i} + 2\vec{j}$
     (g) $\vec{i}$

     (h) $\vec{j}$
9   90° or $\pi/2$
11 $\sqrt{6} \approx 2.4$
13 7.6
15 $21\vec{j} + 35\vec{k}$
17 (a) $50\vec{i}$
     (b) $-50\vec{j}$
     (c) $25\sqrt{2}\vec{i} - 25\sqrt{2}\vec{j}$
     (d) $-25\sqrt{2}\vec{i} + 25\sqrt{2}\vec{j}$
19 $\vec{u}$ and $\vec{w}$; $\vec{v}$ and $\vec{q}$.
21 $\vec{i} + 4\vec{j}$
23 $-\vec{i}$

Section 10.3

1   (a) $\vec{v} = 4.33\vec{i} + 2.5\vec{j}$
            For the second leg of his journey,
            $\vec{w} = x\vec{i}$
     (b) $x = 9.87$
     (c)   14.42
3   $\vec{q}_a = 1.06\vec{i} + 1.97\vec{j}$; $(1.06, 1.97)$
     $\vec{q}_b = 2.70\vec{i} + 3.11\vec{j}$; $(2.70, 3.11)$
     $\vec{q}_c = 2.13\vec{i} + 3.93\vec{j}$; $(2.13, 3.93)$
     $\vec{q}_d = 0.49\vec{i} + 2.79\vec{j}$; $(0.49, 2.79)$
5   48.3° east of north
     744 km/hr

Section 10.4

1   -38
3   14
5   238
7   2100 ft-lbs
9   1.91 radians (109.5°)
11 for both, max = 11, min = 3
13 No
17 Bantus
19 43.297°

Chapter 10 Review

1   $-4.5\vec{i} + 8\vec{j} + 0.5\vec{k}$
3   $\vec{a} = \vec{b} = \vec{c} = 3\vec{k}$
     $\vec{d} = 2\vec{i} + 3\vec{k}$
     $\vec{e} = \vec{j}$
     $\vec{f} = -2\vec{i}$
5   $\|\vec{u}\| = \sqrt{6}$
     $\|\vec{v}\| = \sqrt{5}$
7   (a)   Yes
     (b)   No
9   548.6 km/hr
13 $F = g\sin\theta$
19 $\overrightarrow{AB} = -\vec{u}$; $\overrightarrow{BC} = 3\vec{v}$; $\overrightarrow{AC} = \overrightarrow{AB} + \overrightarrow{BC} = -\vec{u} + 3\vec{v}$; $\overrightarrow{AD} = 3\vec{v}$
21 $3\vec{n} - 3\vec{m}$;
     $3\vec{m} + \vec{n}$;
     $4\vec{m} - \vec{n}$;
     $\vec{m} - 2\vec{n}$

Section 11.1

1   $\sum_{n=1}^{7} 3n$
3   $\sum_{n=1}^{8} (1/2)n$
5   315
7   3825

9 92

11 1600

13 $16n^2$

15 (a) $1, n$

    (b) $n, n(n+1)/2$

17 (b) $c_n = 300 + \frac{1}{2}(n-1)(3n-46)$

       $c_{12} = 245, c_{50} = 2848$

    (c) 19

Section 11.2

1 Yes, $a = 2$, ratio $= 1/2$.

3 No. Ratio between successive terms is not constant

5 Yes, $a = 1$, ratio $= -x$.

7 Yes, $a = y^2$, ratio $= y$.

9 $1/(1+x), |x| < 1$

11 $\sum_{n=0}^{4} 4^n$

13 $\sum_{n=0}^{6} 2(5^n)$

15 889

17 5.997

19 3,985,805

21 1/54

23 8184

25 (a) $0.23 + 0.23(0.01) + 0.23(0.01)^2 + \ldots$

    (b) $0.23/(1-0.01) = 23/99$

27 613/99

29 431/900

31 3781/4950

33 $Q_3 = 260.40$

    $Q_{40} = 260.417$

    Approaches 260.417 mg

37 \$25,503

39 \$1081.11

Section 11.3

1 $y = 3x - 5$

3 $y = 2x^2 + 1$

5 $y = (x+4)^2$

7 $y = x^{3/2}$

9 $y = (2/3)\ln x, x > 0$

11 $(x-3)^2 + (y-2)^2 = 1$

13 Lines from $(0,1)$ to $(1,0)$ to $(0,-1)$ to $(-1,0)$ to $(0,1)$

15 Lines from $(-1,1)$ to $(1,1)$ to $(-1,-1)$ to $(1,-1)$ to $(-1,1)$

17 (a) $x = t, y = t^2$

      $x = t+1, y = (t+1)^2$

    (b) $x = t, y = (t+2)^2 + 1$

      $x = t+1, y = (t+3)^2 + 1$

19 Clockwise for all t

21 Counterclockwise: $t > 0$

23 (a) Right of $(2,4)$

    (b) $(-1,-3)$ to $(2,4)$

    (c) $t < -2/3$

25 (a) Line $y = x$

    (b) Circle, with starting point $(1,0)$ and period 2π

    (c) Ellipse, with starting point $(1,0)$ and period 2π

27 $x = t, y = -4t + 7$

35 (a) $x = t, y = -16t^2 + 48t + 6$

    (c) 6 feet

    (d) 3 seconds

    (e) 42 feet

Section 11.4

1 (a) Center $(2,-4)$, radius $\sqrt{20}$

    (b) Center $(-1,2)$, radius $\sqrt{11}$

3 (a) Center $(-5,2)$; Vertices $(-5 \pm \sqrt{6}, 2)$; Asymptotes $y = \pm(2/\sqrt{6})(x+5) + 2$

    (b) Center $(-1,-2)$; Vertices $(-1 \pm \sqrt{14}, -2)$; Asymptotes $y = \pm(x+1) - 2$

5 Ellipse, centered $(-1,1)$ horizontal axis of length 8 vertical axis of length 4

7 Hyperbola, centered $(-1,-3)$ width of rectangle: 2 length: 6

9 $(x-2)^2 + (y-3)^2 = 36$

11 $\frac{(y-4)^2}{16} - \frac{(x+2)^2}{16} = 1$

13 $\frac{(x+4)^2}{4} + \frac{(y+2)^2}{16} = 1$

15 $x = 2 + 5\cos t, y = 1 + 5\sin t,$ $0 \le t \le 2\pi$

17 $(4 + 4\cos(t-\pi/2), 4 + 4\sin(t-\pi/2))$

19 $x = -3\cos t, y = -7\sin t,$ $0 \le t \le 2\pi$

21 Parabola:

    $y = (x-2)^2, 1 \le x \le 3$

23 Implicit: $xy = 1, x > 0$

    Explicit: $y = \frac{1}{x}, x > 0$

    Parametric: $x = t, y = \frac{1}{t}, t > 0$

25 Implicit:

    $x^2 - 2x + y^2 = 0, y < 0$,

    Explicit:

    $y = -\sqrt{-x^2 + 2x},$

    Parametric:

    $x = 1 + \cos t, y = \sin t,$

    with $\pi \le t \le 2\pi$

27 (a) $a = b = 0, k = 5$ or -5

    (b) $a = 0, b = 5, k = 5$ or -5

    (c) $a = 10, b = -10, k = \sqrt{200}$ or $-\sqrt{200}$

Section 11.5

1 $2e^{i\pi/2}$

3 $5e^{i4.069}$

5 $e^{i3\pi/2}$

7 $-3 - 4i$

9 $-5 + 12i$

11 $3 - 6i$

13 $\frac{\sqrt{3}}{2} + \frac{i}{2}$

15 $\sqrt{2} + i\sqrt{2}$

17 $\frac{\sqrt{3}}{2} + \frac{i}{2}$

19 $\sqrt[3]{2} \cdot \frac{\sqrt{3}}{2} + i\sqrt[3]{2} \cdot \frac{1}{2}$

21 $\frac{1}{\sqrt{2}}\cos(\frac{-\pi}{12}) + i\frac{1}{\sqrt{2}}\sin(\frac{-\pi}{12})$

23 $A_1 = 1 - i, A_2 = 1 + i$

25 (a) $z_1 z_2 = 6 - i2\sqrt{3}$

      $\frac{z_1}{z_2} = i\sqrt{3}$

    (b) $z_1 = 2\sqrt{3}e^{i7\pi/6},$ $z_2 = 2e^{i2\pi/3}$

27 True

29 False

31 True

Section 11.6

3 $\sinh x \to (e^x)/2$ as $x \to \infty$

   $\sinh x \to -(e^{-x})/2$ as $x \to -\infty$

5 $\cosh 2x = \cosh^2 x + \sinh^2 x$

9 $\sin(ix) = i\sinh x$

Chapter 11 Review

1 $\sum_{n=0}^{10}(100 - 10n)$

3 603, 59

5 Yes, $a = 1$, ratio $= 2z$.

7 $1/(1 - 2z), |z| < 1/2$

9 (a) $h_n = 10(3/4)^n$

    (b) $D_1 = 10$ feet

      $D_2 = h_0 + 2h_1 = 25$ feet

      $D_3 = h_0 + 2h_1 + 2h_2 = 36.25$ feet

      $D_4 = h_0 + 2h_1 + 2h_2 + 2h_3 \approx 44.69$ feet

    (c) $D_n = 10 + 60\left(1 - (3/4)^{n-1}\right)$

11 (a) \$1250

    (b) 12.50

13 $x = t, y = 5$

15 $(2\cos(t + \pi/2), 2\sin(t + \pi/2))$

17 Equation of curve is $x = 1 - 2y^2, -1 \le y \le 1$

19 (a) $(x,y) = (t,1)$

    (b) $(x,y) = (t + \cos t, 1 - \sin t)$

21 $13e^{i-1.176}$

23 $-125i$

25 $8i\sqrt[3]{2}$

Appendix A

1 64

2 25

3 121

4 10,000

5 1

6 -1

7 5

8 25

9 1,000

10 1

11 1

12 2

13 4

14 8

15 16

16 4

17 x^2

18 y^4

19 w^4z^2

20 $x^{5/2}y^2$
21 $4x^{3/2}$
22 $7w^{9/2}$
23 $5x^{3/2}z^2$
24 r
25 $r^{3/2}$
26 r^2
27 $6t$
28 $8s^{7/2}$
29 $5\sqrt{2}x^2y^3$
30 $4\sqrt{3}u^5v^6y^{5/2}$
31 $4m^2$
32 $6s^{3/2}t^4v^4$
33 2
34 4
35 2
36 8
37 32
38 1024
39 1/3
40 1/9
41 $1/(3\sqrt{3})$
42 1/25
43 1/625
44 1/125
45 49
46 16
47 -1
48 -121
49 -18
50 1
51 2100
52 -5
53 4
54 -6
55 0.2
56 4
57 3
58 0.5
59 100,000
60 $0.16x^2y^4$
61 3^{x+1}
62 $8L$
63 $70w^{5/6}$
64 $16S^2xt^2$
65 e^x
66 $A^3/\left(3B^3\right)$
67 e^{kt+4}
68 $(M+2)^2$
69 $9x^5$
70 x^{3e}
71 e^{2y}/y^4
72 $4x^{(3\pi-1)}$
73 $64A$
74 $3a$
75 $4u^4v^2w^4$
76 $ab/(b+a)$
77 $25(2b+1)^{20}$
78 -8
79 -8
80 -125

81 Not a real number
82 20,736
83 1/512
84 -512
85 Not a real number
86 243
87 False
88 False
89 True
90 False
91 False
92 False
93 True
94 False
95 False
96 True

Appendix B

1   $x-7$
2   $x-25$
3   $3x+6$
4   $5x-15$
5   $6x-14$
6   $-4y-24$
7   $12x+12y$
8   $-35x+56y$
9   $2x^2+5x$
10 $6xz-27z^2$
11 $-50r^2-60r^2s$
12 $3x^2-2x-16$
13 $5xz-10z-3x+6$
14 x^2+4x+3
15 $x^2+4x-12$
16 $10x^2-17x+3$
17 $3x^2-2x-16$
18 $yz+3y+z+3$
19 $96wy+84y-40w-35$
20 $5xz-10z-3x+6$
21 $4x^2+11x-20$
22 $-200t$
23 $Pp^2-6Ppq+9Pq^2$
24 $A^4-2A^2B^2+B^4$
25 $4x^2-24x+43$
26 $-2x-2\sqrt{2x}-1$
27 2^u+u2^{2u}
28 KRr^2-Kr^3
29 $30+72/x+2x$
30 $(e^{2x}+2+e^{-2x})/4$

Appendix C

1   $2(x+3)$
2   $3(y+5)$
3   $5(z-6)$
4   $2(2t-3)$
5   $5(2w-5)$
6   $u(u-2)$
7   $u^3(3u-4)$
8   $3u^2(u^5+4)$
9   $6x(2x^2y^2-3)$
10 $7rs(2r^3s-3t)$
11 $(x+2)(x+1)$
12 Cannot be factored

13 $(x-2)(x-1)$
14 Cannot be factored
15 Cannot be factored
16 $(x-3)(x+1)$
17 Cannot be factored
18 $(x+3)(x-1)$
19 $(2x+1)(x+2)$
20 $(3x-4)(x+1)$
21 $2(x-3)(x-2)$
22 $(x+7)(x-4)$
23 $x(x-3)(x+1)$
24 $x(x+3)(x-1)$
25 $(x+1.4)(x-2.8)$
26 $(ax-b)(ax+b)$
27 $\pi r(r+2h)$
28 $(B-6)(B-4)$
29 $(x-c)^2$
30 Cannot be factored.
31 $(a-2)(a+2)(a^2+3)$
32 $(t-1)(t+7)$
33 $(hx-3)(x-4)$
34 $(r+2)(r-s)$
35 $(y-2x)(y-x)$
36 $xe^{-3x}(x+2)$
37 $e^{5t}(t+1)(t+2)$
38 $(s+2t+2p)(s+2t-2p)$
39 $P(1+r)^3$
40 $\sin x(x-1)$
41 $(\cos x-1)^2$
42 $(e^x+1)^2$

Appendix D

1   41/35
2   17/30
3   $(3-4x)/6x$
4   $69/7y$
5   $-2(1-2y)/yz$
6   $-2(z-4)/y$
7   $(2-3x)/x^2$
8   $(6y^2+7)/y^3$
9   15/7
10 1/18
11 $18/x^3$
12 $x/2$
13 $20/(x-1)$
14 $41/(2(x-1))$
15 $(4y^3z-3wx)/(x^2y^4)$
16 $7/(y-2)$
17 $(8(y+4))/(y-4)$
18 8
19 $x^3/2$
20 $3(4x+11)/((x+3)(x+2))$
21 $(-27x+44)/((x+1)(3x-4))$
22 $(-18x^2+18x+41)/((x-2)^2(x+1))$
23 $2(11x+27)/((x-3)^2(x+5)^2)$
24 $(x+20)/(x^2-16)$
25 $(x^2+1)/(x-1)$
26 $1/2r$
27 $\left((u+a)^2+u\right)/(u+a)$
28 $(x-1)/(\sqrt{x})^3 = x\sqrt{x}-\sqrt{x}/x^2$
29 $(1+e^x)/e^{2x}$

30 $(4x+1)/(b-a)$

31 $(0.28+3M^3)/(4M)$

32 $(r_2r_3+r_1r_3+r_1r_2)/(r_1r_2r_3)$

33 8

34 $(2a+3)/((a+3)(a-3))$

35 $x(x+2)$

36 $1/(x+y)^2$

37 $1/2$

38 $(-2x-h)/\left(x^2(x+h)^2\right)$

39 $1/(a^2b^2)$

40 Cannot be simplified further

41 $-2x-h$

42 $(b-1)/b^2$

43 $1-(1/a)$

44 $1/(x-1)$

45 $p^3/(p^2+q^2)$

46 $x^2y/(2x+1)$

47 $(1-2\ln x)/\left(3x^3\right)$

48 $(2x-4x^4)/(x^3+1)^3$

49 $(-4x^2+2x+1)/(x^4\sqrt{2x-1})$

50 $13/x^2+1/(2x^3)$

51 $1/3+1/\sqrt{x}$

52 $(2/l^2)+(1/l^3)-4/(3l^4)$

53 $7/(p^2+11)+p/(p^2+11)$

54 $1/6-1/(4x)$

55 $1/t^{5/2}+1/t^{3/2}$

56 $1-7/(x+5)$

57 $1+3/(q-4)$

58 $1+1/R$

59 $1+2/(2u+1)$

60 $1+\sin x/\cos x$

61 $1+1/e^x$

62 False

63 True

64 False

65 False

66 True

67 True

Appendix E

1 $3x+(1/2)x^{-1}+x^2+1/5$

2 $180q^3-60q$

3 $y^2-6y^{-1}+9y^{-4}$

4 x^3+2x+x^{-1}

5 $2P^3+3P^{1/2}$

6 $1/2+3t^{1/2}+(9/2)t$

7 $-3-(1/6)x^2+(1/2)x$

8 $N^{-2}-2+N^2$

9 $-12/7+(3/7)x$

10 $(1/2)x^{7/2}+x^{1/2}+(1/2)x^{-1/2}$

11 $ax+b+cx^{-1}$

12 $1+2p^2+p^4$

13 $12(7+1)^{1/2}$

14 $1000(10-s)^{1/3}$

15 $-0.7(x-1)^4$

16 $(1/2)(x^2+1)^{-3}$

17 $24(6R+2)^3$

18 $2(7x-\pi)^{1/2}$

19 $(1/2)^x$

20 $(1/2)^x$

21 $10{,}000(0.76)^t$

22 $e\left(e^2\right)^x$

23 $2(1/3)^x$

24 $(1/3)6^x$

25 4^t

26 $(1/e)e^x$

27 $-(5/3)^x$

28 $5e(e^x)$

29 $100(e^2)^t$

30 $A(e^k)^t$

31 $x^2+(1/x^3)$

32 $5x^2+x+1$

33 $6y^3-15y$

34 $y^5+2y^4+y^3$

35 z^4+2z^2+1

36 $z^2-(2/z)+6$

37 $(x+4)^2-16$

38 $(y-6)^2-36$

39 $(w+(7/2))^2-(7/2)^2$

40 $2(r+5)^2-50$

41 $(s+3)^2-17$

42 $3(t+4)^2-61$

43 $(x-1)^2-4$

44 $(x-3)^2+1$

45 $-(x-3)^2+7$

46 $3(x-2)^2+1$

47 $-6x/(x^2+7)^4$

48 $(-2^{x+1}\ln 2)/(1+3^x)^3$

49 $\pi\cos(\pi t)/\sin(\pi t)$

50 $-1/\sin^2 z$

51 $-\left(xe^{2x}+2\right)/x^3e^x$

52 $(1-\ln x)/x^2$

53 $\sqrt{5x}$

54 $1/\sqrt{3x-2}$

55 $6y\sqrt[3]{4z-5}$

56 $(3^5z^{5/3})/\sqrt{3w-1}$

57 $x/\sqrt{x^2+16}$

58 $(x+5)/\sqrt{x^2+10x+1}$

59 $\cos(2x)/\sqrt{\sin(2x)}$

60 $2/\left(\sqrt[3]{x^2-e^{3x}}\right)^2$

Appendix F

1 $x=5$

2 $y=-6$

3 $z=(11/2)$

4 $x=7$

5 $y=26$

6 $w=-11$

7 $x=9$

8 $y=7$

9 $z=13/10$

10 $x=2$

11 $y=-(7/41)$

12 $x=-6$ or $x=-1$

13 $y=-1$ or $y=6$

14 $w=(-5/2)$ or $w=2$

15 $x=(-3\pm\sqrt{249})/8$

16 $x=(7/16)$

17 $x=7/4$

18 $y=170$

19 $y=(46/5)$

20 $x=(37/2)$

21 $(18\pm\sqrt{285})/3$

22 $t=3\pm2\sqrt{3}$

23 $r=10$

24 $l=A/w$

25 $r=(C/2\pi)$

26 $P=(I/rt)$

27 $F=(9/5)C+32$

28 $k=(\ln 1.0573/t)$

29 $B=e^{0.079}$

30 $y=(-5x-1)/(3x-2)$

31 $t=(3C-2B)/(A-6B)$

32 $x=-a(b+1)/(ad-c)$

33 $b=\sqrt[5]{C/A}$

34 $x=3, x=-4$

35 $t=\ln\sqrt[k]{2P/Q}$

36 $B=-2$

37 $s=-0.115$

38 $y=-17/16$

39 $t=10/7$

40 $x=2, x=-4/3$

41 $p=3, p=-3, p=-1/2$

42 $N=3, N=1$

43 $t=0, t=\pm8$

44 $x=1\pm\sqrt{2}$

45 $x=4, x=-3/4$

46 $t=3\pm\sqrt{6}$

47 $y=-2\pm\sqrt{6}$

48 $z=-7/2$

49 $x=\pm1$

50 $L=\pm1/2$

51 $q=\pm\sqrt{2}$

52 $r=\pm5$

53 $x=-1/8$

54 $x=0, x=36$

55 $v=700\pi$

56 $x=-4/3, x=2$

57 $x=0.6826$

58 $x=1$

59 $t=\ln 2/\ln 0.97\approx-22.76$

60 $t=\ln 8/2\approx1.04$

61 $x=5$

62 $x=-\log 3.3\approx-0.52$

63 $k=\ln(1/2)/1000\approx-0.000693$

64 $x=\ln(0.5)/\ln(0.8)\approx3.106$

65 $l=gT^2/4\pi^2$

66 $y'=4/(y+2x)$

67 $w=(2/k)(l-l_0)$

68 $y'=2x/(x-y)$

69 $y=(c+d)/(b-a)$

70 $y=(3m-2n)/(m+n)$

71 $r=(S+a)/(S-L)$

72 $a=2(h-v_0t)/t^2$

73 $z=y-(NK)/(MA)$

74 $e=(Rf)/(b+d-R)$

75 $z=(y+wy-xy)/(y-1)$

76 $s=(tw)/(2v-5w)$

77 $x=1/2, -1/4$

78   $x = 4m$

Appendix G

1   $x = 4,\ y = -1$
2   $x = 7,\ y = 4$
3   $x = 4,\ y = 3$
4   $x = (317/13),\ y = (62/13)$
5   $x = (3 \pm \sqrt{63})/2,$
   $y = (-3 \pm \sqrt{63})/2$
6   $x = -55,\ y = 39$
7   $x = 3$ and $y = -3$, or $x = -1$ and $y = -3$
8   $x = -3$ and $y = -5$, or $x = 1$ and $y = 3$
9   $x = 1/2$ and $y = 2$, or $x = -1/2$ and $y = -2$
10   $x = 13.5;\ y = -12$
11   $x = 2.081,\ y = 8.013$
   $x = 4.504,\ y = 90.348$
12   $x = 1,\ y = a$
13   $(0, 0)$ and $(3, 9)$
14   $(3/2, 3/2)$
15   $(4, 3),\ (-3, -4)$
16   $(2, 4)$, or $(4, 16)$, $(-0.7667, 0.5878)$

Appendix H

1   $x \geq 7$
2   $x > 16$
3   $-5 < x < 5$
4   $x > 5$
5   $x \geq -4$
6   $-5 < t < 5$
7   $x \leq -4$ or $x \geq 4$
8   All real numbers
9   $x > 5$
10   All real numbers
11   $-8 \leq \theta \leq 8$
12   No solution
13   $x > 5$
14   $t < -3$
15   $z < -2$
16   $x < 0.001$

17   $-1 < y < 1$
18   $p \neq 5$
19   $k > 0$
20   $r \geq 0$
21   $t \geq 1995$
22   $x > 4$
23   $x > 2$
24   $x < -4$
25   $x > -(6/4)$
26   $x < 2$, or $x > 3$
27   $-1 < x < 8$
28   $x > 3$
29   $5 < x < (17/3)$
30   $x < -2$ or $x > 2$
31   $y < -2$ or $y > 6$
32   $-(2/3) < w < 2$
33   $-(1/2) < z < (15/2)$
34   All real numbers
35   No values
36   (a)   None
   (b) $x = 0,\ x = -2$
   (c) $x > 0,\ x < -2$
   (d) $-2 < x < 0$
37   (a)   None
   (b) $x = 0,\ x = -2$
   (c) $x > 0,\ x < -2$
   (d) $-2 < x < 0$
38   (a)   None
   (b)   None
   (c)   All real numbers
   (d)   None
39   (a)   None
   (b) $t = 2,\ t = 3$
   (c) $t < 2,\ t > 3$
   (d) $2 < t < 3$
40   (a) $x = 0$
   (b)   None
   (c) $x \neq 0$
   (d)   None
41   (a) $p = 0$
   (b)   None

   (c) $p < 0$
   (d) $p > 0$
42   (a)   None
   (b) $x = 0$
   (c) $x > 0$
   (d) $x < 0$
43   (a) $x \leq 0$
   (b) $x = 1$
   (c) $x > 1$
   (d) $0 < x < 1$
44   (a)   None
   (b) $t = 3$
   (c) $t > 3$
   (d) $t < 3$
45   (a) $x = 2$
   (b) $x = 1$
   (c) $x < 1$ or $x > 2$
   (d) $1 < x < 2$
46   (a)   None
   (b) $u = 1/\sqrt{3},\ u = -1/\sqrt{3}$
   (c) $-1/\sqrt{3} < u < 1/\sqrt{3}$
   (d) $u < -1/\sqrt{3},\ u > 1/\sqrt{3}$
47   (a) $x = 0,\ x = 1$
   (b) $x = 1/2$
   (c) $x < 1/2$ and $x \neq 0$
   (d) $x > 1/2$ and $x \neq 1$
48   $-2 < x < 2$
49   $1/2 \leq x \leq 1$
50   $1/2 \geq n > -21/2$
51   $l > 1/48$
52   $t \leq -1,\ t \geq 4$
53   $x < -7/3,\ x > 1$
54   $x \leq \dfrac{45}{2}$
55   $x < 0$ or $2 < x$
56   $r < 2,\ r > 3$
57   $x < -1,\ x > 0$
58   $x < -1,\ x > 0$
59   $-2 < x < 2$

INDEX